QUANTITATIVE
AN[illegible]TICAL
[illegible]STRY

VOLUME I
INTRODUCTION TO PRINCIPLES

VOLUME I
INTRODUCTION
TO
PRINCIPLES

Quantitative Analytical Chemistry

H. A. FLASCHKA
Georgia Institute of Technology

A. J. BARNARD, Jr.
J. T. Baker Chemical Company

P. E. STURROCK
Georgia Institute of Technology

A BARNES & NOBLE BOOK
distributed by
Harper & Row, Publishers, Inc.

QUANTITATIVE ANALYTICAL CHEMISTRY
Volume I – Introduction to Principles

Standard Book Number: 06-042078-2
Library of Congress Catalog Card Number: 77-91298

PREFACE

This is an introductory textbook with a simple central purpose: to provide the student with a basic understanding of the theory and principles of inorganic quantitative analysis, and thereby an appreciation of a major area of quantitative chemistry. The work is directed to the student who has some knowledge of the elementary concepts and descriptive material introduced in first courses in chemistry and physics. Some of these concepts are presented in summary form in this work. Others, however, are developed in some detail since the student probably has been exposed to them from a different viewpoint or to a depth insufficent for an adequate understanding of their application to analytical chemistry. The student thoroughly familiar with these topics may need only to read the relevant text; other students can study the text closely without the need to turn to other books for reference. Last, but not least, *repetitio est mater studiorum.*

The student is urged for each section to undertake the questions and at least the problems for which answers are given—even if this is not specifically requested by the teacher. This effort will provide an appreciation of the mathematical relationships. In many cases the answer to a question should be qualified as to the conditions under which it would be valid; this consideration should further stimulate thinking and understanding.

Since the beginning students seldom, if ever, examine the original journal literature, the authors have not cited papers. It should be understood, however, that the concepts and facts presented are not those of the authors, but rather are the product of research by many workers from many countries extending back into the nineteenth century. The student interested in exploring a particular topic more fully should contact the instructor, who will gladly provide references.

The authors are indebted to many colleagues who kindly read and constructively criticized portions of the manuscript, including Dr. J. C. Gantchoff, Mr. E. F. Joy, and Dr. R. M. Speights. Dr. LeRoy I. Braddock, Dr. J. Butcher, and Dr. J.

Jordan closely studied the entire manuscript; the authors gratefully acknowledge their valuable suggestions.

A Note to the Teacher

What is the first course in quantitative chemistry to bring to the student? The view adopted in this textbook is that such a course should (1) introduce quantitative chemistry in the framework of inorganic quantitative analysis, and (2) afford an understanding of some basic analytical methods, techniques, and operations, and some principal determinations. These goals are probably not controversial or only mildly so. However, three further views are adopted: The course should (3) acquaint the student with the broad spectrum of present-day analysis, ranging from gravimetric and volumetric methods to electrical and optical ones, (4) be presented in two separate, but integrated categories, principles and practice,† and (5) allow these goals to be realized in the short course often required of students majoring in fields other than chemistry.

Analytical chemistry has undergone enormous changes since World War II; consequently, the patterns of collegiate instruction in quantitative chemistry require reevaluation. Often classical analytical methods (that is, gravimetric and titrimetric analysis) are emphasized in the sophomore year and physicochemical methods of analysis and physical chemistry in courses in the junior or senior years. Some teachers still find this approach satisfactory for students majoring in chemistry. However, students concentrating their study in other fields are often required to take only a single course in analytical methods in the sophomore year. What is needed for the student of engineering and for the student in certain preprofessional curricula, such as premedicine, is not a short course directed largely to classical methods, but rather a unified one that affords him a broad appreciation of present-day principles and challenges him with the potentials of analytical chemistry. A chemical engineer, for example, will seldom, if ever, perform a complicated chemical analysis in a research or production capacity. However, he will be in frequent contact with analytical chemistry as he sends samples to an analytical laboratory, studies reports of analyses, and either supervises or performs simple tests and determinations. This example might be extended to

†See the companion work: *Quantitative Analytical Chemistry:* Vol. II, *Short Introduction to Practice*, H. A. Flaschka, A. J. Barnard, Jr., and P. E. Sturrock, Barnes & Noble, Inc., New York, 1969.

other engineers or even to physicians. For these professions, an understanding of the principles, the approaches, and the "language" of quantitative chemical analysis is far more important than a knowledge of procedural details.

Especially with the nonmajor in chemistry, in the limited course time it is not possible to go into practical analysis beyond describing some fundamental determinations and allowing the student to undertake a few in the laboratory sessions. Further, it is impossible to provide a thorough mathematical treatment of all theories and principles, especially for physicochemical methods. Consideration of such methods can only be introduced by keeping the treatment of gravimetry and acid–base and precipitation titrimetry brief. Frequently, these topics, as well as precipitation phenomena and electrolyte equilibria, are treated at a length beyond reasonable benefit to the nonmajor in chemistry. These are the places where the authors have compressed the discussion, hopefully without the loss of any essential, important concept.

Students undertake a course in quantitative inorganic analysis with varied backgrounds and with various degrees of familiarity with the basic principles and facts of chemistry. Consequently, on one hand, the authors have elected to summarize material that the student should have encountered in introductory courses and physics. On the other hand, the authors have elaborated certain topics basic to the presentation of quantitative analysis that may not have been adequately emphasized in previous courses. No attempt has been made to relate the behavior of indicators and chromogenic agents to structural formulas since a sophomore usually first encounters organic chemistry in either a parallel or subsequent course. Only the basic mathematical operations are usually assumed to be known by the student, since he may only encounter calculus in a parallel course. The chemical nomenclature employed is largely consistent with the recommendations of the 1957 report of the Commission on the Nomenclature of Inorganic Chemistry, International Union of Pure and Applied Chemistry [see *J. Am. Chem. Soc.*, **82,** 5523 (1960)]. Although the principles are stressed in the organization of this volume, selected applications of most techniques are briefly described so that the student is made aware of the vitality of the fundamentals and the utility of the techniques.

In the presentation of physical methods, details of design and operation of the instruments have largely been omitted from this volume. Rather, the emphasis has been placed on the analytical use of the instruments, and only simplified descriptions of how they work have been provided. This approach is considered adequate for lectures on principles. If it appears desirable or necessary, more attention to instrumental and operational details can be given in conjunction with the laboratory work. The information then presented can be related to the particular instrument model at hand.

The teaching of quantitative analytical chemistry, as of any subject in which lectures are complemented by laboratory or field work, faces major problems. The student may have only a few hours of lecture before the laboratory work. In addition, special instruments and assemblies may be available only in limited numbers, necessitating scheduling of their use by some students before consideration of the relevant theory in lectures. Mechanical, "cookbook" efforts by the students often result in either of these situations. The authors have attempted to circumvent these difficulties by completely separating the laboratory instruction and placing it in a separate volume.† Each laboratory experiment, or group of experiments, is prefaced by a brief consideration of the underlying theory and principles, often with repetition of facts and equations that are considered in this volume devoted to principles. Obviously the division into principles and practice is somewhat arbitrary. Nevertheless, the authors hope in this way to benefit the student by not confronting him with references throughout his experiment that require him to turn to pages scattered through a textbook in order to gain an appreciation of the immediate task.

This volume is divided into short chapters and sections rather than into lengthy chapters. This makes it feasible for the teacher to employ a somewhat different order of presentation, for example, the treatment of precipitation titrations (Chapter 18) before acid–base titrations (Chapters 11 through 16), and to omit chapters and sections where a different selection or emphasis of topics is desired. Some of the chapters are sufficiently brief and simple in their treatment of a subject that under favorable circumstances they might well be assigned to students for independent self-study. For most chapters, examples of typical calculations will be found in the text, and additionally questions and numerical problems (some with answers) are included. Consideration of these questions and

† *Op. cit.*

problems by the student will aid in his appreciation and understanding of the material presented. Review questions (and problems) have also been provided covering broad subjects (see Chapters 9, 17, 33, and 44).

This textbook, although primarily organized for use in courses for students not majoring in chemistry, is considered suitable for the more intensive course or courses for majors. In such courses the material presented in this volume can be covered more thoroughly. For a subsequent course in instrumental analysis the work can serve as an informational aid.

No work is ever free of errors and shortcomings. The authors will appreciate having mistakes brought to their attention and will welcome comments from teachers as to success or difficulty encountered in the use of this textbook.

CONTENTS

CONTENTS

1 INTRODUCTION

Analytical chemistry, broadly conceived, underlies and contributes to almost all branches of chemistry as an experimental science. The student has already received brief instruction in qualitative analysis. In that field of chemistry, the central question is: *What is present in the sample?* In quantitative analysis, the question is: *How much is present?* Quantitative analysis employs many of the reactions and phenomena utilized in qualitative analysis. However, quantification demands closer controls and more stringent requirements.

Consider a reaction resulting in a precipitate. When such a reaction is used qualitatively as an identification test, it may be desirable, but not necessary, that the precipitation be complete and the precipitate be pure. In contrast, if the reaction is used for quantitative purposes and the amount of the washed and ignited precipitate is to be related to the amount of one of the components of the precipitate, it is imperative that the precipitation of that component be essentially complete and that the precipitate be free of all impurities that cannot be removed by either simple washing or another subsequent treatment, for example, ignition.

The student, in his study of qualitative analysis, was introduced to only a limited number of the reactions and phenomena of value in analytical chemistry. In the study of quantitative analysis, many additional reactions and phenomena will be presented. Consequently, the student's knowledge will also be augmented with respect to qualitative analysis as well as to chemistry generally. Although most of the material in this textbook is directed toward inorganic substances and inorganic analysis, the principles introduced are also applicable to organic analysis.

Quantitative analytical chemistry can, for the purpose of instruction, be separated into two broad areas: (1) the principles and the possibilities they offer, and (2) their application to practice. It is to the first area that this volume is directed. Practice is treated only so far as to show the vitality of the principles and to assure understanding of the basis for the use of equipment, reagents, instruments, and actual procedures.

Many textbooks begin with detailed definition and classification of the subject to be treated. This approach is usually unrewarding since the student is encountering the subject for the first time. Here, such considerations are postponed to the terminal pages of the text, Chapter 52, where they can be studied either after some experience has been gained or at the end of the course with further appreciation and in perspective. It suffices to state that classification by the techniques employed allows orderly development of the subject. Sections devoted to the introduction of basic chemical and physical concepts, such as chemical equilibria and electrochemical and optical principles, are dispersed throughout this development of the topic.

It is important that the student appreciate the distinction between a determination and an analysis. A determination establishes the amount or content of a single component in a sample. An analysis establishes the amounts or contents of a few or all of the components in a material. In an introduction to quantitative analysis, simple determinations must be considered predominantly. Analyses require, beyond a mastery of determinations, additional knowledge, special experience, and attention to the overall composition of the material. Consequently, the resolution of complex samples is beyond the scope of an introductory course.

2 SIGNIFICANT FIGURES AND RULES FOR COMPUTATION

General 2.1

In quantitative analysis, as in most fields of science and technology, numerical values are obtained from measurements. A measurement is restricted in its reliability, and this fact should be reflected in the number of figures retained in the reported value. A figure is a digit that denotes the amount of a quantity in the place in which it stands. A figure that is assigned some reliability is known as a significant figure. The term "figure" includes the zero. But as will be seen below, a zero, depending on its position, can be either a significant figure or merely an indicator of the location of the decimal point.

The rule that may be adopted for most measurements is to retain only one uncertain figure in the value reported. In other words, the final result should be presented in such a manner that the figure before the last one is certain and the last figure is uncertain.† For example, if a certain amount of a substance is weighed on a pan balance that is known to yield results to no better than a few hundredths of a gram, a result might be presented as 5.34 grams. To record 5.342 g after a tedious, but useless, interpolation would be improper, because the second decimal place is already uncertain. Indeed, the result of an additional weighing might be 5.35 or 5.33 g. If the same amount of substance is weighed on an analytical balance with a reproducibility of 0.0003 g, the result might properly be written, for example, as 5.3434 g.

It should be emphasized that the recording of superfluous figures is not only wasted effort, but improper because an incorrect impression is given of the reliability of the reported value.

When the number of significant figures is evaluated, care must be taken with respect to the concept of zero. Assume that an object is found to weigh 3.2 g on a crude balance, and that this number is written with the proper number of significant figures. It may be desired to express this result in milligrams. To write

†For the statistical study of a series of measured values and for intermediate results in various mathematical operations, it is often necessary to retain two or more uncertain figures.

3200 mg, although numerically correct, is ambiguous, because the two zeros, which actually serve to locate the decimal point, may or may not be taken subsequently as significant figures. To circumvent this difficulty, the result is best written in the form 3.2×10^3 mg. In contrast, if it is desired to express the result in kilograms, the expression 0.0032 kg is unambiguous, because the two zeros to the right of the decimal point can only serve to fix its location. Of course, the result can also be represented satisfactorily as 3.2×10^{-3} kg.

If the reliability of a result is known, an even more meaningful way of presenting it can be adopted. Assume that a sample was weighed on a balance that is known to permit no better operation than to ± 0.03 g. Then the result might be reported as, for example, 5.34 ± 0.03 g. Here the term ± 0.03 g represents the *absolute* uncertainty of the reported result. Another way to express the reliability is to utilize the *relative* uncertainty, which is the fraction obtained by dividing the absolute uncertainty by the value of the result. In the present case the relative uncertainty is $0.03/5.34 = 3/534$, or about $1/180$.

The following few examples will serve to illustrate the points made so far. Below each of the data is given the number of significant figures and the approximate relative uncertainty, with the absolute uncertainty assumed to be ± 2 in the last significant figure.

2.34	0.00234	0.0002340	2.34×10^4	23,400	23,400.0
3	*3*	*4*	*3*	*uncertain*	*6*
$\frac{1}{120}$	$\frac{1}{120}$	$\frac{1}{1200}$	$\frac{1}{120}$		$\frac{1}{120{,}000}$

It is sometimes expedient to indicate uncertain figures by special typography, usually by the use of italic or subscript digits, e.g., 3.5*2* or 3.5_2. These practices are especially utilized when a 5 would otherwise be rounded.

2.2 Rules for Rounding

In the adjustment of a measured or calculated value for retention of only the proper number of significant figures, superfluous figures must be rejected and the following rules for "rounding" are recommended.†

†These are the practices recommended by the American Standards Association and the American Society for Testing and Materials, and are paraphrased from the publications of the latter society.

1. When the figure next beyond the last place to be retained is *less than* 5, leave unchanged the figure in the last place retained. (For example, 452.23 would be rounded to 452.2, and 8.03 to 8.0.)
2. When the figure next beyond the last place to be retained is *greater than* 5, increase by 1 the figure in the last place retained. (For example, 23.67 would be rounded to 23.7, and 0.0699 to 0.070.)
3. When the figure next beyond the last place to be retained is *5*: If there are no figures beyond this 5 or only zeros, (a) increase by 1 the figure in the last place retained if it is odd, (b) leave the figure unchanged if it is even. (For example, 235.5 would be rounded to 236, 0.6445 to 0.644, 0.605 to 0.60, and, in rounding to the nearest hundred, 2250.0 and 2350.0 to 2.2×10^2 and 2.4×10^3, respectively.)

   If there are figures other than zeros beyond this 5, increase by 1 the figure in the last place retained. (For example, in rounding to the nearest hundred, 2250.4 and 2354.0 would be rounded to 2.3×10^3 and 2.4×10^3, respectively.)
4. The rounded-off value should be obtained in one step by direct rounding and not in two or more steps of successive roundings. (Thus, 89,490 rounded to the nearest thousand is at once 89,000 and is best written as 8.9×10^4; it would be improper to round off first to the nearest hundred, 89,500, and then to the nearest thousand, giving 90,000, that is, 9.0×10^4.)

Significant Figures in Arithmetic Operations 2.3

The above rules for rounding can be applied readily to the proper expression of a single measured value. The situation, however, is more complicated when the final result is obtained by calculations involving several values, each of which may vary in its own reliability. The maxim applies that "no chain is stronger than its weakest link." In terms of significant figures, this means that no calculated value can be more reliable than its least reliable component. Evaluation of the uncertainty of a computed value will depend on the mathematical operations involved.

Addition and Subtraction. In a result obtained by addition or subtraction, or both, the *absolute* uncertainty of the least reliable component determines the reliability of the result. Consequently, only as many places should be retained to the right of the decimal point as there are in the component with the least number of decimal places.

Example 2-1. A formula weight is calculated from the values for the atomic weights recorded in the atomic weight table (see Table K in the Appendix). Thus for potassium fluoride, KF:

$$\begin{array}{rl} \text{K:} & 39.102 \\ \text{F:} & \underline{18.9984} \\ \text{KF:} & 58.1004 = \text{rounded } 58.100 \end{array}$$

As the atomic weight of potassium is known reliably to only three decimal places, it is pointless to carry any figures beyond that place, and the formula weight of the salt is properly recorded as 58.100.

Multiplication and Division. The following simple "fewest significant figures rule" may be adopted: A product or quotient should not contain more significant figures than are contained in the number with the fewest significant figures used in the multiplication or division. The rule holds also for mixed operations consisting of both multiplication and division. A few examples will illustrate the approach.

Example 2-2. Consider the multiplication 34.2051×3.22. The result is 110.140422. The factor with the fewest, namely three, significant figures is 3.22. Consequently, the result is rounded to three significant figures: 110. However, written in this form it is ambiguous whether the zero is significant or serves only to fix the location of the decimal point. The result is best presented as 1.10×10^2.

It can be seen that superfluous digits have been carried through the calculation and it is common sense to round the first factor from the beginning to the same number of significant figures as is present in the second factor. Then the multiplication simplifies to 34.2×3.22.

Example 2-3. Calculate the result for the following mixed expression:

$$\frac{4.6672 \times 12.4}{53.267 \times 0.13862}$$

The factor with the fewest significant figures has three such figures (12.4). Consequently, all the other factors are rounded to three significant figures: Then

$$\frac{4.67 \times 12.4}{53.3 \times 0.139}$$

The multiplications in the numerator and denominator are next performed with the intermediate results also only written with three significant figures:

$$\frac{57.9}{7.41}$$

And finally the result is obtained as 7.81.

Multiplications and divisions performed on a 10-inch slide rule will have a relative uncertainty no greater than 1 part in 400. Hence such a slide rule can be used in many calculations in science and technology and, of course, is valuable in checking calculations for gross errors at any level of uncertainty.

Logarithms. When converting a number to its logarithm the following rule applies: Retain as many places in the mantissa of the logarithm (that is, to the right of the decimal point in the logarithm) as there are significant figures in the number itself. For example, $\log 24.5 = 1.389$, and $\log 0.34 = 0.53 - 1$.

When logarithms are used to obtain the results of multiplications, divisions, calculation of powers, extraction of roots, or a combination of these operations, the logarithms may be noted down as given in a table of logarithms and the final result adjusted to the proper number of significant figures according to the rules given in the preceding sections. Alternatively, each logarithm in the operations may be written with its proper number of decimals in the mantissa and the calculation be performed with the adjusted logarithms.

Significant Figures in "Mixed" Calculations 2.4

Mixed arithmetic calculations are performed stepwise according to the rules of arithmetic. In additions and subtractions the absolute uncertainties are considered and for multiplication and divisions the rule of "fewest significant figures" is applied. Some examples will illustrate the procedure.

Example 2-4. Consider the following mixed arithmetic calculation:

$$\frac{2.357}{0.26} + \frac{1.265}{4.12}$$

For the first term the numerator is rounded to two significant figures, and as the result of the division $\frac{2.4}{0.26} = 9.2$. Proceeding analogously with the second term yields $\frac{1.26}{4.12} = 0.306$. For the addition according to the rules established, only one decimal place is significant. Consequently, $9.2 + 0.3 = 9.5$.

Example 2-5. Consider the following mixed arithmetic operations:

$$\frac{(248.31 - 248.1) \times 4.012}{(10.01 + 25.12) \times 0.9804}$$

First the subtraction in the numerator is performed. The difference is 0.21 and, written with the appropriate number of significant figures retained, is 0.2. This number is the one with the fewest significant figures. Consequently, all other terms, including the result of 35.13 of the addition term in the denominator, are rounded to one significant figure:

$$\frac{0.2 \times 4}{40 \times 1}$$

The result is 0.02, which also, of course, is recorded with one significant figure. It is noteworthy that in such mixed operations significant figures are lost. From terms with four and five significant figures a result is derived with only one significant figure. Such losses always occur when numbers similar in magnitude appear in a subtraction term, as in this example.

2.5 Remarks

The preceding treatment is simplified but will suffice the student's need. It should be noted that for multiplication and division the "fewest significant figures" rule is a pronounced simplification. By this rule a decision is based solely on the number of significant figures and does not consider the fact that two numbers with the same number of significant figures may differ in their reliability. The number 99, for example, is considered to have two significant figures, but it is so close to 100 that it may well be considered to have a "hidden" third significant figure. In a more rigorous treatment it is not the number of significant figures but the relative uncertainties that should be taken into account. The decision of how many significant figures are to be properly retained in the value resulting from multiplication or division, or both, should be based on the following consideration: The *relative* uncertainty of the result should be of the same order as that of the least certain component. The rule may be adopted that the relative uncertainty of the result should be between twice and two tenths, (that is, one fifth) of the relative uncertainty of the least certain component. One example may illustrate the approach.

Example 2-6. Consider the following multiplication, where each factor is assumed to be uncertain to the extent of ± 1 in the last figure. (The approximate relative uncertainties are expressed as italic fractions.)

$$\underset{1/40}{4.3} \times \underset{1/7000}{6.893} \times \underset{1/5000}{0.5372} =$$

To avoid superfluous numbers being carried through, the numbers are rounded as in the simple rule, but one more figure is retained than in the one with the fewest figures. Thus

$$4.3 \times 6.89 \times 0.537 =$$

Multiplication of the first two factors and again retaining one additional figure yields

$$29.6 \times 0.537 = 15.90$$

The result should have no better certainty than 1/40, which is that of the least certain factor. Then $1/40 \times 15.90 = 0.4$. Consequently, only one decimal place is to be retained, and the result is properly written as 15.9. The uncertainty of the result is 1/160, which is within twice and two tenths of the least certain factor: $2 \times 1/40 = 1/20$ and $0.2 \times 1/40$ is 1/200. Note that the simplified approach would lead to expression of the result as 16, which has a relative uncertainty of 1/16 or approximately 1/20, which falls outside or just at one limit of the uncertainty range required by the rigorous treatment.

This more rigorous treatment is used in all calculations throughout this textbook, and for the sake of simplicity it is assumed that the uncertainty of all given values is ± 1 in the last figure. This practice has no bearing on the mode of calculation adopted.

The student may ask how knowledge is gained as to how many figures are uncertain in a value obtained by a measurement. The answer to this question requires more knowledge of measuring processes and is a significant topic for the practice of analytical chemistry. Here it suffices to state that the accuracy of an instrument can be assessed, and the precision for a series can be evaluated, by means of a statistical treatment, for example from consideration of the so-called standard deviation. (See, for example, H. A. Flaschka, A. J. Barnard, Jr., and P. E. Sturrock, *Quantitative Analytical Chemistry: A Short Introduction to Practice*, Barnes & Noble, Inc., New York, 1969, Section 7.5.)

Problems 2.6

1. The velocity of light in vacuo is 29,979,300,000 cm/sec. How should this value be written in order to express unambiguously that the measured value is known to seven significant figures?

   Answer: 2.997930×10^{10} cm/sec

2. Perform the following arithmetic operations and express the result to the proper number of significant figures assuming that each value is uncertain to the extent of ± 1 in the last figure:

   (a) $3.67 + 34.236 + 863.6$ (c) $3.6 \times 48.1 \times 0.216$
   (b) $521.5 + 3.77 - 8.09$ (d) $(9.8 \times 6.44)/1.21$

   Answers: (a) 901.5; (b) 517.2; (c) 37; (d) 52

3. What is faulty with the following presentation of a value: 646.52 ± 2?

4. Write the logarithms of the following data and retain the correct number of significant figures: (a) 24.6; (b) 2.6×10^{-3}; (c) 0.0643.

   Answers: (a) 1.391; (b) $0.42 - 3$ or -2.58; (c) $0.808 - 2$ or -1.192

2.6 PROBLEMS

2-5. Give the correct number of significant figures for the following numbers.

(a) 231.4
(b) 0.10004
(c) 0.002300
(d) 620.04
(e) 3.23×10^{-3}
(f) 0.044×10^{4}
(g) 202.0
(h) 0.02020
(i) 0.00202
(j) 202

2-6. Give the logarithms of the following numbers with the correct number of significant figures.

(a) 21.4
(b) 22.301
(c) 1.00×10^{4}
(d) 1×10^{-3}
(e) 0.023

2-7. Give the numbers, with the correct number of significant figures, corresponding to the following logarithms.

(a) 0.30
(b) 1.301
(c) 2.3010
(d) −4.66
(e) 8.4563

2-8. Round each of the following to three significant figures.

(a) 88.55
(b) 8.345
(c) 3.2547×10^{-6}
(d) 213.4
(e) 699.9

2-9. Round each of the following to four significant figures.

(a) 33.352
(b) 32.356
(c) 32.3549
(d) 0.49996
(e) 5,632,445.4

2-10. Perform the following additions giving the correct number of significant figures in the results.

(a) $301.0 + 36.54 + 423 =$
(b) $2.56 + 0.732 + 1.6 =$
(c) $4.367 + 0.486 + 16.3 =$
(d) $0.001 + 0.16 + 1.009 =$

2-11. Perform the following arithmetic operations and express the results with the correct number of significant figures assuming that each value has an uncertainty of ±1 in the last place.

(a) $88.3 + 4.22 - 1.457 =$
(b) $(3.21 \times 10^{-4}) \times 21.3 =$
(c) $4.22 \times 9.22 =$
(d) $7.89 + 3.456 - 6.775 =$
(e) $(2.335 + 7.892) \times 3.678 =$
(f) $\dfrac{6.63 \times 2.98}{4.6997} =$
(g) $\dfrac{88.14 + 24.1}{17.99 - 34.64} =$

3 EXPRESSION OF CONCENTRATION AND CONTENT

Introduction 3.1

The analytical chemist is continuously confronted by mixtures of two or more components, and especially by homogeneous mixtures, that is, solutions. Solutions may be liquid, solid, or gaseous; however, in analytical chemistry a *liquid* solution is assumed unless otherwise qualified. In the consideration of a solution of a solid (or a gas) in a liquid, the former is called the solute and the latter the solvent. It is necessary to express unambiguously the composition of a mixture and especially the content of the solute, which for a solution is termed concentration. Concentration and content are expressed in various ways, depending on the data available and the required purposes. An understanding of the diverse modes of expression is mandatory for a full appreciation of the calculations and methods of quantitative analysis. It is possible to group these expressions into two categories: those involving physical units only and those involving chemical units.

Concentration and Content in Physical Units 3.2

Per Cent Weight by Weight. Per cent weight by weight, expressed as % w/w, and often termed "weight per cent," refers to the parts by weight of a component or solute per hundred parts by weight of the mixture or solution. Thus, a binary magnesium–aluminum alloy stated to be 12% w/w in aluminum contains 12 parts by weight (grams, pounds, etc.) of aluminum in 100 parts by weight (grams, pounds, etc.) of alloy. Since it is a binary alloy, it is by difference 88% w/w in magnesium. It is noteworthy that per cent weight by weight is independent of temperature expansion or contraction in contrast to per cent weight by volume. Per cent weight by weight is commonly used in expressing the composition of alloys, the purity (assay) of chemical substances, and the strength of concentrated acids and bases. When the density of a solution is known, that is, the weight per unit volume, a per cent weight by weight value can be converted to per cent weight by volume (see below).

Per Cent Volume by Volume. Per cent volume by volume, expressed as % v/v, and often termed "volume per cent," refers to the parts by volume of a component per hundred parts by volume of the mixture. This form of expression is largely limited to mixtures of gases or of liquids. Thus, a 55% v/v mixture of methanol with water contains 55 volumes (milliliters, pints, etc.) of methanol in 100 volumes (milliliters, pints, etc.) of the methanol–water mixture. Since the volume of a liquid is temperature-dependent, it is usually necessary to specify a temperature when reporting a per cent volume by volume value. It should be emphasized that the mixing of different liquids is often attended by a marked expansion or contraction in volume. A 55% v/v mixture of methanol and water is *not* obtained by mixing 55 ml of methanol with 45 ml of water, but rather by adding water to 55 ml of methanol with mixing to obtain a total solution volume of 100 ml. If 45 ml of water and 55 ml of ethanol were mixed, the volume of the mixture would be slightly less than 100 ml as a result of contraction caused by intermolecular action of the two components.

Per Cent Weight by Volume. Per cent weight by volume, expressed as % w/v, and often termed "weight by volume per cent," refers to the parts by weight of a component in 100 parts by volume of the mixture. The weight and volume units employed should be compatible. Thus, a 10% w/v sodium chloride solution contains 10 g of sodium chloride in 100 ml of solution (not in 100 ml of solvent!). Since the volume of a solution is temperature-dependent, the temperature must usually be specified when reporting a per cent weight by volume value. If the density of a solution is known at the specified temperature, a per cent weight by volume value can be converted to per cent weight by weight and vice versa. For example, the density of a 10.0% w/w aqueous sodium chloride solution at 20.0°C is 1.07 g/ml; hence, the % w/v = 10.0 × 1.07 = 10.7. For very dilute aqueous solutions w/w and w/v percentages do not differ much in value, since the density is close to 1 g/ml, and are for practical purposes often taken as identical.

Parts per Thousand and per Million. All the percentage expressions of concentration have analogous parts per thousand and parts per million expressions. However, a weight by weight basis is usually employed and is assumed unless otherwise stated. As noted above, with very dilute aqueous solutions, the distinction between w/w and w/v is negligible. An alloy 1 ppt in silver contains 1 part by weight of silver in 1000 parts by weight of the alloy. A water sample 2 ppt in sulfate contains, for example, 2 g of sulfate per 1000 g of sample solution or, since a dilute aqueous solution is involved, per 1000 ml (that is, per liter) of sample solution. A value in parts per million is equivalent to milligrams per kilogram and, in the case of dilute aqueous solutions, to milligrams per liter.

Miscellaneous Expressions. In laboratory practice the concentration of aqueous solutions of acids and ammonia is frequently expressed, for example, as 1:8 HNO_3 or 1 + 8 HNO_3; either expression implies that one volume of commercial concentrated nitric acid is mixed with eight volumes of water. This special notation has the advantage that no calculation is necessary and only simple measurement of volumes is required when preparing such a dilute solution. This mode of expression might seem ambiguous at first glance; however, most concentrated acids and ammonia are marketed in the same concentration by all reagent suppliers. Thus, reagent-grade nitric acid contains nominally 70% w/w HNO_3. Where a concentrated acid is commonly available in two different concentrations, this form of expression is best avoided, for example, with perchloric acid, which is available in the reagent grade as 70–72% w/w and 60–62% w/w products. The concentration of solvent mixtures may be expressed in a similar way; thus, 1:1 methanol–water implies that one volume of methanol is mixed with one volume of water.

Density and Specific Gravity. The density of a substance or of a mixture is the weight per unit volume and is usually expressed in grams per milliliter. The density is temperature-dependent since the volume varies with the temperature. The specific gravity of a substance or a mixture is the ratio of its density at a specified temperature to the density of a reference substance, commonly water, at a specified temperature. The two specified temperatures need not be identical. Specific gravity values may be converted to density values if the (absolute) density of the reference substance is known at the specified temperatures. Textbooks on physics and handbooks presenting physical data may be consulted for further details.

For *dilute* aqueous solutions, since the density of water remains close to 1.0 g/ml over a wide range of temperature, specific gravities and densities may be used interchangeably, often with an error less than 1 part in 5000. For simple solutions (that is, one solute and one solvent), the concentration of the solute at a given temperature can be related empirically to the density of the solution. This serves for the assay of some simple solutions, including commercial mineral acids. Common examples include the use of specially calibrated hydrometers in estimating the sulfuric acid content of the electrolyte in lead storage batteries and the ethylene glycol or methanol content of automotive radiator antifreeze solutions.

3.3 Concentration and Content in Chemical Units

The expressions of concentration considered above may be viewed as involving only physical units. For the purposes of quantitative analysis, expression in terms of chemical units is often advantageous.

Definition of the Mole. In this textbook, the atomic weight is defined as the mass ratio of one (average) atom of a particular element to one atom of carbon isotope 12, to which a mass of exactly 12 is assigned by the international atomic weight scale (see Table K in the Appendix). The atomic weight is therefore a dimensionless number. The proper term is "atomic mass"; the traditional term "atomic weight," however, will be maintained in view of the fact that "weight" and "mass" are used interchangeably in chemistry.

The formula weight of a substance is the sum of the atomic weights of all elements in the assigned formula, the atomic weight of each element being taken as many times as indicated by the formula. The formula weight is therefore also a dimensionless number. Unless specifically stated otherwise, the formula weight and molecular weight are taken as identical. A mole (actually a "gram" mole) of a substance is defined as the amount of that substance in grams numerically equal to the formula weight. The mole is frequently too large a unit; hence, one thousandth of it, known as a millimole, finds use. The millimole is then the amount of substance in milligrams numerically equal to the formula weight. One mole of the compound sodium chloride, NaCl, equals 22.9898 + 35.453 = 58.443 g. One millimole of sodium chloride equals 58.443 mg.

Some authors restrict the term "mole" to compounds and employ the terms "gram-atom" and "gram-ion" in applying the mole concept to elements and simple ions, respectively; however, the distinction is not especially fruitful for the purposes of inorganic analysis.

The moles of a substance corresponding to a certain weight of it in grams is obtained from the relationship

$$\text{amount in moles} = \frac{\text{weight in grams}}{\text{formula weight}}$$

Rearrangement of this expression yields

$$\text{formula weight} = \frac{\text{weight in grams}}{\text{amount in moles}}$$

Thus the formula weight has the units of grams per mole. This may seem to contradict the previous statement that a formula weight is a dimensionless number. This contradiction, however, is only an apparent one. It must be realized that the concepts of "dimension" and "unit" are related but not identical. "Gram" and "mole" have the same dimension, mass. Consequently, the formula weight, even with the units "grams per mole" attached, is still dimensionless!

It is important that the student fully appreciate these definitions presented, since he may have been exposed to different ones that have equal validity and possess advantage for some other area of science. It is imperative that a given set of definitions be used consistently and persistently to assure complete dimensional and unit accord of all quantities encountered.

Mole Fraction and Mole Per Cent. If in a given amount of a binary mixture of two substances A and B, a moles of A are present and b moles of B, then the mole fraction of substance A present is given by $a/(a + b)$ and that of B by $b/(a + b)$. The fractions are commonly expressed decimally. The sum of the mole fractions of A and of B is unity. Extension of the concept to mixtures of three or more components is obvious. If a mole fraction is multiplied by 100, a mole per cent is obtained. Consider a mole of sodium chloride: It contains one mole of sodium ion and one mole of chloride ion; the mole fraction and mole per cent of sodium present are, respectively, 0.5 and 50%.

Formality. Formality as an expression of chemical concentration is defined as the moles of a substance per liter of solution (*not* per liter of solvent!). Since the mole is the amount of substance in grams numerically equal to the formula weight, it is necessary to have the formula of the substance either given or understood when stating a formality. With the concentration of a solution expressed in formality, it is possible to relate chemical units to volumes, which are more quickly measured than weights. However, as the volume of a solution is temperature-dependent, in the statement of the formality of a solution the temperature must be specified. If no temperature is given, room temperature is assumed. Formality is denoted by a capital F (in italic in printing, and best underlined in handwritten text and problems). Thus a 1.30 F aqueous sodium chloride solution, which may be further abbreviated as 1.30 F NaCl (water being assumed as the solvent unless otherwise specified), contains $1.30 \times 58.44 = 76.0$ g of sodium chloride dissolved in water and diluted with water to a total volume of exactly 1 liter. (Note that the solution is not obtained by dissolving 76.0 g of sodium chloride *in* 1 liter of water!)

In mathematical expressions formal concentration is in this text denoted by the capital letter C with subscripts, where necessary, to differentiate various species. Thus the expressions C_A and C_{HCl} denote the formal concentration of substance A and hydrochloric acid, respectively.

Molarity and Its Relation to Formality. A solution of a desired formality can be prepared by dissolving the necessary number of moles of the substance and diluting to 1 liter. The statement of the formality in no way implies that this substance is present as such, that is, as added. It may dissociate, associate, form complexes with other substances added or present, or undergo some reaction. However, all these changes do not alter the fact that the solution is of the stated formality.

In many cases it is desirable to have a special label for the concentration of any species that forms as a result of these changes. The concentration of a species *actually* present is conveniently expressed in molarity, that is, as the moles of that species per liter of solution. In mathematical expressions, molarity is denoted by placing the designation for the species within brackets. Thus the expressions [A] and $[Cl^-]$ denote the (actual) molar concentrations of substance A and of chloride ion, respectively.

To bring the distinction between formality and molarity into sharp relief, a simple example will be helpful. Consider that 0.100 F solutions of hydrochloric acid and acetic acid are prepared. These solutions might be prepared by dissolving in water $\frac{1}{10}$ mole of hydrogen chloride and acetic acid, respectively, and diluting each solution to 1 liter with water. From elementary chemistry, the student will appreciate that the hydrogen chloride completely dissociates in the aqueous solution into hydrogen ion and chloride ion and that no undissociated hydrogen chloride is present. The student will also be familiar with the fact that the acetic acid solution contains only small amounts of hydrogen ion and acetate ion and that most of the acetic acid is present in its undissociated form. It should be emphasized that regardless of these chemical facts each solution is 0.100 *F* in its acid. However, the molarity of undissociated hydrogen chloride is *zero*; in contrast, the molarity of the undissociated acetic acid has a finite value. Because of the complete dissociation of hydrogen chloride, the following relations hold for 0.100 *F* HCl:

$$\underset{0.100\,F}{C_{HCl}} = \underset{0.100\,M}{[H^+]} = \underset{0.100\,M}{[Cl^-]}$$

Because of the incomplete dissociation of acetic acid, HOAc, the following relations hold for 0.100 *F* acetic acid:

$$\underset{0.100\,F}{C_{HOAc}} = \underset{98.66 \times 10^{-3}\,M}{[HOAc]} + \underset{1.34 \times 10^{-3}\,M}{[OAc^-]}$$

Since one mole of hydrogen ion is formed for each mole of acetate ion, the molarity of hydrogen ion is also $1.34 \times 10^{-3}\,M$. The molarities given above for the acetic acid solution are calculated by means to be described in a later section.

The distinction between total (analytical) concentrations and actual concentrations expressed in moles per liter, termed, respectively, formality and molarity, is extremely helpful for the unambiguous consideration of chemical equilibria. However, the student's attention is called to the fact that this distinction is not always made and that what is here described as formality is sometimes denoted as molarity. The distinction between the two senses in which the term molarity is then used must be recognized from the context on the basis of the relevant chemical facts.

In physical chemistry, it is convenient to express concentration in moles per 1000 g of *solvent*; this is termed the weight-formality or molality. This mode of expression is almost never employed in chemical analysis.

Equivalents and Normality. When two substances react with each other they need not do so in equal *molar* amounts. One mole of one substance may be "chemically equivalent" to two or more moles of the other substance (e.g., one mole of sulfuric acid requires two moles of sodium hydroxide for complete neutralization). For many purposes it is desirable to have a unit that implicitly incorporates the chemical relationship existing in the reaction. This unit is known as the chemical equivalent or, in brief, the equivalent. The number of equivalents of substance is obtained from the number of moles by multiplication by the equivalence number, which is derived from the molar combination ratio or reaction ratio of the two substances. The equivalent weight of a substance is obtained by dividing the formula weight by the equivalence number. Since the latter is a dimensionless number, the equivalent and equivalent weight have the same dimensions (units) as the mole and formula weight, respectively. The normality of a solution is defined as the equivalents of a substance in 1 liter of solution. Normality is denoted by a capital N (in italic in printing and best underlined in handwritten text and problems).

It should be emphasized that the terms equivalent, equivalent weight, and normality must be related to the reactions to which they apply; otherwise, ambiguity will often result. Where a solution may be used as an active participant in various reactions involving different numbers of reactive species per molecule, it is best to label its concentration in term of formality. Further discussion of these various terms is deferred to the consideration of titrimetry (see Chapters 15, 19, and 26).

Titer. A further approach to chemical expression of concentration is the use of titer. A titer expresses the relation of the solute present in a unit volume of its solution to the amount of a stated substance for a particular reaction of the solute with that substance. Thus, the "strength" of a hydrochloric acid might be stated as 1.00 ml = 4.08 mg of Na_2CO_3. The value 4.08 mg is the sodium carbonate titer of the particular hydrochloric acid under the actual reaction conditions. The use of titers gained early recognition in the evolution of titrimetry, since no knowledge of the chemical reaction (or reactions) involved is needed. The "titer method" simplifies calculations and is especially advantageous in the routine analysis of many similar samples, and indeed where "standard" samples are used to establish the titer under the actual conditions of the determination, some errors may be compensated. When the reaction is known to proceed quantitatively, a titer may be calculated from the stoichiometry of the reaction. The titer concept will be more fully understood when the student is introduced to the theory and practice of titrimetry (see Chapters 19 and 26).

3.4 Illustrative Examples: Concentration and Content

In the calculation of dilution problems, unless otherwise stated, it is assumed that volumes are additive; that is, contraction or expansion in volume on mixing either does not occur or is neglected.

Example 3-1. What weight of NaCl is present in 1.50 kg of a NaCl–KCl mixture 16.2% w/w in NaCl?

Since exactly 100 kg of the mixture contains 16.2 kg of NaCl, then 1.50 kg contains x kg, hence by proportion,

$$\frac{100}{16.2} = \frac{1.50}{x}$$

and $$x = \frac{16.2 \times 1.50}{100} = 0.243 \text{ kg} = 243 \text{ g of NaCl}$$

Example 3-2. What weight of the mixture of Example 3-1 must be taken to provide exactly 400 g of KCl?

Since exactly 100 g of the mixture contains 100 − 16.2 = 83.8 g of KCl, then x g contains 400 g, hence

$$\frac{100}{83.8} = \frac{x}{400}$$

and $$x = \frac{100 \times 400}{83.8} = 477 \text{ g of the mixture}$$

Example 3-3. What weight of water must be used to dissolve 25.0 g NaCl to obtain an 8.00% w/w solution?

Since exactly 100 g of the solution contains 8.00 g of NaCl, then x g contains 25.0 g, hence

$$\frac{100}{8.00} = \frac{x}{25.0}$$

and $$x = \frac{100 \times 25.0}{8.00} = 312 \text{ g of solution}$$

Therefore, the weight of water needed = 312 − 25 = 287 g.

Example 3-4. A 45.0% w/w aqueous solution of substance A has a density of 1.62 g/ml. Calculate the % w/w of this solution.

The weight of 100 ml of the solution is 100 × 1.62 = 162 g; this volume contains 45.0 g of substance A. Hence

$$\% \text{ w/w of substance A} = \frac{45.0 \times 100}{162} = 27.8\%$$

Example 3-5. Express the composition in mole per cent of a mixture of NaCl (*58.4*) and K_2SO_4 (*174.3*) that is 65.0% w/w in NaCl.†

Exactly 100 g of the mixture contains 65.0 g of NaCl or 65.0/58.4 = 1.113 moles of NaCl.

Exactly 100 g of the mixture contains 100 − 65.0 g = 35.0 g of K_2SO_4 or 35.0/174.3 = 0.2008 mole of K_2SO_4.

---

†Here, as throughout this textbook, formula weights (sometimes rounded) are often given parenthetically in the statement of a problem in italic numerals; these values are to be used in the calculations of the *particular* problem so that the arithmetic effort is reduced.

Hence, the total moles in exactly 100 g of the mixture = 1.113 + 0.201 = 1.314 moles and the mole per cent of NaCl = 1.113 × 100/1.314 = 84.7 mole % of NaCl.

By difference, the mole per cent of K_2SO_4 = 100.0 − 84.7 = 15.3 mole %.

Example 3-6. What weight of a 14.0% w/w aqueous solution of substance B must be added to 35.0 g of a 6.0% w/w aqueous solution of B to obtain a 7.0% w/w solution?

Initially the amount of B present = 35.0 × 0.060 = 2.10 g of B. After the addition the weight of solution will be 35.0 + x, where x is the number of grams of the 14.0% solution to be added.

The total amount of B in the final solution is 2.10 + x × 0.14 g. Hence the percentage of B in the final solution is given by

$$\frac{(2.10 + x \times 0.14) \times 100}{35.0 + x} = 7.0\%$$

Solving for x yields x = 5.0 g of the 14.0% w/w solution.

Example 3-7. What volume of 0.12 F H_2SO_4 must be added to exactly 500 ml of 0.09 F H_2SO_4 to obtain 0.10 F H_2SO_4?

By reasoning similar to that used in the solution of Example 3-6, the following equation is obtained, where x stands for the milliliters of 0.12 F H_2SO_4 to be added:

$$\frac{500 \times 0.09 + x \times 0.12}{500 + x} = 0.10$$

Solving for x yields $x = 2.5 \times 10^2$ ml of 0.12 F H_2SO_4.

Example 3-8. What volume of water must be added to exactly 300 ml of 0.250 F HCl to secure 0.200 F HCL?

Reasoning similar to that employed in the solution of Example 3-6 yields

$$\frac{300 \times 0.250 + x \times 0}{300 + x} = 0.200$$

Solving for x yields x = 75.0 ml of water to be added.

Note that the term "$x \times 0$," which is included to make the reasoning fully analogous with Examples 3-6 and 3-7, corresponds to the fact that the concentration of hydrogen chloride in the water is zero.

Example 3-9. What is the formality of a sulfuric acid solution containing 62.0% w/w of H_2SO_4 (*98.0*) and having a density of 1.520 g/ml?

One liter (= 1000 ml) of the acid weighs 1000 × 1.520 = 1520 g and contains 1520 × 0.620 = 942 g of H_2SO_4; this amount of H_2SO_4 represents 942/98.0 = 9.61 moles. Since the calculation is based on a volume of 1 liter, the result is 9.61 F H_2SO_4.

In consolidated form and with the units shown the solution may be presented as

$$\frac{10^3 \dfrac{\text{ml}}{\text{liter}} \times 1.520 \dfrac{\text{g}}{\text{ml}} \times 62.0 \dfrac{\text{g}}{100\text{ g}}}{98.0 \dfrac{\text{g}}{\text{mole}}} = 9.61 \frac{\text{moles}}{\text{liter}} = 9.61\ F$$

Example 3-10. What will be the formality of an ammonia solution secured by diluting 20.0 ml of concentrated aqueous ammonia (density, 0.90 g/ml; 28% w/w NH_3; mol. wt. NH_3, 17.0) to a volume of 50.0 ml with water?

(Note that this dilution represents the addition of 30.0 ml of water.)

By reasoning analogous to that in Example 3-6, the formality of the concentrated aqueous ammonia is calculated:

$$\frac{1000 \times 0.90 \times 0.28}{17.0} = 14.8\ F$$

Then reasoning as in Example 3-8 yields the formality of the final solution:

$$\frac{20.0 \times 14.8 + 30.0 \times 0}{50.0} = 5.9_2\ F \text{ in } NH_3$$

Example 3-11. What volume of 40% w/w HCl (*36.5*; density, 1.198 g/ml) must be diluted with water to obtain exactly 250 ml of a 1.0 F solution?

The formality of the initial acid is

$$\frac{1000 \times 1.198 \times 0.40}{36.5} = 13.1\ F$$

Hence, if x denotes the milliliters of 13.1 F HCl to be taken,

$$\frac{x \times 13.1}{250} = 1.0$$

and x = 19 ml of the 40% w/w HCl to be taken.

Example 3-12. What is the formality of H_2O in pure water? Water has the rounded formula weight of 18.01, and 1 liter of water at 25°C weighs 997.04 g. Hence the formality of H_2O in pure water is

$$\frac{997.04}{18.01} = 55.36\ F$$

Questions 3.5

Define the following terms: mixture, solution, solute, and solvent.

Assume that a 0.100 F aqueous solution has been prepared at 20.0°C. Will the formality be greater or smaller at 30.0°C? Explain your answer.

Discuss the advantages and disadvantages of concentration expressions based on weight by weight as compared with those based on weight by volume.

3-4. Define the following terms: mole, formula weight, formality, and molarity.

3.6 Problems

3-1. An alloy is obtained by melting a mixture of 36.0 g of aluminum, 14.0 g of magnesium, and 2.00 g of manganese. Calculate the percentage composition of the alloy.

Answer: 69.2% Al, 26.9% Mg, and 3.8_5% Mn

3-2. Calculate the composition in mole per cent of the alloy of Problem 3-1, where the atomic weights of Al, Mg, and Mn are taken as 27.0, 24.3, and 54.9, respectively.

Answer: 68.5, 29.6, and 1.9 mole %, respectively

3-3. At a certain temperature, NaCl shows a solubility of 35.0 g in 100 ml of water. Calculate the % w/w of this saturated solution.

Answer: 25.9% w/w

3-4. Solution A is 25% w/w in NaCl and solution B 35% w/w in NaCl. Calculate the per cent of NaCl in the solution obtained by mixing 50 g of solution A and 43 g of solution B.

Answer: $29._6$% w/w

3-5. A solution of a certain substance is prepared by mixing 15 g of a 6.3% w/w solution and 23 g of a 7.2% w/w solution. How much pure water should be added to the mixture to obtain a 5.0% w/w solution?

Answer: 14 g (= 14 ml) of water.

3-6. Calculate the formality of a 26.8% w/w KOH (*56.1*) solution which has a density of 1.255 g/ml.

Answer: 6.00 *F*

3-7. What volume of concentrated HNO_3 (density, 1.513 g/ml; 100% w/w; mol. wt. HNO_3, 63.0) is needed to prepare 400 ml of 1.50 *F* HNO_3?

Answer: 25.0 ml

3-8. Express the concentration in parts per million of chloride (*35.5*) of a 1.6×10^{-5} *F* HCl solution.

Answer: 0.57 ppm of chloride

3-9. Calculate the formality of each of the solutions resulting on mixing solution A with solution B.

	Solution A		Solution B	
	Volume, ml	Formality	Volume, ml	Formality
(a)	300.0	0.120	200.0	0.150
(b)	50.0	0.1015	100.0	0.1250
(c)	80.0	3.2	50.0	2.5
(d)	500.0	1.0	300.0	2.0

Calculate the formality of each of the following solutions.
(a) 5.00 g of H_2SO_4 (*98.08*) in 400.0 ml of solution
(b) 4.00 g of H_3PO_4 (*98.00*) in 100.0 ml of solution
(c) 2.00 g of HCl (*36.46*) in 50.00 ml of solution
(d) 3.00 g of HNO_3 (*63.01*) in 200.0 ml of solution
(e) 33.5 g of H_2SO_4 (*98.08*) in 1.50 liters of solution

What percentage (w/w) of metal is present in each of the following compounds?

	Compound	Metal
(a)	NaCl (*58.44*)	Na (*22.99*)
(b)	$AgNO_3$ (*169.88*)	Ag (*107.87*)
(c)	Na_2SO_4 (*142.04*)	Na (*22.99*)
(d)	K_2CrO_4 (*194.20*)	K (*39.10*)
(e)	K_2CrO_4 (*194.20*)	Cr (*52.00*)
(f)	K_2CrO_4 (*194.20*)	K (*39.10*) and Cr (*52.00*)

Calculate the % w/w of H_2O (*18.02*) in each of the following substances.
(a) $BaCl_2 \cdot 2H_2O$ (*244.28*) (c) $K_2SO_4 \cdot Al_2(SO_4)_3 \cdot 24H_2O$ (*948.8*)
(b) $Na_2CO_3 \cdot 10H_2O$ (*286.14*) (d) $FeCl_3 \cdot 6H_2O$ (*270.3*)

Calculate how many milliliters of pure water must be added to each of the specified solutions to obtain solutions of the desired formalities. Assume that the volumes are additive.

	Original formality	Volume, ml	Desired formality
(a)	0.1012	250.0	0.1000
(b)	0.250	80.0	0.1000
(c)	1.00	125.0	0.250
(d)	0.200	75.0	0.150
(e)	0.200	20.0	0.080

Calculate the grams of solution II that must be added to the indicated amount of solution I to obtain solution III.

	Grams	Solution I, % w/w	Solution II, % w/w	Solution III, % w/w
(a)	500.0	12.0	20.0	16.0
(b)	32.0	30.0	9.5	16.9
(c)	20.0	24.0	18.4	20.0
(d)	120.0	18.0	39.0	25.0
(e)	200.0	25.0	16.0	19.0
(f)	160.0	38.0	29.0	31.0

You have 500.0 ml of a 30.0% v/v solution of glycerol in water. How many milliliters of that solution must you remove and replace with an 80.0% v/v solution to obtain 500.0 ml of a 40.0% v/v solution?

Express the composition of each of the binary alloys in % w/w.
(a) 350 g of Pb, 450.0 g of Cu (c) 620 g of Na, 44 g of K
(b) 36 g of Zn, 88 g of Cd (d) 235 g of Hg, 300.0 g of Ag

3.6 PROBLEMS

3-17. Calculate the number of moles of each of the compounds present in the specified volumes of its solution.

	Compound	Solution density, g/ml	% w/w	Volume, ml
(a)	KOH (*56.11*)	1.43	43.0	250.0
(b)	NaOH (*40.00*)	1.28	25.1	450.0
(c)	HCl (*36.46*)	1.19	37.2	100.0
(d)	HNO_3 (*63.01*)	1.41	70.1	400.0
(e)	H_2SO_4 (*98.08*)	1.83	95.5	150.0
(f)	H_3PO_4 (*98.00*)	1.69	85.0	300.0
(g)	$HClO_4$ (*100.46*)	1.66	69.8	600.0

3-18. Calculate the formality of each of the following solutions of acids and bases.

	Solute	% w/w	Density, g/ml
(a)	HNO_3 (*63.01*)	43.7	1.27
(b)	NaOH (*40.00*)	25.1	1.275
(c)	HCl (*36.46*)	10.5	1.05
(d)	NH_3 (*17.03*)	27.3	0.900
(e)	H_3PO_4 (*98.00*)	85.5	1.70
(f)	$HClO_4$ (*100.46*)	70.5	1.67
(g)	H_2SO_4 (*98.08*)	30.2	1.22
(h)	KOH (*56.11*)	21.4	1.20
(i)	H_3PO_4 (*98.00*)	10.0	1.05

3-19. Calculate the milliliters of each of the concentrated solutions that must be mixed with water to prepare a dilute solution of the specified volume and formality.

		Concentrated solution		Dilute solution	
	Solute	Density, g/ml	% w/w	Volume	Formality
(a)	$HClO_4$ (*100.46*)	1.66	70.0	500.0 ml	0.500
(b)	HNO_3 (*63.01*)	1.42	70.0	1.00 liter	1.00
(c)	HNO_3 (*63.01*)	1.27	43.7	2.00 liters	0.200
(d)	NaOH (*40.00*)	1.43	40.0	500.0 ml	0.286
(e)	NH_3 (*17.03*)	0.900	28.0	2.00 liters	0.100

4 CHEMICAL EQUILIBRIUM

The Law of Mass Action 4.1

Many chemical reactions can be made to proceed in either direction by a suitable choice of conditions; that is, they are reversible. This fact can be expressed in a chemical equation by the use of a double arrow:

$$aA + bB + cC \cdots \rightleftharpoons rR + sS + tT \cdots \qquad (4\text{-}1)$$

Where a reversible reaction has reached the stage at which no further change occurs in the relative amounts of the species involved in the reaction, an equilibrium exists. For such a reaction it can be derived that the concentrations of the constituents in the equilibrium mixture are governed by the relationship

$$K = \frac{[R]^r[S]^s[T]^t \cdots}{[A]^a[B]^b[C]^c \cdots} \qquad (4\text{-}2)$$

where the constant K is called the equilibrium constant. This formulation of a chemical equilibrium is termed the law of mass action, or sometimes the law of Guldberg and Waage, after its discoverers.

In writing the above expressions, certain conventions are observed: The reaction equation is written with the actual or assumed reactants on the left side and the products on the right. The expression for the equilibrium constant is written with the concentrations of the reaction products appearing as factors in the numerator and the concentrations of the reactants in the denominator; in addition, the coefficient of each participant in the balanced chemical equation becomes the exponent of the corresponding concentration.

It is important to realize that the equilibrium is a dynamic one. Although on a gross scale there may be no observable change in the relative amounts of the substances involved, the forward and reverse reactions continue on a molecular and ionic scale but at equal rates. The dynamic situation can be readily demonstrated by "labeling" one of the reactants with a radioactive tracer, as in the following example. Solid silver chloride, in which some of the atoms of silver are radioactive, is introduced into a saturated aqueous solution of inactive silver chloride. After some time, the solid silver chloride is removed by filtration and the filtrate placed in a radiation counter. The previously inactive solution is found to be radioactive. The reaction equilibrium may be represented by the following equation:

$$\underline{AgCl} \rightleftharpoons Ag^+ + Cl^- \tag{4-3}$$

where the underlining of a substance, here silver chloride, indicates that it is present as a solid phase.

Example 4-1. A chemical system is known to react according to the balanced reaction $2A + B \rightleftharpoons 2C$. When an equilibrium reaction mixture was analyzed, the following concentrations were found: $[A] = 1.0 \times 10^{-2}\ M$, $[B] = 4.0 \times 10^{-9}\ M$, and $[C] = 2.0 \times 10^{-5}\ M$. What is the value of the equilibrium constant?

Application of the expression for the law of mass action, (4-2), yields

$$K = \frac{[C]^2}{[A]^2[B]} = \frac{(2.0 \times 10^{-5}\ \text{mole/liter})^2}{(1.0 \times 10^{-2}\ \text{mole/liter})^2 \times (4.0 \times 10^{-9}\ \text{mole/liter})}$$

$$= 1.0 \times 10^3\ \text{liters/mole}$$

4.2 Discussion of the Law of Mass Action

Some important aspects and consequences of the law of mass action that often confuse the student deserve brief mention.

The equilibrium constant describes the *position* of the equilibrium and says nothing about the *rate* at which the equilibrium is attained. Thus, the position of the equilibrium between elemental hydrogen and oxygen, and water, $2H_2 + O_2 \rightleftharpoons 2H_2O$, lies extremely far to the right side, that is, toward the position of water. But it is possible to store a mixture of hydrogen and oxygen gases for months or longer with no significant conversion to water. If, however, a catalyst (e.g., finely divided platinum) is brought into contact with the hydrogen–oxygen mixture, the reaction speeds up tremendously and may even become explosive. Equilibrium is then attained in a fraction of a second.

The expression for the equilibrium constant is valid only for a dilute system and for a *reversible reaction under equilibrium conditions*. The concept of reversibility is a topic of more advanced courses. Here it suffices to state that a reaction is considered reversible if under the condition of interest it can proceed readily in either direction and if equilibrium is rapidly reestablished after a small (actually infinitesimally small) change in the conditions is imposed. Fortunately, the majority of reactions encountered in inorganic quantitative analysis are reversible and proceed rapidly to equilibrium; thus, the law of mass action can be properly applied to their study.

Although it is common practice to write an equilibrium constant without explicit statement of its units, it should be emphasized that usually it is *not* a dimensionless quantity. Thus, for the reaction $2A + B \rightleftharpoons R + W$, the equilibrium constant has the dimensions of $(\text{mass/volume})^{-1}$. In terms of the general form of the law of mass action, given above, (4-2), the equilibrium constant has the dimensions of (mass/volume), raised to the power $(r + s + t \cdots) - (a + b + c \cdots)$. It follows that it is impossible to compare two equilibria on the basis of the values of their respective constants unless the constants have identical dimensions (units).

The *numerical* value of the equilibrium constant will change with the units employed in the expression of concentration. For the purposes of analytical chemistry, molar concentrations are commonly employed. For reactions involving gases, partial pressures, which are proportional to concentrations, are convenient.

Effect of Conditions on Chemical Equilibrium 4.3

Some of the factors that influence either the position of a chemical equilibrium or the value of the equilibrium constant, or both, may be considered briefly.

Nature of the Reaction System. Obviously each reversible reaction involves a particular set of substances and is of unique nature, which is reflected in the numerical value of its equilibrium constant.

Concentration. If the concentration of any component of an equilibrium mixture is changed, the position of the equilibrium will be shifted. However, the numerical value of the equilibrium constant, to a first approximation, will remain unchanged; rather the concentrations of the reactants and products will be so altered that the quotient of the products of their concentrations, with the relevant exponents considered, is unchanged.

Activity. Implicit in the above consideration of the law of mass action is the assumption that each species involved is unaffected by the presence of other substances. This condition is met only in extremely dilute systems. When an equilibrium constant, as defined above, is determined carefully, slightly different values are obtained at different concentration levels, even though temperature and pressure are kept constant. For electrolytes in aqueous solution, which are the principal concern in this textbook, the attraction between oppositely charged ions becomes more pronounced as the concentration is increased and the effective concentration differs from the actual concentration. If the effective concentration, called activity, is substituted for the actual concentration, the so-called activity constant or thermodynamic constant is obtained. The value of this constant is not affected by the concentration level and varies only with temperature. It should also be appreciated that the numerical value of the concentration-based equilibrium constant can change with the concentration of species that do not even participate in the equilibrium.

The activity, a, of a species A can be related to the molar concentration of that species:

$$a_{A} = \gamma[A] \tag{4-4}$$

The activity coefficient is designated by the Greek letter gamma, γ. The values of such coefficients can be determined experimentally or can be estimated from theoretical considerations (Debye–Hückel equation). The latter approach, however, yields only satisfactory results for solutions of low ionic concentration (say, below about 0.5 M). For such solutions the activity coefficients usually have numerical values less than unity and approach unity at infinite dilution.

The value of the activity coefficient of an ion depends on the size and charge of that ion and, most important, on the concentrations and charges of other ions in the solution. The following data will give a rough idea of the magnitude of some activity coefficients. In an aqueous solution 0.1 F in potassium chloride, the activity coefficients of the ions H^{+}, Ca^{2+}, Al^{3+}, and Th^{4+} are approximately 0.9, 0.7, 0.4, and 0.3, respectively. These values are valid for 25°C and under the condition that the concentration of each ion named is negligibly small in comparison to the 0.1 formality of the potassium chloride. The data show that neglection of activity coefficients can lead to sizable errors.

Where equilibrium constants are involved, a practicable way exists to ease the problem. Activity coefficients do not change considerably within the ionic-strength region 0.01 to 1. Therefore, the constants may be determined at an ionic strength within this region, for example, at 0.1. If these "concentration constants" are used in connection with concentrations (rather than activities), the errors will be greatly diminished. Ionic strength in this context is a function of the concentrations and charges of *all* ions present in the solution. All these concepts are considered in detail in physical chemistry courses, and the brief discussion given here serves merely to point out the importance of activity. Throughout this textbook concentrations will be used and the reader should be aware of the fact that the results obtained are often only approximate ones.

Temperature. For a given chemical reaction, temperature is the only factor that may essentially influence the value of the (thermodynamic) equilibrium constant. The value may be either increased or decreased by a change in temperature (see below).

Catalysis. As was pointed out above, the equilibrium constant does not reveal anything about the rate at which the equilibrium is attained. A catalyst affects only the rate of a reaction. In a reversible reaction the rate of the reaction in either direction is affected in the same way; consequently, the position of an equilibrium is unaffected by the presence of a catalyst.

The Principle of Le Châtelier 4.4

The so-called principle or theorem of Le Châtelier permits a qualitative decision as to the direction of the shift in the position of the equilibrium when a change in conditions is imposed. This principle may be expressed in the following form: If a stress (e.g., by a change in temperature, concentration, or pressure) is applied to a system at equilibrium, the equilibrium will be displaced in the direction which tends to diminish the effect of the stress.

Consider the equilibrium $A + B \rightleftharpoons S + T$. If more S is added, the equilibrium position will shift to the left. This can be reasoned from the law of mass action. The same result would be predicted from the principle of Le Châtelier: that the system shifts to the left to diminish the effect of the stress imposed by the addition of more S.

This principle can also serve to indicate the influence of a temperature change if the reaction equation is written with the heat involved as either a reactant or product. (Whether the reaction is exothermic or endothermic, that is, whether heat is evolved or consumed, must be established experimentally.) Consider the reversible, *exothermic* reaction of A and B to yield R. This may be written $A + B \rightleftharpoons R + n$ calories. An increase in temperature may be considered to add the reagent "calories" to the right side. By the Le Châtelier principle, it would be surmised that with an increase in temperature the equilibrium would shift to the left, thus diminishing the applied stress. Since "calories" do not occur in the expression for the equilibrium constant, this shift results in a smaller numerical value of that constant at higher temperature. The use of "calories" as a reactant is only a first approximation, because "caloric" energy may be only a portion of the total energy involved in the reaction. However, this approximation is sufficient for many purposes. For a more rigorous treatment of chemical equilibrium and its thermodynamic basis, textbooks on physical chemistry should be consulted.

4.5 Questions

4-1. What is the meaning of the statement: "A chemical equilibrium is dynamic"?

4-2. In what units may the rate of a chemical reaction be expressed?

4-3. How does temperature affect the rate of attainment of an equilibrium?

4-4. What effect does temperature have on the value of an equilibrium constant?

4-5. What would be the dimensions of a rate constant for a reaction? Give an example.

4-6. Are quantitative predictions possible from the principle of Le Châtelier?

4-7. How will an increase in temperature shift the equilibrium of an endothermic reaction?

4-8. What is the thermodynamic equilibrium constant?

4-9. Will the value of the equilibrium constant of an uncatalyzed reaction change when a catalyst is added?

4-10. How can you prove experimentally that a chemical equilibrium is dynamic?

4-11. When silver chloride, AgCl, is prepared by mixing solutions of its ions, heat is evolved. Would you predict silver chloride to be more soluble in cold water or hot water? Explain your conclusion.

4-12. Write the expression for the equilibrium constant of the reaction $3A + B \rightleftharpoons 2U + V$.

4-13. What is a "concentration constant"? When is it used? What are its advantages?

4. The reaction A + B $\longrightarrow$ C is reversible, endothermic, and at 20°C has an equilibrium constant, $K = 4.0 \times 10^3$. At 46.5°C would you expect the numerical value of the constant to be the same, larger, or smaller? Explain.

5 SOLUBILITY AND SOLUBILITY PRODUCT

Solubility and Its Expression 5.1

All solutions involving solids dissolved in liquids may be thought of as being either saturated, unsaturated, or supersaturated. The solid is called the solute and the liquid the solvent. A saturated solution is one containing the maximum amount of solute soluble in the solvent under equilibrium conditions.

For most solutes the solubility increases with the temperature, although no direct proportionality exists. For example, 100 g of water will dissolve 13 g of potassium nitrate at 0°C and 246 g at 100°C; in contrast, the same quantity of water will dissolve 35.7 g of sodium chloride at 0°C and 39.8 g at 100°C. Substances that exhibit a decrease in solubility with an increase in temperature are less common; well-known examples include aqueous solutions of calcium acetate, calcium hydroxide, and sodium sulfate. The temperature must therefore be specified when a solubility value is given for a particular solute–solvent system. For a dissolved gas the pressure must also be specified.

It is important to appreciate that solubility refers to equilibrium conditions. A solution saturated at a higher temperature may be cooled and, although the solute may be far less soluble at the lower temperature, precipitation may not occur even over an extended period of time. Such a system is not in equilibrium but is said to be supersaturated; that is, it contains a greater amount of solute than is allowed by the solubility equilibrium. In most cases, if a supersaturated solution is agitated by either shaking or stirring, some solute crystallizes out quite promptly; the addition of solute crystals also aids in initiating crystallization.

If sufficient solute is placed in contact with a solvent, eventually a saturated solution will be attained. A dynamic equilibrium is thereby established and the amount of solute dissolving in a given interval of time equals the amount leaving the solution. As stressed in Section 4.1 in the general consideration of chemical equilibria, the rate of attainment of an equilibrium in no way reflects the position of the equilibrium. Because of the inapt use of such expressions as "readily soluble," the beginner often fails to appreciate the difference between the solubility of a substance and its rate of dissolution. A substance may dissolve slowly or rapidly, depending on its state of subdivision (e.g., large crystals or powder) and on the rate of agitation of the solvent, but this has no relation to the solubility of the substance. Substances that are of quite low solubility are often called "insoluble." More precise terms are essentially, sparingly, or practically insoluble.

Solubility may be expressed in many different ways. The mode most commonly encountered in tables of solubility data is the grams of solute per 100 g of *solvent* at a definite temperature, which is usually specified as a superscript number. The statement that the water solubility of sodium chloride is 35.7^{0} and 39.1^{100} is understood to mean 35.7 g of sodium chloride per 100 g of water at 0°C and 39.1 of sodium chloride per 100 g of water at 100°C, respectively.

Solubility may be related to the solution, rather than the solvent, by expressing the concentration of a saturated solution in terms of %w/w, %w/v, %v/v, or, for substances of low solubility, as parts per thousand or parts per million. These modes of expression were discussed in Section 3.2. The solubility may also be expressed as the formality of the saturated solution, that is, moles of solute present in 1 liter of the saturated solution. This is referred to as the formal solubility, S^{f}, and is an especially convenient mode of expression for the theoretical treatment of some problems. (The formal solubility is also often termed the molar solubility.)

5.2 Solubility Product

If the solute is a strong electrolyte, that is, a substance that in aqueous solution is completely (or essentially completely) dissociated into ions, an equilibrium exists that involves only the ions in the saturated solution and the undissolved solute in contact with the solution. For example, for a saturated aqueous solution of silver chloride, the equilibrium may be written as follows, where the underrule denotes the solid state:

$$\underline{AgCl} \rightleftharpoons Ag^{+} + Cl^{-} \tag{5-1}$$

The amount of undissolved solute in contact with the saturated solution has no effect on the concentration of that solution. If the solid silver chloride were separated by filtration, the filtrate (at the same temperature) would still be a saturated solution, and if more solid silver chloride were added, it would remain undissolved in dynamic equilibrium with the dissolved salt. Consequently, the concentrations of the ions of the solute are sufficient to describe the equilibrium. The equilibrium constant in this case is the product of the molar concentrations of the silver ion and chloride ions and is known as the solubility-product constant or simply as the solubility product, K_{sp}, of silver chloride in water:

$$K_{sp} = [Ag^+][Cl^-] \tag{5-2}$$

The concept may be generalized with the restriction of its application to only sparingly soluble, strong electrolytes. Thus, the solubility equilibrium of the strong electrolyte M_mA_a in water is represented by the following (where charges on the ions are omitted to simplify the notation):

$$\underline{M_mA_a} \rightleftharpoons mM + aA \tag{5-3}$$

The solubility-product expression then takes the form†

$$K_{sp} = [M]^m[A]^a \tag{5-4}$$

The solubility-product principle can, of course, be applied to electrolytes that yield more than two ionic species in water solution. For example, the expression for the solubility product of magnesium ammonium phosphate, $Mg(NH_4)PO_4$, is $K_{sp} = [Mg^{2+}][NH_4^+][PO_4^{3-}]$.‡

Ion Product 5.3

Assume that an aqueous solution of silver chloride is unsaturated. The product of the molar concentrations of the silver and chloride ions in the solution, which is known as the ion product, IP, will have a value less than that of the solubility product. If more silver chloride is dissolved, the value of the ion product is increased. Eventually when a saturated solution is reached, the value of the ion product will equal that of the solubility product. In a supersaturated solution the value of the ion product will exceed that of the solubility product. Comparison of the values of the ion product and the solubility product permits a judgment of whether a solution is saturated

---

†Values for the solubility product of some common electrolytes at 25°C are presented in Table C in the Appendix.

‡The solubility of an electrolyte and hence the value of the solubility product may change if the solid phase undergoes modification, including the formation of a different crystalline form, a change in the nature of the hydration, or, where very small particles are involved, by a change in the particle size.

or not and in the former case how far it is removed from that equilibrium condition.

5.4 Formal Solubility and the Solubility Product

It is possible to calculate the formal solubility, S^f, from the solubility product and vice versa.

Example 5-1. The solubility product of silver chloride in water is about 1×10^{-10}. Calculate the formal solubility of this salt. Since the solution contains one silver ion for every chloride ion, the molar concentrations of these two ions are identical and equal the formal solubility of the salt, S^f:

$$[Ag^+] = [Cl^-] = S_f$$
$$1 \times 10^{-10} = [Ag^+][Cl^-] = S^f \times S^f = (S^f)^2$$

and

$$S^f = \sqrt{1 \times 10^{-10}} = 1 \times 10^{-5}\,F$$

Example 5-2. Calculate the formal solubility of silver chromate, Ag_2CrO_4, when its solubility product has the value 2.0×10^{-2}. The solubility-product expression is

$$K_{sp} = 2.0 \times 10^{-12} = [Ag^+]^2[CrO_4^{2-}]$$

Since on dissolution of the salt two silver ions are formed for each chromate ion, and since the formal solubility of the salt, S^f, equals the molar concentration of the chromate ion, it follows that

$$[CrO_4^{2-}] = S^f$$
$$[Ag^+] = 2[CrO_4^{2-}] = 2S^f$$
$$2.0 \times 10^{-12} = (2S^f)^2 \times S^f$$
$$2.0 \times 10^{-12} = 4(S^f)^3$$
$$S^f = \sqrt[3]{\frac{2.0 \times 10^{-12}}{4}} = \frac{10^{-4}}{\sqrt[3]{2.0}} = 8 \times 10^{-5}\,F$$

The general relation between the formal solubility, S^f, and the solubility product, K_{sp}, of a strong electrolyte M_mA_a in water is given by

$$S^f = \sqrt[m+a]{\frac{K_{sp}}{[M]^m[A]^a}} \qquad (5\text{-}5)$$

The student should derive this formula by generalization of the reasoning of Example 5-2.

Superficial comparison of the value of the solubility product of silver chloride ($K_{sp} = 1 \times 10^{-10}$) with that of silver chromate ($K_{sp} = 2.0 \times 10^{-12}$) might lead to the conclusion that silver chloride is more soluble than silver chromate. However, comparison of the corresponding formal solubilities ($1 \times 10^{-5}\,F$ and $7.9 \times 10^{-5}\,F$, respectively) reveals that the opposite is true.

The explanation resides in the fact that solubility products of electrolytes of different types are not comparable. As noted in Section 4.1, values of equilibrium constants are in general recorded without their units being explicitly stated. The solubility products of silver chloride and silver chromate have different units—(moles/liter)^2 and (moles/liter)^3, respectively—hence, their direct comparison is impossible.

For both silver chloride and barium sulfate, the solubility product has a value of about 1×10^{-10}. Since the two substances are of the same type, MA, their formal solubility is identical, even though one involves the singly charged ions Ag^+ and Cl^- and the other the doubly charged ions Ba^{2+} and SO_4^{2-}.

Common-Ion Effect 5.5

The addition of a participant to a chemical system at equilibrium shifts the position of the equilibrium (see Section 4.3). Consequently the addition of a further amount of an ion involved in the solubility equilibrium should lead to further precipitation of the solute from the saturated solution. This phenomenon, known as the common-ion effect, finds important application either for reducing the concentration of an ionic species or for maintaining its concentration at a low level. The effect of the addition of a common ion can be evaluated from the expression for the solubility product.

Example 5-3. In 1 liter of a saturated water solution of AgCl ($K_{sp} = 1 \times 10^{-10}$), KCl is dissolved to the extent of 10 millimoles. What is the effect on the silver ion concentration?

The initial silver ion concentration is obtained from the expression for the solubility product:

$$K_{sp} = 1 \times 10^{-10} = [Ag^+][Cl^-]$$

The reasoning and calculation related to the solution with the KCl added takes the following form. The potassium ion is not a common ion and hence has no effect on the equilibrium. Chloride ion now stems from two sources, from the AgCl and from the KCl. Thus, we can write for the total chloride concentration

$$[Cl^-] = [Cl^-]_{AgCl} + [Cl^-]_{KCl}$$

The concentration of chloride from the KCl equals 10 millimoles/liter or $1.0 \times 10^{-2}\,M$. The concentration of the chloride from the AgCl equals that of the silver ion, since one silver ion is produced from each chloride ion. Hence

$$[Cl^-] = [Ag^+] + 1.0 \times 10^{-2}$$

Substituting this into the solubility-constant expression yields

$$1 \times 10^{-10} = [Ag^+]([Ag^+] + 1.0 \times 10^{-2})$$

This is a quadratic equation and an exact value can be obtained by solving it. However, it is possible to simplify matters. It is known that the addition of KCl causes a repression of the silver concentration. Since this concentration without repression is 10^{-5} *M*, it is permissible to neglect the term $[Ag^+]$ in the parentheses because 10^{-5} is small in comparison with 10^{-2}.† Then

$$1 \times 10^{-10} = [Ag^+] \times 1 \times 10^{-2}$$

and

$$[Ag^+] = 1 \times 10^{-8}\,M$$

It is seen that by increasing the chloride ion concentration by a factor of 10^3, the silver ion concentration is decreased by a factor of 10^3.

Example 5-4. To 1 liter of a saturated solution of AgCl, KCl is added until its concentration is 1×10^{-5} *F*. What will the silver ion concentration be when equilibrium is reestablished?

Proceeding as in Example 5-3, the following equation is obtained:

$$1 \times 10^{-10} = [Ag^+]([Ag^+] + 1 \times 10^{-5})$$

Neglecting the term $[Ag^+]$ in the parentheses and solving for $[Ag^+]$ as in the preceding example yields

$$[Ag^+] = \frac{1 \times 10^{-10}}{1 \times 10^{-5}} = 1 \times 10^{-5}\,M$$

Upon checking the result it becomes obvious that the neglected term is not much smaller than 10^{-5} and that an approximation analogous to that in the preceding example is not allowed here. Consequently, the full equation must be solved. The quadratic equation, as obtained on rearrangement, may be solved by application of the quadratic formula:

$$[Ag^+]^2 + 1 \times 10^{-5}[Ag^+] - 1 \times 10^{-10} = 0$$

$$[Ag^+] = \tfrac{1}{2}(-1 \times 10^{-5} \pm \sqrt{1 \times 10^{-10} + 4 \times 10^{-10}})$$

$$= 6 \times 10^{-6}\,M$$

Only the positive root is taken, because a negative concentration has no physical meaning.

5.6 Diverse-Ion Effect

The solubility product, as considered above, is a "concentration" constant and its numerical value will change somewhat even at constant temperature when further electrolytes are

†As a rule in equilibrium calculations a term may be neglected if it is 5% or less of the numerical value of the term to which it is to be added or from which it is to be subtracted. If the value is unknown or cannot be estimated, the simplified calculation should be performed. If examination of the result thus obtained indicates that the simplification is invalid, the more exact expression must be solved.

dissolved that have no ion in common with the insoluble substance. If activities (Section 4.3) were used in place of molar concentrations, the solubility-product expression would yield a true constant, influenced only by temperature. In general, the solubility of a sparingly soluble, strong electrolyte in water will increase with increasing concentrations of other electrolytes. This phenomenon is variously known as the inert-electrolyte effect, diverse-ion effect, or salt effect. For example, barium sulfate is increasingly soluble at increasing concentrations of such salts as KNO_3 and $Mg(NO_3)_2$. This enhanced solubility not only depends on the concentration of the diverse ions but also on their charges. Thus magnesium chloride, $MgCl_2$, will increase the solubility of barium sulfate in water more than an equimolar amount of potassium chloride.

Complications in Use of Solubility-Product Principle 5.7

The solubility-product expression considered above assumes that the ions of the dissolved strong electrolyte undergo no secondary reactions other than the formation of hydrated species, that is, aquo complexes. However, the anion or cation or both may undergo hydrolysis. An example of this is a solution of metal sulfide, MS, for which $K_{sp} = [M^{2+}][S^{2-}]$. The sulfide ion, S^{2-}, undergoes hydrolysis to the hydrogen sulfide ion, HS^-, according to

$$S^{2-} + H_2O \rightleftharpoons HS^- + OH^- \qquad (5\text{-}6)$$

The solubility-product expression involves the S^{2-} ion, but much of the ionic sulfur present in the solution is in the form of the HS^- ion. If this hydrolysis is neglected, there is no agreement between the formal solubility calculated from the solubility product and that established experimentally. Consideration of the influence of hydrolysis on solubility is deferred until Section 8.10.

Complex formation is a further type of secondary reaction that may affect a solubility equilibrium. For example, silver ion forms strong, soluble complexes with ammonia, and hence silver chloride is soluble in ammoniacal solution; the student is familiar with this reaction, which is used in the qualitative analysis scheme.

The formation of a complex may even occur with an ion directly involved in the solution equilibrium. Thus, silver ion forms the soluble chloro complex, $AgCl_2^-$, and hence silver chloride is soluble in solutions containing a high concentration of chloride ion. If chloride is added, as hydrochloric acid, for example, to a saturated solution of silver chloride in water, initially silver chloride precipitates due to the operation of the common-ion effect. Then as more hydrochloric acid is added, the effect of the formation of soluble complexes predominates, and the precipitate redissolves progressively and ultimately completely. For the present purpose it suffices to mention that hydrolysis and complex formation may occur and to a degree that they cannot be neglected in calculations based on the solubility-product principle.

5.8 Calculations Using Solubility-Product Principle

Some additional examples will serve to familiarize the student with calculations based on the solubility-product principle.

Example 5-5. A saturated aqueous solution of AgI ($K_{sp} = 1 \times 10^{-16}$) is given. (a) What is the concentration of silver ion in this solution? (b) What concentration of iodide ion must be established (e.g., by addition of KI) to reduce the silver ion concentration to 1×10^{-12} M?

(a)
$$1 \times 10^{-16} = K_{sp} = [Ag^+][I^-]$$

and since $[Ag^+] = [I^-]$,

$$[Ag^+] = \sqrt{1 \times 10^{-16}} = 1 \times 10^{-8}\,M$$

(b) Because the concentration of iodide ion furnished by the AgI may be safely neglected, it follows that

$$1 \times 10^{-16} = 1 \times 10^{-12}[I^-]$$

and

$$[I^-] = \frac{1 \times 10^{-16}}{1 \times 10^{-12}} = 1 \times 10^{-4}\,M$$

Example 5-6. Solid Ag_2CrO_4 ($K_{sp} = 2 \times 10^{-12}$) is equilibrated with $1 \times 10^{-3}\,F$ $AgNO_3$. What is the equilibrium concentration of chromate ion in the solution?

$$2 \times 10^{-12} = K_{sp} = [Ag^+]^2[CrO_4^{2-}]$$

The concentration of silver ion is $1 \times 10^{-3}\,M$ (from the $AgNO_3$) plus twice the chromate concentration (since two silver ions are produced for each chromate ion):

$$[Ag^+] = 1 \times 10^{-3} + 2[CrO_4^{2-}]$$

Substitution into the solubility-product expression yields

$$2 \times 10^{-12} = (1 \times 10^{-3} + 2[CrO_4^{2-}])^2[CrO_4^{2-}]$$

Neglecting $2[CrO_4^{2-}]$ within the parentheses gives the approximation

$$2 \times 10^{-12} \simeq (1 \times 10^{-3})^2[CrO_4^{2-}]$$

and

$$[CrO_4^{2-}] = \frac{2 \times 10^{-12}}{1 \times 10^{-6}} = 2 \times 10^{-6}\,M$$

Finally the approximation must be checked. 1×10^{-3} is indeed much larger than $2 \times 2 \times 10^{-6}$, so neglecting the parenthetical term is permissible.

Example 5-7. The value of the ion product of an unsaturated solution of $BaSO_4$ ($K_{sp} = 1 \times 10^{-10}$) is 1×10^{-12}. (a) What is the concentration of barium and sulfate ion in this solution? (b) What weight (in milligrams) of 100% w/w of H_2SO_4 (*98*) can be added to 2 liters of this solution before precipitation of sulfate commences?

(a)
$$\text{IP} = 1 \times 10^{-12} = [Ba^{2+}][SO_4^{2-}]$$
$$[Ba^{2+}] = [SO_4^{2-}] = \sqrt{1 \times 10^{-12}} = 1 \times 10^{-6}\,M$$

(b) The concentration of barium remains $1 \times 10^{-6}\,M$ up to the point of saturation. Therefore,

$$1 \times 10^{-10} = K_{sp} = [Ba^{2+}][SO_4^{2-}] = 1 \times 10^{-6} \times [SO_4^{2-}]$$

and
$$[SO_4^{2-}] = \frac{1 \times 10^{-10}}{1 \times 10^{-6}} = 1 \times 10^{-4}\,M$$

The difference in $[SO_4^{2-}]$ between the saturated and unsaturated solutions corresponds to

$$1 \times 10^{-4} - 1 \times 10^{-6} = 99 \times 10^{-6} \simeq 1 \times 10^{-4}\,M$$

Hence for 2 liters, 2×10^{-4} mole or 0.2 millimole of sulfate can be added before precipitation commences. This amount is furnished by $0.2 \times 98 = 20$ mg of 100% w/w H_2SO_4.

The next example demonstrates one possible way in which a formal solubility and the corresponding solubility product may be determined.

Example 5-8. Solid $SrSO_4$ *(184)* was added to about 2 liters of distilled water. The mixture was shaken for several hours at constant temperature and then was filtered. Exactly 1.5 liters of the clear filtrate was evaporated in a platinum dish to dryness. The residue of $SrSO_4$, after ignition and cooling, was found to weigh 0.174 g. Calculate the solubility product of $SrSO_4$.

The saturated solution contains 0.174 g of $SrSO_4$ in 1.50 liters. The formal solubility of this salt is therefore

$$S^f = \frac{0.174}{184} \times \frac{1}{1.50} = 6.30 \times 10^{-4}\,F$$

Hence

$$K_{sp} = [Sr^{2+}][SO_4^{2-}] = (6.30 \times 10^{-4})^2 = 39.7 \times 10^{-8}$$
$$= 3.97 \times 10^{-7}$$

Questions 5.9

1. Distinguish between the formal solubility and the solubility product of a strong electrolyte and explain their relationship.

5.10 PROBLEMS

5-2. Explain why the solubility-product concept can be applied only to strong electrolytes of low solubility.

5-3. Why is it unnecessary to include the concentration of the undissolved solute in the solubility-product expression?

5-4. Explain what is meant by the statement that a solubility equilibrium is dynamic?

5-5. In equilibrium calculations a concentration term is frequently neglected in additions or subtractions. What general rule was adopted for deciding whether the resulting approximation is valid?

5-6. What is the common-ion effect? Explain it in terms of the solubility equilibrium.

5-7. How is the solubility equilibrium of an electrolyte affected by the addition of a salt which does not contain a common ion?

5-8. Describe one way in which the value of a solubility product might be determined.

5-9. Elaborate on how the solubility of an electrolyte may be influenced by the presence of other substances.

5-10. What is the inert-electrolyte effect?

5.10 Problems

5-1. Calculate the formal solubilities of the salts MA($K_{sp} = 7.0 \times 10^{-10}$) and M_3A_2($K_{sp} = 1.0 \times 10^{-31}$) and state which is more soluble.

Answers: 2.6×10^{-5} *F* and 2.5×10^{-7} *F*, respectively; MA is more soluble

5-2. When solid AgBr ($K_{sp} = 5.0 \times 10^{-13}$) is equilibrated with 0.020 *F* KBr, what will be the silver ion concentration in the solution?

Answer: 2.5×10^{-11} *F*

5-3. Calculate the formal solubility of the salt M_2A_3 for which K_{sp} has the value 5.0×10^{-96}.

Answer: 3.4×10^{-20} *F*

5-4. How many milligrams of barium (*137*) will be present in 500 ml of a solution obtained by equilibrating solid $BaSO_4$ ($K_{sp} = 1 \times 10^{-10}$) with 0.1 *F* H_2SO_4?

Answer: 7×10^{-5} mg

5-5. Calculate the value of the solubility product of the salt from its given formal solubility. Neglect hydrolysis.

	Salt	Formal solubility
(a)	AgCl	1.3×10^{-5}
(b)	AgBr	7.2×10^{-7}
(c)	$BaSO_4$	1.1×10^{-6}
(d)	$Ca(C_2O_4)_2$	4.8×10^{-5}
(e)	Ag_2SO_4	1.1×10^{-2}
(f)	Ag_3PO_4	1.6×10^{-5}

•. Calculate the formal solubility of the salt from the given value of its solubility product. Neglect hydrolysis.

	Salt	Solubility product
(a)	AgI	8.3×10^{-17}
(b)	$SrSO_4$	3.2×10^{-7}
(c)	$PbCl_2$	1.6×10^{-5}
(d)	AgSCN	1×10^{-12}

. From the given value for the solubility product, calculate the amount, in milligrams, of the salt present in 50.0 ml of its saturated solution.

	Salt		Solubility product
(a)	$BaSO_4$	*(233.4)*	1.3×10^{-10}
(b)	$Ba(IO_3)_2$	*(487.2)*	2×10^{-9}
(c)	AgI	*(234.8)*	3×10^{-7}
(d)	SrF_2	*(125.6)*	1.2×10^{-5}

. The solubility of a salt AB_2 (FW = 100.0) is 0.60 g per 100 ml of water. A volume of 30.0 ml of a 0.300 M solution of A^{2+} is mixed with 30.0 ml of a 0.200 M solution of B^-. Show by calculation whether or not a precipitate of AB_2 will form.

•. What concentration of KCl *(74.6)* in milligrams per liter must be established in a saturated solution of AgCl ($K_{sp} = 1.8 \times 10^{-10}$) to adjust the silver ion concentration to 5.0×10^{-7} M?

•. A volume of 16.0 ml of 0.040 F $AgNO_3$ is mixed with 30.0 ml of 0.030 F NaCl. Calculate the amount, in milligrams, of Ag^+ *(107.9)* remaining in solution. K_{sp} for AgCl is 1.8×10^{-10}.

. To 50.0 ml of a saturated solution of AgCl ($K_{sp} = 1.8 \times 10^{-10}$) is added 50.0 ml of a solution containing 1.70 g of $AgNO_3$ *(169.9)*. Calculate the chloride ion molarity in the resulting solution.

•. To 10.0 ml of 0.010 F $AgNO_3$ is added 10.0 ml of 0.020 F Na_2SO_4. Show by calculation whether or not a precipitate of Ag_2SO_4 (K_{sp} = 2.1×10^{-5}) will form.

. What molar concentration of chloride ion is required to initiate the precipitation of the metal chloride of the given solubility product if the solution contains 10 mg of the metal per milliliter?

	Metal chloride	Solubility product
(a)	AgCl	1.8×10^{-10}
(b)	$PbCl_2$	1.6×10^{-5}
(c)	Hg_2Cl_2	1.3×10^{-18}

. A saturated solution of AgBr is shaken with an excess of AgCl until equilibrium is established. Calculate the molar concentration of bromide ion in the solution.

. To a saturated solution of AgI is added an excess of solid AgCl. The system is shaken until equilibrium is attained. What is the ratio of the molar concentration of iodide ion in the resulting solution to that originally present?

6 PRINCIPLES OF GRAVIMETRIC ANALYSIS

General Considerations 6.1

Gravimetric analysis depends on measuring weights or weight changes and using these data as the basis for the calculation of the amount of a sought-for substance. Although in analytical chemistry the expression *weight* is used, it is actually *mass* that is determined. The usage stems from the fact that a balance is employed, which establishes the mass of an object by comparing the weight of the object with the known weight of a reference mass. It is convenient to distinguish three different types of gravimetric determinations: precipitation methods, electrogravimetric methods, and evolution methods.

Precipitation Methods. Precipitation methods typically involve the following steps:

1. Weighing of a certain amount of material to be analyzed.
2. Dissolution of the weighed sample.
3. Addition of an appropriate reagent to form a sparingly soluble compound with the sought-for substance.
4. Isolation of the precipitate formed.
5. Purification of the precipitate.
6. Weighing of either the precipitate after drying or a compound formed from it by an appropriate conversion.

From the final weight obtained, the amount of sought-for substance can be calculated or, in conjunction with the sample weight, the content of sought-for substance in the sample material can be established.

For example, if sodium chloride is to be determined by a gravimetric method in a mixture with sodium carbonate, a weighed amount of the mixture is dissolved in dilute nitric acid and a solution of silver nitrate is added. The silver chloride precipitated is separated by filtration, washed, dried, and finally weighed. From the weight of the silver chloride, the corresponding weight of sodium chloride can be calculated and consequently the percentage of this substance in the original mixture.

Electrogravimetric Methods. Electrogravimetric methods are based on the deposition of a substance on an electrode. The weight of the deposit is obtained as the difference in the weight of the electrode with and without the (dried) deposit. Electrogravimetric methods are considered in Chapter 30.

Evolution Methods. Evolution methods are based on the volatilization of a component of the sample, usually by heating. The amount of substance volatilized is obtained from the difference in the sample weights before and after the volatilization step. This simple type of a gravimetric method is frequently employed for the determination of water by heating the sample for some time at 105 to 110°C and of carbon dioxide by ignition at a higher temperature. It is also possible to absorb the volatilized compound (water or carbon dioxide in the examples above) in an appropriate medium, which is weighed before and after the absorption process. The amount of the volatilized compound is obtained from the weight increase. Various evolution methods are considered in Chapter 48.

6.2 Computations in Gravimetric Analysis

The student is already familiar with "chemical arithmetic" from an introductory course in chemistry; consequently, the calculations in gravimetric analysis require only brief treatment.

The balanced reaction equation

$$Ag^+ + Cl^- \rightarrow \underline{AgCl} \tag{6-1}$$

not only expresses qualitatively that silver and chloride ions react to form silver chloride, but quantitatively that one silver ion reacts with one chloride ion to yield one molecule of silver chloride. Since one mole of silver ion contains the same number of particles as one mole of chloride ion or one mole of silver chloride, the equation also implies that one mole of silver ion reacts with one mole of chloride to yield one mole of silver chloride. Since one mole of a substance is its formula weight expressed in grams, the reaction equation also implies that 107.868 g of silver ion reacts with 35.453 g of chloride ion to yield 107.868 + 35.453 = 143.321 g of silver chloride, provided, of course, the substances are pure and the reaction proceeds to completion.

This quantitative relationship permits the calculation of the amount of silver or chloride in a given amount of silver chloride that is obtainable from a given amount of silver or chloride ion on complete reaction with chloride or silver ion, respectively.

Example 6-1. What weight of Ag (*107.87*) is present in 100.0 g of AgCl (*143.32*)?

One mole of AgCl contains one mole of Ag; consequently 143.32 g of AgCl contains 107.87 g of Ag, and 100.0 g of AgCl contains x g of Ag. By proportion:

$$\frac{143.32}{107.87} = \frac{100.0}{x}$$

and

$$x = \frac{107.87}{143.32} \times 100.0 = 75.27 \text{ g of Ag}$$

The student oriented to solving problems by "ratio" rather than by "proportion" will achieve the result directly by writing the last line of the above solution of the problem (and similarly in the illustrative problems below).

Example 6-2. What amount of Na (*22.99*) is present in 50.00 g of Na_2SO_4 (*142.04*)?

Since one mole of Na_2SO_4 contains two moles of Na, 142.04 g of Na_2SO_4 contains 2 × 22.99 g of Na and 50.00 g of Na_2SO_4 contains x g of Na. By proportion,

$$\frac{142.04}{2 \times 22.99} = \frac{50.00}{x}$$

and

$$x = \frac{2 \times 22.99}{142.04} \times 50.00 = 16.19 \text{ g of Na}$$

Example 6-3. What amount of NaCl (*58.44*) is present in a sample that yields on precipitation of the chloride with $AgNO_3$ 0.8342 g of AgCl (*143.32*)?

Since one mole of NaCl yields on mole of AgCl, 58.44 g of NaCl corresponds to 143.32 g of AgCl and x g of NaCl to 0.8342 g of AgCl.

$$\frac{58.44}{143.32} = \frac{x}{0.8342}$$

and

$$x = \frac{58.44}{143.32} \times 0.8342 = 0.3402 \text{ g of NaCl}$$

Example 6-4. How much $BaCl_2$ (*208.24*) is present in a solution if on addition of a sufficient amount of $AgNO_3$, the weight of the isolated AgCl (*143.32*) is 1.3456 g?

One mole of $BaCl_2$ contains two moles of Cl and therefore yields two moles of AgCl; consequently 208.24 g of $BaCl_2$ corresponds to 2 × 143.32 g of AgCl, and x g of $BaCl_2$ to 1.3456 g of AgCl.

$$\frac{208.24}{2 \times 143.32} = \frac{x}{1.3456}$$

and $$x = \frac{208.24}{2 \times 143.32} \times 1.3456 = 0.9776 \text{ g of } BaCl_2$$

The student should appreciate that in Examples 6-3 and 6-4 it is not necessary to calculate the amount of chloride ion and then to convert this result into sodium chloride and barium chloride, respectively. The calculations proceed directly via the overall weight relationships.

6.3 Gravimetric Factors

In each of the preceding examples the left side of the proportion or the ratio immediately following the equals sign in the last line of the calculation scheme involves only formula weights and is independent of the actual amounts of the substances involved in the particular example. These terms are known as chemical factors and in reference to gravimetric analysis as gravimetric factors. Such a factor may be represented by simply writing the relevant chemical formulas (commonly in italic in printed works), which are then understood to imply the formula weights.

For the determination of sodium chloride by precipitation as silver chloride (Example 6-3), the following general proportion may be written:

$$\frac{NaCl}{AgCl} = \frac{x}{w} \tag{6-2}$$

where w is the weight of the precipitate and x is the corresponding weight of sodium chloride. By rearrangement,

$$x = \frac{NaCl}{AgCl} \times w \tag{6-3}$$

Here the quotient $NaCl/AgCl$ is a gravimetric factor and need only be calculated once. This factor actually expresses the amount of sodium chloride corresponding to a unit amount of silver chloride and is a dimensionless number; w may therefore be expressed in any desired unit of weight (grams, milligrams, pounds, etc.) and x will and must (!) have the same unit.

The gravimetric factor for sulfate determined by precipitating and weighing barium sulfate corresponds to $SO_4^{2-}/BaSO_4$. For certain applied purposes, the sulfate content is expressed as sulfur trioxide; the factor then is $SO_3/BaSO_4$.

Phosphate can be determined by precipitating magnesium ammonium phosphate hexahydrate, which can be dried and weighed. The gravimetric factor for the orthophosphate ion will be $PO_4^{3-}/MgNH_4PO_4 \cdot 6H_2O$. If the phosphate content is to be given in terms of phosphorus pentoxide (as is common in the analysis of ores, fertilizers, etc.), the factor is

$$P_2O_5/2MgNH_4PO_4 \cdot 6H_2O$$

Observe that the coefficient 2 in the denominator compensates for the fact that two phosphorus atoms appear in the formula of the oxide but only one in that of the phosphate salt.

By ignition, magnesium ammonium phosphate can be readily converted to magnesium pyrophosphate, which can then be weighed. In this case the factors for the orthophosphate ion and for the pentoxide become

$$2PO_4^{2-}/Mg_2P_2O_7 \qquad \text{and} \qquad P_2O_5/Mg_2P_2O_7$$

respectively. In the first of these factors, a coefficient 2 is required in the numerator to compensate for the fact that only one phosphorus atom appears in the formula of the phosphate ion but two in that of the pyrophosphate salt. In the second factor, two phosphorus atoms appear in both formulas; consequently, both coefficients are unity and need not be written.

The student should realize that the coefficients are governed solely by the relation between the "linking" species (the phosphorus atom in the above examples) in the sought-for substance and the weighed substance. Mistakes of the novice are often associated with failure to recognize the linking atom.

The result of a determination is frequently expressed as the percentage of a constituent in the sample. The calculations then involved are best explained by examples.

Example 6-5. A mixture of NaCl (*58.44*) and Na_2SO_4 is to be analyzed for its NaCl content. A sample weighing 0.9532 g is dissolved in water and the chloride precipitated by the addition of an excess of $AgNO_3$. The dried precipitate of AgCl (*143.32*) weighs 0.7033 g. What is the per cent NaCl in the sample?

The calculation is evident from the following (compare Example 6-3):

$$x = \frac{NaCl}{AgCl} \times w$$

↑	↑	↑
g of NaCl present in sample	gravimetric factor	g of AgCl found in precipitate

$$= \frac{58.44}{143.32} \times 0.7033$$

The percent NaCl in the mixture is given by the following, where W is the sample weight in grams.

$$\% = \frac{x \times 100}{w} = \frac{58.44 \times 0.7033 \times 100}{143.32 \times 0.9532} = 30.09\%\ \text{NaCl}$$

Example 6-6. When a 1.3260-g sample of a substance was heated at 110°C, water was volatilized; after cooling under anhydrous conditions, the weight was 1.0112 g. What is the percentage of volatizable water in the substance?

$$\%H_2O\ \text{lost at } 110°C = \frac{\overbrace{(1.3260 - 1.0112)}^{\text{loss in weight} = \text{g of } H_2O} \times 100}{\underbrace{1.3260}_{\text{sample weight in g}}} = 23.74\%\ H_2O$$

A common error made by the beginner is to use different units in expressing the weight of the final substance and the weight of the sample. If the sample weight is expressed in grams or milligrams, then the weight of the final substance must also be expressed in grams or milligrams, respectively.

6.4 Calculation of Results on a "Dry Basis"

Different samples of a material often vary considerably in their moisture content, and these differences, of course, cause changes in the percentage composition with respect to other constituents. Sometimes analyses are performed to compare various samples, to check a material to learn whether it complies with certain specifications, or for similar purposes. It then becomes necessary to refer to the composition of the material on a "dry basis." Detailed consideration of the determination of moisture is deferred to other sections. However, the calculations can be appreciated from examples.

Example 6-7. A sample of an iron ore is found to have a moisture content of 1.56% and an "as is" iron content of 26.24%. What is its iron content on a "dry basis"?

A 100.00-g sample of ore contains 1.56 g of H_2O and 26.24 g of Fe. When dried, the weight remaining is 100.00 − 1.56 = 98.44 g, of which 26.24 g is still Fe. Hence, $\frac{26.24 \times 100}{98.44} = 26.66\%$ Fe (dry basis).

Example 6-8. One portion of a sample of a copper ore weighing 1.5653 g is dried to a constant weight of 1.4920 g. A second portion weighing 1.0075 g is analyzed for copper, and an amount of 320.3 mg of copper is found. Calculate the moisture content of the ore and the copper content on both the "as is" basis and dry basis.

The moisture content is

$$\frac{(1.5653 - 1.4920) \times 100}{1.5653} = 4.68\%\ H_2O$$

The copper content is

$$\frac{0.3203 \times 100}{1.0075} = 31.79\%\ \text{Cu ("as is" basis)}$$

and

$$\frac{31.79 \times 100}{100.00 - 4.68} = 33.35\%\ \text{Cu (dry basis)}$$

Precipitation 6.5

Precipitation is employed in quantitative analysis for various reasons. It serves for the separation of one substance from other substances, as the basis for a precipitation titration (Chapter 18), and as one of the initial operations in a gravimetric determination. For the last purpose, it is unnecessary that the composition of the precipitate be the same as that of the compound finally weighed. Hence in gravimetric analysis it is convenient to differentiate between the precipitation form and the weighing form. The requirements to be fulfilled by the two forms are different.

Precipitation Form. The precipitation form should have a low solubility in the medium employed; it should be pure, or at least contain only impurities that can be readily removed before the weighing step, and preferably it should be coarsely crystalline. The compound precipitated should also be readily dried, ignited, or otherwise converted to an appropriate weighing form.

Weighing Form. It is highly desirable that the compound representing the weighing form be strictly stoichiometric in composition, because the gravimetric factor can then be derived readily from stoichiometric considerations. If the composition is not stoichiometric, an empirical factor may be applied, but this practice is not favored and, of course, requires that the composition be the same in every determination. An additional practical requirement is that the weighing form not be readily attacked by moisture or by carbon dioxide or oxygen of the air. These attacks can be prevented by weighing in a closed container, but such a procedure is inconvenient. It is desirable that the compound used as a weighing form affords a gravimetric factor small in its numerical value; that is, the amount of the weighing form corresponding to a unit weight of the sought-for substance should be large. In such a case, for a given amount of the sought-for substance, the influence of weighing errors, impurities, uptake of small amounts of moisture or carbon dioxide, etc., is greatly decreased.

6.6 Mechanism of Precipitation

Sooner or later a precipitate must form when a solution contains more of a particular solute than is allowed by the solubility under the prevailing conditions. Consequently, the first step in the precipitation is the establishment of supersaturation. Then, as the second step, crystal nuclei make their appearance; that is, nucleation takes place. Crystal nuclei are exceedingly minute and may even consist of only a few unit cells of the crystal lattice. The rate at which such nuclei form increases with the degree of supersaturation existing.

In the third and final step these nuclei serve as centers for the deposition of more material and larger crystalline aggregates are formed. In addition to the growth of individual aggregates, two or more established crystals may cluster together to form a larger particle. These processes act to achieve what is called crystal growth.

Of course, more nuclei may form while crystal growth proceeds. Consequently, the rate of nucleation must be kept low and crystal growth encouraged in order to obtain a small number of large crystals rather than a large number of small ones. To achieve this goal the supersaturation must be kept low, because as it becomes greater the rate of nucleation is increased to a much greater extent than the rate of crystal growth. Supersaturation can be kept low by adding the precipitant slowly, under vigorous stirring, and as a dilute solution. But the degree of dilution allowed for the precipitant solution is limited. The more dilute this solution, the more of it must be used to bring the necessary amount of precipitant into the sample solution. Thus the latter solution will be diluted, possibly to such a degree that precipitation becomes incomplete. Even when a dilute precipitant solution is employed, a region of local supersaturation developes at the site where a drop enters the sample solution. While local supersaturation can be reduced to a great extent by adequate stirring, it is never avoided completely.

Even where the precipitation is effected according to these findings, the improvement in the character of the precipitate for some systems is inadequate. Such unfavorable cases include various metal sulfides and the hydrous oxides of aluminum, iron(III), and certain other polyvalent metal ions. In these cases the rate of nucleation or other unfavorable factors predominate and crystal growth is inadequate. As a result a gelatinous network of barely crystalline material forms. In some of these stubborn cases and in others, crystallinity can be improved, or at least the particle size be increased, by aging of the precipitate (see below).

Contamination of Precipitates 6.7

A precipitate, as it forms, is rarely pure! For gravimetric purposes it should be possible to remove the impurities readily or at least to reduce them to an extent such that the accuracy of the determination is not affected. This removal should take place during the washing of the precipitate or during its subsequent conversion to the weighing form.

Contamination of a precipitate can occur through various processes; a brief consideration of the important ones will throw some light on the conduct of a precipitation for gravimetric purposes.

Coprecipitation. In the broadest sense, coprecipitation is the concurrent precipitation of two or more substances. Thus, for example, the addition of silver nitrate to a solution containing both sodium chloride and sodium bromide leads to the precipitation of both silver chloride and silver bromide. In analytical chemistry and especially in the consideration of the contamination of a precipitate, the term coprecipitation is conveniently used in a more restricted sense. Here it refers to the carrying down of one or more substances that if present alone would not be precipitated by the reagent used. For example, calcium is partially coprecipitated when iron(III) is precipitated as its hydrous oxide by neutralization of an acidic solution to pH 4 to 5. Under the same conditions in the absence of iron, calcium remains in solution.

Coprecipitation can proceed through a number of mechanisms, including solid solution formation, surface adsorption, and occlusion. These mechanisms are considered separately below, but it should be appreciated that contamination often results from the concurrent action of more than one of these processes.

Solid Solutions. Just as two liquids may be miscible to form a liquid solution, two solids may be sufficiently soluble in each other to form a solid solution, which should be differentiated from a mixture obtained by merely grinding the substances together. Two substances may form "mixed crystals"; that is, one substance is incorporated into the crystal lattice of the other. Such mixed crystals are frequently, but not exclusively, formed when the two substances are isomorphous, that is, when their crystals are of the same crystallographic system. Such incorporation occurs even if the contaminating compound is soluble in the liquid solution under the conditions of the precipitation. For example, owing to the close similarity of structure, size, charge, and electronic configuration of chromate and sulfate ion, barium chromate may be incorporated into the precipitate of barium sulfate. This coprecipitation occurs even when barium sulfate is precipitated from a strongly acidic solution in which the barium chromate is soluble. The precipitate is then more or less yellow in color, depending on the chromate content. Permanganate ion can be incorporated to a certain extent into barium sulfate, although barium permanganate is quite soluble and the permanganate ion has a charge differing from that of the sulfate ion. The barium sulfate precipitate obtained is then pink in color. The minor constituent of a solid solution is scattered throughout the crystal and held there quite firmly by the lattice forces; washing of the precipitate is therefore ineffective in removing the contaminant.

Adsorption. On the surface of the particles of a precipitate, "active sites" are present, capable of attracting and holding species that by themselves are not precipitated. Obviously, contamination by this mechanism increases with the surface area of the precipitate. Consequently, in a gravimetric determination a coarse, crystalline precipitate with its smaller relative surface area is preferred over a finely divided one. Contamination by adsorption is most frequently encountered with gelatinous precipitates. Although adequate washing of a precipitate contaminated via adsorption is often a satisfactory remedy, it seldom succeeds with a gelatinous precipitate because its filtration proceeds very slowly and the wash liquid fails to penetrate it fully within a reasonable period of time.

Occlusion. The carrying down of impurities in the interior of primary particles is known as occlusion. This process would include in a broad sense the formation of solid solutions, considered above, but can be used in the more restricted sense of a mechanical occlusion, including the inclusion of mother liquor and ion entrapment, that is, the growth of a precipitate around an adsorbed ion. Retention of mother liquor is especially pronounced with gelatinous precipitates. The degree to which occlusion takes place depends to some extent on the rate of the precipitation. It can be appreciated that washing of the precipitate will never remove occluded impurities adequately.

Postprecipitation. In postprecipitation, an initially pure precipitate is contaminated by the subsequent deposition of a second solid phase formed by another substance that is only slightly soluble in the solution. Frequently this deposition proceeds from a supersaturated solution of the contaminant. Consider the precipitation of calcium as the oxalate from a magnesium-containing solution. If the calcium oxalate precipitate is not separated promptly by filtration, some magnesium oxalate is adsorbed on the surface and induces the precipitation of more magnesium oxalate, which once deposited does not redissolve. However, in the absence of calcium oxalate a supersaturated solution of magnesium oxalate can be kept for a considerable period of time without precipitation taking place.

Various metal sulfides offer interesting examples of postprecipitation. When copper(II) is precipitated from a zinc-containing solution by saturation of an acidic solution with hydrogen sulfide, zinc-free copper(II) sulfide is formed initially. When left in contact with the mother liquor, the copper(II) sulfide increases progressively in its zinc content. It may be reasoned that sulfide ion is adsorbed on the surface of the precipitate, thus raising the sulfide ion concentration locally to an extent that zinc sulfide deposits. Postprecipitation can usually be avoided, or at least be reduced, by prompt separation of the precipitate from the mother liquor.

Purification of Contaminated Precipitates 6.8

Every precipitate is washed before its conversion to the weighing form. The wash liquid should at least remove the mother liquor adhering to the precipitate. As noted above, washing does not always result in a complete removal of impurities; in such a case additional steps must be taken to achieve adequate purity.

Reprecipitation (Double Precipitation). Reprecipitation is often a satisfactory, but tedious, method to obtain a pure precipitate. The extent of the contamination of a precipitate depends largely on the concentration of the contaminating material in the solution. Hence, if an impure precipitate is redissolved and the precipitation repeated, the resulting precipitate is often far less contaminated. As mentioned above, when hydrous iron (III) oxide is precipitated from a calcium-containing solution, some calcium is coprecipitated. The contaminated hydrous iron(III) oxide can be separated by filtration, be washed and dissolved in dilute hydrochloric acid, and precipitated again by the addition of ammonia. Since the concentration of calcium in the second solution is far less than in the first, the calcium content of the second precipitate is usually negligible. Reprecipitation is impossible, or at least extremely difficult and tedious, when the precipitate cannot be redissolved readily. Barium sulfate represents a common example of such a case. In addition, reprecipitation may be of little value if the contamination is caused by an impurity of low solubility or by one that is isomorphous, or if the contamination is produced by the precipitant itself.

From the facts and phenomena considered above, it can be appreciated that it is not a simple task to obtain a precipitate of a purity suitable for a gravimetric determination. Since it is especially difficult to remove a contaminant, once introduced, it is desirable to perform the precipitation in such a way that a sufficiently pure product is obtained initially.

6.9 Aging of Precipitates (Digestion)

When a precipitate is allowed to stand in contact with its mother liquor, changes often take place that lead to a product of larger particle size and thereby improved filterability and having possibly less contamination. This overall transformation is called aging and may be the result of the operation of a number of processes. Small crystals have a slightly greater solubility than large ones; consequently they dissolve and their material is then deposited on the larger crystals. Small clusters of crystals may be joined and imperfections in crystals may be "smoothed." In addition, colloids undergo coagulation; that is, small colloidal particles join and the larger particles formed precipitate.

In practice, aging is often effected at an elevated temperature, for example, on a steam bath, so that the processes involved are accelerated. The operation is called aging by digestion or simply digestion of the precipitate.

Aging or digestion may also reduce the contamination of a precipitate, especially that due to adsorption of substances, since the large particles formed have a relatively smaller surface area. The precipitation of barium sulfate is an outstanding example where a steam bath digestion effects a great improvement in the filterability of the precipitate and reduces some of the contamination significantly. Aging, of course, fails and is inappropriate when the contamination is due to postprecipitation.

Precipitation from Homogeneous Solution 6.10

If the precipitant solution is added to the sample solution, undesirably high supersaturation results at least locally at the site where the drops of the precipitant solution enter. Here extensive nucleation may take place even upon slow addition and under vigorous stirring. This difficulty can be avoided if the precipitant is not added from without but rather is formed within the sample solution itself. This elegant technique is appropriately termed precipitation from homogeneous solution; its principles are best understood from a discussion of some practical examples. In the precipitation of hydrous iron(III) oxide, ammonia is often employed as the precipitating reagent. More realistically the precipitant is the hydroxide ion, which is formed according to

$$NH_3 + H_2O \rightleftharpoons NH_4^+ + OH^- \qquad (6\text{-}4)$$

Even if a dilute aqueous solution of ammonia is added slowly and with vigorous stirring, a local excess of hydroxide is unavoidable and a slimy precipitate results. However, if the hydroxide ion is generated in the solution itself at a low rate, the supersaturation is controlled, any local overconcentration is avoided, and a coarse, crystalline precipitate results. Urea is one substance that allows this favorable sequence of events. When heated in slightly acidic aqueous solution, urea reacts slowly with water, according to the reaction

$$CO(NH_2)_2 + H_2O \rightarrow (NH_4)_2CO_3 \rightarrow CO_2\uparrow + 2NH_3 \qquad (6\text{-}5)$$

The carbon dioxide is evolved from the boiling solution and the ammonia causes formation of hydroxide ion, as indicated above. Consequently, throughout the solution, hydroxide ion is generated evenly and slowly and no local supersaturation occurs. A very coarse product is obtained in the formation of the hydrous oxides of iron(III), aluminum, and certain other metals by this method.

Urea can also be employed for the precipitation of calcium oxalate in a coarse form. The latter compound is insoluble in neutral or alkaline aqueous solution but soluble in acidic solution. Thus, if an aqueous solution of a calcium salt is made acidic and oxalate ion is added, no precipitation occurs. If urea is now added and the solution is boiled, hydroxide ion is generated, and calcium oxalate is precipitated progressively as the solution is neutralized. In this manner, crystals more than 0.5 mm in length can be readily obtained.

The sulfides of nickel and cobalt are "problem" precipitates in the sulfide separation of cations employing gaseous hydrogen sulfide. Colloidal suspensions are obtained that do not coagulate and that are difficult to separate by filtration. Thioacetamide can serve for the homogeneous precipitation of the sulfides of these and other metals. In alkaline solution, the formation of sulfide ion from thioacetamide may be represented by

$$CH_3CSNH_2 + 2OH^- \rightarrow CH_3COO^- + NH_3 + HS^- \quad (6\text{-}6)$$

$$HS^- + OH^- \rightleftharpoons S^{2-} + H_2O \quad (6\text{-}7)$$

The decomposition of thioacetamide proceeds slowly even in boiling solution; consequently, the generation of sulfide ion is gradual and leads to the homogeneous precipitation of the metal sulfide. The products, thus obtained, are coarsely crystalline and are readily separated by filtration.

Precipitation from homogeneous solution is a tedious process, because boiling for 30 minutes to several hours may be required. The time spent, however, is often compensated by the ready filtration of the product and its lower degree of contamination, which make reprecipitation unnecessary. Contamination is reduced because the larger particles have a relatively smaller surface area and therefore adsorption is decreased. In addition, contamination caused by all types of occlusion and entrapment is greatly reduced, since the low rate of crystal formation allows time for the displacement of any foreign material from the crystal lattice.

6.11 Separation and Washing of Precipitates

The precipitate, whether separated from its mother liquor by filtration, centrifugation, or even decantation, must be properly washed, and certain criteria for the selection of a wash liquid merit mention. The (final) wash liquid should leave in the precipitate no foreign substance that cannot be removed in the subsequent treatment (usually drying or ignition) before the weighing.

Unless the precipitate has an extremely low solubility, it is important to keep the amount of wash liquid as small as possible. The washing will be more efficient if each small portion is completely removed from the precipitate before the addition of the next portion. If a significantly soluble precipitate is involved, its solubility in the wash liquid may often be reduced by resort to the common-ion effect (Section 5.5), that is, by having present in the wash liquid an ion common with the precipitate (but, of course, not the one to be determined!). Another possibility is to use at first a wash liquid saturated with the compound precipitated and in a final washing step to remove the saturated liquid with a minimum amount of water. Organic solvents such as alcohol or acetone, in which most ionic compounds are less soluble, may be employed to reduce losses during the washing procedure.

Where the precipitate is a coagulated colloid, for example, hydrous tin(IV) oxide, the use of pure water as the wash liquid may lead to losses by reversion of the precipitate to a colloidal solution that passes through the filter. This reversion, known as peptization, is avoided if the water contains a small amount of an electrolyte. To avoid a permanent contamination of the precipitate, this electrolyte must be one that can be volatilized in the subsequent ignition (e.g., ammonium salts).

Conversion of Precipitation Form to Weighing Form 6.12

Where the precipitation form and the weighing form are essentially identical, the precipitated and washed compound may be merely dried and weighed. Simple drying needs no discussion except to state that in some cases the temperature must be maintained within a specified range to assure the formation of a compound of definite composition without allowing partial decomposition or volatilization to take place. More involved operations are often necessary to obtain an appropriate and suitable weighing form, ranging from ignition to treatment with one or more chemicals. The object of ignition is to obtain a stable substance of definite composition. The optimum conditions will depend on the particular determination. If filter paper is used in the filtration of the precipitate, it must be burned, and often most carefully, to avoid any reduction of the filtered substance. For example, the carbon formed in the initial, incomplete combustion of the paper may reduce barium sulfate to barium sulfide according to the reaction

$$BaSO_4 + 4C \rightarrow BaS + 4CO \uparrow \qquad (6\text{-}8)$$

Sometimes for various reasons the ignition must be performed at a much higher temperature than that required to attain a desired stoichiometric composition. For example, hydrous aluminum oxide when heated to 200 to 300°C loses its water completely; however, the aluminum oxide thus obtained proves to be highly hygroscopic at room temperature. On heating for 10 minutes at 1100°C or higher, the aluminum oxide is converted to a nonhygroscopic form that after cooling can be weighed in an open crucible.

The determination of calcium involving calcium oxalate as the precipitation form offers instructive examples of weighing forms. The precipitate can be dried at 100°C to the monohydrate, but extreme care is required in the drying process to attain a product of the stoichiometric composition of the monohydrate. Calcium oxalate can be ignited to calcium carbonate, which is a suitable weighing form and is commonly used as such. Controlled ignition (475 to 525°C) is mandatory since at too high a temperature the calcium carbonate decomposes to calcium oxide. Complete conversion to this oxide, which is also a possible weighing form, can be accomplished at a sufficiently high temperature (above 880°C). However, the oxide is extremely hygroscopic, absorbs carbon dioxide, and consequently it is not a recommended weighing form.

6.13 Remarks on General Use of Precipitation

Although the considerations above have been directed principally to precipitation as a step for a gravimetric determination, it should be recognized that this method of separation has many other applications in both qualitative and quantitative analysis. The phenomena and facts considered above are relevant to this broader use of precipitation.

It is interesting to note that coprecipitation can be used to advantage to achieve the concentration and recovery of trace amounts of a substance. This substance may be deliberately coprecipitated with another compound, which is then known as a trace carrier, collector, or scavenger. As would be expected from the above considerations, gelatinous (colloidal) precipitates are especially suitable for this purpose.

6.14 Questions†

6-1. Define "weight" and "mass" and elaborate on their relation with emphasis to the common usage of the terms in gravimetric analysis.

6-2. In your own words define gravimetric analysis; outline the general

†Problems related to gravimetry appear in Section 7.5.

principles on which this technique is based; give diverse classifications of gravimetric analysis as they are possible from various points of view.

What is "coprecipitation" in the broad and restricted senses of the term?

List the major steps commonly involved in a gravimetric determination.

Differentiate between "precipitation" and "weighing" forms and name the requirements for the two forms that are similar, different, mandatory, and desirable.

What three steps lead to the formation of a precipitate?

What is "digestion" of a precipitate and what purpose does it serve?

By what mechanisms can a precipitate be contaminated? How may such contaminations be avoided or at least reduced?

What is a gravimetric factor? Should its numerical value be high or low? Explain.

How can crystal growth be favored over nucleation?

Compare the usual precipitation technique with precipitation from homogeneous solution. Discuss requirements in time and apparatus; purity, composition, and appearance of the precipitates.

For what purpose other than gravimetry might precipitation be performed in analytical laboratories?

What is coagulation?

What is a colloid?

Discuss advantages and disadvantages of separation techniques such as filtration, decantation, and centrifugation.

What means can you name that can improve the situation in a gravimetric analysis when a precipitate is involved that is not essentially insoluble?

7 APPLIED GRAVIMETRIC ANALYSIS

Selectivity and Sensitivity 7.1

One or more gravimetric methods are known for the accurate determination of almost all the major elements. Although the methods vary in their chemistry and complexity, in general it is not difficult to determine one element when it is present alone. However, in practical analysis the sought-for substance is usually accompanied by other substances. Hence, the problem of securing sufficient sensitivity in the determination is overshadowed by the far greater problem of securing adequate selectivity. In other words, the precipitant and the procedure selected must not only guarantee a complete precipitation but also a "clean" separation from other constituents of the solution. Not recognizing this difficulty, the novice frequently has the impression that gravimetric analysis is an easy task requiring for each element or ion nothing more than the mere application of one method. In actual analytical practice, a large variety of methods must be at hand and an appropriate selection must be made according to the amount and nature of the substances to be determined as well as the amount and nature of accompanying materials.

For example, the gravimetric determination of barium via the precipitation of barium sulfate is quite satisfactory when barium is present in a solution as the only cation. In contrast, the method is useless in the presence of strontium or calcium, or both, because these other alkaline earths also form sparingly soluble sulfates. In such a case barium is precipitated under carefully adjusted conditions as barium chromate. The precipitation of iron(III) as the hydrous oxide by the addition of ammonia followed by ignition and weighing as iron(III) oxide, Fe_2O_3, furnishes an additional example of the problem of analytical selectivity. When iron is present as the only cation, the method is satisfactory. In the presence of aluminum, chromium(III), and titanium(IV), the method fails because these metals (and some others) are also precipitated on the addition of ammonia. It is rather difficult to achieve a sufficiently "clean" separation of iron from these metals so that a gravimetric determination is made feasible. As mentioned in Section 6.7, even the presence of calcium may complicate the determination of iron because of coprecipitation. The difficulties increase when a gravimetric determination is contemplated for a substance that is present either in small or trace amounts. In a complex mixture, separations become then even much more involved. The introduction of organic precipitants has in many cases greatly improved the situation. In contrast to inorganic precipitants a greater variety of compounds is available and many of the reagents per se exhibit a higher degree of selectivity. In addition, the properties of organic compounds can be changed within certain limits by modifications of their structure. Often, with organic reagents of large formula weights, favorable gravimetric factors are obtained. The student may be familiar with the use of dimethylglyoxime in the qualitative test for nickel. This oxime is also employed in quantitative analysis and under appropriate conditions serves as a highly selective precipitant for nickel.

An introductory course is not the place to discuss the many possibilities and practical considerations for gravimetric determinations. The student, however, should bear the abovementioned limitations in mind when studying the following brief consideration of some important and instructive examples of gravimetric determinations for inorganic substances.

7.2 Selected Examples, Gravimetric Determinations

Many trivalent and tetravalent cations can be readily precipitated as hydrous oxides by ammonia; this affords a possibility for their separation from monovalent and some divalent ions, but reprecipitation is often necessary because of coprecipitation. The conversion to the anhydrous oxide as a weighing form is achieved simply by ignition. The sum of the sesquioxides of iron and aluminum (and chromium, where present) is often designated as R_2O_3 (i.e., "residual" oxides) when determined in this way and is often reported as such in the analysis of rocks, minerals, glass, etc., especially where further differentiation is not required. The precipitate will also contain any titanium present as TiO_2, and also manganese as MnO_2 if an oxidant such as hydrogen peroxide is present during the ammonia precipitation.

The precipitation of hydrous oxides with ammonia will be familiar to the student from his knowledge of the qualitative separation of inorganic cations for their detection. Many other precipitation reactions of the qualitative scheme are applied for quantitative purposes. Some cations can be precipitated and weighed as sulfides, including those of lead, mercury(II), cadmium, zinc, arsenic(V), antimony(V), bismuth, cobalt(II), nickel, and manganese(II). However, since an excess of sulfur is present in certain of the precipitates, special measures are necessary to achieve a suitable weighing form. Because of the low selectivity of the sulfide precipitation, it is seldom employed as the basis of a gravimetric determination, but hydrogen sulfide still serves as an important reagent for "group" separations.

Chloride ion is a familiar precipitant for silver ion and serves for the gravimetric determination of that metal. This precipitation reaction can also be applied to the determination of chloride, employing silver nitrate as the precipitant. Such dual use of a given precipitation reaction is a rather general practice. The precipitation of barium sulfate serves for the separation and gravimetric determination of both barium and sulfate, and that of barium chromate for both barium and chromate.

Nickel ion can be precipitated with dimethylglyoxime and the precipitate dried to the stoichiometric composition $Ni(C_8H_{14}H_4O_4)_2$, which affords a most favorable gravimetric factor (0.2032). The gravimetric determination of even quite small amounts of nickel is possible.

The gravimetric determination of calcium based on the precipitation of calcium oxalate has been described in Section 6.12, including some of the possible weighing forms. Magnesium can be precipitated from ammoniacal solution as magnesium ammonium phosphate hexahydrate, $MgNH_4PO_4 \cdot 6H_2O$, and can be weighed as such after drying. Solutions of this salt show a pronounced tendency for supersaturation. Scratching of the vessel walls is employed to initiate the precipitation and prolonged standing is necessary to assure its completeness, especially when only a small amount of magnesium is present. Well-crystallized precipitates are obtained in practice by adding orthophosphate to the acidic sample solution and then adding a considerable excess of ammonia to the boiling solution slowly and with stirring. The precipitate may be ignited at a high temperature, whereby according to the reaction

$$2MgNH_4PO_4 \cdot 6H_2O \rightarrow Mg_2P_2O_7 + 2NH_3\uparrow + 7H_2O\uparrow \tag{7-1}$$

magnesium pyrophosphate is formed. This compound is preferred as the weighing form since its stoichiometric formation is more readily assured than that of the hexahydrate salt.

The same precipitation reaction may be used for the gravimetric determination of orthophosphate, magnesium ion serving as the precipitant. The addition of an ammonium salt (chloride or nitrate) is necessary to prevent such a high alkalinity being attained that magnesium hydroxide is precipitated.

The important alkali metals, sodium and potassium, form only a very few compounds that are sparingly soluble in water. Their gravimetric determination is therefore often accomplished after complete removal of all other cations via precipitations with reagents, such as ammonia, sulfide ion, and oxalate ion, that are readily volatilized or destroyed. The final solution, after addition of sulfuric acid, is evaporated to dryness and the residue ignited. The alkali metal sulfates remaining are weighed. Frequently, as in the analysis of rocks and minerals, only the sum of the alkali metals is thus established.

Potassium may be determined in the sulfate mixture by dissolving the latter and precipitating potassium as the sparingly soluble perchlorate, $KClO_4$. Ethanol is added to decrease the solubility sufficiently and is also used in the washing of the precipitate. Sodium is then found by difference. Sodium tetraphenylborate, $NaB(C_6H_5)_4$, introduced in recent years, is an excellent precipitant for potassium. Potassium tetraphenylborate, $KB(C_6H_5)_4$, of all known potassium compounds, has the lowest solubility in water. It is stable in air, can be dried to the stoichiometric composition, and has a most favorable gravimetric factor.

7.3 Indirect Gravimetric Analysis

When both components of a binary mixture are to be determined, it is preferable to determine each of them individually, performing a separation if necessary. This is called direct analysis. Another approach involves determination of one of the components and establishment of the other by difference. For example, the percentage of aluminum in a binary aluminum–magnesium alloy may be determined and the magnesium content found by subtraction from 100. A third approach is known as indirect analysis. Two measurements, of which one is usually the weight of the sample, are secured and used to set up a pair of simultaneous equations. Their algebraic solution establishes the amount of each component.

One type of indirect analysis involves a mixture in which the components have an element or radical in common. From the determination of this element or radical and the weight of the sample, the amounts of the individual components can be established.

Example 7-1. A sample, known to contain only AgCl *(143.32)* and AgI *(234.77)*, weighs 1.5000 g. It is reduced quantitatively to metallic silver *(107.87)*, which amounted to 0.8500 g. What is the weight of AgCl and AgI in the sample?

Let

$$x = \text{grams of AgCl in the sample}$$

$$y = \text{grams of AgI in the sample}$$

Then by the description of the sample

$$x + y = 1.5000$$

Now

$$\text{grams of Ag from the AgCl} + \text{grams of Ag from the AgI} = 0.8500$$

and by the applications of the chemical factors

$$\text{grams of Ag from the AgCl} = x \times \frac{Ag}{AgCl}$$

$$\text{grams of Ag from the AgI} = y \times \frac{Ag}{AgI}$$

Hence,

$$x \times \frac{Ag}{AgCl} + y \times \frac{Ag}{AgI} = 0.8500$$

Substitution of the values for the chemical factors gives

$$0.7527x + 0.4595y = 0.8500$$

This constitutes the second equation of the required pair. Multiplication of the first equation of the pair by 0.7527 yields

$$0.7527x + 0.7527y = 1.1290$$

Subtraction of the second equation of the pair from this equation gives

$$0.2932y = 0.2790$$

and $y = 0.952$ g of AgI in sample

From the first equation of the pair

$$x = 1.5000 - y = 0.548 \text{ g of AgCl in sample}$$

A second type of indirect analysis involves the conversion of a mixture in which the components have an element or radical in common to a second mixture in which the components have a different element or radical in common.

Example 7-2. The mineral dolomite is a mixture of $CaCO_3$ *(100.07)* and $MgCO_3$ *(84.32)*, with small amounts of impurities neglected for the present purposes. A 1.000-g sample of a dolomite is ignited, volatilizing CO_2 and leaving a residue of CaO *(56.08)* and MgO *(40.31)*, which is found to weigh 0.5200 g. What is the approximate percentage composition of the dolomite?

Let

$$x = \text{grams of } CaCO_3 \text{ in the 1-g sample}$$

$$y = \text{grams of } MgCO_3 \text{ in the 1-g sample}$$

Then $x + y = 1.000$ as the first equation of the pair. Application of chemical factors yields

$$x \times \frac{CaO}{CaCO_3} + y \times \frac{MgO}{MgCO_3} = 0.5200$$

and substitution of the values for the factors gives

$$0.5604x + 0.4781\,y = 0.5200$$

as the second equation of the pair. Multiplication of the first equation by 0.5604 results in

$$0.5604x + 0.5604\,y = 0.5604$$

and on solution

$$x = 0.509 \text{ g} \qquad \text{and} \qquad y = 0.491 \text{ g}$$

Hence the approximate composition is 51% $CaCO_3$ and 49% $MgCO_3$.

Other types of indirect analysis might be illustrated. However, it will suffice to state that if n constituents are to be determined, n independent measurements are required, and n simultaneous equations must be somehow formulated and solved. Where more than two constituents are to be determined, indirect analysis is seldom of practical value because the accuracy and precision are poor. Indeed, even when applied to only two components, the necessary mathematical treatment leads to the loss of significant figures and hence a lowering of the precision.

Best results are obtained in an indirect analysis if all the substances involved are present in reasonable amounts. If only a small amount of one constituent is present, then even a slight absolute deviation in the measurement will lead to a rather large relative deviation in the result for that constituent. It will be noted that in Example 7-2, in the indirect analysis of dolomite, the conditions are favorable since the components of the mixtures are present in somewhat similar amounts. From an inspection of the examples it is evident that the tendency to lose significant figures is the smaller the greater the difference in the numbers on the right side of the simultaneous equations and the greater the difference in the conversion factors appearing on the left side of the equations.

Questions 7.4

1. What two forms of the separated product may be differentiated in gravimetric analysis?

2. In your own words, summarize the mechanisms and phenomena that lead to coprecipitation.

3. In gravimetric analysis, why are ammonium salts present in many wash liquids?

4. Elaborate on the terms solid solution, occlusion, peptization, post-precipitation, and precipitation from homogeneous solution.

5. What is the significance of a gravimetric factor? Of an empirical gravimetric factor?

6. How would you weigh a hygroscopic substance?

7. By what means can the solubility of a precipitate be reduced?

8. Is a large or small value for a gravimetric factor preferred? Explain your answer.

9. For a sought-for substance A, there are two possible precipitants B and C. The formula weights of these precipitants are 100 and 300, respectively. The precipitates obtained are AB and A_2C. Assuming all other factors equal (solubility, ease of handling, etc.), which precipitant would you select?

0. In your own words state some factors to be considered when selecting a wash liquid for a precipitate.

1. What is indirect analysis? What are its advantages and disadvantages?

Problems 7.5

1. A volume of 50.00 ml of a solution containing 0.460 millimole of Na_2SO_4 (*142.04*) per milliliter is evaporated to dryness. What is the weight of the residue in grams?

   Answer: 3.27 g

2. How many grams of AgCl (*143.32*) will be obtained when the chloride present in a solution containing 0.2435 g of NaCl (*58.44*) is completely precipitated?

   Answer: 0.5972 g

3. A sample of KI (*166.01*) weighing 0.2463 g is treated with H_2SO_4, the excess of acid evaporated, and the residue ignited. What is the weight in milligrams of the residue of K_2SO_4 (*174.27*)?

   Answer: 129.3 mg

4. A sample of Ag_2CrO_4 (*331.73*) weighing 0.1000 g is so treated with an excess of $BaCl_2$ (*157.34*) that the precipitation reaction $Ag_2CrO_4 + BaCl_2 \rightarrow \underline{2AgCl} + \underline{BaCrO_4}$ proceeds to completion. The precipi-

tate containing AgCl (*143.32*) and $BaCrO_4$ (*253.33*) is filtered, washed, dried, and weighed. What will be its weight in grams?

Answer: 0.1628 g

7-5. Calculate the gravimetric factors for

Substance sought	Substance weighed	Answer
SO_3 (*80.06*)	$BaSO_4$ (*233.40*)	0.34302
As_2O_3 (*197.84*)	Ag_3AsO_4 (*462.53*)	0.21387
$HClO_4$ (*100.459*)	AgCl (*143.323*)	0.70093
$FeSO_4$ (*152.908*)	Fe_2O_3 (*159.692*)	1.91504

7-6. In a mixture, $CaCO_3$ (*100.09*) is the only component that loses weight on ignition, by volatilization of CO_2 (*44.01*). When a sample of this mixture weighing 0.4532 g was ignited, the weight of the residue was 428.9 mg. (a) What is the per cent loss in weight on ignition? (b) What is the per cent of $CaCO_3$ in the mixture?

Answers: (a) 5.36%; (b) 12.19% $CaCO_3$

7-7. The analysis of a mixture established that it contains 32.55% Fe_2O_3 and that the per cent loss in weight on drying is 1.25%. Calculate the per cent Fe_2O_3 on a "dry" basis.

Answer: 32.96% Fe_2O_3 (dry basis)

7-8. The formal solubility of AgCl in water is 1.0×10^{-5} *F*. A precipitate of AgCl (*143*) is washed with 50 ml of water. Assuming that the wash liquid is 80% saturated in AgCl, how many milligrams of the precipitate will be "lost" by this washing?

Answer: 0.057 mg

7-9. A mixture, containing only NaCl (*58.44*) and KCl (*74.56*) and weighing 1.2328 g, was treated with H_2SO_4 and then heated to dryness. The residue of Na_2SO_4 (*142.04*) and K_2SO_4 (*174.27*) was found to weigh 1.4631 g. What is the percentage composition of the original mixture? (*Caution*! Note that only one mole of an alkali metal sulfate is obtained from two moles of the corresponding chloride.)

Answer: 39% NaCl, 61% KCl

7-10. With the data of Example 7-1, consider that all possible pairings of 0.5-mg errors are made in the two weighings, that is, +0.5, +0.5; +0.5, −0.5; −0.5, +0.5; −0.5, −0.5 mg. Observe the effect of these errors on the calculated results. Proceed analogously with the data of Problem 7-9.

7-11. Exactly 1 g of an alloy gave a weight of 0.24 g of Al_2O_3 (*102*) after proper treatment. Calculate the % w/w of Al (*27*) in the alloy.

7-12. A sample of 1.116 g of an alloy gave a weight of 1.280 g of Fe_2O_3 (*159.7*) after proper treatment. Calculate the % w/w of iron in the alloy.

7-13. What amount in grams of a 1.00% w/w solution of $BaCl_2$ (*208.25*) is required to quantitatively react with 0.100 mg of Na_2SO_4 (*142.04*)?

7-14. The sulfur in 0.2280 g of thiourea, $(NH_2)_2CS$ (*76.13*), is converted to sulfate and precipitated as barium sulfate (*233.4*). How many grams of $BaSO_4$ will be obtained?

5. From qualitative analysis a coin was known to contain only silver (*107.87*) and copper (*63.54*). The coin was dissolved in nitric acid and the silver precipitated as AgCl (*143.32*), of which an amount of 0.2868 g was obtained. The sample taken weighed 0.3083 g. Calculate the % w/w of silver and copper in the coin.

6. Sodium tetraphenylborate, $NaB(C_6H_5)_4$, is used as a precipitant for potassium. What is the % w/w of K_2O (*94.20*) in a fertilizer when a 0.4320-g sample gives 0.2138 g of $KB(C_6H_5)_4$ (*358.2*)?

7. The arsenate present in a solution is precipitated as Ag_3AsO_4 by adding 40.00 ml of 0.100 *F* $AgNO_3$. The precipitate is separated by filtration and the silver in the filtrate is precipitated as AgCl, yielding 0.1434 g. Calculate the amount of arsenate in the original solution expressed as milligrams of As_2O_3.

8. Calculate the per cent loss on ignition for the following substances from which CO_2 (*44.01*) is volatilized.
   (a) $CaCO_3$ (*100.09*)
   (b) $MgCO_3$ (*84.32*)
   (c) $BaCO_3$ (*197.35*)
   (d) $MnCO_3$ (*114.95*)
   (e) $SrCO_3$ (*147.63*)

9. A pure dolomite [only $CaCO_3$ (*100.0*) and $MgCO_3$ (*84.32*) present] shows a 47.29% w/w loss on ignition. Calculate the % w/w of $CaCO_3$ in the dolomite.

8 ACID-BASE EQUILIBRIA

Introduction 8.1

According to the classical concepts, developed by Arrhenius, acids and bases are substances that yield in aqueous solution hydrogen ion, H^+, and hydroxide ion, OH^-, respectively. An acid–base reaction is therefore the combination of hydrogen ion with hydroxide ion to yield water. These classical concepts are useful ones, chiefly in their application to acid–base reactions in aqueous solution. They allow a mathematical treatment of the equilibria involved but fail to explain adequately many of the phenomena encountered, especially in nonaqueous solutions.

In 1923 Brønsted introduced a more general acid–base theory. According to his concepts, an acid is a substance, electrically neutral or ionic, that can give up a proton, H^+. A base is a substance, electrically neutral or ionic, that can take up a proton. In other words, acids and bases are defined as proton donors and proton acceptors, respectively. Lowry independently recognized the ideas underlying Brønsted's definitions; consequently, the phrase "Brønsted–Lowry acid–base concepts" is often used.

Consider hydrogen chloride, for which the following ionization equilibrium may be written:

$$HCl \rightleftharpoons H^+ + Cl^- \qquad (8\text{-}1)$$

As the reaction proceeds from left to right, hydrogen chloride gives up a proton and consequently is an acid. In the reverse reaction, chloride ion accepts a proton and consequently is a base. Hydrogen chloride and chloride ion are said to be a conjugate acid–base pair: HCl is the conjugate acid of Cl^-, and Cl^- is the conjugate base of HCl.

Such an equilibrium may be written in general terms:

$$\text{acid} \rightleftharpoons \text{proton} + \text{base} \qquad (8\text{-}2)$$

Acids need not be uncharged molecules (molecular or electrically neutral acid), such as hydrogen chloride; they can equally well be a cation (cationic acid) or an anion (anionic acid). Identical considerations apply to bases. The following selected examples illustrate the situation:

$$HBr \text{ (molecular acid)} \rightleftharpoons \text{proton} + Br^- \text{ (anionic base)} \quad (8\text{-}3a)$$

$$NH_4^+ \text{ (cationic acid)} \rightleftharpoons \text{proton} + NH_3 \text{ (molecular base)} \quad (8\text{-}3b)$$

$$H_2SO_4 \text{ (molecular acid)} \rightleftharpoons \text{proton} + HSO_4^- \text{ (anionic base)} \quad (8\text{-}3c)$$

$$HSO_4^- \text{ (anionic acid)} \rightleftharpoons \text{proton} + SO_4^{2-} \text{ (anionic base)} \quad (8\text{-}3d)$$

$$Al(H_2O)_6^{3+} \text{ (cationic acid)} \rightleftharpoons \text{proton} + Al(H_2O)_5(OH)^{2+} \text{ (cationic base)} \quad (8\text{-}3e)$$

In this list the hydrogen sulfate ion, HSO_4^-, appears both as an anionic acid and an anionic base; species, such as this, that can both donate and accept a proton are called ampholytes and are said to be amphiprotic or amphoteric.

Acids and bases may also be classified according to the number of protons they can donate or accept per molecule. If an acid can donate only one proton it is variously known as monoprotic, monofunctional, monobasic, or monoequivalent. The prefixes di-, tri-, tetra-, etc., are substituted for mono- if two, three, four, etc., protons can be donated or accepted. For an acid (or base) capable of donating (or accepting) more than one proton but with no particular number of protons specified, the prefix poly- is used.

The electrical conductance of benzene is virtually zero and remains so when hydrogen chloride is dissolved in this solvent. This finding indicates that neither the solvent nor the solution contains any appreciable concentration of ions. In contrast, the electrical conductance of pure water is extremely low, but it increases tremendously if hydrogen chloride is dissolved in it. Consequently, it must be concluded that the aqueous solution, that is, hydrochloric acid, contains a considerable concentration of ions; in other words, the originally molecular hydrogen chloride has ionized.

The extreme difference in the behavior of hydrogen chloride in the two solvents indicates that whether or not ionization takes place does not depend on the acid alone but rather that the solvent plays an active role.

According to Brønsted's concepts, a proton is only given up by an acid if a base is present that can accept the proton. Since benzene does not possess basic properties, no proton transfer takes place and consequently no ions are formed. Water, however, can act as a base and accept protons. In water, therefore, ionization of the acid occurs. Consequently, the classical "dissociation" of an acid in terms of Brønsted's concepts is actually an acid-base reaction between the solute and the solvent.

The present example may be formulated as

$$HCl + H_2O \rightleftharpoons H_3O^+ + Cl^- \tag{8-4}$$

The "dissociation" of hydrogen chloride in water is a protolytic reaction involving two conjugate acid–base pairs, $HCl—Cl^-$ and $H_3O^+—H_2O$. The monohydrated proton, H_3O^+, written above is known as the oxonium ion; however, where an indefinite degree of hydration is actually intended the term hydronium ion may be used. The "dissociation" of an acid may be formulated as a proton-transfer reaction of the following general type:

$$\text{acid}_1 + \text{base}_2 \rightleftharpoons \text{base}_1 + \text{acid}_2 \tag{8-5}$$

where base_2 is commonly the solvent.

Acid–Base Strength 8.2

The strength of an acid or base may be described in terms of its intrinsic strength, that is, its tendency to donate or accept a proton, respectively. However, this tendency can only be measured on a relative basis because the acid donates a proton only if a base is present to accept it and the base can accept a proton only if an acid is present that donates it. Consequently, the strength of an acid (or base) can be evaluated only in relation to a base (or acid), commonly the solvent itself. In aqueous solution hydrochloric, nitric, and perchloric acids, for example, all behave as strong acids of essentially equal strength. All three acids donate their protons virtually completely to the base water to form the hydronium ion, which is the strongest acid that can exist in water. In water-free acetic acid, however, perchloric acid acts as a far stronger acid than nitric or hydrochloric acids. The greater intrinsic strength of perchloric acid (that is, its greater tendency to donate protons) comes into fuller play in the acetic acid medium because this solvent is less basic than water, that is, it accepts a proton less readily.

It can be concluded that the position of the equilibrium in reaction (8-5) will be shifted more to the right the stronger acid_1 and base_2 and the weaker base_1 and acid_2. In other words, if an acid-base reaction takes place, the weaker acid and the weaker base of the two conjugate acid-base pairs will always form.

If an acid is strong, that is, if it has a great tendency to donate a proton, then obviously its conjugate base has a rather small tendency to accept a proton. Since hydrochloric acid (in water) is a strong acid, chloride ion is an extremely weak base. This conclusion may be presented as a general statement: The stronger an acid (or base), the weaker its conjugate base (or acid). The strengths of an acid and its conjugate base can be related mathematically (see Section 8.5).

8.3 Autoprotolysis of Water

The "dissociation" of water, after Brønsted, is actually an acid-base reaction, involving two conjugate acid–base pairs which, in this special case, have one component in common. The process may be written as follows:

$$\underset{\text{acid}_1}{H_2O} + \underset{\text{base}_2}{H_2O} \rightleftharpoons \underset{\text{base}_1}{OH^-} + \underset{\text{acid}_2}{H_3O^+} \qquad (8\text{-}6)$$

One molecule of water acts as the proton donor (acid) and the other as the proton acceptor (base). The hydronium ion is the strongest acid that can exist in aqueous solution and the hydroxide ion the strongest base. Since, as already stated, an acid–base reaction always proceeds toward the formation of the weaker acid and base, it can be surmised that the position of the equilibrium is such as to favor water. Indeed, the experimental evidence shows that the equilibrium position is extremely far to the left. In pure water the concentrations of hydronium and hydroxide ions are each only 10^{-7} mole/liter.

As stated earlier, the writing of H_3O^+ is not to be taken to imply that this species is actually present as such. Solvation definitely takes place and species such as $H(H_2O)_2^+$ or $H(H_2O)_3^+$ may persist. The "hydrogen ion" molar concentration always equals the total molar concentration of solvated protons, regardless of the species considered to be present. Consequently, it is possible, whenever explicit recognition of the solvation of protons is not required, to simplify the notation by writing H^+. This common practice is followed in the present textbook.

Application of the law of mass action to the autoprotolysis reaction of water, reaction (8-6), leads to the expression for the equilibrium constant:

$$K_{eq} = \frac{[H_3O^+][OH^-]}{[H_2O]^2} = \frac{[H^+][OH^-]}{[H_2O]^2} \quad (8\text{-}7)$$

Since the extent of ionization is exceedingly small, the concentration of water can be considered to remain essentially unaltered and can be incorporated into the equilibrium constant. Then

$$K_w = [H_3O^+][OH^-] = [H^+][OH^-] \quad (8\text{-}8)$$

This new constant, K_w, is known as the ion product of water. Its numerical value at 25°C is 1.00×10^{-14}. Since the autoprotolysis of water is strongly endothermic, the value of the constant increases greatly with temperature, reaching, at 60°C, 10 times the value for 25°C.

The autoprotolysis reaction yields equimolar concentrations of hydrogen and hydroxide ions. Consequently, for pure water

$$[H^+] = [OH^-] \quad (8\text{-}9)$$

Hence

$$[H^+]^2 = [OH^-]^2 = K_w = 1.00 \times 10^{-14} \quad (8\text{-}10)$$

$$[H^+] = [OH^-] = \sqrt{K_w} = 1.00 \times 10^{-7}\,M \quad \text{(at 25°C)} \quad (8\text{-}11)$$

Pure water or any aqueous solution that affords equimolar concentrations of hydrogen and hydroxide ions is said to be neutral. Pure water in this context means chemically pure water, in contrast to distilled or deionized water in equilibrium with the air. The latter, so-called equilibrium water, contains a small amount of carbon dioxide, which forms carbonic acid. Consequently, such water will be slightly acidic; that is, the hydrogen ion concentration will be greater than 1.00×10^{-7} *M*. Depending on the carbon dioxide content of the air, this concentration ranges from 1×10^{-6} to 3×10^{-6} *M*.

The Concept of pH 8.4

It is sometimes inconvenient to present small molar concentrations in an exponential manner. Sørensen, therefore, proposed that such concentrations and the small values of equilibrium constants be expressed in terms of the negative of the logarithm of the numerical value. This mode of expression is indicated by a prefixed letter p.

Accordingly, for the molar concentrations of hydrogen and hydroxide ions, the expressions pH and pOH are employed and defined by

$$pH = -\log [H^+] \quad \text{and} \quad [H^+] = 10^{-pH} \tag{8-12}$$

$$pOH = -\log [OH^-] \quad \text{and} \quad [OH^-] = 10^{-pOH} \tag{8-13}$$

The present-day definition of pH, however, involves the hydrogen ion activity rather than concentration. But, as in previous cases, for dilute solutions substitution of concentrations for activities is made as an approximation. For the establishment of a practical, that is, an operationally defined pH scale, see Chapter 23.

The expression for the ion product of water, using Sørensen's convention, can be written as

$$pK_w = pH + pOH = 14.00 \tag{8-14}$$

Consequently, for pure water or a neutral solution, from equation (8-11), pH = pOH = 7.00.

Unless stated otherwise, a temperature of 25°C is assumed in connection with pH values. In the numerical expression of either pH or pOH, usually only two decimal figures are given, because in the common experimental determination of the hydrogen or hydroxide ion concentration it is difficult to attain a precision greater than that corresponding to two significant figures (see Section 2.3 for significant figures in logarithmic expressions).

The relation of the hydrogen and hydroxide ion concentrations to the acidity or alkalinity of an aqueous solution at 25°C can be summarized:

$$\text{Alkaline solution:} \quad pH > 7.00 > pOH \tag{8-15a}$$

$$\text{Neutral solution:} \quad pH = 7.00 = pOH \tag{8-15b}$$

$$\text{Acidic solution:} \quad pH < 7.00 < pOH \tag{8-15c}$$

It should be realized that the numerical value of the ion product is not changed by the addition of an acid or base. Only the hydrogen and hydroxide ion molar concentrations are altered and in such a manner that their product still equals K_w.

A few numerical examples will illustrate practical aspects of the concepts developed.

Example 8-1. What is the pH of a solution 0.030 M in hydrogen ion?

$$[H^+] = 0.030 = 3.0 \times 10^{-2}\,M$$

$$pH = -\log[H^+] = -\log(3.0 \times 10^{-2}) = -\log 10^{-2} - \log 3.0$$
$$= 2 - 0.48 = 1.52$$

Example 8-2. Calculate the pH and pOH of a solution 0.0020 M in hydroxide ion.

$$[OH^-] = 0.0020 = 2.0 \times 10^{-3}\,M$$

$$pOH = -\log[OH^-] = -\log(2.0 \times 10^{-3}) = -\log 10^{-3} - \log 2.0$$
$$= 3 - 0.30 = 2.70$$

$$pH = 14.00 - pOH = 14.00 - 2.70 = 11.30$$

This last calculation may be checked by application of equation (8-8).

$$[H^+] = \frac{1.00 \times 10^{-14}}{[OH^-]} = \frac{1.00 \times 10^{-14}}{2.0 \times 10^{-3}} = 0.50 \times 10^{-11}$$
$$= 5.0 \times 10^{-12}\,M$$

$$pH = -\log(5.0 \times 10^{-12}) = 12 - \log 5.0$$
$$= 12 - 0.70 = 11.30$$

Example 8-3. What is the molar concentration of hydrogen ion in a solution having a pH value of 4.30?

$$[H^+] = 10^{-pH} = 10^{-4.30} = 10^{-5.00+0.70} = 10^{-5} \times 10^{0.70}$$
$$= 10^{-5} \times \text{antilog}\,0.70 = 5.0 \times 10^{-5}\,M$$

The expression of the negative exponent as the algebraic sum of a positive and a negative sum is necessary because an antilogarithm must be a positive decimal number to be found in a table of logarithms.

Example 8-4. What is the hydroxide ion concentration in a solution of pH 10.15?

$$pOH = 14.00 - pH = 14.00 - 10.15 = 3.85$$
$$[OH^-] = 10^{-pOH} = 10^{-3.85} = 10^{-4.00+0.15} = 10^{-4} \times 10^{0.15}$$
$$= 10^{-4} \times \text{antilog}\,0.15 = 1.4 \times 10^{-4}\,M$$

Acidity and Basicity Constants 8.5

Equation (8-5) can be written in a brief form

$$a_1 + b_2 \rightleftharpoons b_1 + a_2 \qquad (8\text{-}16)$$

where a and b designate an acid and a base, respectively. If the molar concentrations of the various species are denoted by brackets, the equilibrium-constant expression for reaction (8-16) can be written as

$$K_{prot} = \frac{[b_1][a_2]}{[a_1][b_2]} \qquad (8\text{-}17)$$

The equilibrium constant, K_{prot}, is called the protolysis constant.

8.5 ACIDITY AND BASICITY CONSTANTS

If the discussion is restricted to the solution of an acid in water, b_2 and a_2 become H_2O and H_3O^+, respectively. The reaction taking place when an acid, a, is dissolved in water may be written as

$$a + H_2O \rightleftharpoons b + H_3O^+ \tag{8-18}$$

The equilibrium constant for this reaction is given by

$$K_{\text{prot}} = \frac{[b][H_3O^+]}{[a][H_2O]} \tag{8-19}$$

Since the concentration of water remains essentially unchanged in a dilute solution and since $[H^+]$ can be substituted for $[H_3O^+]$, the above relation reduces to

$$K_a = [H_2O]\,K_{\text{prot}} = \frac{[H^+][b]}{[a]} \tag{8-20}$$

The constant K_a is variously known as the dissociation, ionization, or acidity constant of the acid a. The larger the numerical value of this constant, the stronger the acid.

It should be noted that equation (8-18) seems to be unbalanced with respect to charge. As has been pointed out, the essential criterion for an acid is its ability to denote protons, regardless of whether the acid is electrically neutral, anionic, or cationic. Consequently, in the general formulation, charges are omitted on a and b. With actual species written, either a or b or both may carry charges and in such a way that the charges on the left side of the equation balance those on the right side.

If a base b is dissolved in water, the protolysis reaction taking place is

$$b + H_2O \rightleftharpoons a + OH^- \tag{8-21}$$

The protolysis constant for this equilibrium is

$$K_{\text{prot}} = \frac{[a][OH^-]}{[b][H_2O]} \tag{8-22}$$

Again the concentration of water remains essentially unchanged and can be incorporated into a new constant, which is variously known as hydrolysis, dissociation, ionization, or basicity constant K_b, of base b.

$$K_b = [H_2O]\,K_{\text{prot}} = \frac{[a][OH^-]}{[b]} \tag{8-23}$$

Multiplication of equation (8-23) by (8-20) yields

$$K_a K_b = \frac{[H^+][b]}{[a]} \times \frac{[a][OH^-]}{[b]} = [H^+][OH^-] = K_w \qquad (8\text{-}24)$$

The statement made previously that the stronger an acid the weaker its conjugate base can now be made quantitative: The acidity constant of an acid is inversely proportional to the basicity constant of its conjugate base and vice versa. The proportionality constant is the ion product of the solvent. Since hydrochloric acid is a strong acid with a large acidity constant, its conjugate, chloride ion, is an extremely weak base. In fact, chloride is so weak a base that it does not influence the acidity of a solution (for example, a sodium chloride solution is neutral because neither the chloride nor sodium ion affects the pH). In general, the base character of the conjugate of a strong acid need not be considered in calculations related to acid–base equilibria.

Monofunctional Strong Acids and Bases 8.6

An acid or base is defined as being strong if its protolytic reaction with the solvent (water in the present context) goes essentially to completion even in moderately concentrated solutions. The calculation of the pH value of the solution of a strong monofunctional acid (or base) is simple. The molar concentration of hydrogen ion (or hydroxide ion) equals the formality of the acid (or base) as long as the concentration of the solute is sufficiently large to make the contribution of hydrogen (or hydroxide) ion from the autoprotolysis of water insignificant. Some examples will illustrate the approach.

Example 8-5. What is the pH of a 0.0040 F solution of the strong acid HCl?

$$C_{\mathrm{HCl}} = [H^+] = 0.0040 = 4.0 \times 10^{-3}\,F$$
$$\mathrm{pH} = 3 - \log 4.0 = 3.00 - 0.60 = 2.40$$

Example 8-6. What is the pH of a 0.020 F solution of the strong base NaOH?

$$C_{\mathrm{NaOH}} = [OH^-] = 2.0 \times 10^{-2}\,F$$
$$\mathrm{pOH} = 2 - \log 2.0 = 1.70$$
$$\mathrm{pH} = 14.00 - 1.70 = 12.30$$

It should be pointed out that sodium hydroxide, NaOH, according to the Brønsted concepts, is not strictly a base but a salt. This compound has an ionic lattice and upon dissolution in water simply dissociates to yield sodium ion, Na^+, and hydroxide ion, OH^-. The sodium ion undergoes solvation, but this is not a protolytic reaction and is of no further consequence. The hydroxide ion is actually the base. According to older concepts, sodium hydroxide was described as a base, and even when the Brønsted concepts are applied this compound is still loosely termed a base.

The simple approach employed in Examples 8-5 and 8-6 fails if the concentration of either of the ions produced in the autoprotolysis of water cannot be neglected. For example, if the simple approach were applied to 1.0×10^{-7} *F* hydrochloric acid, the calculated result would be pH = 7.00. This value would imply that addition of a strong acid to pure water yields a neutral solution! This is nonsense. The following example shows how to proceed in this case.

Example 8-7. What is the pH of a solution 1.0×10^{-7} *F* in HCl? Hydrogen ion is produced by two reactions, the autoprotolysis of water and the protolysis of the hydrogen chloride. Consequently,

$$[H^+] = [H^+]_{water} + [H^+]_{HCl}$$

From the HCl, one hydrogen ion is obtained for every chloride ion formed. Since the protolysis of the HCl goes to completion, the molar concentration of Cl^- equals the formality of the acid, and this in turn equals the molar concentration of the hydrogen ion produced from the HCl. For the autoprotolysis of water it is known that for each hydrogen ion one hydroxide ion is formed. Consequently, the molar concentration of hydrogen ion must equal the sum of the molar concentrations of the hydroxide ion and chloride ion:

$$[H^+] = [OH^-] + [Cl^-]$$

The chloride concentration is known to equal 1.0×10^{-7} *M*. The hydroxide ion concentration can be expressed in terms of $[H^+]$ from the ion product of water; this leads to

$$[H^+] = \frac{1.0 \times 10^{-14}}{[H^+]} + 1.0 \times 10^{-7}$$

On rearrangement the following quadratic equation is obtained:

$$[H^+]^2 - 1.0 \times 10^{-7}[H^+] - 1.0 \times 10^{-14} = 0$$

Application of the quadratic formula yields

$$[H^+] = \tfrac{1}{2}[1.0 \times 10^{-7} \pm \sqrt{(1.0 \times 10^{-7})^2 + 4.0 \times 10^{-14}}]$$
$$= 1.6 \times 10^{-7}\ M$$

Only the positive root has physical meaning. The pH is then

$$\text{pH} = 7 - \log 1.6 = 7 - 0.20 = 6.80$$

It should be added that the preceding example serves only to demonstrate that under certain conditions the autoprotolysis of water cannot be neglected and to indicate how to proceed in such a case. The numerical value obtained has little practical significance, because so dilute an acid solution is extremely susceptible to small amounts of impurities (for example, carbon dioxide from the air).

This refinement in the calculation need be undertaken only if very small concentrations of an acid or a base are involved; that is, the pH resulting is close to that of the neutral point. Slightly removed from that point, the differences between the concentrations of hydrogen and hydroxide ions become so great that the concentrations of one of these species can be neglected. For example, if the acidity of 1.00×10^{-6} *F* HCl is calculated according to the refined approach, the result is $[H^+] = 1.01 \times 10^{-6}$ *M* (the actual calculation is left to the student), while the approximate approach yields 1.00×10^{-6} *M*. Consequently, the simple calculation is quite satisfactory for concentrations of monofunctional acids (and bases) greater than 10^{-6} *M*.

Charge Balance 8.7

The equation $[H^+] = [OH^-] + [Cl^-]$ of Example 8-7 was derived by appropriate reasoning; this equation can be obtained, however, more directly.

A solution must be electrically neutral; that is, the number of positive charges must equal the number of negative charges. Consequently, for any solution the sum of the molar concentrations of all positive charge must equal the sum of the molar concentrations of all negative charge. The mathematical formulation of this fact is the "charge-balance equation" or simply the "charge balance." Thus the charge balance for Example 8-7 can be immediately written as given above.

The novice frequently fails to differentiate between the concentration of *charge* and the concentration of *species*. For example, the molar concentration of barium ion in a solution is written $[Ba^{2+}]$; the molar concentration of the positive charge for this ion requires the term $2[Ba^{2+}]$, since each ion of barium carries two positive charges and consequently one mole of barium ion corresponds to two moles of positive charge. Similarly, the orthophosphate ion in a charge balance requires the term $3[PO_4^{3-}]$, since the ion carries three negative charges. Another common mistake is the failure to realize that the concentration term for a species occurs only once, even when the species is produced by different solutes. For example, the charge-balance equation for a solution obtained by dissolving sodium sulfate and sodium chloride in water is

$$[H^+] + [Na^+] = [OH^-] + [Cl^-] + [HSO_4^-] + 2[SO_4^{2-}]$$

The coefficient of the sodium term is unity, although sodium ion stems from two sources, one of which even provides two sodium ions per molecule. Also note that the concentration term for the hydrogen sulfate ion occurs but not that of the electrically neutral sulfuric acid. Of course, every charge balance for an aqueous solution contains the autoprotolysis products of water. The mistakes mentioned result in part from the failure of the beginner to differentiate clearly between a charge-balance equation and the equation describing a dissociation equilibrium.

The charge balance is not restricted to problems of the type considered here but is of general value. It often provides the additional equation required to make the number of equations equal the number of unknowns.

To reduce the mathematical effort in solving problems it is always good practice to determine whether any terms in the charge-balance equation are significantly smaller than others and may therefore be neglected. For example, the solution of Example 8-5 could well be started by writing the charge balance $[H^+] = [OH^-] + [Cl^-]$. It could then be reasoned that in a strongly acidic solution the hydroxide concentration is considerably smaller than the hydrogen ion concentration. Thereby the equality $[H^+] = [Cl^-] = C_{HCl}$ could be obtained directly.

One additional example of pH calculations should serve to clarify the point being made.

Example 8-8. What is the pH of a solution 2.0×10^{-7} *F* in the strong diacidic base $Ba(OH)_2$? The charge balance is

$$[H^+] + 2[Ba^{2+}] = [OH^-]$$

Appropriate substitutions yield

$$[H^+] + 2 \times 2.0 \times 10^{-7} = \frac{1.0 \times 10^{-14}}{[H^+]}$$

On rearrangement

$$[H^+]^2 + 4.0 \times 10^{-7}[H^+] - 1.0 \times 10^{-14} = 0$$

Solution of the quadratic equation yields

$$[H^+] = \tfrac{1}{2}(-4.0 \times 10^{-7} \pm \sqrt{16 \times 10^{-14} + 4 \times 10^{-14}})$$
$$= 2.5 \times 10^{-8}\ M$$

and $pH = 8 - \log 2.5 = 7.60$

The remark made after Example 8-7 also applies here.

8.8 General Treatment, Monofunctional Strong Acids and Bases

The derivation of a general formula that allows the acidity (alkalinity) of a solution of a strong acid (base) to be calculated from its concentration is simple. The reasoning involved is merely a generalization of that employed in Example 8-7. The hydrogen ions stem from two sources, the ionization of the acid and the autoprotolysis of water. Since the acid is a strong one and thus ionizes completely and since the acid is monofunctional, the hydrogen ion concentration from this source equals the formal concentration of the acid C_a. The hydrogen ion concentration from the autoprotolysis of the water equals the hydroxide concentrations. Consequently,

$$[H^+] = C_a + [OH^-] \tag{8-25}$$

From the expression of K_w, $[OH^-]$ is given as

$$[OH^-] = \frac{K_w}{[H^+]} \tag{8-26}$$

Substitution of this yields

$$[H^+] = C_a + \frac{K_w}{[H^+]} \tag{8-27}$$

Since the value of K_w is known, this equation suffices to calculate the acidity. The quadratic equation obtained on rearrangement of equation (8-27) is

$$[H^+]^2 - [H^+]C_a - K_w = 0 \tag{8-28}$$

Solution yields

$$[H^+] = \tfrac{1}{2}(C_a + \sqrt{C_a^2 + 4K_w}) \tag{8-29}$$

In almost all practical cases the solution of a strong acid has an acidity such that $[H^+]$ is considerably larger than $[OH^-]$. The latter term in equation (8-25) can then be neglected and the simple relation obtained is

$$[H^+] \simeq C_a \tag{8-30}$$

and

$$pH = -\log C_a \tag{8-31}$$

For a strong monofunctional base at formal concentration C_b, the reasoning is analogous and can be given in summary form:

$$[OH^-] = C_b + [H^+] \tag{8-32}$$

Proceeding as before, the following equation for $[OH^-]$ is obtained:

$$[OH^-] = \tfrac{1}{2}(C_b + \sqrt{C_b^2 + 4K_w}) \tag{8-32}$$

For a strong base, $[OH^-]$ in almost all practical cases is much larger than $[H^+]$, and the latter term can be neglected in equation (8-32):

$$[OH^-] \cong C_b \tag{8-34}$$

Substitution of $K_w/[H^+]$ for $[OH^-]$ and rearrangement yields

$$[H^+] = \frac{K_w}{C_b} \tag{8-35}$$

and
$$pH = pK_w - \log C_b = 14 - \log C_b \tag{8-36}$$

It is of interest to investigate the situation where a strong monofunctional acid and a strong monofunctional base are dissolved simultaneously. At equilibrium, the sum of the concentrations of *all* bases must equal the sum of the concentrations of *all* acids. Consequently, with the equalities already previously stated,

$$C_a + [OH^-] = C_b + [H^+] \tag{8-37}$$

Substitution of $K_w/[H^+]$ for $[OH^-]$ and solution of the quadratic equation obtained gives

$$[H^+] = \tfrac{1}{2}[(C_a - C_b) + \sqrt{(C_a - C_b)^2 + 4K_w}] \tag{8-38}$$

If $C_a = C_b$ the equation yields $[H^+] = \sqrt{K_w} = 10^{-7}$; that is, as expected, the solution is neutral. If either the acid or the base is in slight excess, the pH is far removed from that of the neutral point and then either $[OH^+]$ or $[H^+]$ can be neglected in equation (8-37). For an acidic solution (acid in excess, $C_a > C_b$)

$$[H^+] = C_a - C_b \tag{8-39}$$

For an alkaline solution (base in excess, $C_b > C_a$)

$$[OH^-] = C_b - C_a \tag{8-40}$$

or
$$[H^+] = \frac{K_w}{C_b - C_a} \tag{8-41}$$

It should be mentioned that the formulas for a strong acid, a strong base, or both are valid also if more than one acid or base is present if C_a and C_b is replaced by the sum of concentrations, that is, ΣC_a and ΣC_b, respectively.

Monofunctional Weak Acids and Bases 8.9

An acid or base is defined as being weak if the protolysis reaction with the solvent, here water, does not proceed essentially to completion, unless the solution is extremely dilute. The calculation of the acidity and alkalinity, that is, of the pH, of solutions containing weak protolytes can best be explained in terms of examples. To simplify the notation the weak acid may be denoted as HA. For example, for acetic acid, CH_3COOH, A^- is the anion CH_3COO^-. The protolysis reaction may now be written

$$HA + H_2O \rightleftharpoons A^- + H_3O^+ \qquad (8\text{-}42)$$

The acidity constant of the acid according to equation (8-20) is given as

$$K_a = \frac{[A^-][H^+]}{[HA]} \qquad (8\text{-}43)$$

Since the ionization does not go to completion, the acid, when equilibrium is established, will be present partially as nonionized acid and partially as the anion, that is, as the conjugate base. The formal concentration of the acid, C_a, will equal the sum of the molar concentrations of the nonionized and ionized portions,

$$C_a = [HA] + [A^-] \qquad (8\text{-}44)$$

This equation is called a material-balance equation or simply a material balance. The charge balance provides one more equation,

$$[H^+] = [A^-] + [OH^-] \qquad (8\text{-}45)$$

Rearrangement yields

$$[A^-] = [H^+] - [OH^-] \qquad (8\text{-}46)$$

Combination of this equation with equation (8-44) yields an expression for [HA]:

$$[HA] = C_a - ([H^+] - [OH^-]) \qquad (8\text{-}47)$$

Substitution of the relevant terms in the expression of the constant (8-43) gives

$$K_a = \frac{[H^+]([H^+] - [OH^-])}{C_a - ([H^+] - [OH^-])} \qquad (8\text{-}48)$$

When $[OH^-]$ is replaced by $K_w/[H^+]$, an expression is obtained that can be solved for $[H^+]$; with K_a, K_w, and C_a known, rigorous calculation of the acidity of the solution is possible. Unfortunately, the equation is third-order in $[H^+]$ and thus difficult to solve. However, for practical cases simplification is almost always possible. Unless the acid is extremely weak and the solution is very dilute, the acidity will be always sufficiently high to permit neglection of $[OH^-]$ with respect to $[H^+]$; this amounts to neglecting the contribution due to the autoprotolysis of water. Then the expression will simplify to

$$K_a \cong \frac{[H^+]^2}{C_a - [H^+]} \tag{8-49}$$

For many practical cases only a moderate acidity will result, and then the term $[H^+]$ in the denominator can be dropped, since it is much smaller than C_a. Then equation (8-49) simplifies to

$$K_a \cong \frac{[H^+]^2}{C_a} \tag{8-50}$$

The hydrogen ion concentration can therefore be obtained from the simple equation

$$[H^+] \cong \sqrt{K_a C_A} \tag{8-51}$$

The critical question arises as to when it is allowed to neglect a term. The values of the acidity constants are commonly reliable only within a few per cent. Thus the rule of thumb can be established that a term can be dropped if it is 5% or less than the term with which it is compared. Hence the first approximation, that is, neglecting of $[OH^-]$ in equation (8-48), can be made if $[OH^-]$ is 5% or less of $[H^+]$. The second approximation, that is, neglecting $[H^+]$ in the denominator of equation (8-49), is permissible if $[H^+]$ is 5% or less of C_a. It should be emphasized that approximations of this type are possible only with sums and differences and *never* with products or quotients; beginners often overlook this restriction.

Of course, it cannot be decided offhand whether an approximation is allowed or not. For example, the decision as to whether $[H^+]$ can be dropped in the denominator of equation (8-49) requires knowledge of the value of $[H^+]$, and it is this value that is to be obtained by the calculations. This difficulty is resolved in the following way. The approximation is made and the approximate value thus obtained is used for a check. If the 5% requirement (or whatever other limit is considered appropriate) is not fulfilled, the calculation must be repeated employing the more involved formula without the approximation that has proved unallowable.

Several examples will illustrate the approach.

Example 8-9. What is the pH of a 0.10 F solution of acetic acid when $K_a = 1.8 \times 10^{-5}$? Application of the approximate formula (8-51) yields

$$[H^+] \cong \sqrt{1.8 \times 10^{-5} \times 1.0 \times 10^{-1}} = \sqrt{1.8 \times 10^{-6}} = 1.3 \times 10^{-3}\ M$$

This value is now used to test the approximation. The hydrogen ion concentration is definitely large enough to drop $[OH^-]$, that is, to neglect the contribution of H^+ by the autoprotolysis of water. Thus the first approximation is valid. The value of $[H^+]$ obtained amounts to $\dfrac{1.3 \times 10^{-3} \times 100}{10^{-1}} = 1.3\%$ that of C_a. Consequently, the second approximation is also permitted. It is now safe to proceed with the calculation of the pH of the solution.

$$\text{pH} = -\log(1.3 \times 10^{-3}) = 3 - \log 1.3 = 3 - 0.11 = 2.89$$

Example 8-10. What is the pH of a 0.0050 F solution of formic acid with $K_a = 1.8 \times 10^{-4}$?

Application of the approximate formula (8-51) yields

$$[H^+] \cong \sqrt{1.8 \times 10^{-4} \times 5 \times 10^{-3}} = \sqrt{9 \times 10^{-7}}$$
$$= \sqrt{90 \times 10^{-8}} = 9.5 \times 10^{-4}\ M$$

and $\qquad \text{pH} = 3.02$

Testing the approximations reveals that the ionization of water does not contribute significantly to the acidity. However, a comparison of $[H^+]$ with C_a shows that the former is $\dfrac{9.5 \times 10^{-4} \times 100}{5 \times 10^{-3}} \cong 20\%$ of the latter. Consequently, the second approximation is not permitted according to the 5% requirement, and the calculation must proceed via the quadratic equation (8-49).

Substituting the numerical values yields

$$1.8 \times 10^{-4} = \frac{[H^+]^2}{5 \times 10^{-3} - [H^+]}$$

Rearrangement gives

$$[H^+]^2 + 1.8 \times 10^{-4}[H^+] - 9 \times 10^{-7} = 0$$

and solution

$$[H^+] = \tfrac{1}{2}(-1.8 \times 10^{-4} + \sqrt{3.2 \times 10^{-8} + 36 \times 10^{-7}})$$
$$= 8.6 \times 10^{-4}\ M$$

Finally,

$$\text{pH} = 3.07$$

Comparison of this result with the pH obtained by the approximate formula shows the difference to be only 0.05 pH unit, indicating that the 5% requirement is a fairly stringent one.

Example 8-11. What is the pH of a 5×10^{-6} F solution of hydrocyanic acid with $K_a = 5 \times 10^{-10}$?

Application of the approximate formula (8-51) yields

$$[H^+] \cong \sqrt{5 \times 10^{-6} \times 5 \times 10^{-10}} = \sqrt{25 \times 10^{-16}}$$
$$= 5 \times 10^{-8}\ M$$

The result is obviously wrong. This hydrogen ion concentration corresponds to an alkaline solution. Weak as an acid may be, its solution can never be alkaline. Obviously, with so weak an acid and at so small a concentration the contribution of ions by the water cannot be neglected and even the first approximation is not permitted. It might seem necessary to apply the third-order equation (8-48). In fact, however, the calculation is quite simple. The pH is obviously quite close to 7. Consequently, $[H^+]$ and $[OH^-]$ are therefore very small in comparison to C_a and can be neglected in the denominator of (8-48). Then

$$[H^+] \cong \sqrt{K_aC_a + K_w} = \sqrt{25 \times 10^{-16} + 10^{-14}}$$
$$= \sqrt{1.25 \times 10^{-14}} = 1.12 \times 10^{-7}\ M$$

and $\quad$ $pH = 6.95$

However, the calculation is practically meaningless, because such a solution would be so susceptible to influences of trace impurities (e.g., carbon dioxide from the air) that the pH actually measured would probably differ quite significantly from that calculated. The example serves the sole purpose of showing the importance of testing approximations and how to proceed with evaluating possibilities of dropping terms. Later, examples will be encountered that are of practical significance.

The derivation of a formula for calculating the alkalinity of a solution containing a weak monofunctional base, B, is fully analogous. The base protolyzes according to

$$B + H_2O \rightleftharpoons BH^+ + OH^- \qquad (8\text{-}52)$$

The basicity constant is given as

$$K_b = \frac{[OH^-][BH^+]}{[B]} \qquad (8\text{-}53)$$

Material balance yields

$$C_b = [B] + [BH^+] \qquad (8\text{-}54)$$

Charge balance yields

$$[H^+] + [BH^+] = [OH^-] \qquad (8\text{-}55)$$

By rearrangement and substitution into the expression for the basicity constant,

$$K_b = \frac{[OH^-]([OH^-] - [H^+])}{C_b - ([OH^-] - [H^+])} \tag{8-56}$$

This formula could have been written without derivation by simply replacing in the formula for a weak acid, (8-48), $[H^+]$ by $[OH^-]$ and vice versa, and substituting C_b and K_b for C_a and K_a, respectively. The expression obtained upon neglecting the contributions of ions by water (here equivalent to dropping the terms $[H^+]$) is

$$K_b \cong \frac{[OH^-]^2}{C_b - [OH^-]} \tag{8-57}$$

If the base is not too strong and the solution not too dilute, $[OH^-]$ is small enough to be neglected in comparison with C_b and the further approximate formula,

$$[OH^-] \cong \sqrt{K_b C_b} \tag{8-58}$$

is obtained. Application of these equations, the reasoning, and the testing of approximations are analogous to those for a weak acid. An example will delineate the point.

Example 8-12. What is the pH of a 0.020 F solution of ammonia (protolysis corresponding to $NH_3 + H_2O \rightleftharpoons NH_4^+ + OH^-$) if $K_b = 1.8 \times 10^{-5}$?

Application of the approximate formula (8-58) yields

$$[OH^-] = \sqrt{1.8 \times 10^{-5} \times 2 \times 10^{-2}} = \sqrt{3.6 \times 10^{-7}}$$
$$= \sqrt{36 \times 10^{-8}} = 6.0 \times 10^{-4}\,M$$

Testing the approximations shows that neglecting the protolysis of water is permitted. Since $[OH^-]$ is only $\frac{6 \times 10^{-4} \times 100}{2 \times 10^{-2}} = 3\%$ of C_b, the second approximation is also permitted. Proceeding to the calculation of the pH,

$$\text{pOH} = 4 - \log 6.0 = 3.22$$

and

$$\text{pH} = 14 - 3.22 = 10.78$$

General Treatment, Monofunctional Weak Acids and Bases 8.10

The derivations given in Section 8.9 involved molecular acids and bases only and may have recalled facts to which the student had already been exposed in previous courses. Analogous derivations are possible for ionic acids and bases. However, it is more convenient to have general formulas available applicable to all types of monofunctional acids and bases.

Regardless of whether an acid is electrically neutral or ionic, the protolysis equilibrium in water yielding the conjugate base proceeds according to

$$a + H_2O \rightleftharpoons b + H_3O^+ \tag{8-59}$$

The formal concentration of the acid, C_a, is, according to material balance,

$$C_a = [a] + [b] \tag{8-60}$$

Next we may write a scheme that shows the reactants and products of all protolytic reactions:

Initial protolytes	Bases formed	Acids formed
a of formal concentration C_a	b of concentration $[b]$	
H_2O	OH^- of concentration $[OH^-]$	H_3O^+ of concentration $[H^+]$

The sum of the concentrations of all acids formed must equal the sum of the concentration of all bases formed. (This condition is analogous to the charge balance used in the previous derivation.) Consequently,

$$[H^+] = [OH^-] + [b] \tag{8-61}$$

and, on rearrangement,

$$[b] = [H^+] - [OH^-] \tag{8-62}$$

Combination of (8-60) and (8-62) yields

$$[a] = C_a - [b] = C_a - ([H^+] - [OH^-]) \tag{8-63}$$

Substitution into the expression for the acidity constant, (8-20), yields the desired formula:

$$K_a = \frac{[H^+]([H^+] - [OH^-])}{C_a - ([H^+] - [OH^-])} \tag{8-64}$$

This expression is identical with that derived for a molecular acid (8-48). The arguments leading to approximations apply identically here and the simplified formulas are

$$K_a \simeq \frac{[H^+]^2}{C_a - [H^+]} \tag{8-65}$$

and

$$[H^+] \simeq \sqrt{K_a C_a} \tag{8-66}$$

The derivation for weak bases proceeds analogously and is left as an exercise to the student. The formulas obtained are

$$K_b = \frac{[OH^-]([OH^-] - [H^+])}{C_b - ([OH^-] - [H^+])} \tag{8-67}$$

$$K_b \cong \frac{[OH^-]^2}{C_b - [OH^-]} \tag{8-68}$$

and $$[OH^-] \cong \sqrt{K_b C_b} \tag{8-69}$$

A few examples will further strengthen the points.

Example 8-13. What is the pH of a 0.015 F solution of ammonium chloride if K_b for ammonia is 1.8×10^{-5}?

Ammonium chloride is a strong electrolyte and upon dissolution in water completely dissociates according to

$$NH_4Cl \rightarrow NH_4^+ + Cl^-$$

The chloride ion is the conjugate base of the strong acid HCl and is such a weak base that its basicity is of no consequence. The ammonium ion is a cationic acid. Thus the problem reduces to finding the pH of a solution containing the weak acid NH_4^+ in a concentration of 0.015 F. We recall that NH_4^+ is the conjugate acid of the base ammonia. Employing the relationship (8-24) ($K_a K_b = K_w$) the acidity constant of NH_4^+ is found to be

$$K_a = \frac{1.0 \times 10^{-14}}{1.8 \times 10^{-5}} = 5.6 \times 10^{-10}$$

Now the formula for a weak acid (8-66) is applied:

$$\begin{aligned}[H^+] &= \sqrt{5.6 \times 10^{-10} \times 1.5 \times 10^{-2}} = \sqrt{8.4 \times 10^{-10}} \\ &= 2.9 \times 10^{-5}\, M\end{aligned}$$

and $\quad pH = 4.54$

The tests for the validity of the approximations are left to the student.

Example 8-14. What is the pH of a 0.080 F solution of sodium benzoate if K_a for benzoic acid is 6.3×10^{-5}?

If the benzoate ion is denoted by R^-, the dissociation of the strong electrolyte sodium benzoate upon dissolution in water may be described as $NaR \rightarrow Na^+ + R^-$. The sodium ion solvates but does not show any acid–base properties and is therefore of no further interest. The benzoate ion is the conjugate base of benzoic acid and thus the problem reduces to finding the pH of a solution containing this weak base in a concentration of 0.080 F. The basicity constant of the benzoate ion is readily found to be

$$K_b = \frac{1.0 \times 10^{-14}}{6.3 \times 10^{-5}} = 1.6 \times 10^{-10}$$

Application of the approximate formula (8-69) yields

$$\begin{aligned}[OH^-] &= \sqrt{1.6 \times 10^{-10} \times 8 \times 10^{-2}} = \sqrt{12._8 \times 10^{-12}} \\ &= 3.6 \times 10^{-6}\, M\end{aligned}$$

$$\text{pOH} = 5.44$$

and $$\text{pH} = 8.56$$

The student should test the validity of the approximations.

Example 8-15. Calculate the pH of a solution obtained by dissolving 4.92 g of sodium acetate (*82.0*) in water to a volume of 400 ml. K_a for acetic acid is 1.8×10^{-5}.

Sodium acetate upon dissolution in water fully dissociates into sodium ion (which is of no further consequence) and acetate ion. The latter is the conjugate base of acetic acid and therefore has a basicity constant of

$$K_b = \frac{1.0 \times 10^{-14}}{1.8 \times 10^{-5}} = 5.6 \times 10^{-10}$$

To apply the approximate formula (8-69) the formal concentration of the base must be known, which in this case is equal to that of the salt. This concentration is obtained as

$$C_b = C_s = \frac{4.92}{82.0} \times \frac{1000}{400} = 0.150\ F$$

By substitution,

$$[\text{OH}^-] = \sqrt{5.6 \times 10^{-10} \times 1.5 \times 10^{-1}} = \sqrt{8.4 \times 10^{-11}}$$
$$= 9.2 \times 10^{-6}\ M$$

$$\text{pOH} = 5.04$$

and $$\text{pH} = 8.96$$

Example 8-16. The solution of the sodium salt of a weak monofunctional acid has a pH of 9.00. What will be the pH if the solution is diluted with pure water to twice its volume?

The anion of the weak acid is a weak base with K_b and at concentration C_b. Application of the approximate formula to the original solution yields

$$[\text{OH}^-]_{\text{original}} = \sqrt{K_b C_b}$$

Upon dilution to twice the volume the relationship holds:

$$[\text{OH}^-]_{\text{dilute}} = \sqrt{K_b C_b / 2}$$

Division of one equation by the other gives

$$\frac{[\text{OH}^-]_{\text{original}}}{[\text{OH}^-]_{\text{dilute}}} = \sqrt{2}$$

Substitution of $K_w/[\text{H}^+]$ for $[\text{OH}^-]$ yields

$$\frac{K_w/[\text{H}^+]_{\text{original}}}{K_w/[\text{H}^+]_{\text{dilute}}} = \sqrt{2}$$

and finally

$$[\text{H}^+]_{\text{dilute}} = [\text{H}^+]_{\text{original}} \sqrt{2}$$

Substitution of the numerical value gives

$$[H^+]_{dilute} = 1.0 \times 10^{-9} \times \sqrt{2} = 1.4 \times 10^{-9}\,M$$

and

$$pH = 8.85$$

Solution of a Monofunctional Weak Acid (or Base) and Its Salt 8.11

The situation becomes somewhat more involved if a weak acid (or base) is present in a solution together with its salt, that is, its conjugate. While it is always possible to establish enough independent relationships to obtain a number of equations equal to the number of unknowns, the final general expression for $[H^+]$ (or $[OH^-]$) is always of third or higher order. Fortunately, in most of the cases that are of practical importance reasonable approximations are allowed, and if applied lead to expressions that can be handled more readily. At least an approximate value can always be obtained that permits the rough evaluation of the situation at hand.

No full derivations will be given in this section. Rather, the approximations will be made a priori and only approximate expressions will be derived. However, the student is warned to take a cautious look at the numerical results whenever such expressions are employed and especially so in cases where the acid or base is either extremely weak or relatively strong, where either a very dilute or a relatively concentrated solution is involved, or where the pH value is very high or very low.

First the situation will be examined for a weak acid present in a solution with its sodium salt. The acid undergoes protolysis according to

$$a + H_2O \rightleftharpoons b + H_3O^+ \qquad (8\text{-}70)$$

The acidity constant is

$$K_a = \frac{[H^+][b]}{[a]} \qquad (8\text{-}71)$$

The salt dissociates and the sodium ion is of no further consequence. The conjugate base, however, undergoes hydrolysis according to

$$b + H_2O \rightleftharpoons a + OH^- \qquad (8\text{-}72)$$

Neither the protolytic reaction of the acid or of the conjugate base is assumed to proceed to any great extent. Protolysis of the acid is restricted because its conjugate base, b, is present in significant concentration from the addition of the salt.

Analogously, the hydrolysis of the base is restricted because its conjugate acid, a, is present in significant amounts. Consequently, as an approximation the molar concentration of the acid, $[a]$, can be taken as equal to the concentration of the acid added, [acid]. Analogously, the molar concentration of the conjugate base, $[b]$, can be set equal to the concentration of the salt, [salt]. Substitution into the expression for the acidity constant yields

$$K_a = \frac{[H^+][\text{salt}]}{[\text{acid}]} \tag{8-73}$$

On rearrangement the approximate formula for calculating the acidity of the solution is secured:

$$[H^+] \cong K_a \frac{[\text{acid}]}{[\text{salt}]} \tag{8-74}$$

It can be seen that for a given system, that is, with K_a given, the acidity is solely a function of the ratio of the formal concentrations of acid and salt. It should be pointed out that this simple approach holds only for acids with K_a in the approximate range 10^{-3} to 10^{-10} and at reasonable concentrations, say not below 10^{-3} to 10^{-4} F. It cannot be applied to very highly concentrated solutions, but here the calculations are invalid anyway because of the failure to consider activities. In addition, it should be pointed out that the simple approach is not applicable if the salt involved introduces any species other than the conjugate base that exhibits significant acid–base properties.

These restrictions hold analogously for solutions containing a weak base and its conjugate acid. From the analogy between acids and bases already stressed several times, the formula for the calculation of the hydroxide in concentration of a solution containing a weak base and its salt can be written immediately as

$$[OH^-] \cong K_b \frac{[\text{base}]}{[\text{salt}]} \tag{8-75}$$

The derivation is recommended to the student as an exercise.

Some examples will further familiarize the student with the topic.

Example 8-17. What is the pH of a solution 0.050 F in sodium acetate and 0.025 F in acetic acid ($K_a = 1.8 \times 10^{-5}$)?

Substitution in (8-74) yields

$$[H^+] = \frac{1.8 \times 10^{-5} \times 2.5 \times 10^{-2}}{5.0 \times 10^{-2}} = 9.0 \times 10^{-6}\ F$$

and $$pH = 6.00 - \log 9.0 = 6.00 - 0.95 = 5.05$$

Example 8-18. Calculate the pH of a solution prepared by dissolving 3.0 g of acetic acid (*60.0*; $K_a = 1.8 \times 10^{-5}$) and 4.9 g of sodium acetate (*82.0*) in water and diluting with water to a volume of exactly 200 ml.

The moles of acid and salt present are:

$$\text{moles of acetic acid} = \frac{3.0}{60.0} = 0.050$$

$$\text{moles of sodium acetate} = \frac{4.9}{82} = 0.060$$

The corresponding concentrations are:

$$[\text{acid}] = 0.050 \times \frac{1000}{200}$$

$$[\text{salt}] = 0.060 \times \frac{1000}{200}$$

Substitution in (8-74) yields

$$[H^+] = 1.8 \times 10^{-5} \times \frac{0.050 \times 1000/200}{0.060 \times 1000/200} = 1.5 \times 10^{-5}$$

and $$pH = 4.82$$

It can be seen in Example 8-18 that the volume of the solution cancels. In many applications of formulas (8-74) and (8-75) this is the case, and it is unnecessary to carry the volume through the calculations; that is, the millimoles of the species can be used instead of their concentrations.

Polyfunctional Acids, Bases, and Their Salts 8.12

Polyfunctional acids (bases) are capable of donating (accepting) more than one proton per molecule. The discussion is here limited to diprotic acids. Diacidic bases can be treated analogously replacing $[H^+]$ by $[OH^-]$ and the acidity constant by the basicity constant.

A molecular diprotic acid, H_2A, in water undergoes protolysis in steps:

$$H_2A + H_2O \rightleftharpoons HA^- + H_3O^+ \qquad (8\text{-}76)$$

$$HA^- + H_2O \rightleftharpoons A^{2-} + H_3O^+ \qquad (8\text{-}77)$$

The corresponding expressions for the two acidity constants

are

$$K_{a,1} = \frac{[H^+][HA^-]}{[H_2A]} \tag{8-78}$$

and

$$K_{a,2} = \frac{[H^+][A^{2-}]}{[HA^-]} \tag{8-79}$$

It may be noted that in general the value of the acidity constant for each successive protolytic reaction is smaller; that is, $K_{a,1} > K_{a,2}$. The fact is largely explained by the following. In the first step, the acid loses a proton and the conjugate base has one less positive charge. This base is the acid in the second step and with one negative charge less it holds remaining protons more strongly; in other words, it has less tendency to donate a proton.

For sulfuric acid the first step corresponds to the protolysis of a strong acid and goes essentially to completion. The acidity constant for the second step is fairly large (1.1×10^{-2}). Thus frequently the approximation is made that ionization is also complete here. Hence, the hydrogen ion molar concentration is taken as twice the formality of the acid. However, it is interesting to test this approximation by calculating the actual acidity of a sulfuric acid solution.

Example 8-19. (a) Derive a formula for the calculation of the acidity of a sulfuric acid solution assuming that the first protolytic step goes to completion and that the second is that of an acid having $K_a = 1.1 \times 10^{-2}$. (b) Use the formula to calculate the hydrogen ion concentration of a 0.010 *F* sulfuric acid.

To derive the requested formula a number of equations will be stated that equal the number of unknowns. These equations will then be combined according to the rules of algebra so that eventually $[H^+]$ is obtained as a sole function of C_a and K_a.

The first ionization step, as that of a strong acid, goes to completion. For the second step the acidity constant expression may be written

$$K_a = \frac{[H^+][SO_4^{2-}]}{[HSO_4^-]}$$

The charge balance is

$$[H^+] = 2[SO_4^{2-}] + [HSO_4^-] + [OH^-]$$

which can definitely be simplified by neglecting $[OH^-]$ since the solution is strongly acidic.

$$[H^+] = 2[SO_4^{2-}] + [HSO_4^-]$$

If the formal concentration of the sulfuric acid is denoted by C_a, the material balance yields

$$C_a = [SO_4^{2-}] + [HSO_4^-]$$

By rearrangement,

$$[HSO_4^-] = C_a - [SO_4^{2-}]$$

Combination of this equation with the simplified charge balance yields

$$[H^+] = 2[SO_4^{2-}] + C_a - [SO_4^{2-}] = [SO_4^{2-}] + C_a$$

or, after rearrangement,

$$[SO_4^{2-}] = [H^+] - C_a$$

Combination of the two preceding equations gives

$$[HSO_4^-] = C_a - [H^+] + C_a = 2C_a - [H^+]$$

Substitution into the acidity constant expression gives

$$K_a = \frac{[H^+]([H^+] - C_a)}{2C_a - [H^+]}$$

Rearrangement gives

$$[H^+]^2 - [H^+](C_a - K_a) - 2C_aK_a = 0$$

and solution secures the requested formula,

$$[H^+] = \tfrac{1}{2}[(C_a - K_a) + \sqrt{(C_a - K_a)^2 + 8C_aK_a}]$$

(b) Substitution of the numerical values gives

$$[H^+] = \tfrac{1}{2}(1.0 \times 10^{-2} - 1.1 \times 10^{-2} + \sqrt{1.0 \times 10^{-6} + 8 \times 1.1 \times 10^{-4}}) = 1.4 \times 10^{-2}\ M$$

Thus the hydrogen ion concentration is $1.4 \times 10^{-2}\ M$, which is 30% less than the value of $2.0 \times 10^{-2}\ M$ obtained under the assumption that both steps go to completion. The student should evaluate results for more concentrated and more dilute solutions, say, 0.10 and 0.0010 F, and observe the differences.

If both protolytic steps correspond to weak acids, the following reasoning is applied when evaluating the acidity of a solution containing the acid. Since the second acidity constant is usually considerably smaller than the first, ionization according to the second step can be assumed to be completely repressed by the relatively large hydrogen ion concentration produced in the first step. In other words, all of the hydrogen ion is assumed to stem from the first step, and calculation of the acidity of the solution reduces to the case of a weak monoprotic acid having an acidity constant equal to $K_{a,1}$.

The calculation of the pH of solutions containing salts of difunctional acid and bases is deferred until Section 11.9.

Mixed Equilibria: Acid–Base and Salt Solubility 8.13

The dissolution of a sparingly soluble salt has been discussed in Chapter 5 in terms of the solubility equilibria only. This approach is inadequate or at most serves only as a first approximation if either the cation or the anion, or both, undergo protolytic reactions. If protolysis is extensive, and is neglected in the calculation, the values of the formal solubility thus obtained can be grossly in error (low). How correct results can be obtained by taking protolysis into account will be explained for a few simple cases in terms of examples.

Example 8-20. Calculate the formal solubility in pure water of the sparingly soluble salt MA ($K_{sp} = 2.5 \times 10^{-27}$), composed of the metal M^+ and the anion, A^-, which is the conjugate base of weak acid HA ($K_a = 1.2 \times 10^{-14}$). The metal ion does not undergo any protolytic reaction.

The competitive solubility and protolysis equilibria may be displayed as

$$\begin{array}{lll} \underline{MA} \rightleftharpoons M^+ + & A^- & \\ & + & \\ & H_2O \rightleftharpoons & HA + OH^- \end{array}$$

From this reaction scheme it can be seen that the protolysis of A^- shifts the solubility equilibrium to the right, that is, increases the solubility of the salt.

It is evident that the following material balance holds:

$$[M^+] = [A^-] + [HA]$$

Rearrangement of the expression for the acidity constant yields

$$[HA] = \frac{[A^-][H^+]}{K_a}$$

By substitution,

$$[M^+] = [A^-] + \frac{[A^-][H^+]}{K_a} = [A^-]\left(1 + \frac{[H^+]}{K_a}\right)$$

Regardless of protolysis, for a saturated solution the solubility-product expression for the salt (5-2) still holds. However, it must be realized that this expression involves only the concentrations of M^+ and A^- and does not consider the portion of the anion that is transferred by hydrolysis to HA. Hence

$$[A^-] = \frac{K_{sp}}{[M^+]}$$

Combination of the last two equations yields

$$[M^+] = \frac{K_{sp}}{[M^+]}\left(1 + \frac{[H^+]}{K_a}\right)$$

And the formal solubility, since it here equals the molar concentration of M^+, becomes

$$S^f = [M^+] = \sqrt{K_{sp}\left(1 + \frac{[H^+]}{K_a}\right)}$$

To calculate the formal solubility the pH of the solution must be known. Since the salt is of very low solubility the approximation can be introduced that the hydroxide ion concentration due to the hydrolysis of the anion is negligibly small in comparison with that due to the autoprotolysis of water. Then the hydrogen ion concentration is 10^{-7} M. Substitution of this and the other numerical values yields

$$S^f = \sqrt{2.5 \times 10^{-27}\left(1 + \frac{1.0 \times 10^{-7}}{1.2 \times 10^{-14}}\right)} = 1.4 \times 10^{-10}\,F$$

Testing of the approximation involves the following reasoning. Since hydroxide ion is produced by the dissolution and protolysis of molecular MA, the hydroxide ion concentration stemming from this process can at most be equal to that of the formal solubility and is thus negligible in comparison to the value 10^{-7} from the ionization of water. It is of interest to contrast the values of the solubility obtained with and without considering hydrolysis. The latter value is

$$S^f = \sqrt{2.5 \times 10^{-27}} = 5.0 \times 10^{-14}\,F$$

As can be seen, the protolysis of the anion causes the formal solubility to increase by a factor of about 10,000.

In the equation for the solubility of a sparingly soluble salt of type M^+A^- developed in Example 8-20,

$$S^f = [M^+] = \sqrt{K_{sp}\left(1 + \frac{[H^+]}{K_a}\right)} \qquad (8\text{-}80)$$

the term $1 + ([H^+]/K_a)$ may be denoted as α_H. For a particular acid the value of this term is a function only of the acidity of the solution. The product $K_{sp} \times \alpha_H$ may be denoted as K'_{sp} and be termed the effective or conditional solubility product. Either term is appropriate since this constant is the one *effective* under the particular acidity *condition* prevailing. The values of α_H or of $\log \alpha_H$ may be tabulated or plotted as a function of pH. Then for any sparingly soluble salt of the particular acid, the effective solubility product can be calculated readily and may be used in any other type of calculations by employing the effective constant exactly in the same manner as K_{sp} itself.

By reasoning in a manner analogous to that in Example 8-20, the student may derive the general formula for the formal solubility of a salt $M^{2+}A^{2-}$, where the anion A^{2-} of a weak diprotic acid H_2A undergoes hydrolysis. The derivation starts with the equality

$$[M^{2+}] = [A^{2-}] + [HA^-] + [H_2A]$$

The result is

$$S^f = \sqrt{K_{sp}\left(1 + \frac{[H^+]}{K_{a,2}} + \frac{[H^+]^2}{K_{a,1}K_{a,2}}\right)} = \sqrt{K_{sp}\alpha_H} \quad (8\text{-}81)$$

To gain deeper insight, the student is strongly advised to calculate α_H in equation (8-81) for the weak diprotic acid hydrogen sulfide ($K_{a,1} = 1.0 \times 10^{-7}$, $K_{a,2} = 1.2 \times 10^{-14}$) as a function of pH for pH values of values of 0, 1, 2, 3, and so forth up to pH 10, and to plot log α_H versus pH. The student should observe when the various terms in the expression for α_H contribute significantly and when they can be neglected. Further application of the values obtained is recommended to the calculation of the solubilities of the sulfides of lead, cadmium, and zinc in acidic solutions ranging from pH 6 to pH 0. The results should be related to the qualitative sulfide separation scheme for cations.

A further example will demonstrate additional reasoning involved in this type of problem.

Example 8-21. Calculate the formal solubility of calcium oxalate (a) in pure water and (b) in an aqueous solution buffered to pH 2.00. The following data are given: $K_{sp} = [Ca^{2+}][C_2O_4^{2-}] = 2.3 \times 10^{-9}$; $K_{a,1} = 5.4 \times 10^{-2}$; $K_{a,2} = 5.1 \times 10^{-5}$.

(a) Since both dissociation constants of oxalic acid are relatively large, hydrolysis of oxalate ion in pure water can be neglected and the formal solubility of the salt can be calculated from the simple expression for the solubility product,

$$S^f = [Ca^{2+}] = \sqrt{K_{sp}} = \sqrt{2.3 \times 10^{-9}} = 4.8 \times 10^{-5}\,F$$

By substituting the value $[H^+] = 1.0 \times 10^{-7}\,M$ in the full expression (8-81) it can be confirmed that protolysis can indeed be neglected in pure water.

(b) For pH = 2.00 the hydrogen ion concentration $[H^+] = 1.0 \times 10^{-2}\,M$. Substitution of this and the other relevant numerical values into equation (8-81) yields

$$S^f = \sqrt{2.3 \times 10^{-9}\left(1 + \frac{1.0 \times 10^{-2}}{5.1 \times 10^{-5}} + \frac{(1.0 \times 10^{-2})^2}{5.4 \times 10^{-2} \times 5.1 \times 10^{-5}}\right)}$$
$$= 7.3 \times 10^{-4}\,F$$

Calcium oxalate is therefore about 15 times more soluble at pH 2 than at pH 7.

8.14 Questions

8-1. What is an acid? A strong acid? A weak acid? A salt?

8-2. What fundamental relation exists between the hydrogen ion concentration and formality of a monoprotic strong acid?

3. What are the approximate formulas describing the acidity of a weak acid solution and the basicity of a weak base solution?

4. Elaborate on the restriction existing for the application of the approximate formulas mentioned in Question 8-3.

5. What is stated by the electroneutrality rule?

6. Discuss the advantages of the pH concept.

7. How does the value of the ion product of water change with temperature?

8. Elaborate on the meaning of a negative pH value.

9. What does "protolysis of an ion" mean?

0. What is an amphiprotic substance? Give examples.

1. Explain why SO_4^{2-} is a weak base in aqueous medium.

2. State whether the solution of the following substances in pure water will be acidic, neutral, or alkaline and explain your answers: (a) NaCl; (b) NH_4Cl; (c) ammonium acetate ($K_a = K_b$); (d) potassium acetate.

3. Is the statement "HCl is a strong acid" complete? If not, what might be added to make it complete?

4. Elaborate on the fact that according to the strict Brønsted concept KOH is a salt. Why may it loosely be called a base?

5. What relationship exists between the acidity constant of a weak acid and the basicity constant of its conjugate base?

6. Why is the second dissociation constant of a diacidic weak base always smaller in value than the first dissociation constant?

7. What is an effective solubility-product constant?

8. Why does a solution of KCN show an alkalinity close to that of NaOH?

9. Elaborate on the conditions under which the autoprotolysis of water may be neglected when dealing with acid–base problems.

0. Two acids are of equal strength in aqueous solution. Are these two acids necessarily of equal strength in another solvent? Elaborate.

1. In the *strict* sense of the Brønsted concepts, which of the following substances would be considered acids, bases, ampholytes, and salts in aqueous medium?

HBO_2	KCN	CN^-	HSO_4^-	$Ba(OH)_2$	CH_3COONa
1	*2*	*3*	*4*	*5*	*6*

$Fe(H_2O)_6^{3-}$	KCl	HCN	NH_3	NH_4^+
7	*8*	*9*	*10*	*11*

Answers: acids: *1*, *4*, *7*, *9*, and *11*; bases: *3*, *4*, and *10*; ampholyte: *4*; salts: *2*, *5*, *6*, and *8*

2. Write for each of the substances in Question 8-21 the protolysis reaction occurring on their contact with water and discuss the type of species formed in terms of the Brønsted concepts.

8.15 Problems

8-1. Calculate the pH corresponding to the following conditions in an aqueous solution at room temperature:

(a) $[H^+] = 1.0 \times 10^{-9}\ M$ (d) $[OH^-] = 1.2 \times 10^{-3}\ M$
(b) $[H^+] = 2.35 \times 10^{-2}\ M$ (e) $[OH^-] = 0.012\ M$
(c) $[H^+] = 0.00321\ M$ (f) $[OH^-] = 1 \times 10^{-5}\ M$

Answers: (a) 9.0; (b) 1.63; (c) 2.49; (d) 11.08; (e) 12.08; (f) 9.0

8-2. Convert the following pH values to the corresponding values of $[H^+]$ and $[OH^-]$: (a) 7.63; (b) 1.48.

Answers: (a) $[H^+] = 2.3 \times 10^{-8}\ M$ and $[OH^-] = 4.3 \times 10^{-7}\ M$;
(b) $[H^+] = 3.3 \times 10^{-2}\ M$ and $[OH^-] = 3.0 \times 10^{-13}\ M$

8-3. Calculate the pH of $2.0 \times 10^{-7}\ F$ HCl.

Answer: pH 6.62

8-4. Calculate the pH of $1.5 \times 10^{-7}\ F$ NaOH.

Answer: pH 7.30

8-5. Calculate the pH of a 0.050 F solution of a weak monoprotic acid when $K_a = 1.0 \times 10^{-7}$.

Answer: pH 4.15

8-6. What is the pH of a 0.020 F solution of a weak monoprotic base when $K_b = 5 \times 10^{-6}$?

Answer: pH 10.5

8-7. What is the pH of a 0.040 F solution of a weak monoprotic acid when $K_a = 1.0 \times 10^{-3}$?

Answer: pH 2.23

8-8. What is the pH of a 0.10 F solution of a weak monoprotic base when $K_b = 5.0 \times 10^{-3}$?

Answer: pH 12.30

8-9. What is the pH of an equimolar mixture of a weak base, B^+, of $K_b = 1.8 \times 10^{-5}$ and its nitrate salt, BNO_3?

Answer: pH 9.26

8-10. Calculate the pH of a solution obtained by dissolving 50 ml of concentrated aqueous ammonia (*17.0*) (density 0.90 g/ml; 28% *w/w* NH_3) and 2.0 g of NH_4Cl (*53.5*) in water and diluting to exactly 250 ml ($K_b = 1.8 \times 10^{-5}$).

Answer: pH 10.55

8-11. What is the pH of the solution obtained by mixing 50 ml of a 0.15 M solution of the sodium salt of a weak monoprotic acid ($K_a = 4.0 \times 10^{-6}$) with 40 ml of 0.10 M HCl?

Answer: pH 5.34

8-12. The pH of a 0.050 F solution of sodium acetate was found to be 8.70. What is the value of the dissociation constant of acetic acid based on this experiment?

Answer: $K_a = 2.0 \times 10^{-5}$

3. A volume of 50 ml of 1.120 *F* HCl was diluted with 120 ml of pure water. What is the pH of the resulting solution?

Answer: pH 0.48

4. Calculate the formal solubility of a salt MR in (a) pure water, (b) in a solution buffered to pH 3.00, and (c) in 0.100 *F* HCl, when $K_{sp} = 4.0 \times 10^{-8}$ and $K_{a,HR} = 8.0 \times 10^{-4}$.

Answers: (a) 2.0×10^{-4} *F*; (b) 3.0×10^{-4} *F*; (c) 2.2×10^{-3} *F*

5. Write the complete charge-balance equation for a solution obtained by dissolving NaCl, Na_2HPO_4, and $ZnSO_4$, and HCl in water.

Answer: $[H^+] + [Na^+] + 2[Zn^{2+}] = [OH^-] + [Cl^-] + [HSO_4^-] + 2[SO_4^{2-}] + [H_2PO_4^-] + 2[HPO_4^{2-}] + 3[PO_4^{3-}]$

6. Which of the terms in the charge-balance equation of Problem 8-15 can be neglected in good approximation if the concentrations of each of the solutes present is about 0.1 *F*?

Answer: $[OH^-]$, $3[PO_4^{3-}]$, $2[H_2PO_4^{2-}]$

7. What is the basicity constant in water of the anion of a weak acid HA with $K_a = 1 \times 10^{-11}$?

Answer: 1×10^{-3}

8. Aniline is an organic base, $C_6H_5NH_2$. It undergoes protolysis in aqueous media to form the anilinium ion, $C_6H_5NH_3^+$. K_b for aniline is 4.0×10^{-10}. Calculate the acidity constant of the anilinium ion.

Answer: 2.5×10^{-5}

9. Calculate the pH value of a 0.040 *M* solution of anilinium chloride in water. (*Hint:* The anilinium ion is an acid; the chloride ion is such a weak base that its influence can be neglected.)

Answer: pH 3.0

0. The hydroxopentaaquoaluminate ion, $Al(H_2O)_5(OH)^{2+}$, is a base with a K_b value in water of 3.5×10^{-10} at 15°C. Calculate the acidity constant of the hexaaquoaluminate ion at this temperature at which pK_w has the value 14.35.

Answer: 1.3×10^{-5} (Note that this ion has about the same acid strength as acetic acid!)

1. Calculate the pH value of the solutions of the following molar hydrogen ion concentrations.

(a) 6.2×10^{-3}
(b) 3.0×10^{-9}
(c) 6.5×10^{-5}
(d) 5.0×10^{-9}
(e) 9.3×10^{-6}
(f) 1.2
(g) 2.9
(h) 4.5×10^{-14}
(i) 1.9×10^{-15}
(j) 4.0×10^{-15}

2. Calculate the molar hydrogen ion concentrations of the solutions with the following pH values.

(a) 6.53
(b) 9.82
(c) 4.85
(d) −1.22
(e) 7.20
(f) 14.57
(g) −1.52
(h) 14.20
(i) 8.98
(j) 7.91

8.15 PROBLEMS

8-23. Calculate the molar hydroxide ion concentrations of the solutions with the following pH values.

(a) 1.51
(b) 13.62
(c) 3.67
(d) 6.29
(e) 2.70
(f) 1.32
(g) 3.89
(h) 6.73

8-24. The following solutions of strong acids are mixed. Calculate the pH of each solution resulting.

(a) 70.0 ml of 0.010 F HCl and 30.0 ml of 0.015 F $HClO_4$
(b) 25.0 ml of 0.050 F HNO_3 and 50.0 ml of 0.025 F HCl
(c) 25.0 ml of 0.0080 F $HClO_4$ and 20.0 ml of 0.011 F HNO_3

8-25. What is the pH of 3.00×10^{-7} F HCl?

8-26. What is the pH of 4.00×10^{-7} F HCl?

8-27. What is the pH of 1.00×10^{-7} F $Ba(OH)_2$?

8-28. What is the pH value of each of the following solutions of a weak acid or base having the indicated K_a or K_b value:

	Solution	Dissociation constant
(a)	0.40 F $HC_2H_3O_2$	1.8×10^{-5}
(b)	0.15 F HClO	2.8×10^{-8}
(c)	0.01 F HSCN	1.4×10^{-1}
(d)	0.05 F HF	6.0×10^{-4}
(e)	0.02 F weak base	4.2×10^{-3}
(f)	0.18 F weak base	6.0×10^{-4}

8-29. Calculate the value of the dissociation constant of each of the weak monobasic acids listed below from the pH values of their respective solutions:

	Solution	pH
(a)	0.010 F $HC_2H_3O_2$	3.37
(b)	0.050 F HCN	5.35
(c)	0.20 F $HCHO_2$	2.22
(d)	0.40 F HNO_2	1.85

8-30. Calculate the pH value of each of the following solutions. The solution is

(a) 0.0200 F in acetic acid ($K_a = 1.8 \times 10^{-5}$) and 0.100 F in sodium acetate
(b) 0.0200 F in ammonia ($K_b = 1.8 \times 10^{-5}$) and 0.600 F in ammonium chloride
(c) 0.100 F in formic acid ($K_a = 1.8 \times 10^{-4}$) and 0.150 F in sodium formate
(d) 0.0100 F in boric acid ($K_a = 5.9 \times 10^{-10}$) and 0.0500 F in sodium borate
(e) 0.0500 F in hydrofluoric acid ($K_a = 6.0 \times 10^{-4}$) and 0.150 F in sodium fluoride
(f) 0.300 F in tartaric acid ($K_{a,1} = 9.2 \times 10^{-4}$) and 0.0200 F in potassium acid tartrate

1. Calculate the pH value of each of the solutions obtained by mixing
   (a) 25.0 ml of 0.100 F acetic acid ($K_a = 1.8 \times 10^{-5}$) and 20.0 ml of 0.100 F NaOH
   (b) 30.0 ml of 0.100 F KCN and 25.0 ml of 0.100 F HCl (for HCN, $K_a = 5.0 \times 10^{-10}$)
   (c) 20.0 ml of 0.100 F formic acid ($K_a = 1.8 \times 10^{-4}$) and 20.0 ml of 0.150 F sodium formate
   (d) 50.0 ml of 0.200 F acetic acid ($K_a = 1.8 \times 10^{-5}$) and 50.0 ml of 0.100 F NaOH
   (e) 30.0 ml of 0.100 F ammonia ($K_b = 1.8 \times 10^{-5}$) and 20.0 ml of 0.100 F HCl
   (f) 35.0 ml of 0.300 F hydrofluoric acid ($K_a = 6.0 \times 10^{-4}$) and 65.0 ml of 0.100 F sodium fluoride

2. What is the formal solubility of LiF ($K_{sp} = 4.0 \times 10^{-3}$) in (a) pure water and (b) 1.0 F HF ($K_a = 6.0 \times 10^{-4}$)?

3. At what pH value will the formal solubility of CaF_2 ($K_{sp} = 4.0 \times 10^{-11}$) be 100 times that in pure water? $K_{a,HF} = 6.0 \times 10^{-4}$.

4. The dissociation constant of HF has the value 6.0×10^{-4}. What will be the solubility, in moles per liter, of CaF_2 ($K_{sp} = 3.7 \times 10^{-8}$) (a) in pure water, and in a solution buffered to (b) pH 3.50, (c) pH 2.50, and (d) pH 1.50?

5. What amount, in milligrams, of lead (*207.2*) is present in 500 ml of a saturated solution of PbF_2 ($K_{sp} = 3.7 \times 10^{-8}$) (a) in pure water, in a solution buffered to (b) pH 2.00 and (c) pH 1.00, and (d) in a solution buffered to pH 1.50 and made 0.50 F in NaF?

9 REVIEW QUESTIONS, CHEMICAL EQUILIBRIA AND GRAVIMETRIC ANALYSIS

1. Elaborate on the difference between an ion product and a solubility product.
2. Elaborate on the factors that govern the rate of a precipitation.
3. What is peptization? What difficulty may it cause? What can be done to avoid it?
4. Elaborate on the possibilities of obtaining a coarsely crystalline precipitate relatively free of impurities.
5. What advantages and disadvantages does reprecipitation have?
6. When a certain substance is dissolved in water, a temperature decrease is observed. From this fact predict whether its solubility will increase or decrease with increasing temperature.
7. Give the units of the solubility products for lead(II) sulfate and lead(II) iodide.
8. Strontium oxalate is slightly soluble in water. Name several ions which would influence the solubility of this salt via the common-ion or diverse-ion effects.
9. By application of the solubility-product concept establish whether a medium 0.1 *F* in $AgNO_3$ or one 0.1 *F* in K_2CrO_4 would be more effective in repressing the solubility of Ag_2CrO_4.
10. The following statement is made: The formal solubility of a salt MA is 1×10^{-6}; consequently, the solubility product is 1×10^{-12}. Elaborate on this statement and discuss what qualifications must be made before the statement can be considered to be correct.
11. Is it safe to assume that drying a substance at 110°C would remove all the water? Elaborate.
12. In quantitative analysis, samples are usually dried before they are weighed. What is the merit of this procedure?
13. How may the common-ion effect be used to assure essentially complete precipitation?
14. In gravimetric analysis, it is necessary to ignite some precipitates rather than to simply dry them. Elaborate on this statement and give examples.
15. How does addition of a catalyst affect the position of an equilibrium?
16. When barium sulfate is precipitated in the presence of sodium ions, some of the sodium is coprecipitated as sodium sulfate. Will this

cause a low or a high result in the gravimetric determination of sulfate?

9-17. In the gravimetric determination of lead, the $PbSO_4$ precipitate is first washed with a dilute solution of sulfuric acid rather than water. Explain the merit of this practice.

9-18. Elaborate on the mechanisms by which impurities can be carried down with a precipitate.

9-19. Two substances have identical numerical values for their respective solubility products. One substance is of the type MA; the other, MA_2. Hydrolysis being negligible, which substance is more soluble?

9-20. Consider the situation of Question 9-19 when the substances are of the types M_2A and MA_2.

9-21. How would you treat a sample solution containing NaCl and Na_2S prior to the gravimetric determination of chloride as AgCl? What would happen if the solution were used without a pretreatment?

9-22. Why is it important in many cases to refer to the "dry basis" in reporting an analytical result?

9-23. During the ignition of hydrous iron(III) oxide to Fe_2O_3, it is possible that some Fe_3O_4 forms if improper conditions prevail. Will the result of the gravimetric determination of iron then be low or high?

9-24. Elaborate on the requirements for a substance to be used as (a) a precipitation form and (b) as a weighing form.

9-25. Consider the precipitation of the following substances: $BaSO_4$, CaC_2O_4, CaF_2, Ag_2CrO_4, $BaCO_3$, and AgCl. For which would you assume that the acidity of the solution during the precipitation is an important factor? Explain your answer.

9-26. Explain the experimental finding that sparingly soluble salts of weak acids, but not of strong ones, are commonly dissolved by strong acids.

9-27. What is the meaning assigned the "p" in terms such as pK?

9-28. Explain the alkaline reaction of a solution of the sodium salt of a weak acid.

9-29. What is coagulation?

9-30. In a calculation related to the pH of an acid-base equilibrium, approximate formulas were used and the value for the pH was found to be close to 7. What should you check in that case?

9-31. An acid–base equilibrium is considered at pH 3. Is it necessary to take the hydroxide ion concentration into account when formulating the charge balance? Explain your answer.

9-32. What is the effect of temperature on the ionization equilibrium of water?

9-33. Distinguish between strong and weak electrolytes.

9-34. Distinguish between monoprotic and polyprotic acids.

What will be the pH of a solution of a salt formed from a weak acid and a weak base if $K_a = K_b$?

Compare the pH values of aqueous solutions of hydrochloric acid, sulfuric acid, and acetic acid equal in *formality*. Explain the differences.

Distinguish between the following terms: common-ion effect and foreign-ion effect; coprecipitation and postprecipitation; molarity and formality; protolyte and ampholyte; dry basis and "as is" basis; weak base and strong base; and charge balance and material balance.

Is the pAg value in a saturated silver chloride solution larger or smaller than in a saturated silver bromide solution? Explain.

10 TITRIMETRY: GENERAL CONSIDERATIONS

This chapter will serve only to introduce the subject of titrimetry, to unify it, and to relate it to facts discussed in previous courses. The full meaning and nuances of many of the statements will be better appreciated as the student gains a more detailed knowledge of titrimetric methods through the study of subsequent sections of this volume and through laboratory experience. It is strongly recommended that the student return to this section and relate its content to each type of titration presented.

Concepts and Definitions 10.1

Titrimetric analysis or titrimetry is a term for quantitative methods in which the amount of sought-for substance is calculated from the known concentration and measured amount (usually volume) of reagent solution added to the sample solution until essentially all the sought-for substance has been consumed in a definite reaction with this reagent. The process is known as a titration. The reagent is called the titrant and its solution is known as the titrant solution or standard solution. The concentration of this solution is either calculated from the known amount of titrant dissolved and diluted to an exact volume or established by a determination known as a standardization (see Section 10.3). In the latter case the solution may be called a standardized solution where this distinction is of significance. The species (or solution) titrated is sometimes termed the titrand. However, usage of this term is not encouraged because the close similarity between titran*t* and titran*d* readily leads to confusion.

The following requirements must be fulfilled in order that a reaction can serve as the basis of a titrimetric determination: (1) The reaction must proceed rapidly so that the titration can be performed within a reasonable period of time; (2) the reaction must be unambiguous, or at least sufficiently "well-behaved," so that a definite equivalence of the reactants exists; and (3) the reaction should go essentially to completion.

There are titrimetric procedures where the reaction is not fully complete, where side reactions occur, or where other factors adversely affect the stoichiometry. In such cases an empirical factor is employed in the calculation of the result. However, the preferred situation is that the computation be based on the strictly stoichoimetric relation between the sought-for substance and the titrant; this requires that the relevant balanced reaction equation be known. The reaction on which the titration is based is the titration reaction and may be expressed by a chemical equation which is termed the titration equation.

The point in the process of a titration at which the titrated species and the titrant are present in exactly equivalent amounts is called the equivalence point. Unless an empirical factor is employed in the computation (see above) this point is unambigously defined by the relations implicit in the titration equation. However, it should be appreciated that some substances can be titrated with the same titrant according to different titration equations, and then two or more equivalence points are defined.

It is necessary to establish the point at which sought-for substance and titrant are present in equivalent amounts. This process is called indication. Ideally the indication should locate the equivalence point. In practice, however, this is difficult to accomplish and the point actually established will be near but not coincident with the equivalence point. This experimental point is termed the end point of the titration. Indication can be achieved visually or by instrumental methods. In some cases the end point may be calculated from data taken during the titration or be established from a graphical presentation of such data.

Some titrations proceed so that visual indication is afforded by the titration system itself; this is termed self-indication. If a system is not self-indicating, an indicator may be added to the titration medium. Visual indicators, here under consideration, show a change in color related to the change of some property of the titration system.

The titration curve is a curve obtained as a plot on a rectilinear coordinate system. On the abscissa commonly milliliters of titrant is plotted. On the ordinate is plotted the numerical value of some parameter that is related to the concentration of one or more species participating in the titration reaction. If the value of this parameter varies during the titration over many orders of magnitude, it is often plotted as a logarithmic function. The titration curve may be calculated from the stoichiometry of the titration equation and is then called the theoretical titration curve. Alternatively it may be obtained from plotting actual experimental data and may then be called the experimental titration curve, whenever such a distinction is warranted.

Classification of Titrations 10.2

Titrimetric methods may be classified from different points of view. They may be grouped into four broad categories according to the nature of the titration reaction: acid–base titrations, precipitation titrations, compleximetric titrations, and redox titrations. They may also be classified according to the titrant used, for example, acidimetry, alkalimetry, permanganatometry, argentimetry, and iodimetry. Another approach is to arrange titrations according to the method of end-point detection; thus visual titrations, electrometric titrations, photometric titrations, etc., may be differentiated. An additional criterion is the size of the sample taken, that is, macro, semimicro, micro, or submicro titrations. All these classification schemes have merit, and the one selected will depend on the purpose to be served. In the present book the considerations are arranged according to the nature of the titration reaction involved, with visual titrations being dealt with before those involving instrumental methods of end-point detection.

One important description of titrations is according to their mode of performance. When the standard solution is run directly into the sample solution until the end point is established, the process is called a direct titration. Sometimes a known amount of titrant solution in excess of that required to react completely with the sought-for substance is added to the sample solution and the excess is titrated with a second titrant, called the back-titrant. The process is called a back-titration.

Standardization and Primary Standards 10.3

It is not always possible to prepare a standard solution in such a way that its concentration is known to the necessary degree of reliability (usually four significant figures) from the amount of titrant dissolved and diluted to definite volume. This occurs, for example, where the starting material is of unknown purity, contains variable amounts of water, or presents difficulties in its handling (e.g., it absorbs moisture or carbon dioxide from the air). In such a case a solution of approximately the desired concentration is prepared and the exact concentration is then established by a determination known as a standardization. The reference substance used in the standardization is often known as the primary standard. Such a substance must fulfill certain requirements: (1) It must be of known purity, preferably 100% or very close to it; (2) its reaction with the relevant component of the solution to be standardized must be unequivocal and stoichiometric so that the reaction can be used as the basis of the calculations; (3) it should present no difficulties in handling; and (4) it should have a high equivalent weight, so that the weighing error has a negligible effect. In addition, it is desirable that the substance be readily available.

11 ACID-BASE TITRATION CURVES

A titration curve, as noted in Section 10.1, is obtained by plotting the numerical values of some parameter of the titration system that changes as a function of the amount of titrant added versus the volume of titrant solution. For an acid–base titration in an aqueous medium, the hydrogen ion concentration may be selected as this parameter. However, since this concentration changes over many orders of magnitude, it is expedient to employ a logarithmic function and to plot pH.

Titration curves aid in the establishment of the feasibility of a titration, in the prediction of the sharpness of the end point, and in the selection of a suitable indicator. These curves are often derived from calculations based on the equilibria involved. Such calculations and the evaluation of the curves obtained are the subject of this chapter.

Titration of a Strong Acid with a Strong Base 11.1

The calculations necessary to obtain the data for the construction of a titration curve of a strong acid with a strong base, both being monofunctional, are straightforward because the acid, the base, and the salt formed are all completely dissociated. The mode of calculation will best be explained in terms of examples.

Assume that a volume of exactly 100 ml of a 0.100 F solution of a monoprotic strong acid is titrated with a 0.100 F solution of a strong monoequivalent base (for example, hydrochloric acid is titrated with sodium hydroxide).

The pH at the start of the titration is calculated as given in Section 8.4:

$$[H^+] = \text{formality of the acid} = 0.100\ M$$

and $$pH = 1.00$$

On addition of 5.00 ml of the base, for example, the calculation of the pH takes the following form. The amount of acid initially present is 100 × 0.100 = 10.0 millimoles.† The amount of base added is 5.00 × 0.100 = 0.500 millimole; consequently, the amount of acid remaining unreacted is 10.0 − 0.5 = 9.5 millimoles. The solution volume is now 100 + 5 = 105 ml. Consequently, the pH of the solution is calculated as follows:

$$[H^+] = \frac{9.5}{105} M$$

and $$pH = \log 105 - \log 9.5 = 2.02 - 0.98 = 1.04$$

Analogous calculations are applicable as further volumes of base are added, and the pH values obtained are plotted on the ordinate and the milliliters of base added on the abscissa (see Figure 11-1).

The mode of calculation can readily be put in general terms by use of the following formula, which should be understandable without detailed derivation:

$$\frac{\overbrace{\underbrace{ml_a \times F_a}_{\text{initial millimoles acid}} - \underbrace{ml_b \times F_b}_{\text{millimoles base added}}}^{\text{millimoles acid remaining unreacted}}}{\underbrace{ml_a + ml_b}_{\text{total volume}}} = \text{formality of unreacted acid}$$

$$= \text{molarity of hydrogen ion}$$
$$= [H^+] \qquad (11\text{-}1)$$

Here ml_a and ml_b are the volumes and F_a and F_b the formalities of the acid titrated and the base added, respectively.

Application of the formula may be exemplified by calculating the pH after the addition of 99.99 ml of base.

$$\frac{100.0 \times 0.100 - 99.99 \times 0.100}{100.0 + 99.99}$$

$$= \frac{10.00 - 9.999}{199.99} \cong \frac{0.001}{200} \cong 5 \times 10^{-6} M$$

and $$pH \cong 5.3$$

For the equivalence point, equation (11-1) yields $[H^+] = 0$, which is an impossible value. At this point, the pH value is, of course, 7.00. The unacceptable value of zero results from

†The relation between the amounts of acid and base reacting could be expressed in milliequivalents; however, since both reactants are monofunctional, the number of milliequivalents and millimoles are identical.

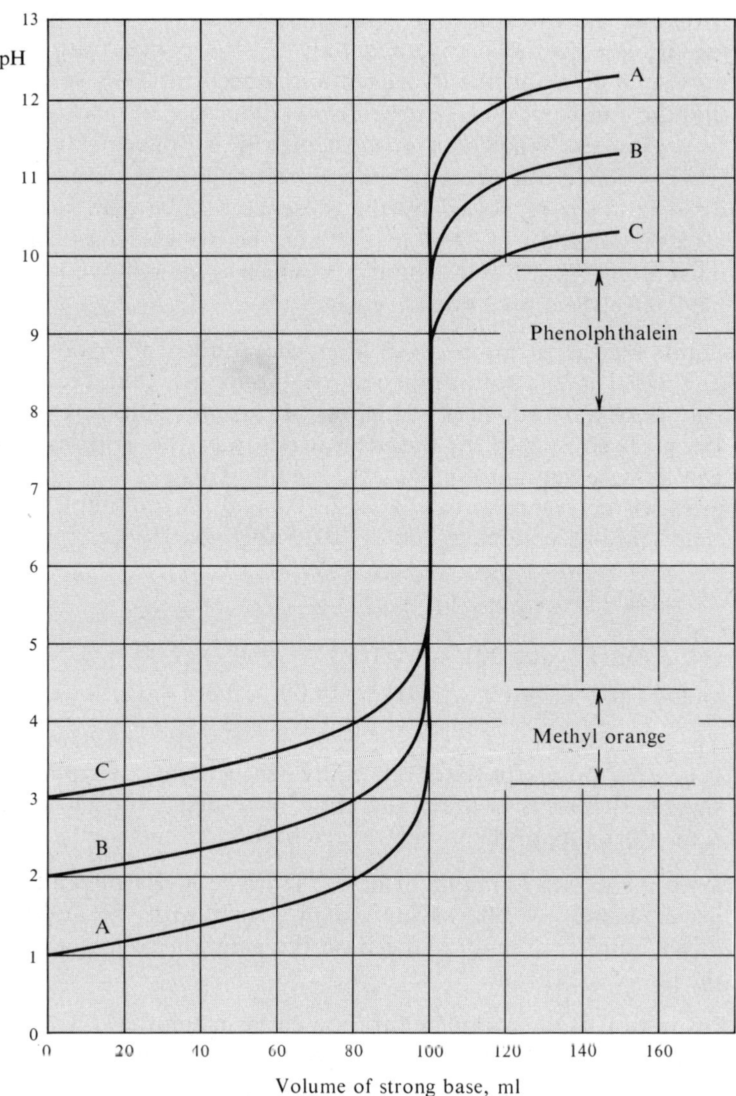

Figure 11-1. Calculated titration curves for the titration of a monoprotic strong acid with a monoequivalent strong base, with the pH transition intervals of methyl orange and phenolphthalein shown. Curve A: 100 ml of 0.100 *F* acid titrated with 0.100 *F* base. Curve B: 100 ml of 0.0100 *F* acid titrated with 0.0100 *F* base. Curve C: 100 ml of 0.00100 *F* acid titrated with 0.00100 *F* base.

the fact that (11-1) is only an approximation. The formula neglects the hydrogen ion resulting from the dissociation of water. However, application of this formula to points removed from the equivalence point was appropriate. At such points the pH is essentially governed only by the concentration of unreacted acid and the hydrogen ion concentration is so large that the contribution from the dissociation of water can safely be neglected. Water can at most provide a concentration of $[H^+] = 10^{-7}$, but actually contributes much less because its dissociation is repressed by the presence of free acid. Even after the addition of 99.99 ml of base, the pH value, as calculated above, is still only about 5.3, which is far removed from pH 7.0 (see Section 8.4).

Points on the titration curve after the equivalence point are calculated in the following manner. Suppose that the total volume of base added is 101.00 ml. From this volume exactly 100 ml is consumed for the neutralization of the acid; exactly 1 ml of base remains as unreacted excess. This amount of base in excess corresponds to $1.00 \times 0.100 = 0.100$ millimole, and is present in a volume of $100 + 101 = 201$ ml. Hence,

$$[OH^-] = \frac{0.100}{201}\ M$$

$$pOH = \log 201 - \log 0.100 = 2.30 + 1.00 = 3.30$$

and $$pH = 14.00 - pOH = 14.00 - 3.30 = 10.70$$

The student should derive a general formula analogous to (11-1) for points beyond the equivalence point. (*Hint:* The signs of the terms in the numerator are reversed and the result is the formality of the unneutralized base.)

With increasing amounts of base, the curve levels off and approaches pH = 13.00 asymptotically, since with the addition of an infinite volume of the base the solution would assume the pH of the titrant.

Inspection of the complete titration curve in Figure 11-1, curve A, reveals that the addition of about 90% of the volume of titrant necessary to reach the equivalence points fails to change the pH by as much as the addition of only a fraction of a drop of base in the vicinity of that point. It is this sharp rise in pH upon addition of a very small amount of titrant that makes indication of the end point possible.

Curves B and C given in Figure 11-1 apply to the analogous titration of 0.0100 F and 0.00100 F strong acid with a strong base of corresponding formality. It can be seen that the shape of the curve remains unchanged, but the "break," that is, the nearly vertical rise in the region around the equivalence point, becomes progressively smaller as the concentrations of the acid and the titrant are decreased.

Titration of a Strong Base with a Strong Acid 11.2

The titration curve of a monoequivalent base with a strong monoprotic acid is the reverse of that given above and is calculated in the same manner but replacing $[H^+]$ by $[OH^-]$. It is good practice for the student to derive the general formula, to perform the calculations, and to plot such a curve. It can be readily appreciated that each point on the titration curve for the base can be obtained by subtracting the corresponding pH values of the titration curve for the acid from 14.00. In other words, the titration curve for the base is the mirror image of the corresponding curve for the acid, and is obtained by reflection of the latter curve along the line passing through pH 7.00 and parallel to the abscissa.

Titration of a Weak Acid with a Strong Base 11.3

The formulas derived in Sections 8.9, 8.10, and 8.11 for the acid–base equilibria involving a weak acid and its conjugate base, present as its salt, may be applied to the calculation of the curve for the titration of a weak monoprotic acid with a strong monoequivalent base.

Assume that a volume of exactly 100 ml of a 0.100 *F* solution of a weak monoprotic acid, having an acidity constant, $K_a = 1.0 \times 10^{-5}$, is titrated with a 0.100 *F* solution of a strong monoequivalent base. (The data closely correspond to the titration of acetic acid with sodium hydroxide.) The pH at the start of the titration is calculated from equation (8-66) as follows:

$$[H^+] \simeq \sqrt{K_a C_a} = \sqrt{1.0 \times 10^{-5} \times 1.00 \times 10^{-1}}$$
$$= \sqrt{1.0 \times 10^{-6}} = 1.0 \times 10^{-3}\ M$$

and $\quad \text{pH} = 3.00$

After the addition of 10.00 ml of the strong base, for example, the calculation of the pH is based on equation (8-74) for the hydrogen ion concentration of a solution containing a weak acid and its conjugate base, present as the alkali metal salt. As was pointed out, in this approximate formula the concentrations of the acid and its salt appear in a fraction, the solution volume cancels, and the calculation can simply be based on the millimoles of acid and salt present. The base amounts to 10.00 × 0.100 = 1.00 millimole.

Initially the acid present amounted to 100 × 0.100 = 10.0 millimoles. The base added amounts to 10.00 × 0.100 = 1.00 millimole and transforms an equivalent amount of acid to the conjugate base, that is, to the salt; consequently, the amount of salt present is 1.00 millimole and that of the acid, 10.0 − 1.0 = 9.0 millimoles. Hence

$$[H^+] = K_a \frac{[\text{acid}]}{[\text{salt}]} = 1.0 \times 10^{-5} \frac{9.0}{1.0} = 9.0 \times 10^{-5}\,M$$

and $\quad pH = 5 - \log 9.0 = 5 - 0.95 = 4.05$

Using this mode of calculation, the pH values corresponding to the addition of 20, 30, . . . ml of strong base may be calculated and plotted (see curve A, Figure 11-2).

At the equivalence point the acid is completely neutralized and the solution is identical to that containing the equivalent amount of conjugate base as the alkali salt in the prevailing volume. The concentration of the base or the salt can be calculated from the original concentration of the acid by recognition of the fact that the volume changed during the titration. The amount of salt is 100 × 0.100 = 10.0 millimoles present in a total of 100 + 100 = 200 ml. Hence C_s at the equivalence point is 10.0/200. The basicity constant of the conjugate base of the acid is $K_b = K_w/K_a$. Application of equation (8-58) and substitution of the numerical values yields

$$[OH^-] = \sqrt{\frac{1.0 \times 10^{-14} \times 10.0}{1.0 \times 10^{-5} \times 200}} = \sqrt{5.0 \times 10^{-11}}$$

and $\quad pH = 14.00 - \frac{1}{2}(11 - \log 5.0) = 8.85$

It should be emphasized that the number of millimoles of acid initially present and the number of millimoles of salt present at the equivalence point are identical. But the concentration of the acid at the beginning of the titration and the concentration of the salt at the equivalence point are different because of the dilution effected during the titration. The novice often fails to appreciate this fact!

Beyond the equivalence point, the strong base is solely responsible for the hydroxide ion concentration and governs the further shape of the titration curve; consequently, this portion of the curve is identical to that obtained in the titration of a strong acid with a strong base involving identical volumes and concentrations. The complete titration curve is shown as curve

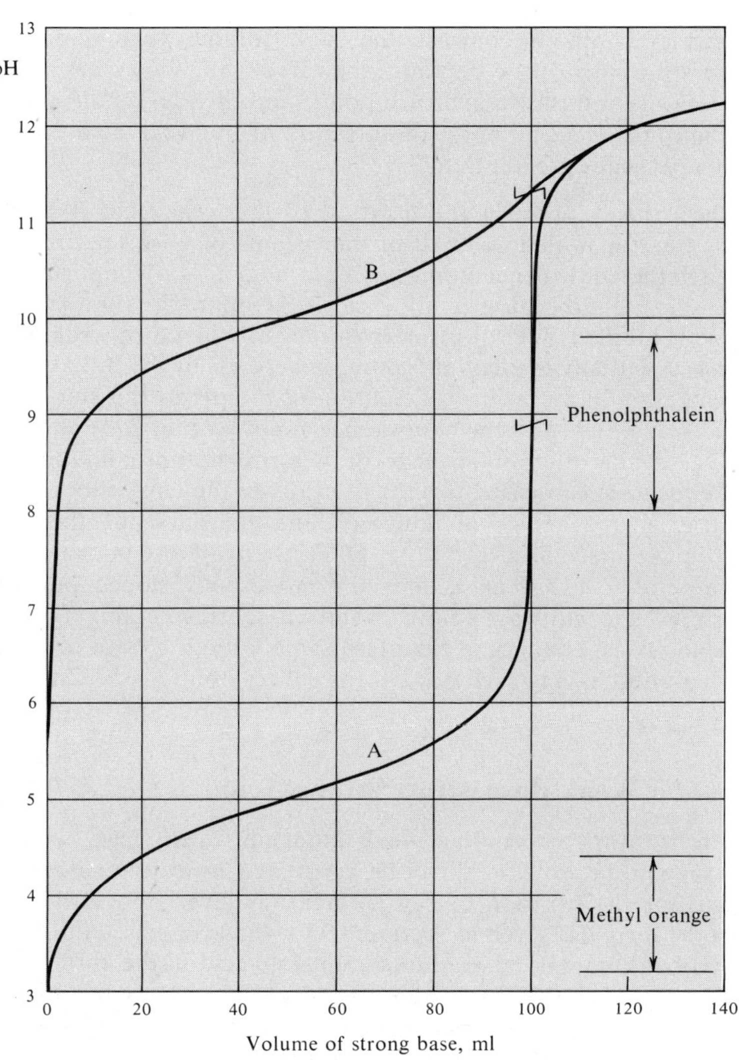

Figure 11-2. Calculated titration curves for the titration of a monoprotic weak acid with a monoequivalent strong base, with the pH transition intervals of methyl orange and phenolphthalein shown. Curve A: 100 ml of 0.100 *F* acid ($K_a = 1.0 \times 10^{-5}$) titrated with 0.100 base. Curve B: 100 ml of 0.100 *F* acid ($K_a = 1.0 \times 10^{-10}$) titrated with 0.100 *F* base.

A in Figure 11-2 and should be compared with the corresponding titration curve for a strong acid (curve A, Figure 11-1). It will be seen that the curve for the latter remains at a low pH for a large part of the titration, then inclines only moderately, exhibits a large break around the equivalence point, and then levels off. In contrast, the curve for the weak acid rises from the start of the titration, then levels off, shows a smaller break around the equivalence point, and then levels off again. The equivalence point for the titration of the weak acid occurs at a pH value greater than 7.

The halfway point in the titration of the weak acid, that is, the point at which one half of the volume of base required to reach the equivalence point has been added, is of importance. At this point (because [acid] = [salt]) the approximation pH $\simeq$ pK_a is fulfilled, and all titration curves for the given weak acid pass essentially through this point, regardless of the initial concentration of the acid. Calculation of the pH at the start, the halfway point, and the equivalence point in the titration of a weak acid is often sufficient to draw a rough approximation of the titration curve and from it to evaluate the feasibility of the titration, the choice of indicator, the sharpness of the end point, etc. (see Section 11.6). Of further interest in the titration curve of a weak acid is the moderately sloped portion around the halfway point. Solutions corresponding to this region resist changes in the pH upon addition of base or acid. A solution having such behavior is called a buffer (see Chapter 13).

11.4 Titration of a Weak Base with a Strong Acid

The titration curve of a weak monoequivalent base with a strong monoprotic acid can be calculated in exact analogy to that for the titration of a weak acid, as given above, by use of the formulas given in Section 8.9 for the acid–base equilibria involving a weak base and its conjugate acid in the form of a salt. The titration curve can also be obtained by a mirror reflection of the curve for the corresponding titration of a weak acid along the line passing through pH 7.00 and parallel to the abscissa provided K_b equals K_a and the concentrations involved are the same.

11.5 Feasibility of an Acid–Base Titration

The nature of a titration curve in the vicinity of the equivalence point is of importance when judging the feasibility of an acid–base titration, and is especially helpful in the selection of an appropriate acid–base indicator. Such indicators are considered in Chapter 12. For the present purpose it need only be stated that an acid–base indicator changes in color within a definite pH interval characteristic of the particular indicator. In Figures 11-1, 11-2, and 11-4, the pH transition intervals of two such indicators, methyl orange and phenolphthalein, are designated.

Ideally the titration should be performed to the particular color exhibited by the indicator at the exact pH of the equivalence point. However, since the human eye is incapable of such subtle color discrimination, the experimental end point usually does not coincide exactly with the equivalence point even with a satisfactory indicator. With either of the two indicators mentioned, the resulting deviation from the correct result is quite small in the titration of a strong acid (Figure 11-1), since the titration curve passes nearly perpendicularly through the transition intervals of both indicators. Consequently, either indicator is applicable in the titration of a 0.1 *F* solution of a strong acid with a 0.1 *F* solution of a strong base, both being monofunctional, even though the end point signaled is displaced from the pH of the equivalence point.

The situation is different in the titration of a weak acid. From curve A in Figure 11-2 it can be appreciated that with methyl orange as the indicator the change in color begins at the start of the titration and is complete when only about one half of the weak acid has been titrated; for this titration methyl orange is not suitable as an indicator. In contrast, the steepest part of the titration curve passes through the transition interval of phenolphthalein, and a satisfactory result is obtained when this indicator is used. The student's attention is drawn to the importance of making a clear distinction between the equivalence point and the experimentally established end point.

Thus a titration curve permits a judgment as to what indicators might be satisfactory and also as to whether a sharp end point can be secured. In this context, sharpness denotes the extent of the color change exhibited upon addition of a certain, small amount of titrant around the end point. Inspection of the titration curve, in connection with the transition interval and limiting colors of an indicator, also allows an estimate of what color shade should best be taken as the end point.

Titration of Very Weak Acids and Bases 11.6

The mode of calculation outlined above can also be applied to approximate the curve for the titration of an extremely weak acid or base, that is, where K_a or K_b is smaller than about 10^{-9}. Inspection of such a curve reveals that a titration is not feasible, as can be appreciated from an example.

Assume that a volume of exactly 100 ml of a 0.100 F solution of a weak monoprotic acid having an acidity constant, K_a, of 1.0×10^{-10} is titrated with a 0.100 F solution of a strong monofunctional base.

The halfway point of the titration corresponds to pH $\simeq$ pK_a $\simeq$ 10. The pH at the equivalence point is found from equation (8-58) with C_s being 10.0/200 and the basicity constant of the conjugate base given by $K_b = K_w/K_a = 1.0 \times 10^{-14}/1.0 \times 10^{-10} = 1.0 \times 10^{-4}$:

$$[OH^-] \simeq \sqrt{1.0 \times 10^{-4} \times \frac{10.0}{200}} = \sqrt{5.0 \times 10^{-6}}$$

and $$pH = 14.00 - \tfrac{1}{2}(6 - \log 5.0) = 11.35$$

The titration curve must approach pH 13.00 asymoptotically, since this is the pH of the 0.100 F titrant. Hence, between the halfway point of the titration and the ultimate pH the interval is only $13 \times 10 = 3$ pH units. Most two-color acid–base indicators have an actual transition interval of about 1.8 pH units (Section 12.1). Even without the calculation of the pH of the equivalence point, it can be appreciated that the titration is not feasible. Assume that the midpoint of the transition interval of the indicator approximates the equivalence point; then the color change is most gradual. The basic limit of the color transition is only reached after the titrant is added in an excess of about 50%. Hence, the color change extends over the addition of about 70 ml of titrant! The calculated titration curve, given as curve B in Figure 11-2, bears out these conclusions.

It should be pointed out that the conjugate base of a very weak acid is a base of considerable basicity.

In the above example the basicity constant of the conjugate base is 10^{-4}, which is stronger than ammonia. Consequently, if the conjugate base is present as a solution, for example, of the alkali metal salt of the acid, such a solution can be titrated with a strong acid. The traditional term for such a titration is a "replacement titration," because the anion of the very weak acid upon addition of the strong acid titrant reacts with hydrogen ion to form the undissociated acid and is "replaced" by the anion of the strong acid.

Analogously the solution of a salt of a very weak base contains its corresponding conjugate acid, which is fairly strong and can be titrated readily with a strong base.

Titration of Polyfunctional Acids and Bases 11.7

Polyprotic acids furnish more than one hydrogen ion per molecule. It is of interest to investigate the curve for their titration with a monofunctional strong base, such as sodium hydroxide. The treatment will be limited here to a diprotic acid, H_2A; however, extension to acids furnishing more than two hydrogen ions per molecule is readily possible. Titrations involving triprotic phosphoric acid are considered in Section 14.5.

The dissociation of a diprotic acid, H_2A, was considered in Section 8.12. The relevant expressions (8-78) and (8-79) are repeated here:

$$H_2A \rightleftharpoons H^+ + HA^-: \quad K_{a,1} = \frac{[H^+][HA^-]}{[H_2A]} \qquad (11\text{-}2)$$

$$HA^- \rightleftharpoons H^+ + A^{2-}: \quad K_{a,2} = \frac{[H^+][A^{2-}]}{[HA^-]} \qquad (11\text{-}3)$$

In principle, each of the dissociable hydrogens should be capable of being titrated with a strong base:

$$H_2A + OH^- \rightleftharpoons HA^- + H_2O \qquad (11\text{-}4)$$

$$H_2A + 2OH^- \rightleftharpoons A^{2} + 2H_2O \qquad (11\text{-}5)$$

To obtain two usable equivalence points, that is, two well-defined breaks in the titration curve, the values of $K_{a,1}$ and $K_{a,2}$ must differ sufficiently. If the two constants are too close, the first break is obliterated or poorly defined. In addition, since an acid–base indicator has a color transition interval of about 1.8 pH units (Section 12.1), if the two acidity constants are too close in value, the two equivalence points may fall within a single transition interval. Mathematical treatment of the situation reveals that, to secure two adequately defined breaks in the titration of 0.1 to 0.001 F solutions, the necessary condition is that $K_{a,1}/K_{a,2} > 10^3$ to 10^4. In addition, the value of the second acidity constant must be sufficiently large to make the titration of the second hydrogen feasible; this limitation is identical to that already considered for a weak monoprotic acid.

Assume that a volume of exactly 100 ml of a 0.100 F solution of a diprotic acid H_2A, having acidity constants $K_{a,1}$ =

1.0×10^{-4} and $K_{a,2} = 1.0 \times 10^{-8}$, is titrated with a 0.100 F solution of sodium hydroxide.

At the start, it is "safe" to assume that the second dissociation step will be completely repressed by the high acidity resulting from the first dissociation step. Hence, the expression for a monoprotic acid, (8-51), may be applied:

$$[H^+] \simeq \sqrt{K_{a,1}C_a} = \sqrt{1.0 \times 10^{-4} \times 0.100}$$
$$= \sqrt{1.0 \times 10^{-5}}$$

and $$pH = 2.50$$

Even in the further course of the titration of the first hydrogen, the effect of the second dissociation can be neglected up to the vicinity of the first equivalence point. The curve is therefore calculated in the manner described for a monoprotic weak acid. At the (first) halfway point, pH $= pK_{a,1} = 4.00$.

The situation at the first equivalence point corresponds to that of a solution of the acid salt, NaHA. Since this salt is a strong electrolyte, it will dissociate into Na^+ and HA^-. The sodium ion is of no further interest. The anion HA^- can undergo dissociation according to

$$HA^- \rightleftharpoons H^+ + A^{2-} \qquad (11\text{-}6)$$

If no further equilibrium were involved, as in the case of the salt of a monoprotic weak acid, the concentration of the anion A^{2-} would equal the hydrogen ion concentration. But the anion HA^- can also combine with hydrogen ion and form the undissociated acid: $HA^- + H^+ \rightarrow H_2A$. Therefore the concentration of the anion A^{2-} will equal the sum of the concentration of the "free" hydrogen ion and that of the hydrogen ion "hidden" in H_2A. Hence,

$$[A^{2-}] = [H^+] + [H_2A] \qquad (11\text{-}7)$$

Replacement of terms in this equality by the expressions obtained from equations (11-3) and (11-2) by rearrangement yields

$$\frac{K_{a,2}[HA^-]}{[H^+]} = [H^+] + \frac{[HA^-][H^+]}{K_{a,1}}$$

The result on rearrangement is

$$[H^+]^2\left(1 + \frac{[HA^-]}{K_{a,1}}\right) = K_{a,2}[HA^-]$$

Solution for $[H^+]$ yields

$$[H^+] = \sqrt{\frac{K_{a,1}K_{a,2}[HA^-]}{K_{a,1} + [HA^-]}} \tag{11-8}$$

If the amount of H_2A formed is small, and if $K_{a,1}$ is not too great, it is possible as an approximation to take $[HA^-]$ as equal to $C_{s,1}$, the formal concentration of the salt NaHA. It then follows that

$$[H^+] \simeq \sqrt{\frac{K_{a,1}K_{a,2}C_{s,1}}{K_{a,1} + C_{s,1}}} \tag{11-9}$$

If the concentration of the acid being titrated is not extremely low (and consequently that $C_{s,1}$ is not too small) and the acid is a relatively weak one, then $K_{a,1} \ll C_{s,1}$; with this condition $K_{a,1}$ can be neglected in the denominator, and the factor $C_{s,1}$ cancels to yield the approximate formula

$$[H^+] \simeq \sqrt{K_{a,1}K_{a,2}} \tag{11-10}$$

or, in logarithmic form,

$$pH_1 \simeq \tfrac{1}{2}(pK_{a,1} + pK_{a,2}) \tag{11-11}$$

As a first approximation, therefore, the hydrogen ion concentration is independent of the concentration of the salt. It is necessary to check whether the approximations in the above derivation are allowed when calculating the exact pH of a solution of an acid salt. In any event, the expression permits a rough evaluation of the situation and, hence, a judgment as to whether the titration is feasible. In the example, the pH at the first equivalence point is given by

$$pH_1 \simeq \tfrac{1}{2}(4.00 + 8.00) = 6.00$$

The calculations for points beyond the first equivalence point are based solely on the second dissociation equilibrium; that is, one proceeds as if a monoprotic acid with $K_{a,2}$ were titrated. Thus at the second halfway point, $pH \simeq pK_{a,2} = 8.00$.

The pH of the second equivalence point is obtained as follows. At this point the acid is completely neutralized and present as its conjugate A^{2-} in the form of the sodium salt. Consequently, equation (8-58) may be applied. The concentration of the salt, $C_{s,2}$, required for the calculation is readily obtained. The initial amount of acid was $100 \times 0.100 = 10.0$ millimoles. Since all the acid is transferred to the salt, Na_2A, this is also the amount of salt. This amount is present in a volume of 300 ml (100 ml of original acid, 100 ml of base to reach the first equivalence point, and an additional 100 ml to reach the second equivalence point.) Consequently, the concentration of the salt is 10.0/300 F. The basicity constant of base A^{2-} is $K_b = K_w/K_{a,2}$. Substitution of the numerical values into equation (8-58) yields

$$[OH^-] = \sqrt{\frac{1.0 \times 10^{-14}}{1.0 \times 10^{-8}} \times \frac{10.0}{300}} = \sqrt{\frac{1.0 \times 10^{-7}}{3.0}}$$

and $$pH = 14.00 - \tfrac{1}{2}(7 + \log 3.0) = 10.26$$

Beyond the second equivalence point, the pH is governed solely by the excess of sodium hydroxide. This portion of the curve takes the same shape as in the titration of a strong acid with a strong base. The entire titration curve for the diprotic acid is given in Figure 11-3.

The titration of a dibasic acid can be considered as a special case of the titration of the mixture of two weak acids of dif-

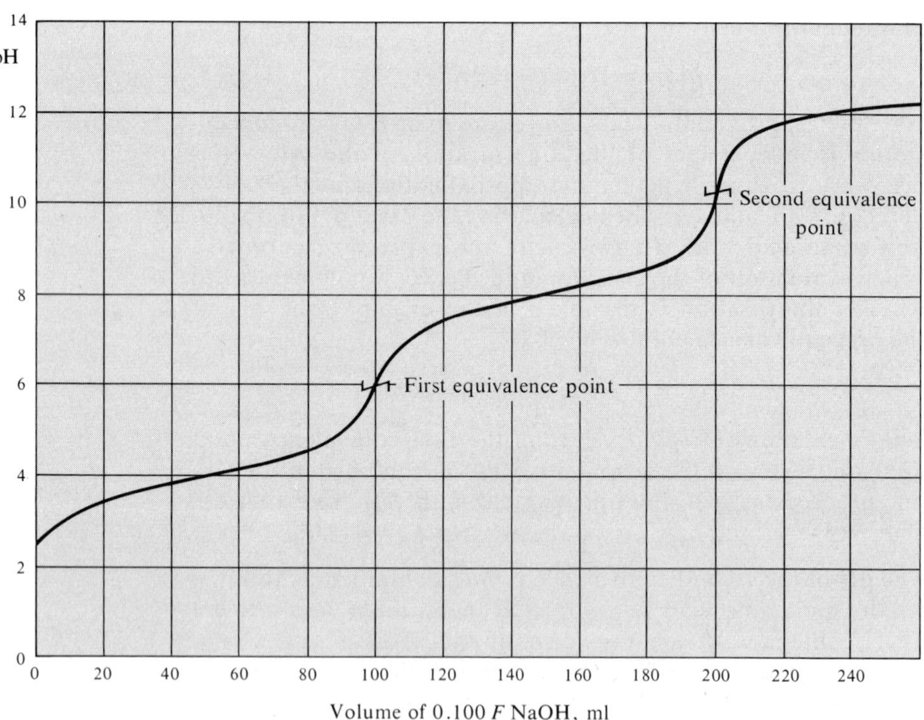

Figure 11-3. Calculated titration curve for the titration of 100 ml of a 0.100 F solution of a diprotic acid ($K_{a,1} = 1.0 \times 10^{-4}$; $K_{a,2} = 1.0 \times 10^{-8}$) with a 0.100 F solution of a monoequivalent strong base.

ferent acidity constants and present at identical initial concentrations. The calculations in the above example reflect this viewpoint.

The titration of a polyfunctional base can be treated in an analogous manner by replacing OH^- for H^+ and K_b for K_a.

Of special practical interest is the titration of the difunctional base carbonate ion, CO_3^{2-}, present in the solution as an alkali metal or alkaline earth salt. The titration of carbonate with strong acid proceeds in two steps, first to hydrogen carbonate ion, HCO_3^-, and then to carbonic acid:

$$CO_3^{2-} + H^+ \rightarrow HCO_3^-$$
$$HCO_3^- + H^+ \rightarrow H_2CO_3$$

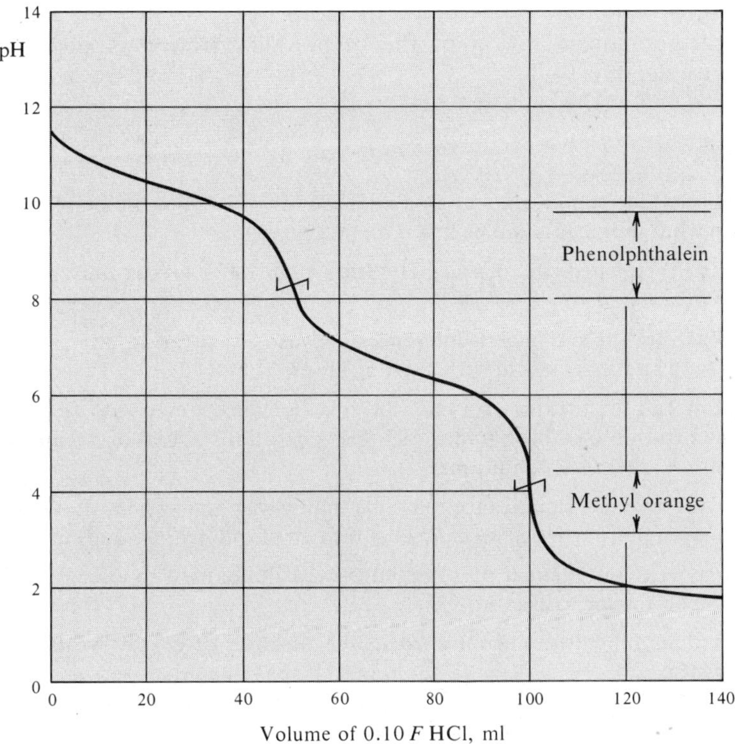

Figure 11-4. Calculated titration curve for the titration of 100 ml of a 0.050 *F* solution of sodium carbonate with 0.10 *F* hydrochloric acid, with the pH transition intervals of methyl orange and phenolphthalein shown.

The latter decomposes according to the equilibrium

$$H_2CO_3 \rightleftharpoons H_2O + CO_2$$

From the acidity constants of carbonic acid, $K_{a,1} = 4.5 \times 10^{-7}$ and $K_{a,2} = 5.6 \times 10^{-11}$, the basicity constants of carbonate ion are readily obtained as $K_{b,1} = 1.8 \times 10^{-4}$ and $K_{b,2} = 2.2 \times 10^{-8}$. Note the reverse order of numbering: $K_{a,1}$ and $K_{a,2}$ correspond to $K_{b,2}$ and $K_{b,1}$, respectively.†

The calculated titration curve is shown in Figure 11-4. The two breaks are clearly defined since $K_{b,1}/K_{b,2} \simeq 10^4$. The first equivalence point occurs at pH about 8.3 (the calculation is left to the student) and phenolphthalein may be used as the indicator. The second break at pH about 4.0 is not extremely sharp because the second basicity constant is that of a fairly weak base. However, the situation can be greatly improved by heating the solution to volatilize carbon dioxide and thereby shift the position of the second titration reaction to the right. Further consideration of the carbonate titration is deferred until Section 14.7.

11.8 Questions

11-1. Discuss the significance of an acid–base titration curve and elaborate on what judgments can be based on this curve.

11-2. Why is it expedient to plot pH rather than $[H^+]$ versus the volume of titrant added?

11-3. What are the significant differences between the titration curves of a strong and a weak acid with a strong base?

11-4. How can the titration curve of an acid be transformed into that for the titration of a base under identical conditions? Elaborate on the phrase "identical conditions."

11-5. What special significance has the halfway point when the same amount of a particular weak acid is titrated at different concentrations?

11-6. Why is it impractical or sometimes even impossible to use a weak acid or a weak base as a titrant?

11-7. Explain difficulties encountered in the titration of a very weak acid or base.

11-8. What is a replacement titration?

---

†The first acidity constant for carbonic acid and consequently the second basicity constant for carbonate ion are "apparent" constants. It is assumed that all the carbonic acid is present as such, that is, as undissociated acid. Actually, however, it holds that $C_{H_2CO_3} = [H_2CO_3] + [CO_2]$ according to the above equilibrium.

-9. Elaborate on the conditions that must be fulfilled to permit the visual titration of a diprotic acid to two equivalence points.

10. Elaborate on the possibility of titrating each of two acids in their mixture.

Problems 11.9

-1. A solution in a volume of 250 ml contains 5.0 millimoles of HCl and 2.0 millimoles of H_2SO_4. What is the pH of the solution? Assume that both hydrogen ions of sulfuric acid are completely dissociated.

Answer: pH 1.44

-2. Calculate the pH of a solution obtained by mixing 100 ml of 0.10 *F* HCl and 150 ml of 0.080 *F* NaOH.

Answer: pH 11.90

-3. Calculate the pH of a solution obtained by mixing 50 ml of 0.00010 *F* HCl and 50 ml of 0.10 *F* HA ($K_a = 1.0 \times 10^{-4}$).

Answer: pH 2.66

-4. A 25.0-ml portion of a 0.0800 *F* solution of a strong monoprotic acid is diluted with water to exactly 100 ml and titrated with 0.100 *F* NaOH. Calculate the pH at the starting point and after the addition of 2.0, 5.0, 10.0, 20.0, and 22.0 ml of titrant.

Answers: pH 1.70, 1.75, 1.84, 2.04, 7.00, 11.21

-5. A 25.0-ml portion of 0.0800 *F* NaOH is diluted to exactly 100 ml with water and titrated with a 0.100 *F* solution of a monoprotic strong acid. Calculate the pH at the same points as in Problem 11-4.

Answers: pH 12.30, 12.25, 12.16, 11.96, 7.00, 2.79

-6. A 25-ml portion of a 0.0800 *F* solution of a weak monoprotic acid ($K_a = 5.0 \times 10^{-5}$) is diluted to exactly 100 ml with water and titrated with 0.100 *F* NaOH. Calculate the pH at the same points as in Problem 11-4.

Answers: pH 3.00, 3.35, 3.82, 4.30, 8.26, 11.21

-7. A 25-ml portion of a 0.0800 *F* solution of a weak monofunctional base ($K_b = 5.0 \times 10^{-5}$) is diluted to exactly 100 ml with water and titrated with 0.100 *F* HCl. Calculate the pH at the same points as in Problem 11-4.

Answers: pH 11.00, 10.65, 10.18, 9.70, 5.74, 2.79

-8. A 50.0-ml portion of a 0.1020 *F* solution of a diprotic acid ($K_{a,1} = 4.0 \times 10^{-4}$; $K_{a,2} = 6.0 \times 10^{-8}$) is diluted to exactly 150 ml with water and titrated with 0.1200 *F* NaOH. Calculate the pH at the starting point, first halfway point, first equivalence point, second halfway point, and the second equivalence point.

Answers: pH 2.43, 3.40, 5.31, 7.22, 9.82

-9. Assume the same numerical data as in Problem 11-8, but a diprotic base is titrated with a strong acid.

Answers: pH 11.57, 10.60, 8.69, 6.78, 4.18

11-10. How many milliliters of 0.0500 F H_2SO_4 must be diluted to exactly 1000 ml with water so that the pH of the resulting solution will equal that of a 0.080 F solution of acetic acid ($K_a = 1.8 \times 10^{-5}$)?

Answer: 12 ml

11-11. Calculate the pH at the start, halfway to the equivalence point, and at the equivalence point when the solutions of weak acids specified below are titrated with a strong base.

(a) 25.0 ml of 0.100 F acid ($K_a = 8.0 \times 10^{-6}$) with 0.050 F NaOH
(b) 50.0 ml of 0.050 F acid ($K_a = 5.0 \times 10^{-7}$) with 0.100 F NaOH
(c) 100.0 ml of 0.010 F acid ($K_a = 5.7 \times 10^{-9}$) with 0.050 F NaOH
(d) 20.0 ml of 0.030 F acid ($K_a = 3.6 \times 10^{-7}$) with 0.040 F NaOH

11-12. A weak acid, HA, has a dissociation constant of 1.0×10^{-6}. A volume of 30.0 ml of 0.100 F HA is titrated with 0.050 F NaOH. Calculate the pH after 0.0, 30.0, 60.0, and 90.0 ml of the NaOH solution have been added.

11-13. Calculate the pH at the start, halfway to the equivalence point, and at the equivalence point when solutions of the weak bases as specified below are titrated with a strong acid.

(a) 25.0 ml of 0.150 F base ($K_b = 4.0 \times 10^{-6}$) with 0.100 F HCl
(b) 30.0 ml of 0.040 F base ($K_b = 1.0 \times 10^{-7}$) with 0.040 F HCl
(c) 100.0 ml of 0.010 F base ($K_b = 5.0 \times 10^{-8}$) with 0.050 F HCl
(d) 50.0 ml of 0.050 F base ($K_b = 6.5 \times 10^{-6}$) with 0.070 F HCl

11-14. Calculate the approximate pH values at the start, halfway to the first equivalence point, at the first equivalence point, midway between the two equivalence points, and at the second equivalence point when the following solutions of difunctional weak acids (or bases) are titrated with strong base (or acid).

(a) 10.0 ml of 0.010 F acid ($K_{a,1} = 1.0 \times 10^{-6}$, $K_{a,2} = 1.0 \times 10^{-8}$) with 0.020 F NaOH
(b) 25.0 ml of 0.400 F acid ($K_{a,1} = 1.0 \times 10^{-4}$, $K_{a,2} = 1.0 \times 10^{-9}$) with 0.200 F NaOH
(c) 25.0 ml of 0.100 F base ($K_{b,1} = 1.8 \times 10^{-4}$, $K_{b,2} = 2.2 \times 10^{-8}$) with 0.100 F HCl
(d) 30.0 ml of 0.050 F base ($K_{b,1} = 2.0 \times 10^{-10}$, $K_{b,2} = 1.9 \times 10^{-13}$) with 0.100 F HCl

12 ACID–BASE INDICATORS

For a visual indication of the end point in a neutralization titration a so-called acid–base indicator is employed. Such an indicator is a weak organic acid (or base) having the special feature that its color differs from that of the corresponding conjugate base (or acid). At least one form of the indicator must be colored. Where one form is colorless the substance is described as a one-color indicator in contrast to a two-color indicator. Each indicator changes in color within a definite pH range, called its visual transition interval or its transition range. Some indicators possess more than one such range (see Table F in the Appendix). The location of this range on the pH scale depends on the acidity (or basicity) constant of the indicator. At pH values below and above this range the indicator is present predominantly in its so-called acid form and base form, respectively. Within the transition range both forms are present in substantial proportions.

It is possible to relate the color phenomena to changes in the structure of the indicator, but the discussion of such changes requires more knowledge of organic chemistry than the student possesses at this point of his study. However, such knowledge is unnecessary for an understanding of the functioning of an indicator and of the requirements for its practical application. The more important requirements may be summarized as follows.

At least one form of an indicator must be intensely colored so that a clearly visible coloration is imparted to the solution to be titrated even at the extremely low indicator concentrations usually employed (commonly 10^{-4} to 10^{-5} F, or lower). The transition range should be small, so that an abrupt color change results from the addition of a very small amount of titrant. In the case of a two-color indicator the two limiting colors (that is, the colors of the acid and base forms) should differ markedly in hue so that good color contrast is realized; ideally the colors should be complementary.

An indicator should be selected so that the pH of the end point signaled is close, ideally identical, to that of the equivalence point. The transition range with reference to the titration curve should lie on the steepest portion. Since an indicator is an acid or base, it may require titrant or may neutralize a portion of the substance to be titrated. Consequently, the smallest amount of indicator necessary should be used and approximately the same indicator concentration should be employed in the titration of standards and unknowns. Adequate reproduction of the indicator concentration is assured by addition of a definite number of drops of a dilute indicator solution to a stated volume of solution to be titrated.

These and several other points will be discussed in detail and become more clear in the following development of the topic.

12.1 Indicator Constant

The ionization reaction of an indicator may be presented as follows:

$$\text{acid form} \rightleftharpoons \text{H}^+ + \text{base form} \tag{12-1}$$

This generalized form has the advantage that it applies to indicator acids as well as to indicator bases and regardless of whether they are uncharged, cationic, or anionic. The equilibrium constant, here called the indicator constant,† is then given as

$$K_{in} = \frac{[\text{base form}][\text{H}^+]}{[\text{acid form}]} \tag{12-2}$$

This equation may be rearranged to yield

$$[\text{H}^+] = K_{In} \frac{[\text{acid form}]}{[\text{base form}]} \tag{12-3}$$

or, written in logarithmic form,

$$\text{pH} = \text{p}K_{In} + \log \frac{[\text{base form}]}{[\text{acid form}]} \tag{12-4}$$

This formula in connection with an assumption as to the intensities of indicator colors and to the color discrimination of the observer's eye permits a theoretical treatment that should

†It may be pointed out that for an indicator acid (e.g., $\text{H}In \rightleftharpoons \text{H}^+ + In^-$) the indicator constant equals the acidity constant of the acid, K_a. For an indicator base (e.g., $In + \text{H}_2\text{O} \rightleftharpoons In\text{H}^+ + \text{OH}^-$) the relation between the indicator constant and the basicity constant is $K_{In} = K_w/K_b$.

provide an insight into the functioning of indicators. Since there are certain distinctive differences in the theory of two-color and one-color indicators, they will be considered separately.

Two-Color Indicators 12.2

An inspection of equation (12-4) reveals that with changes in pH the ratio [base form]/[acid form] changes. In other words, since both forms are colored, a certain hue, due to the mixture of the two colorants, is related to a certain pH value. For a closer assessment of the situation the following assumptions, admittedly approximations, are made:

1. To the eye the color intensity is proportional to the concentration of the colored species, and both forms of the indicator have about the same color intensity.
2. If two colors are mixed, the eye is no longer able to discern a change in hue (that is, color shade) if the concentration ratio of the two colorants exceeds a value of 8:1 to 9:1.

By combination of statement 2 and equation (12-4) it follows that the eye is able to observe color changes, due to change in the ratio [base form]/[acid form], within a pH range of approximately 1.8 pH units (see Example 12-1). Beyond this visual transition range no changes in hue can be differentiated, and the eye sees the limiting colors of the indicator. The span of 1.8 pH units can also be obtained from the data in Table F in the Appendix, by averaging the transition intervals of the indicators listed. If [base form] = [acid form] the logarithmic term in equation (12-4) vanishes and pH = pK_{In}. This pH value approximates the midpoint of the transition interval that stretches about 0.9 pH unit above and below the midpoint. If the contrast of the two limiting colors of two two-color indicators is of the same quality, the indicator with the smaller transition range is the better one, because with a smaller transition range a larger change in hue is obtained for a certain change in pH.

Some examples will advance the student's appreciation of the topic.

Example 12-1. A two-color indicator has pK_{In} = 5.3. (a) Within what approximate pH range will it be possible to discern a color change? Will it be possible to use this indicator to indicate a pH of about (b) 4.5 and (c) 8.5?

(a) As an average the values 8.5:1 and 1:8.5 may be taken for the limiting ratio [base form]/[acid form]. Substituting these values into equation (12-4) yields

$$\text{pH} = 5.3 \pm \log 8.5 = 5.3 \pm 0.93$$

The transition range therefore stretches from pH 4.4 to 6.2.

(b) The required pH of 4.5 is within the transition range and by adjusting the color very closely to that of the pure acid form the pH of 4.5 can be well approximated.

(c) The pH of 8.5 cannot be approximated. At any pH above 6.2 only the limiting color of the base form is viewed. Consequently, once this color is reached no conclusion can be drawn other than that a pH higher than about 6.2 prevails.

Example 12-2. At what pH will a two-color indicator with $K_{In} = 1.0 \times 10^{-10}$ be present 60% in the base form?

Sixty per cent base form corresponds to a base–acid form ratio of 60:40. Substitution into equation (12-4) yields

$$\text{pH} = 10.00 + \log \frac{60}{40} = 10.18$$

It is important to appreciate that for a two-color indicator, although the intensity of the color viewed increases with increasing indicator concentration, it is only a certain hue to which the titration is performed. This hue depends only on the ratio [base form]/[acid form] and is independent of the indicator concentration, at least for indicator concentrations employed in actual titrations. Thus the indicator concentration for a two-color indicator is not overly critical. Of course, if the concentration is too great the solution will be so dark that changes in hue can no longer be precisely observed, and if too low there will be insufficient color for visual differentiation.

12.3 One-Color Indicators

The situation with respect to indicator concentration is quite different with a one-color indicator. The titration with such an indicator is performed to a certain color *intensity* rather than to a particular hue. This intensity occurs at a certain pH for a given indicator concentration. For an indicator with the base form being colored the same color intensity occurs at a lower pH, if the indicator concentration is raised. This is, for example, the case with phenolphthalein, the most important one-color indicator. When this indicator is used, the titration of an acid is performed to a certain intensity of pink. In practice the titration is often stopped at the first appearance of a permanent trace of pink, corresponding to a pH of 8.3 to 8.5. This procedure is permissible in the titration of a strong acid with a strong base because this pH range clearly falls on the steep part of the titration curve. In the titration of a rather weak acid, with an equivalent point, say, at pH 9.0 to 9.5 a deeper pink color should be taken.

For such a one-color indicator, the transition interval is not a region within which one color changes to another, but rather where colorless changes to colored with a progressive increase in intensity until a final color intensity is reached when essentially all the indicator is in the colored form.

An example will illustrate some of the relationships.

Example 12-3. A one-color indicator has $pK_{In} = 8.0$ and a molecular weight of 120. It is known that a coloration is just discernible if the indicator is present in its base form at a concentration of 2×10^{-5} *M*. How many drops of an exactly 0.1% solution of this indicator must be added to the solution to be titrated so that the appearance of the first discernible color indicates a pH of 8.5? Assume the volume at the end point to be 100 ml.

Material balance for the indicator yields

$$C_{In} = [\text{base form}] + [\text{acid form}]$$

Combination of this equation with equation (12-4) gives

$$\text{pH} = pK_{In} + \log \frac{[\text{base form}]}{C_{In} - [\text{base form}]}$$

Substitution of the numerical values yields

$$8.5 = 8.0 + \log \frac{2 \times 10^{-5}}{C_{In} - 2 \times 10^{-5}}$$

Solution for C_{In} gives

$$10^{0.5} = 3.2 = \frac{2 \times 10^{-5}}{C_{In} - 2 \times 10^{-5}}$$

and

$$C_{In} = 2.6 \times 10^{-5}\, F$$

The formality of the 0.1% indicator solution is

$$\frac{0.1}{120} \times \frac{1000}{100} = 8.3 \times 10^{-3}\, F$$

To establish the above-calculated indicator concentration in 100 ml of the solution to be titrated, the number of milliliters of indicator solution required is

$$\frac{100 \times 2.6 \times 10^{-5}}{8.3 \times 10^{-3}} = 0.31 \text{ ml}$$

If one drop is assumed to be 0.05 ml, the number of drops required is

$$\frac{0.31}{0.05} = 6 \text{ drops}$$

12.4 Indicator Blank

Since an acid–base indicator is either a weak acid or a weak base, it consumes either sample or titrant. For titrations involving relatively concentrated sample solutions and titrants, this influence can be neglected because only a minute amount of the indicator is added. Where dilute solutions and small amounts of unknown are involved, an indicator blank should be determined and used to correct the titration result. Assume that an acid is titrated with a base and an acid–base indicator in its acid form is used. The blank in this case is secured by titration of an amount of the indicator in a volume of water, both identical to those employed in the principal titration. The small volume of base consumed is then subtracted from the volume required in the principal titration. If in the same titration an indicator in its base form is used, the blank is secured by a similar titration but using a standard acid solution. The volume of acid consumed is then expressed as the equivalent volume of the base used in the principal titration, and this volume is then added numerically to the volume required in that titration.

The value of an indicator constant may be markedly influenced by the presence of salts. Hence, in the determination of an indicator blank, it is often desirable to employ a salt solution of appropriate concentration rather than water in place of the sample solution. Temperature in some cases also has a pronounced influence on an indicator constant, especially for a two-color indicator.

For a two-color indicator an indicator correction can often be avoided by "pretitrating" the indicator. For this purpose the indicator in the amount to be used in the principal titration is added to a few milliliters of water and by addition of very dilute acid or base is brought to the exact shade of the end-point color. This solution is then added to the solution to be titrated.

Screened and Mixed Indicators 12.5

The purpose of a screened or mixed indicator is to produce a more pronounced color change at the end point. These types of indicators consist of either a mixture of an indicator and an inert dye or a mixture of two indicators. Their functioning will be understood from the following considerations.

Ideally the two limiting colors of any two-color indicator should be fully complementary, so that a gray develops at some pH within the transition range. Since the indicator concentration is small, the gray content will be so feeble that the solution is virtually colorless. In other words, under ideal conditions the end point corresponds to the change from one color to its complementary color, passing through colorless. This condition would be realized, for example, if the limiting colors were a red and its exactly complementary green, because red + green yields black, which on dilution is gray. In practice, however, the limiting colors are usually far from complementary, with red and yellow the most frequent pair (see Table F in the Appendix). If an inert dye, that is, one showing no acid–base properties (at least not in the pH range of interest) is mixed with the indicator, it acts as a screen against which or through which the actual color change is observed. Such a mixed indicator is therefore appropriately termed a screened indicator. If the inert dye is properly selected with respect to its color and mixed in the right proportion with the indicator, the color change then observed is from one color to its complementary color or close to it. For example, addition of an appropriate blue dye will screen an end-point color change from red to yellow in the following manner: Red + blue yields violet and yellow + blue yields green. The end-point color is an intermediate gray, since violet + green yields gray. The detection of the end point is then easier and its location more precise. For some actual examples of screened indicators, see Table F in the Appendix.

Screening can also be established by mixing two indicators that undergo color transition in approximately the same pH range. When the two indicators are appropriately selected with regard to hue, and mixed in correct proportions, the solution becomes gray or colorless at the end point and the color change is sharper than with either indicator alone.

Several indicators selected so that their transition intervals overlap can be mixed, and such a mixture gives continuous color changes as the pH is changed over a wide range. Such mixtures can be used to establish the pH of a solution by comparison of the color developed in the test solution with a color chart. By impregnation of filter paper with such an indicator mixture so-called pH paper is obtained.

12.6 Color-Comparison End Point

Even when the limiting colors of an indicator afford good color contrast, it is difficult to locate the end point if the slope of the titration curve is small within the transition range of the indicator. In such a case the color change is so gradual that the change after each addition of a drop of titrant is not pronounced. The operator is limited in his ability to remember the color existing before the addition of an increment of titrant solution and to distinguish this color from that observed after the addition. The operator can improve his color recall as he gains practice with a given end point but only to a limited degree. However, the human eye is capable of detecting minute differences in hues or shades when they are presented side by side under identical ilumination. This capability is utilized in the performance of a titration to a color-comparison end point. This technique can be explained by an example—the titration of orthophosphoric acid, H_3PO_4, with a standard base to the methyl orange end point. Here it is difficult to locate this point exactly and the remedy proceeds as follows. At the equivalence point of this titration the salt NaH_2PO_4 is present. Therefore, two titration vessels of identical size and shape are selected. In one is placed the sample solution, in the other a solution of sodium dihydrogen phosphate of a concentration and volume close to that which will exist at the end point in the titration of the sample. If necessary, a rough preliminary titration is performed of an aliquot of the sample solution to establish the approximate concentration and volume involved. Then the same number of drops of the indicator solution is added to both vessels and the sample is titrated until the colors of the two solutions appear identical. By use of this technique, the precision in the location of the end point can be increased by a factor of 10.

12.7 Questions

12-1. Define the terms: indicator, acid–base indicator, and screened indicator.

12-2. What substances are used as acid–base indicators?

What is the significant difference observed in the behavior of a one-color and a two-color acid–base indicator as the indicator concentration is changed?

Why can only *weak* acids and bases be used as acid–base indicators?

State two causes for a sluggish color change: one related to the acid–base indicator, the other to the system being titrated.

Why does a titration to a color-comparison end point give better results?

What is the transition range of a two-color indicator?

Formerly, pH values were often established colorimetrically. Can you suggest a procedure? Elaborate.

Elaborate on the possibility of improving color contrast of indicators by the wearing of colored eyeglasses or the use of colored light to illuminate the titration vessel.

Explain the fact that any indicator changing its color in the interval pH 4 to 9 gives acceptable results when hydrochloric acid is titrated with sodium hydroxide unless extremely dilute solutions are involved.

Explain the fact that methyl orange is an unsatisfactory indicator for the titration of acetic acid with sodium hydroxide.

Present in your own words the significant factors involved in the selection of an indicator for a particular acid–base titration.

What significant advantage do you find in the use of the indicator constant rather than the acidity or basicity constant of an acid–base indicator?

One-color indicators can be screened, too, and the screened preparation behaves as a two-color indicator. If phenolphthalein is to be screened, what color would you suggest should be shown by the screening dye? What will be the color change by the appropriately screened indicator?

Some acid–base indicators have two transition ranges (for example, thymol blue; see Table F in the Appendix). Suggest a general ionization scheme and write expressions for the two indicator constants, $K_{In,1}$ and $K_{In,2}$.

Problems 12.8

An indicator has $pK_{In} = 5.3$. In a certain solution the indicator is found to be 80% in its acid form. What is the pH of the solution?

Answer: pH 4.7

An indicator with $pK_{In} = 9.0$ has a color transition from yellow (acid form) to blue (base form). At what pH will the indicator be 75% in the blue form?

Answer: pH 9.5

12.8 PROBLEMS

12-3. A volume of 100 ml of a 0.100 *F* solution of a triprotic acid is titrated with 0.100 *F* NaOH. The acidity constants of this acid have the values 10^{-3}, 10^{-8}, and 10^{-12}. Two-color indicators with the following pK_{In} values are available: *A*, 3.1; *B* 4.9; *C*, 7.8; *D*, 9.9. Which indicator will be satisfactory for the titration of the acid to the (a) first, (b) second, and (c) third equivalence point? State also whether the titration should be performed to the midcolor or to a more alkaline or acid color of the indicator.

Answers: (a) *B*, quite alkaline color; (b) D, close to midcolor; (c) the acidity constant is too small to permit a titration using an acid–base indicator

12-4. A one-color indicator (base form colored) with $pK_{In} = 9.30$ is used in a titration performed to exactly pH 9.00 and to a certain color intensity. The indicator was present at a concentration of 1.0×10^{-5} *F*. What will be the pH of the end point when the same titration is performed to the same color intensity but at an indicator concentration of 1.0×10^{-4} *F*? (*Hint:* Apply the equation derived in Example 12-3 of the text to both titrations; combine the two equations via the term [base form] and solve for $[H^+]$.)

Answer: pH 7.84

12-5. At what pH will a weak-acid type of indicator ($K_{In} = 1.0 \times 10^{-8}$) be 60% dissociated?

Answer: pH = 8.18

12-6. A weak-acid type of indicator was found to be 60% dissociated at pH 9.20. What will be the per cent dissociation at pH 9.00?

12-7. Calculate the indicator constant for each of the following two-color, weak-acid-type acid–base indicators from the per cent dissociation and the solution pH.

(a) 80% at pH 9.00
(b) 55% at pH 7.90
(c) 60% at pH 5.80
(d) 40% at pH 6.20

12-8. The following two-color indicators (with the pK_{In} values given) are available: *A*(3.2), *B*(4.1), *C*(4.8), *D*(5.5), *E*(6.7), *F*(7.6), *G*(8.4), *H*(9.2). Establish which indicator (or indicators) is suitable for the titration of each of the following acids or bases. Assume a 1.0×10^{-2} *F* salt concentration at each end point.

	K_1	K_2	K_3
(a) acid	1.0×10^{-6}	2.0×10^{-8}	—
(b) acid	6.0×10^{-3}	6.0×10^{-8}	4.4×10^{-13}
(c) acid	2.0×10^{-3}	—	—
(d) acid	5.4×10^{-2}	5.1×10^{-5}	—
(e) base	1.0×10^{-4}	—	—
(f) base	1.8×10^{-4}	2.2×10^{-8}	—
(g) base	1.0×10^{-6}	5.0×10^{-8}	—
(h) base	4.0×10^{-4}	4.0×12^{2}	—

12-9. The color of a one-color indicator is just perceptible in a 4.0×10^{-5} *F* solution when the indicator is 25% dissociated. If the pK_{In} value is 8.30, at what pH will the color first be visible?

13 BUFFERS

Buffering Action and Buffer Capacity 13.1

In many chemical operations, it is necessary to establish a certain pH or to maintain the pH at a definite level, or both. A solution that resists changes in pH is called a buffered solution. Buffering action in a solution is achieved by adding a reagent (or a mixture of reagents) called buffer substance(s). The reagent(s) are often added in the form of a solution called the buffer solution or simply the buffer. Two types of buffered solutions may be differentiated with respect to their action. Solutions of the first type resist changes in pH upon changes in concentration, commonly dilution. Solutions of the second type are not only indifferent to concentration changes, but also resist changes in pH upon addition of acid or base. The degree of the latter resistance is the *buffer capacity,* which is commonly expressed as the number of moles of strong base that must be added to or removed (by addition of acid) from 1 liter of solution to change the pH by 1 unit. The buffer capacity depends on the type and concentration of the buffer. As will be shown in the following paragraphs, the discussion of buffering action and buffer capacity can be based on acid–base titration curves.†

Strong Acids or Strong Bases as Buffers 13.2

Inspection of the titration curve of either a strong acid or a strong base (Figure 11-1) shows that its initial portion is nearly horizontal. Consequently, it can be expected that a solution of a strong acid or base would not exhibit a marked change in pH upon addition of a small amount of a base or acid. Since the buffer capacity depends on the concentration, strong acids and bases are effective buffers only at low or high pH values, that is, below about pH 2 and above about pH 12, respectively. However, unless very concentrated buffers are involved, dilu-

†Students familiar with calculus will appreciate the mathematical definition of the buffer capacity, given as $dC_b/d(\text{pH})$, where dC_b is the change in concentration of base. Where the buffer composition is described by a point on a titration curve, the buffer capacity is proportional to the reciprocal of the slope of the curve at that point.

tion affects any strong acid or strong base buffer to the same extent, regardless of the buffer concentration. An example will illustrate the point.

Example 13-1. Compare a buffer consisting of 0.10 *F* HCl with one consisting of 0.0010 *F* HCl. (a) Calculate the pH values of the original solutions, (b) calculate the pH increase due to dilution of 100 ml of buffer with 100 ml of water, and (c) calculate the increase in pH when to 100 ml of each buffer is added a volume of 20 ml of 0.0010 *F* NaOH.

The calculations involved are straightforward and given in consolidated form only.

For the 0.10 *F* solution:

$$\text{(a)} \quad \text{pH} = -\log 1.0 \times 10^{-1} = 1.00$$

$$\text{(b)} \quad [\text{H}^+] = \frac{100 \times 1.0 \times 10^{-1}}{200} = 5.0 \times 10^{-2}\,M$$

$$\text{pH} = 1.30$$

$$\Delta\text{pH} = 1.30 - 1.00 = 0.30 \text{ pH unit}$$

For the 0.0010 *F* solution:

$$\text{(a)} \quad \text{pH} = -\log 1.0 \times 10^{-3} = 3.00$$

$$\text{(b)} \quad [\text{H}^+] = \frac{100 \times 1.0 \times 10^{-3}}{200} = 5.0 \times 10^{-4}\,M$$

$$\text{pH} = 3.30$$

$$\Delta\text{pH} = 3.30 - 3.00 = 0.30 \text{ pH unit}$$

Note that the dilution affects both buffers to the same degree.

For the 0.10 *F* solution:

$$\text{(c)} \quad [\text{H}^+] = \frac{100 \times 1.0 \times 10^{-1} - 20 \times 1.0 \times 10^{-3}}{120}$$

$$= 8.33 \times 10^{-2}\,M$$

$$\text{pH} = 1.08$$

$$\Delta\text{pH} = 1.08 - 1.00 = 0.08 \text{ pH unit}$$

Note that here the increase in pH is essentially due to the dilution.

For the 0.0010 *F* solution:

$$\text{(c)} \quad [\text{H}^+] = \frac{100 \times 1.0 \times 10^{-3} - 20 \times 1.0 \times 10^{-3}}{120}$$

$$= 6.7 \times 10^{-4}\,M$$

$$\text{pH} = 3.17$$

$$\Delta\text{pH} = 3.17 - 3.00 = 0.17 \text{ pH unit}$$

13.3 Mixture of Weak Acid and Conjugate as Buffer

Inspection of the titration curve of a weak monoprotic acid with a strong base (Figure 11-2) reveals a portion of small slope around the halfway point. In this region the addition of small amounts of base or of acid results in only a slight change in pH. The halfway point corresponds to a solution containing equimolar amounts of the weak acid and its conjugate base. The pH of such a solution is obtained by the approximation pH $\simeq$ pK_a. For a given total concentration of the acid and its conjugate base, this composition exhibits the greatest buffer capacity.

The buffer system acetic acid–acetate is frequently encountered in analytical chemistry. The solution is prepared by mixing appropriate amounts of acetic acid and sodium acetate, acetic acid and sodium hydroxide, or sodium acetate and hydrochloric acid. It has a "practical" buffer range of about pH 3.7 to 5.7, which corresponds to approximately p$K_a \pm 1$; beyond this region the buffer capacity decreases rapidly, as can be deduced from the shape of the titration curve (Figure 11-2).

Some calculations involving mixtures of a monoprotic acid and its conjugate have already been illustrated by Examples 8-10 through 8-13. Several additional examples are worthy of study.

Example 13-2. A buffer solution is prepared by mixing 100 ml of 0.100 F HA ($K_a = 1.0 \times 10^{-5}$) and 100 ml of 0.100 F NaA. (a) Calculate the pH of the mixture. (b) Calculate the change in pH when a volume of 20 ml of 0.100 F HCl is added. (c) Calculate the change in pH when the same volume of 0.100 F HCl is added to 200 ml of pure water.

(a) The approximate expression for the pH of a mixture of a weak acid and its conjugate base, added as an alkali metal salt of the acid, can be obtained by converting equation (8-74) to the logarithmic form:

$$\text{pH} = \text{p}K_a + \log \frac{[\text{salt}]}{[\text{acid}]}$$

where [salt] and [acid] are the concentration of the salt (or the conjugate base) and the weak acid, respectively. Since here [salt] = [acid],

$$\text{pH} \simeq \text{p}K_a = 5.00$$

(b) Addition of the strong acid converts an equivalent amount of the conjugate base to the weak acid by the reaction

$$A^- + H^+ \rightarrow HA$$

The amount of strong acid added is $20 \times 0.100 = 2.0$ millimoles. Prior to this addition, the amounts of weak acid and of its conjugate base present each were $100 \times 0.100 = 10.0$ millimoles. Hence, the amount of that base now present is $10.0 - 2.0 = 8.0$ millimoles, and of weak acid, $10.0 + 2.0 = 12.0$ millimoles. As observed in Example

8-11, the value for the ratio [salt]/[acid] may be computed using millimoles rather than concentrations, since the volume cancels:

$$\mathrm{pH} = 5.00 + \log\frac{8.0}{12.0} = 5.00 + \log 8.0 - \log 12.0$$
$$= 5.00 + 0.903 - 1.079 = 4.82$$

Hence the change in pH on addition of the strong acid is 5.00 − 4.82 = 0.18 pH unit.

(c) The addition of the strong acid to 200 ml of pure water would yield a hydrogen ion concentration of

$$[\mathrm{H}^+] = \frac{20 \times 0.100}{200 + 20} = 0.0091\ M$$

Hence,

$$\mathrm{pH} = 2.04$$

and the change in pH is 7.00 − 2.04 = 4.96 pH units.

Comparison of the results in (b) and (c) delineates the difference between the behavior of a well-buffered and an unbuffered system.

Example 13-3. The same volumes as in Example 13-2 are employed, but the solutions of the weak acid and its alkali metal salt are 0.0500 *F* each. What is the change in the pH on the addition of 20 ml of 0.100 *F* HCl?

The initial pH is again 5.00, since [salt] = [acid]. The calculation of the final pH parallels that of Example 13-2 and is shown in consolidated form.

$$\mathrm{pH} = 5.00 + \log\frac{100 \times 0.0500 - 20 \times 0.100}{100 \times 0.0500 + 20 \times 0.100}$$
$$= 5.00 + \log\frac{5.00 - 2.0}{5.00 + 2.0}$$
$$= 5.00 + \log\frac{3.0}{7.0} = 4.63$$

The change in pH is therefore 5.00 − 4.63 = 0.37 unit.

Note that although the buffer of Example 13-3 is one half the concentration of that of Example 13-2, the addition of the same amount of strong acid still leads to only a relatively small change in pH (0.37 versus 0.18 pH unit). Compare the results of Examples 3-2 and 3-3 with those of Example 3-1.

In actual practice, concentrated solutions, containing a weak acid and an alkali metal salt of it in the proper proportion, are commonly held in stock and an appropriate volume of this solution is added to the solution to be buffered.

When the solution to be buffered has a pH far removed from the desired value, it is best to adjust the solution to about that value by the addition of a strong acid or base, and only then to add the buffer solution. If the buffer were introduced without the previous adjustment, its capacity might be insufficient to secure the desired pH unless a very large amount were added. Too large an addition would be undesirable or, in many instances, intolerable.

Mixture of Weak Base and Conjugate as Buffer 13.4

A mixture of a weak base and its conjugate acid, added as an appropriate salt, exhibits buffer action in the alkaline range. The situation and mathematical treatment are analogous to that in Section 13.3. The buffer system ammonia–ammonium chloride is frequently used in analytical chemistry and has a "practical" buffer range of about pH 8.3 to 10.3.

Buffers of Extended Range 13.5

When polyfunctional weak acids or bases and their conjugates are used or more than one weak acid and base are involved, buffers can be obtained that show high capacity in several pH ranges. In addition, when the dissociation constants are close together, the regions of high buffer capacity overlap and buffers can be obtained that show high capacity over a large pH range. For example, buffers prepared by mixing solutions of sodium hydroxide and (tribasic) citric acid show useful buffer capacity over the interval pH 2 to 6.

Acid Salts of Dibasic Acids 13.6

The acid salt of the type MHA of a dibasic acid H_2A, with M being an alkali metal, receives special attention for the preparation of a solution of known pH. According to equation (11-11), the pH of a solution of such an acid salt is given as $\text{pH} \simeq \frac{1}{2}(\text{p}K_{a,1} + \text{p}K_{a,2})$. The pH remains essentially constant over a relatively wide range of salt concentration, but the buffer capacity is extremely low. Such a solution corresponds to that at the first equivalence point in the titration of a weak dibasic acid (Section 11.9) and at this point the titration curve is far from horizontal! Consequently, addition of small amounts of an acid or base will result in a substantial change in pH. Such salt solutions are used as pH standards ("buffers") to standardize pH meters (Section 23.7 and Table E in the Appendix).

13.7 Summary of Buffer Types

Table 13-1 summarizes the various types of buffers and compares their behavior with various changes.

Table 13-1 Buffer Types and Behavior

	Strong acid or strong base	Weak acid or weak base, and its conjugate	Ampholyte
Effect of dilution	Slight. For solutions up to about 1 *F*, about 1 pH unit for a change in concentration by a factor of 10	Essentially negligible	Essentially negligible
Effect of addition of small amounts of strong acid or strong base	Slight, depending on the buffer concentration	Slight, depending on concentration and composition of the buffer	Quite large, regardless of concentration

13.8 Questions

13-1. Define "buffer."

13-2. Elaborate on the relation of the various types of buffers to the relevant acid–base titration curves.

13-3. Describe the buffer capacity qualitatively. On what does it depend?

13-4. In what sense can a solution of sodium hydrogen carbonate, $NaHCO_3$, be called a buffer?

13-5. Elaborate on the possibility of preparing buffer solutions by the addition of hydrochloric acid to a solution of sodium acetate and to a solution of ammonia.

13-6. What acid, base, salt, or mixture of these would you select to secure a buffer of pH 9.0? Explain your answer.

13.9 Problems

13-1. A buffer solution is prepared by mixing 20.0 ml of 0.10 *F* sodium acetate and 8.0 ml of 0.12 *F* HCl (K_a for acetic acid, 1.8×10^{-5}).

(a) Calculate the pH of this solution. (b) Calculate the change in pH caused by addition of 5.0 ml of 0.10 *F* HCl.

Answers: (a) pH 4.78; (b) 0.47 pH unit

2. An NH_3–NH_4Cl buffer is prepared by mixing equal volumes of 0.200 *F* NH_3 ($K_b = 1.8 \times 10^{-5}$) and 0.200 *F* NH_4Cl. Calculate the pH of this buffer.

Answer: pH 9.26

3. Calculate the pH of the solution obtained if to 100 ml of buffer described in Problem 13-2 is added 10.0 ml of (a) 0.10 *F* NH_3, (b) 0.10 *F* NaOH, and (c) 0.10 *F* HCl.

Answers: (a) pH 9.30; (b) pH 9.34; (c) pH 9.18

4. What minimum volume (in millimeters) of a buffer solution which is 0.100 *F* in both lactic acid ($K_a = 1.4 \times 10^{-4}$) and sodium lactate must be diluted to exactly 100 ml so that the resulting buffered solution will change its pH by not more than 0.50 pH unit upon the addition of 10.0 ml of 0.010 *F* HCl?

Answer: 1.9 ml

5. How many grams of sodium acetate (*82.0*) must be added to 50.0 ml of 0.080 *F* acetic acid ($K_a = 1.8 \times 10^{-5}$) to attain a pH of 4.20?

Answer: 0.094 g

6. Calculate how many grams of ammonium chloride (*53.49*) must be added to each of the ammonia solutions of specified volume and formality to establish the desired pH.

	Ammonia solution	Desired pH
(a)	100.0 ml of 0.100 *F*	9.00
(b)	50.0 ml of 0.120 *F*	9.25
(c)	30.0 ml of 0.195 *F*	9.05
(d)	25.0 ml of 0.679 *F*	10.00
(e)	50.0 ml of 0.500 *F*	10.00
(f)	200.0 ml of 0.200 *F*	9.75

7. The following solutions of weak acids are mixed with solutions of the sodium salt of the same weak acid. Calculate the pH of each mixture.

	Acid solution	K_a	Salt solution
(a)	30.0 ml of 0.020 *F*	3.2×10^{-7}	40.0 ml of 0.015 *F*
(b)	50.0 ml of 0.015 *F*	4.0×10^{-5}	25.0 ml of 0.025 *F*
(c)	20.0 ml of 0.100 *F*	6.5×10^{-6}	150.0 ml of 0.080 *F*
(d)	35.0 ml of 0.040 *F*	1.8×10^{-5}	45.0 ml of 0.050 *F*

8. How many milliliters of the acetic acid solution of indicated formality must be added to each of the NaOH solutions of specified volume and formality to obtain the desired pH?

	NaOH solution	Acetic acid solution	Desired pH
(a)	100.0 ml of 0.100 *F*	0.200 *F*	4.74
(b)	150.0 ml of 0.215 *F*	0.100 *F*	5.00
(c)	200.0 ml of 0.075 *F*	0.095 *F*	4.50
(d)	75.0 ml of 0.095 *F*	0.350 *F*	4.90

13-9. A buffer is made by adding 25.0 ml of 0.100 *F* sodium acetate to 50.0 ml of 0.100 *F* acetic acid ($K_a = 1.8 \times 10^{-5}$). What would be the pH if to this buffer were added: (a) 25.0 ml of distilled water? (b) 10.0 ml of 0.100 *F* HCl? (c) 10.0 ml of 0.100 *F* NaOH?

13-10. Three buffer solutions are prepared from 0.100 *F* ammonia ($K_b = 1.8 \times 10^{-5}$) and 0.100 *F* NH_4Cl solutions.

Buffer	Ammonia solution, ml taken	NH_4Cl solution, ml taken
A	50.0	50.0
B	25.0	75.0
C	75.0	25.0

(a) Calculate the pH of each buffer. (b) Compare the change of pH resulting on the addition of 1.0 ml of 1.0 *F* HCl to each buffer.

13-11. What must be the formal concentrations of HA ($K_a = 1.0 \times 10^{-6}$) and NaA in a buffer of pH 5.30 so that the pH will be lowered to 5.00 upon the addition of 10.0 ml of 0.100 *F* HCl to 100.0 ml of the buffer solution?

13-12. A buffer solution has been prepared by making the solution 0.100 *F* in acetic acid ($K_a = 1.8 \times 10^{-5}$) and 0.100 *F* in sodium acetate. What volume of this buffer solution must be used so that the addition of 50.0 ml of 0.050 *F* H_2SO_4 will change the pH by exactly 1 unit?

14 APPLIED ACID-BASE TITRATIONS

Standard Acid Solutions 14.1

Hydrochloric acid is the acid most frequently used as a titrant in acid–base titrations in aqueous media. Occasionally sulfuric acid and perchloric acid are employed. Nitric acid finds use in some special determinations. Possibilities exist for the direct preparation of a standard acid solution by weighing a known amount of acid and diluting it to a known volume with water. The use of constant-boiling hydrochloric acid for this purpose is a notable example. The composition of the distillate of the constant-boiling mixture is dependent only on the atmospheric pressure. Consequently it is possible to distill this acid under exactly known pressure and to read the concentration of the distillate from a table. In practice, however, it is simpler to prepare a solution of approximately the desired strength and to establish its exact concentration by means of an acid–base titration employing a primary standard, or by evaluation against a standard solution of a strong base.

Sulfuric acid and perchloric acid are not volatilized when their aqueous solutions of even considerable strength are boiled. Hydrochloric acid and nitric acid are more volatile, and it is important to appreciate this fact where boiling of the solution is required during a titration. It may be noted, however, that a 0.1 *F* hydrochloric acid solution can be boiled for over 30 minutes with no detectable loss of hydrogen chloride provided the water evaporated is replaced at intervals.

The storage of standard acid solutions presents no major problems. Their strength is maintained in glass vessels that are tightly closed to prevent the evaporation of water or the absorption of basic substances from the laboratory air.

Primary Standards for Acids 14.2

The requirements for primary standards are summarized in Section 10.3. Although many primary standards for acids are known, they quite often do not function as well as those for bases. Hence, frequently a solution of a strong base is first standardized, and then the strength of the acid is evaluated against this base.

14.2 PRIMARY STANDARDS FOR ACIDS

Sodium Carbonate. Sodium carbonate, Na_2CO_3, is suitable as a primary standard for acids. For accurate work it is necessary to heat the reagent-grade powdered form of the anhydrous compound to about 285°C for at least 30 minutes. This treatment removes all water and converts any traces of hydrogen carbonate to carbonate. Since the product is somewhat hygroscopic, it should be weighed promptly after cooling in a dessicator, and its storage for a protracted period is not recommended. A weighed amount of the sodium carbonate is dissolved in water and is directly titrated with the acid to be standardized to the methyl orange end point. The details of this titration are considered in Section 14.7. In this titration the equivalent weight of sodium carbonate is one half of its formula weight $= \frac{1}{2} \times 105.989 = 52.994$.

Sodium Tetraborate Decahydrate. Sodium tetraborate decahydrate (borax), $Na_2B_4O_7 \cdot 10H_2O$, is a salt of a very weak acid, and in aqueous solution the tetraborate ion reacts with a strong acid as follows:

$$B_4O_7^{2-} + 2H^+ + 5H_2O \rightarrow 4H_3BO_3 \qquad (14\text{-}1)$$

The equivalent weight of borax is therefore one half of its formula weight $= \frac{1}{2} \times 381.373 = 190.686$. The equivalence point in the standardization of a 0.1 *F* solution of strong acid with borax is at about pH 5.1; hence, methyl red is an appropriate indicator. The advantages of borax as a primary standard include its high equivalent weight and its ease of preparation in a highly pure state, simply by recrystallization from water below 55°C. The product requires special storage to assure that the exact composition of the decahydrate is maintained; this is best accomplished by storage over deliquescent sodium bromide, which affords the proper humidity.

2-Amino-2-(hydroxymethyl)-1,3-propanediol. 2-Amino-2-(hydroxymethyl)-1,3-propanediol, $(HOCH_2)_3CNH_2$, is a weak monoequivalent base having a formula (and equivalent) weight of 121.137. It is often assigned the trivial designation *tris*. The high-purity material available can be dried at 100 to 105°C. The equivalence point in its titration with a strong acid occurs at about pH 4.7. Methyl red is a suitable indicator for this titration; bromocresol green is even better. For general analytical work, this base is possibly the most suitable one for the standardization of strong acid solutions.

Standard Base Solutions 14.3

Solutions of bases must be prepared, stored, and used in such a way that they are protected from absorption of carbon dioxide from the air, because the presence of carbonate is a frequent source of difficulties in titrations (see Section 14.7). For ordinary work a small amount of carbonate can be tolerated, but it must be kept as small as possible. Sodium hydroxide is the base most frequently applied as a titrant in acid-base titrations in aqueous media. Brief rinsing of the reagent-grade pellets of this base with water before dissolution in water is one possible approach to the removal of carbonate. The preferred classical method is to dissolve 50 parts by weight of sodium hydroxide in pellet form in 50 parts of water. In such a solution, sodium carbonate is insoluble and settles out. A portion of the clear supernatant liquid, or of the filtrate obtained by vacuum filtration through a sintered glass filter, is used in the preparation of the dilute solution. This preparation has been simplified by the availability of reagent-grade 50% sodium hydroxide solution low in carbonate furnished in polyethylene bottles; clear portions of this solution may be directly diluted. The water used to prepare the standard solution is either boiled to remove carbon dioxide and then cooled or is taken directly from a mixed bed of ion-exchange resins (Chapter 47). The amount of carbon dioxide in water in equilibrium with the general atmosphere is small (1.5×10^{-5} F). Laboratory air, however, may contain substantially more carbon dioxide and hence so may distilled water that has been in contact with such air. Solutions of sodium hydroxide should be stored in containers that are either airtight or fitted with a tube filled with soda-lime or soda-asbestos. Since alkaline solutions attack glass, sodium hydroxide solutions are best stored in polyethylene bottles, although for short periods borosilicate glass bottles may be employed.

Sodium hydroxide solutions may be freed of carbonate by passage through a column of a strongly basic anion-exchange resin in the hydroxide form (Chapter 47).

For very accurate work, barium hydroxide is often recommended, since barium carbonate is insoluble. A barium hydroxide solution of about the desired strength is prepared and allowed to stand in an airtight bottle for several days. The clear supernatant liquid is siphoned off and stored in polyethylene bottle with precautions to prevent absorption of carbon dioxide.

Potassium hydroxide is sometimes used as a base in acid–base titrimetry. However, potassium carbonate is relatively soluble in concentrated solutions of potassium hydroxide; hence, the classical technique employed to remove carbonate from sodium hydroxide is not applicable. Where an ethanolic solution of a base is required, potassium hydroxide is often used because it is readily soluble in this organic solvent.

14.4 Primary Standards for Bases

Potassium Hydrogen Phthalate. Potassium hydrogen phthalate is probably the most frequently employed primary standard in acid–base titrimetry. This compound is the acid salt of the weak diprotic acid, phthalic acid ($K_{a,2} = 3.1 \times 10^{-6}$), and acts as a weak monoprotic acid. Its equivalent weight in the titration with a strong base equals its formula weight, 204.229. The equivalence point is in the alkaline region and phenolphthalein is generally employed as the indicator. Potassium hydrogen phthalate is stable, may be heated up to 130°C without decomposition, and is virtually nonhydroscopic. This salt is commercially available in a primary standard grade.

Sulfamic Acid. Sulfamic acid, NH_2SO_3H, is a strong monoprotic acid of formula weight 97.093, and in its titration with a strong base any indicator changing in color in the region pH 4 to 9 may be employed. The substance is stable in air, but decomposes slowly in aqueous solution to yield ammonium hydrogen sulfate, NH_4HSO_4. Consequently, the titration is best performed soon after dissolution of the weighed amount of sulfamic acid.

14.5 Determination of Phosphoric Acid

Orthophosphoric acid, H_3PO_4, is a tribasic acid ($K_{a,1} = 6.0 \times 10^{-3}$, $K_{a,2} = 6.3 \times 10^{-8}$, $K_{a,3} = 4.4 \times 10^{-13}$; $pK_{a,1} = 2.15$, $pK_{a,2} = 7.20$, $pK_{a,3} = 12.36$).

Only the first two of the dissociable hydrogen ions can be titrated directly with a strong base, such as sodium hydroxide, since the third constant is too small in magnitude. The approximate pH values for the equivalence points are $\frac{1}{2}(2.15 + 7.20) = 4.68$ and $\frac{1}{2}(7.20 + 12.36) = 9.73$. These pH values correspond to the pH of a solution of monosodium phosphate, NaH_2PO_4, and disodium phosphate, Na_2HPO_4, respectively.

The titration to the first equivalence point can be performed using methyl orange and titrating almost to its alkaline color.

Location of this color is facilitated by comparison with a solution containing the indicator and monosodium phosphate in concentrations equal to those in the principal titration (see Section 12.4). Bromocresol green is a better indicator, but again use of a comparison solution is preferable. The second equivalence point is at a pH value slightly more alkaline than can readily be indicated by phenolphthalein; hence, thymolphthalein is a better indicator.

Even though a direct, simple titration of the third dissociable hydrogen ion in phosphoric acid is impracticable, phosphoric acid can be titrated as a tribasic acid if the phosphate ion is precipitated under proper conditions. For this purpose, a neutralized solution of calcium chloride is added to the sample solution, and the titration with sodium hydroxide solution is started. As the titration proceeds, the pH increases progressively until a value is reached at which calcium orthophosphate starts to precipitate. Its occurrence, however, is disregarded and the titration completed to a persisting pink color of phenolphthalein. The precipitate has hardly any effect on the recognition of the end-point color but can coprecipitate unreacted material, and thus errors as great as 1 or 2% may be encountered.

Boric Acid 14.6

Borate is often separated from interfering substances by distillation of trimethyl borate, $B(OCH_3)_3$, which is formed when the borate-containing sample is boiled with concentrated hydrochloric acid, anhydrous calcium chloride (which binds the water present and formed in the esterification), and methanol, CH_3OH. This borate ester is condensed and collected in a sodium hydroxide solution. Next, this solution is boiled to decompose the ester and to volatilize the methanol. The remaining solution is then neutralized to methyl orange and boiled to remove carbon dioxide. Free boric acid is thus present in the solution, acting as a very weak monoprotic acid ($K_a = 5.9 \times 10^{-10}$) that cannot be titrated.† Boric acid, however, forms complex entities with organic polyhydroxy compounds, such as glycerol or mannitol. These complex compounds are far stronger acids than boric acid itself. With mannitol the acidity constant of the complex acid formed has

†Metaboric acid, HBO_2, has one less water molecule than orthoboric acid, H_3BO_3. In aqueous media, the two cannot be distinguished and hence boric acid may be written either as H_3BO_3 or HBO_2.

a value of about 1×10^{-4}; thus the complex acid is stronger than acetic acid and can readily be titrated with a strong base to a sharp end point using phenolphthalein as indicator. The reaction with mannitol may be represented by the following equation, where mannitol is denoted by Ma:

$$HBO_2 + 2Ma \rightarrow H^+ + Ma_2BO_2^- \qquad (14\text{-}2)$$

14.7 Determination of Carbonates

The principles underlying the "replacement" titration of a carbonate salt with a strong acid have been outlined in Section 11.11.

Carbonate can be titrated to hydrogen carbonate ion (pH 8.3). However, the "break" at this equivalence point is not well defined, as can be seen from the curve in Figure 11-4; consequently, where highly reliable results are desired, a comparison solution should be used containing amounts of indicator and sodium hydrogen carbonate similar to those in the titration solution. This is especially necessary when phenolphthalein is used as the indicator, since its complete decoloration occurs at about pH 8.0, which is beyond the equivalence point and in a region where the titration curve is far from perpendicular.

The hydrogen carbonate can be further titrated to carbonic acid. The titrant is added until the methyl orange indicator color begins to change from yellow to orange. Of course, it is possible to perform the titration of carbonate ion directly to carbonic acid, omitting phenolphthalein and adding methyl orange only. The solution is then boiled to expel carbon dioxide, and after cooling the titration is continued to a red-orange color. Cooling is necessary because the dissociation equilibrium of methyl orange and consequently of the color exhibited at a certain pH strongly depends on the temperature. The small amount of carbon dioxide formed during the addition of the last portions of titrant does not affect the color change and a relatively sharp color change occurs.

Alternatively, it is often more convenient to add a known amount of acid in excess, to heat the solution to boiling, then to cool, and to back-titrate with a standard sodium hydroxide solution to the methyl orange end point. Indeed, since all carbonate has been evolved as carbon dioxide and only strong acid remains, it is possible to back-titrate to the phenolphthalein end point. In this case it is unnecessary to wait until the solution is cooled. This back-titration approach is mandatory where a water-insoluble carbonate such as calcium carbonate is to be titrated.

Hydroxide ion can be determined in the presence of carbonate ion if the latter is first precipitated as barium carbonate. An excess amount of a neutralized solution of barium chloride is added to the hydroxide-carbonate sample; the hydroxide ion is then titrated with hydrochloric acid to the phenolphthalein end point in the presence of the precipitate. (The methyl orange end point cannot be used since the barium carbonate would dissolve in the acidic solution.) This procedure has limitations because the precipitate may carry down some hydroxide ion.

Various titrimetric processes may be combined to effect the resolution of mixtures of carbonate and hydrogen carbonate, or of carbonate and hydroxide. If a solution contains substantial amounts of both carbonate and hydroxide ions, titration with hydrochloric acid first to the phenolphthalein end point neutralizes the hydroxide and converts carbonate to hydrogen carbonate; then, after addition of methyl orange, continued titration to a second end point (with the heating step discussed above) converts hydrogen carbonate to carbonic acid. The volume of hydrochloric acid required in the second titration step corresponds to one half of that required to convert carbonate to carbonic acid. Hence, both the hydroxide and carbonate contents can be calculated. A similar stepwise titration permits the determination of both hydrogen carbonate and carbonate when present together. When only a small amount of carbonate is present, the method fails, since then the volume of titrant required to proceed from the phenolphthalein end point to the methyl orange end point is small and unrealiable results are obtained.

In such a case a different approach is used. In one aliquot of the sample solution, the total content of base, that is, the sum of hydroxide and carbonate, is determined by titration to the methyl orange end point. In a second aliquot the carbonate is precipitated by addition of barium chloride and the titration performed to the phenolphthalein end point. The difference in consumption of titrant in the two titrations corresponds to the carbonate. This approach also permits the determination of a small amount of hydrogen carbonate in a carbonate sample. To one aliquot of the sample solution a known volume of standard sodium hydroxide solution is added in excess of that required to convert hydrogen carbonate to carbonate ion. Then barium chloride is added, and the mixture is titrated with hydrochloric acid to the phenolphthalein end point; an identical volume of sodium hydroxide is similarly titrated. The difference in titrant volumes required in the two titrations corresponds to hydrogen carbonate. In a second aliquot of the sample solution the total of carbonate and hydrogen carbonate is determined by titration to the methyl orange end point; carbonate is then obtained by difference.

It is convenient to discuss at this point the effect of the presence of carbonate in a base used for the titration of a strong acid. From the above considerations, it is obvious that there will be a difference in the titration results, depending on whether phenolphthalein or methyl orange is used as the indicator. The greater the carbonate content, the greater the difference. When only a small amount of carbonate is present, it is possible to standardize the base using either indicator and to apply the corresponding "phenolphthalein" (P.P.) or "methyl orange" (M.O.) formality as appropriate. However, if large amounts of carbonate are present, this approach fails because further difficulties arise.

When the carbonate-containing base is added to the acidic sample solution, carbon dioxide is formed. As long as this formation is not excessive the titration to the methyl orange end point will not be affected. However, if the titration is conducted to the phenolphthalein end point, the carbon dioxide is titrated to hydrogen carbonate. But some of it may have already escaped from the solution and, since the amounts lost are uncontrolled, the titrations are not closely reproducible. Furthermore, the reconversion of carbon dioxide to carbonic acid (which is the species that reacts with the base) is not very rapid and leads to a fading end point; that is, the solution becomes pink on addition of one drop of the base, but the color fades within a few seconds. This fading is not to be confused with the slower fading that occurs after possibly 30 seconds or more due to the absorption of atmospheric carbon dioxide by the alkaline solution. For the experienced analyst, the rapid fading and sluggish character of an end point in a titration with a strong base to the phenolphthalein end point is a clear signal that an undesirably high amount of carbonate is present in either the standard base or the sample solution or both.

14.8 Determination of Temporary Water Hardness

Natural water may contain significant amounts of the soluble hydrogen carbonate salts of the alkaline earths (principally calcium and magnesium) as well as the sulfates and chlorides of these elements. These are the major constituents that contribute to the hardness of water. Upon boiling of hard water, precipitation of carbonate occurs according to the reactions

$$Ca^{2+} + 2HCO_3^- \rightleftharpoons \underline{CaCO_3} + H_2O + CO_2 \quad (14\text{-}3)$$

$$Mg^{2+} + 2HCO_3^- \rightleftharpoons \underline{MgCO_3} + H_2O + CO_2 \quad (14\text{-}4)$$

The portion of the hardness corresponding to the hydrogen carbonate salts is termed temporary hardness or carbonate hardness. The remaining hardness is due to the presence of alkaline earth sulfates or chlorides, which do not precipitate on boiling, and is termed the permanent hardness or noncarbonate hardness. The sum of the two types of hardness, expressed in identical units, is known as total hardness. Hardness is usually expressed in terms of parts per million of calcium, calcium oxide, or calcium carbonate, regardless of whether calcium or magnesium is actually present. The determination of water hardness is of practical importance, for example, in boiler-feed water and in water intended for laundry purposes. On heating of hard water in a boiler, calcium and magnesium carbonates deposit in the boiler tubes as so-called boiler scale. Calcium and magnesium ions also form insoluble soaps, and consequently hard water requires the use of much soap before adequate suds can be developed.

Temporary hardness is determined by simply adding methyl orange to a known volume of the water sample and titrating the hydrogen carbonate with hydrochloric acid. Since a low concentration of hydrogen carbonate ion is involved, a blank is run employing an identical volume of carbonate-free distilled water, and the volume of acid required in the blank is subtracted from that in the titration of the water sample.

Total hardness was formerly determined by a tedious acid–base titration procedure, but the result can now be attained more rapidly and with higher reliability by a compleximetric titration (Section 20.10).

Determination of Nitrogen 14.9

Where the nitrogen in a substance is present as ammonium ion or can be converted to it, a determination of nitrogen based on an acid–base titration is feasible. The solution of an ammonium salt may be made strongly alkaline and boiled, thereby volatilizing ammonia with water. Ammonia in the distillate is determined by an acid–base titration. Kjeldahl found that the nitrogen in many organic substances can be converted to ammonium ion by heating with concentrated sulfuric acid. After alkalization of the digested mixture, the ammonia can be distilled and determined. Inorganic nitrates, which are not reduced to ammonia by this acid treatment, may be converted to ammonia by an alkaline digestion with a metal or an alloy, such as Devarda's alloy, having reducing properties under such conditions.

Kjeldahl Digestion. The reactions occurring when an organic substance containing nitrogen is heated with concentrated sulfuric acid are complex and not fully understood. The overall process may be described as the oxidation of the substance by sulfuric acid, which is reduced to sulfuric dioxide, and reduction of nitrogen by the sulfur dioxide to ammonium ion. (With organic matter having oxidized nitrogen, such as nitro groups, a preliminary reduction is often appropriate.) To hasten the digestion, potassium sulfate is added to elevate the boiling point of the digestion mixture. Various substances may be added to catalyze the reactions occurring during the digestion process, including salts and oxides of copper, mercury, or selenium. The digestion is usually carried out in special long-necked flasks, known as Kjeldahl flasks, in a hood or a collecting manifold, since fumes of sulfuric acid and sulfur dioxide are evolved. The digestion is continued until the solution is colorless, indicating the complete destruction of organic matter. The final decolorization may be hastened by addition of a few drops of hydrogen peroxide.

Distillation after Kjeldahl Digestion. When the Kjeldahl digestion is complete, the solution is allowed to cool. A distillation assembly may be attached to the flask, or the solution may be transferred to a separate distillation apparatus. An excess of sodium hydroxide is added and the solution is brought to a boil. Ammonia and water vapor are driven off and condensed, and are caught in a flask containing a measured amount of standard acid. After all the ammonia has been distilled, the excess of the acid is back-titrated with standard base to the methyl red end point (since ammonia is a weak base, the equivalence point is in the acidic region).

Treatment after Devarda. The reduction of inorganic nitrate to ammonia may be effected in strongly alkaline solution by Devarda's alloy (45% aluminum, 5% zinc, and 50% copper):

$$3NO_3^- + 8Al + 5OH^- \rightarrow 8AlO_2^- + 3NH_3 \qquad (14\text{-}5)$$

The ammonia distills with water and is determined as described above.

Formol Method for Ammonia 14.10

Ammonia in ammonium salts may be determined by a more rapid method involving the reaction of formaldehyde, CH_2O, with ammonia to form the extremely weak base hexamethylenetetramine, $(CH_2)_6N_4$, and hydrogen ion, which can be titrated with a strong base. This reaction may be represented by

$$4NH_4^+ + 6CH_2O \rightarrow (CH_2)_6N_4 + 6H_2O + 4H^+ \quad (14\text{-}6)$$

One hydrogen ion is produced for each ammonium ion reacting. The procedure, known as the "formol titration," takes the following course. The sample solution containing ammonium ion and only the anions of strong acids is adjusted to the transition color of methyl red. Formaldehyde, which has been neutralized to phenolphthalein, is now added in excess and then a few drops of this indicator. After several minutes, the solution is titrated with a standard base to the alkaline color of phenolphthalein (here screened by the alkaline color of methyl red to orange, that is, pink + yellow). If anions of weak acids are present, the initial pH adjustment is made with phenol red.

Errors in Acid–Base Titrations 14.11

The correctness of the result of a titration is influenced by many factors and to varying degrees, including uncertainties in the weighing of the sample, delivery of a solution by a pipet, the reading and nonreproducible drainage of a buret, impurities in the sample solution, changes in temperature, failure to titrate always to the same indicator hue (visual discrimination error), and others. All these factors produce small deviations so that the individual results in a series of titrations performed under "identical" conditions and with "identical" samples fluctuate around their average. These deviations are known as random errors. Depending on their magnitude, the repeatability, that is, the precision of a titration, is high or low. Although random errors can never be completely avoided, they can be reduced by careful operation.

Another type of error, known as a systematic error, is inherent in the method employed. Such an error causes a departure of the result from the correct value in a single direction only, and its magnitude is a measurement of the accuracy of a titration result. It may be emphasized that this classification of errors is of theoretical interest and of importance in the treatment of experimental data.†

In practice, however, it is difficult to separate random and systematic errors, because every systematic error is accompanied by a random component. This is especially true of the drop error in titrimetry. When the end point is near, one more drop of titrant is added and in the visual titration the color change is observed. The closeness of the approach to the end point before addition of the final drop will vary in a series of titrations, and hence the small amount of titrant added in excess varies from titration to titration. This error may be negligible if a large volume of titrant is required but is significant with a small volume of titrant. The drop error usually includes the visual discrimination error. Careful work, the use of a color comparison solution (Sections 12.4 and 14.5), if necessary, and completion of the titration with fractions of a drop will reduce the drop error.

The indicator error is a further, largely systematic error. The indicator as a weak acid or base consumes some material to be titrated or titrant. Since only a minute amount of indicator is added, this error is usually negligible. However, with small samples and dilute titrants, it may become significant and require correction via an indicator blank or exclusion by other means (see Section 12.2). Further systematic errors may be introduced, for example, by inaccurate standardization of the titrant and incorrect calibration of the volumetric glassware.

With all these possibilities for error, the impression might be gained that a correct titration result is hard to be attained. However, the random errors from various sources vary in magnitude and sign: One may be large when another is small; one may be positive when another is negative. Hence, partial compensation occurs and the net random error may be small. With good equipment and appropriate care, the net random error, that is, the precision of the titration, may readily be within one or two drops of titrant, that is, within ± 0.05 to ± 0.10 ml.

Of special importance is the titration error or chemical error, which corresponds to the difference between the observed end

†See Chapter 7 in *Quantitative Analytical Chemistry: A Short Introduction to Practice.*

point and the equivalence point. The calculation of the magnitude of this error will be considered for the titration of strong monofunctional acids and bases and will give a further insight into titration processes. The calculation for titrations involving weak acids or bases is possible, but involved.

Example 14-1. A volume of exactly 50 ml of exactly 0.1 F HCl is titrated with exactly 0.1 F NaOH to a certain hue of methyl orange corresponding to exactly pH 4. What is the titration error?

As in the calculation of titration curves [Section 11.2, equation (11-1)], the molar concentration of hydrogen is obtained by subtracting the millimoles of base added from the millimoles of acid initially present and dividing by the total solution volume in milliliters. The volume of base added may be denoted by V_b. The expression becomes

$$10^{-4} = \frac{50 \times 0.1 - 0.1\,V_b}{50 + V_b}$$

$$50 \times 10^{-4} + 1 \times 10^{-4}\,V_b = 5 - 0.1\,V_b$$

$$0.1001\,V_b = 4.995$$

$$V_b = 49.90 \text{ ml}$$

The equivalence point corresponds to exactly 50 ml of base; hence, the titration error, ϵ, expressed in per cent, is

$$\epsilon = \frac{(49.90 - 50) \times 100}{50} = -0.2\%$$

Since in conventional titrimetry a result accurate to about 0.5 to 1% is acceptable, it may be concluded that methyl orange is a suitable indicator for this titration. Titration to a more yellow hue of the indicator might be considered, thereby decreasing the titration error.

Example 14-2. A volume of exactly 50 ml of exactly 0.01 F NaOH is titrated with exactly 0.01 F HCl to exactly pH 4 (methyl orange). What is the titration error?

The solution at the end point has a pH below 7 and hence is overtitrated; that is, a small amount of acid in excess is added. The calculation is analogous to Example 14-1, with V_a denoting the volume of acid added.

$$10^{-4} = \frac{0.01\,V_a - 50 \times 0.01}{V_a + 50}$$

$$10^{-4}\,V_a + 50 \times 10^{-4} = 0.01\,V_a - 0.5$$

$$0.0099\,V_a = 0.505$$

$$V_a = 51.01 \text{ ml}$$

The titration error in per cent is, therefore,

$$\epsilon = \frac{(51.01 - 50) \times 100}{50} = +2\%$$

Comparison of the data in Examples 14-1 and 14-2 shows, as

expected, that with a 10 times more dilute titrant being used, a 10 times greater error results. The overtitration corresponds to a positive sign for the error. For this titration, methyl orange is not a suitable indicator, and one changing color at a higher pH should be used (e.g., methyl red).

Example 14-3. A volume of exactly 25 ml of exactly 0.1 F HCl is titrated with exactly 0.05 F NaOH to exactly pH 9 (phenolphthalein). What is the titration error?

Since an overtitration obviously occurs, let V_b denote the volume of base added.

$$\text{pH } 9 = \text{pOH } 5 \qquad \text{and} \qquad [OH^-] = 10^{-5}\,M$$

Hence,

$$10^{-5} = \frac{0.05\,V_b - 25 \times 0.1}{25 + V_b}$$

$$25 \times 10^{-5} + 10^{-5}\,V_b = 0.05\,V_b - 2.5$$

$$0.04999\,V_b = 2.50025$$

$$V_b = 50.015 \text{ ml}$$

The titration error in per cent is therefore

$$\epsilon = \frac{(50.015 - 50) \times 100}{50} = +0.03\%$$

These few examples show the importance of selecting an appropriate indicator in order to obtain a correct result. In the calculations, the amount of hydrogen or hydroxide ion furnished by water itself has been neglected and permittedly so, since these pH values are far from 7. Where the end point is close to pH 7, this amount may no longer be neglected, and a more involved formula must be used. If, however, in a strong acid–strong base titration, the end point is close to pH 7 (that is, close to the equivalence point!), the titration error is obviously negligible and no calculation is needed.

14.12 Questions

14-1. What is a primary standard? Describe its use and requirements.

14-2. What are the difficulties encountered with primary standards for the standardization of acid solutions?

14-3. How would you proceed to analyze a mixture containing hydrochloric and acetic acids? What indicators would you select?

14-4. Why is a back-titration recommended in the acid–base titration of calcium carbonate?

14-5. How might you determine zinc oxide by means of an acid–base titration?

4-6. Discuss in your own words the terms temporary and permanent hardness.

4-7. Explain why phenolphthalein is not employed in the determination of temporary hardness. What would be the result if this indicator were used?

4-8. How can nitrate be determined via an acid–base titration procedure?

4-9. Elaborate on the possibility of titrimetrically determining in a mixture both hydrochloric acid and nitric acid employing silver acetate as an auxiliary reagent.

-10. What chemical processes are involved in the distillation method for the determination of borate?

-11. What chemical reactions are involved in the formol method of ammonia?

-12. What difficulties will be encountered if a sample of hydrochloric acid containing some iron(III) chloride is titrated with sodium hydroxide to determine the free acid?

-13. What is the purpose of adding a calcium chloride solution when phosphoric acid is titrated as a triprotic acid?

-14. A certain amount of sulfur is burned under conditions yielding both SO_2 and SO_3. The gases are absorbed in water, yielding a mixture of H_2SO_3 and H_2SO_4. Would it be possible to differentiate between the two acids on the basis of a simple acid–base titration?

-15. Sodium dihydrogen phosphate, NaH_2PO_4, can be titrated with a sodium hydroxide solution. Write the equation for the titration reaction. What indicator should be used? What is the equivalent weight of the salt in this titration?

-16. What types of errors can be differentiated?

-17. Describe the indicator error and discuss ways of avoiding or reducing it.

-18. Name some sources of systematic errors in acid–base titrimetry.

-19. What percentage error in the result is caused by a drop error of 0.05 ml of titrant when the volume of titrant required is (a) 5.00 ml and (b) 50.00 ml?

-20. Name some sources of random errors encountered in titrimetry.

-21. How would you proceed to reduce the visual discrimination error in a titration?

-22. The use of an inaccurately standardized titrant solution introduces what type of error? Random or systematic?

-23. A general formula for the titration error in the strong acid–strong base titration may be derived in which the volume of the solution being titrated and the volume of the titrant solution do *not* occur; only the concentrations of the two solutions affect the size of the titration error. Verify this fact by repeating the calculations of

Examples 14-1 and 14-2 by employing volumes other than exactly 50 ml.

14.13 Problems†

14-1. A volume of 100 ml of 0.1 *F* HCl is titrated with 0.05 *F* NaOH to pH 8.5 (phenolphthalein). What is the titration error?

Answer: +0.01%

14-2. Calculate the titration error when a volume of 20 ml of 0.01 *F* HCl is titrated with 0.001 *F* NaOH to an end point at pH 3.3.

Answer: −37%

14-3. Calculate the titration error when a volume of 50 ml of 0.01 *F* HCl is titrated with 0.01 KOH to an end point at pH 5.3 (methyl red).

Answer: −0.1%

†Assume in the solution of these problems that all volumes, formalities, and pH values are exactly known.

15 CALCULATIONS OF RESULTS IN ACID–BASE TITRATIONS

Basis of Calculations 15.1

Regardless of the acids or bases involved, the principal titration equation is always the transfer of a proton: $HA + B \rightarrow A + HB$ (with charges omitted for simplicity). Consequently, the question as to what amount of an acid is equivalent to what amount of a base reduces to how many protons are donated per molecule or ion of the acid and accepted per molecule or ion of the base. Therefore, one equivalent of an acid is defined as the amount of acid that donates under the actual titration conditions one mole of protons. Analogously, one equivalent of a base is defined as the amount of base that under the actual titration condition accepts one mole of protons. Then, with regard to titrations, the equivalent weight EW, of an acid is obtained, by dividing the formula weight FW, by the equivalence number, f:

$$EW = \frac{FW}{f} \qquad (15\text{-}1)$$

The equivalence number denotes the moles of protons donated per mole of acid or accepted per mole of base and *actually transferred under the conditions of the titration.*

The italicized portion of the preceding sentence is of utmost importance. It is *not* the number of protons that *potentially can* be donated or accepted but rather the number that *is* actually donated or accepted. This definition of equivalence may be different from that the student has encountered in previous courses and which was more convenient for the purposes then at hand. For titrimetry, however, the definitions given above are the most suitable ones. It should be pointed out that the equivalence number will be an integer or, in rare cases, a ratio of integers. An equivalence number, in the sense here used, is not an experimentally measured value and hence it will have no influence on the number of significant figures to be retained in the presentation of the result. The equivalence number is dimensionless, but depending on reference to an acid or base has the units "moles of protons donated per mole of acid" or "moles of protons accepted per mole of base," respectively.

As was previously shown, the amount of a substance in moles

is obtained by dividing the grams of substance, g, by the formula weight:

$$\text{moles} = \frac{\text{g}}{\text{FW}} \tag{15-2}$$

Analogously the amount of a substance in equivalents is obtained by dividing the grams by the equivalent weight:

$$\text{eq} = \frac{\text{g}}{\text{EW}} \tag{15-3}$$

Combination of equations (15-1) and (15-3) yields

$$\text{eq} = \frac{\text{g}}{\text{FW}/f} \tag{15-4}$$

The definition of formality is

$$F = \frac{\text{moles}}{\text{liters}} \tag{15-5}$$

Combination of this equation with equation (15-2) gives

$$F = \frac{\text{g}}{\text{FW} \times \text{liters}} \tag{15-6}$$

Substitution of FW from equation (15-4) yields

$$F = \frac{\text{eq}}{f \times \text{liters}} \tag{15-7}$$

Rearrangement of the last equation gives

$$\text{liters} \times F \times f = \text{eq} \tag{15-8}$$

The equivalent is usually too large a unit, and one thousandth of it, that is, the milliequivalent, meq, is commonly employed. Dividing both sides of equation (15-8) by 1000 yields

$$\text{ml} \times F \times f = \text{meq} \tag{15-9}$$

The last two equations show the relationship between the volume and formality of a solution and the equivalents or milliequivalents of reacting substance provided. The concentration of a solution can also be expressed in equivalents per liter of solution, termed the normality of the solution, and denoted by N. Hence, $F \times f = N$. Although the concept of normality has merit in certain cases of practical titrimetry, it is not stressed in the present textbook because the normality in general does not have a fixed value but rather depends on the conditions under which the titration is performed. For example, phosphoric acid can react with hydroxide ion (added as NaOH) in the following three ways:

$$H_3PO_4 + OH^- \rightarrow H_2PO_4^- + H_2O \qquad N = F \times 1$$
$$H_3PO_4 + 2OH^- \rightarrow HPO_4^{2-} + 2H_2O \qquad N = F \times 2$$
$$H_3PO_4 + 3OH^- \rightarrow PO_4^{3-} + 3H_2O \qquad N = F \times 3$$

As indicated at the right of the reaction equations, the normality of a phosphoric acid solution of given formality can have three different values! The titrants most commonly employed in acid–base titrimetry—HCl, $HClO_4$, HNO_3, NaOH, and KOH—are monofunctional and hence their equivalence number is always unity. Only two common acid–base titrants, H_2SO_4 and $Ba(OH)_2$, have an equivalence number of 2 and *always* 2, since for aqueous media their two acidity and basicity constants, respectively, are not sufficiently separated to allow a titration to two distinct end points.

At the equivalence point the equivalents of titrant added exactly equals the equivalents of substance titrated. Using the subscripts t and s to refer to titrant and substance titrated, respectively, this fact can be expressed as

$$\left.\begin{aligned} \text{eq}_t &= \text{eq}_s \\ \text{meq}_t &= \text{meq}_s \end{aligned}\right\} \text{ at the equivalence point only} \qquad \begin{aligned}(15\text{-}10a)\\(15\text{-}10b)\end{aligned}$$

Combination of (15-10b) with (15-9) yields the first fundamental equation for titrimetry:

$$\text{ml}_t \times F_t \times f_t = \text{ml}_s \times F_s \times f_s \qquad (15\text{-}11)$$

The right-hand side of equation (15-11) is the milliequivalents of substance titrated, and can be expressed by

$$\text{meq}_s = \frac{\text{mg}_s}{\text{FW}_s/f_s}$$

Substitution of this term yields

$$\text{ml}_t \times F_t \times f_t = \frac{\text{mg}_s}{\text{FW}_s/f_s} \qquad (15\text{-}12)$$

On rearrangement the second fundamental equation is obtained:

$$\text{ml}_t \times F_t \times f_t \times \frac{\text{FW}_s}{f_s} = \text{mg}_s \qquad (15\text{-}13)$$

This expression relates volume and concentration of the titrant solution with the formula weight, equivalence number, and amount of the substance titrated. Again the relationship holds *only* at the equivalence point. A dimensional analysis applied to equation (15-13) is of interest:

$$\text{ml}_t \times F_t \times f_t \times \left[\frac{\text{FW}_s}{f_s}\right] = \text{mg}_s$$

$$\text{volume} \times \frac{\text{mass}}{\text{volume}} \times \text{none} \times \left[\frac{\text{none}}{\text{none}}\right] = \text{mass}$$

Since "volume" cancels, the identity "mass = mass" shows dimensional satisfaction for the equation.

Of more practical interest is a units analysis of this equation:

$$\text{ml} \times \frac{\text{moles}_t}{\text{liter}} \times \overbrace{10^{-3}\,\frac{\text{liters}}{\text{ml}}}^{\text{to convert liter to milliliter}}$$

$$\times \frac{\text{moles, protons donated (accepted)}}{\text{moles}_t}$$

$$\times \frac{\dfrac{\text{g}_s}{\text{mole}_s}}{\dfrac{\text{moles, protons accepted (donated)}}{\text{mole}_s}}$$

$$= 10^{-3} \times \text{g}_s = \text{mg}_s$$

Only the term $10^{-3} \times \text{g}_s$ remains on the left, which corresponds to mg_s, proving the units to be correct. All other terms cancel. including the terms referring to the moles of protons accepted or donated. The latter two terms cancel, because the number of protons donated (or accepted) by the titrant *must* exactly equal the number of protons accepted (or donated) by the substance titrated.

In many cases it is not the amount of substance titrated that is required but rather its content in the sample material in per cent. Obviously this percentage is obtained as follows, with W denoting the sample weight:

$$\frac{\text{mg}_s \times 100}{W\,(\text{in mg})} = \%_s \qquad (15\text{-}14)$$

Combination of this expression with (15-13) yields

$$\frac{\text{ml}_t \times F_t \times f_t \times \text{FW}_s \times 100}{W\,(\text{in mg}) \times f_s} = \%_s \qquad (15\text{-}15)$$

It is of utmost importance to realize that the sample weight must be expressed in milligrams.

Often a substance cannot be determined by a direct titration but a back-titration can be successfully applied. Calculation of the results in such a case is based on the following formula:

$$(\text{ml}_t \times F_t \times f_t - \text{ml}_r \times F_r \times f_r) \times \frac{\text{FW}_s}{f_s} = \text{mg}_s \qquad (15\text{-}16)$$

where the subscript r refers to the back-titrant. Here the expression within parentheses corresponds to the milliequivalents of the titrant actually reacted with the substance titrated, and is, of course, the difference between the milliequivalents of titrant added and the milliequivalents of back-titrant required in the back-titration.

The analog to equation (15-15) is obviously

$$\frac{(\text{ml}_t \times F_t \times f_t - \text{ml}_r \times F_r \times f_r) \times \text{FW}_s \times 100}{W\,(\text{in mg}) \times f_s} = \%_s \qquad (15\text{-}17)$$

Where it appears in a formula, the term FW_s refers to the substance considered to be titrated. Frequently, however, the result of the titration is expressed in terms of another substance. For example, phosphoric acid may be titrated, but the result is to be expressed in terms of P_2O_5. In such cases additionally one (or more) chemical factors are applied.

Example 15-1. A standard solution of HCl, 0.1000 F, is used to standardize a NaOH solution. A volume of exactly 25 ml of the NaOH solution is transferred to a beaker, some phenolphthalein indicator added, and the titration performed. To reach the end point a volume of 26.03 ml of the acid is required. What is the formality of the NaOH solution? Here HCl is the titrant and NaOH the substance titrated. Application of equation (15-11) yields

$$\underbrace{\text{ml}_{\text{HCl}} \times F_{\text{HCl}} \times f_{\text{HCl}}}_{\text{meq}_{\text{HCl}}} = \underbrace{\text{ml}_{\text{NaOH}} \times F_{\text{NaOH}} \times f_{\text{NaOH}}}_{\text{meq}_{\text{NaOH}}}$$

Substitution of the numerical values given above and recognition that the equivalence number for both the acid and base is unity yields

$$26.03 \times 0.1000 \times 1 = 25.00 \times F_{\text{NaOH}} \times 1$$

and

$$F_{\text{NaOH}} = \frac{26.03 \times 0.1000}{25.00} = 0.1041\,F$$

Since $f_{\text{NaOH}} = 1$, the normality of the base is 0.1041 N.

Example 15-2. On the titration of 20.00 ml of a H_2SO_4 solution with 0.1030 F NaOH to the phenolphthalein end point, a volume of 38.30 ml of the base is required. What is the formality of the H_2SO_4 solution?

The reasoning employed in Example 15-1 is also relevant here and equation (15-11) is applicable. The equivalence number of H_2SO_4, whatever the indicator used, is 2; hence

$$38.30 \times 0.1030 \times 1 = 20.00 \times F_s \times 2$$

and
$$F_{H_2SO_4} = \frac{38.30 \times 0.1030}{20.00 \times 2} = 0.0986\ F$$

Since $f_{H_2SO_4} = 2$, the acid is also $2 \times 0.0986 = 0.1972\ N$.

Example 15-3. A NaOH solution is to be standardized using potassium hydrogen phthalate (*204.14*), KHP, which is a monoprotic acid. Of this primary standard, an amount of 0.8632 g is dissolved in water and titrated with the NaOH solution. To reach the phenolphthalein end point, a volume of 38.64 ml of the base is required. What is the formality of the base?

Application of equation (15-13) is appropriate, with the subscripts t and s referring to NaOH and KHP, as the titrant and substance titrated, respectively. The equivalence number of both NaOH and KHP is unity. Substitution of the numerical data yields

$$38.64 \times F_t \times 1 \times \frac{204.14}{1} = 863.2$$

Solving for the formality yields

$$F_t = 0.1094\ F\ \text{NaOH}$$

Since $f_{NaOH} = 1$, the base is also $0.1094\ N$.

Example 15-4. A HNO_3 solution is to be standardized using Na_2CO_3 (*105.99*). A 0.1562-g sample of this standard is dissolved in water and titrated with the HNO_3 solution. To reach the methyl orange end point, a volume of 31.22 ml of the acid is required. What is the formality of the HNO_3 solution?

Again equation (15-13) is applicable. The equivalence number of HNO_3 is unity; that of Na_2CO_3 is 2, since the CO_3^{2-} ion accepts two protons on titration to the methyl orange end point. Insertion of the numerical data yields

$$31.22 \times F_t \times 1 \times \frac{105.99}{2} = 156.2$$

Solving for the formality yields

$$F_t = 0.0944\ F\ HNO_3$$

Since $f_{HNO_3} = 1$, the acid is also $0.0944\ N$.

Example 15-5. A sample, weighing 0.1935 g, of a weak organic acid is titrated with 0.1034 F NaOH to the phenolphthalein end point. The volume of base required is 29.44 ml. What is the equivalent weight of the acid? Application of equation (15-13) and recognition of FW_s/f_s as the equivalent weight of the substance titrated, here the acid, yields

$$29.44 \times 0.1024 \times 1 \times EW_s = 193.5$$

Solving,

$$EW_s = 64.19$$

Example 15-6. A 0.5000-g sample of a material containing $CaCO_3$ (*100.09*) as the only active substance is dissolved in 50.00 ml of 0.1080 *F* HCl, CO_2 is boiled off, and the excess of acid is back-titrated with 0.0904 *F* NaOH, of which a volume of 4.22 ml is required. What is the % $CaCO_3$ in the material?

To the sample was added $50.00 \times 0.1080 \times 1 = 5.400$ meq of HCl. The excess of acid required $4.22 \times 0.0904 \times 1 = 0.381_5$ meq of NaOH. The number of milliequivalents of HCl consumed by the $CaCO_3$ is $5.400 - 0.381_5 = 5.018_5$ and that is also the number of milliequivalents of $CaCO_3$ present. The equivalence number of $CaCO_3$ is 2, since two protons are accepted by one molecule of $CaCO_3$. (Write the reaction equation.) Consequently, the amount of $CaCO_3$ present is

$$5.018_5 \times \frac{100.09}{2} = 251.2 \text{ mg of } CaCO_3 \text{ in the sample weight taken}$$

The % $CaCO_3$ in the sample is then given by

$$\frac{251.2 \times 100}{500.0} = 50.23\% \ CaCO_3$$

Note that the sample weight has to be expressed in milligrams because the weight of the $CaCO_3$ is in milligrams. The entire calculation can be presented in an integrated fashion according to equation (15-17):

$$\frac{\overbrace{\overbrace{(50.00 \times 0.1080 \times 1)}^{\text{meq acid added}} - \overbrace{(4.22 \times 0.0904 \times 1)}^{\text{meq base for back-titration}}}^{\text{meq acid reacting with } CaCO_3 = \text{ meq of } CaCO_3} \times \overset{\text{formula weight of } CaCO_3}{100.09} \times \overset{\text{conversion factor for } \%}{100}}{\underset{\text{sample wt. in mg}}{500.0} \times \underset{f_{CaCO_3}}{2}}$$

$$= 50.23\% \ CaCO_3$$

Example 15-7. A 2.52-g sample of a mixture containing Na_2CO_3 (*105.99*), $NaHCO_3$ (*84.01*), and inert material is dissolved in water and diluted with water to exactly 250 ml. A 50.00-ml aliquot of this solution is titrated with 0.1020 *F* HCl to the phenolphthalein end point, a volume of 16.44 ml being required. Next a 25.00-ml aliquot is titrated with the same acid to the methyl orange end point, a volume of 22.46 ml being required. Calculate the % Na_2CO_3 and % $NaHCO_3$ in the mixture.

The principles involved in this determination of hydrogen carbonate and carbonate, when present together, have been considered in Section 14.7. In the first titration the carbonate present is titrated to hydrogen carbonate ion and $f_{Na_2CO_3} = 1$. The calculation takes the following form:

$$\frac{\overbrace{16.44 \times 0.1020}^{\text{meq acid added}} \times \underset{\uparrow}{1} \times \overset{\substack{\text{formula wt. of } Na_2CO_3\\ \downarrow}}{105.99} \times \overset{\substack{\text{conversion factor for \%}\\ \downarrow}}{100} \times \overset{\substack{\text{only } \frac{1}{5} \text{ of total sample titrated}\\ \downarrow}}{5}}{\underset{\substack{\uparrow\\ \text{sample wt. in mg}}}{2520.3} \times \underset{\substack{\uparrow\\ f_{Na_2CO_3}}}{1}} = 35.26\%\ Na_2CO_3$$

For the second titration, the calculation for the hydrogen carbonate percentage is based on the difference in the titration to the methyl orange end point and to the phenolphthalein end point. The calculation takes the following form:

$$\frac{\overbrace{(22.46 - 16.44)}^{\substack{\text{net ml of acid}\\ \text{(see text below)}}} \times \overset{\substack{F_{HCl}\\ \downarrow}}{0.1020} \times \overset{\substack{f_{HCl}\\ \downarrow}}{1} \times \overset{\substack{\text{formula wt. of } NaHCO_3\\ \downarrow}}{84.01} \times \overset{\substack{\text{conversion factor for \%}\\ \downarrow}}{100} \times \overset{\substack{\text{only } \frac{1}{10} \text{ of total sample taken}\\ \downarrow}}{10}}{\underset{\substack{\uparrow\\ \text{sample wt. in mg}}}{2520.3} \times \underset{\substack{\uparrow\\ f_{NaHCO_3}}}{1}}$$

$$= 20.5\%\ NaHCO_3$$

The acid required to reach the methyl orange end point in the second titration is consumed in three processes:

1. The acid converts all carbonate present to hydrogen carbonate ion. In the first titration it was established that for a 50.00-ml aliquot a volume of 16.44 ml of acid is required for this purpose; hence, for a 25.00-ml aliquot, a volume of $0.5 \times 16.44 = 8.22$ ml.
2. The acid converts the hydrogen carbonate ion thus formed to carbonic acid. In the 25-ml aliquot, obviously a volume of 8.22 ml of acid is required for this conversion of the hydrogen carbonate that was originally present as carbonate. In other words, to completely convert the carbonate present in a 25.00-ml aliquot to carbonic acid, a volume of $2 \times 8.22 = 16.44$ ml is required.
3. The acid converts the hydrogen carbonate initially present to carbonic acid, and for this purpose a volume of $22.46 - 16.44$ ml of acid is required for the 25.00-ml aliquot. This reasoning explains the factor labeled "net ml of acid" in the above calculation.

15.2 Indirect Analysis and Acid–Base Titrations

Indirect analysis, discussed in Section 7.3 for gravimetric methods, is also possible with titrimetric methods. The principles of calculation are similar to those previously outlined and are readily understood from an example.

Example 15-8. A mixture is known to contain only calcium carbonate (*100.09*) and magnesium carbonate (*84.32*). A 0.4000-g sample of this mixture is treated with 50.00 ml of 0.2000 F HCl, CO_2 is boiled off, and the excess of acid is back-titrated with 0.1000 F NaOH to the phenolphthalein end point, requiring 12.00 ml of base. Calculate the % $CaCO_3$ and % $MgCO_3$ in the mixture.

Let x and y represent the % $CaCO_3$ and % $MgCO_3$, respectively, in the mixture. The first equation of the required pair follows from the description of the mixture:

$$x + y = 100.00$$

The result of the titration of total carbonate content in the sample may be expressed in terms of $CaCO_3$:

$$\frac{(50.00 \times 0.2000 - 12.00 \times 0.1000) \times 100.09 \times 100}{400.0 \times 2}$$
$$= 110.10\% \text{ total carbonate expressed as } CaCO_3$$

The finding of a percentage over 100% is associated with the expression of the result in terms of the substance with the *higher* equivalent weight, $CaCO_3$. (The example can, of course, be solved alternatively by expressing the result as $MgCO_3$, an exercise left to the student.)

The second equation of the required pair is then based on the expression of $MgCO_3$ in terms of $CaCO_3$ and applying the titration result:

$$x + y\frac{CaCO_3}{MgCO_3} = \% \text{ total carbonate expressed as } CaCO_3$$
$$x + 1.1870y = 110.10$$

Solving the pair of simultaneous equations by subtracting one from the other yields

$$0.1870y = 10.10$$
$$y = 54.01 = 54.0\% \; MgCO_3$$
$$x = 100.00 - 54.0 = 46.0\% \; CaCO_3$$

Because of the limited accuracy of indirect analysis, it is best to report 54% $MgCO_3$ and 46% $CaCO_3$.

The discussion of the advantages and disadvantages of indirect analysis given in Section 7.3 is also applicable to titrations and needs no further elaboration. The student may, as an exercise and for the purpose of gaining a better understanding of the limitations of indirect analysis, recalculate Example 15-8 with the assumption that an error of +0.10 ml was made in establishing the volume of acid needed and an error of −0.10 ml in the base required.

15.3 Problems

15-1. A 0.8326-g sample of potassium hydrogen phthalate (*204.23*), KHP, is titrated with a NaOH solution, of which a volume of 33.33 ml is required. Calculate the formality of the base.

Answer: 0.1223 *F*

15-2. In order to standardize a HCl solution, a volume of 25.00 ml of the acid is titrated with 0.1035 *F* NaOH, of which a volume of 24.30 ml is required. What is the formality of the acid?

Answer: 0.1006 *F*

15-3. A 50.00-ml volume of a solution of Na_2CO_3 (*105.99*) is titrated to the methyl orange end point with 0.1000 *F* HCl. A volume of 8.20 ml of this acid is required. Calculate the concentration of the Na_2CO_3 solution expressed as milligrams per liter.

Answer: 869 mg/liter

15-4. A 10.00-ml sample of aqueous ammonia is titrated with 0.0488 *F* H_2SO_4, of which a volume of 7.35 ml is required. What amount, in milligrams, of nitrogen (*14.01*) is present in the volume of the base taken?

Answer: 10.0_5 mg of nitrogen

15-5. A 0.5000-g sample of a material containing $CaCO_3$ (*100.09*) as the only active ingredient is treated with 50.00 ml of 0.0936 *F* HNO_3. The excess acid is then back-titrated with 0.1045 *F* NaOH, of which a volume of 4.25 ml is required. What is the % $CaCO_3$ in the material?

Answer: 42.40%

15-6. A solution was prepared by mixing 25.00 ml of 0.0882 *F* H_3PO_4 and 20.00 ml of 0.1030 *F* NaH_2PO_4. What volume of 0.1130 *F* NaOH is required to titrate this solution (a) to the methyl orange end point and (b) to the phenolphthalein end point?

Answers: (a) 19.51 ml; (b) 57.26 ml

15-7. To 50.00 ml of a HCl solution, an excess of $AgNO_3$ solution is added. The AgCl (*143.32*) formed is separated, dried, and found to weigh 0.1205 g. What is the formality of the acid?

Answer: 0.01682 *F*

15-8. A 0.6030-g sample of a material containing Na_3PO_4 as the only active component is titrated to the methyl orange end point with 0.1035 *F* $HClO_4$, of which a volume of 12.24 ml is required. (a) What is the % P_2O_5 (*141.94*) in the material? (b) What volume of a H_2SO_4 solution of the same formality as the $HClO_4$ solution would be required for the same titration?

Answers: (a) 7.46%; (b) 6.12 ml

15-9. Pure $BaCO_3$ (*197.35*) is used for the standardization of a HCl solution in the following way. A 0.8320-g sample of the carbonate is dissolved in 50.00 ml of the acid. The solution is boiled and cooled. Then methyl orange indicator is added and the excess of acid back-titrated with a NaOH solution, of which a volume of 5.25 ml is required.

Then a 10.00-ml portion of the acid is added to the solution and titrated with the same base, of which now a volume of 12.05 ml is required. Calculate the formality of the HCl solution.

Answer: 0.1847 *F*

. A 0.3500-g sample of a material containing Na_2CO_3 (*105.99*) and $NaHCO_3$ (*84.01*) as the only reactive substances is dissolved in water, phenolphthalein is added, and a titration performed with 0.0955 *F* HCl, of which a volume of 25.40 ml is required. Methyl orange indicator is now added, the buret is refilled, and a further titration performed to the methyl orange end point, now requiring 34.60 ml of the acid. Calculate the % Na_2CO_3 and the % $NaHCO_3$ in the material.

Answer: 73.46% Na_2CO_3, 21.09% $NaHCO_3$

. A material is known to contain only $CaCO_3$ (*100.1*) and $MgCO_3$ (*84.3*). A 0.500-g sample of this material is dissolved in 60.00 ml of 0.200 *F* HCl, CO_2 is boiled off, and the excess of acid back-titrated with 0.100 *F* NaOH to the phenolphthalein end point, requiring 10.48 ml of the base. Calculate the % $CaCO_3$ and % $MgCO_3$ in the material.

Answer: 49% $CaCO_3$, 51% $MgCO_3$

. The weights of potassium hydrogen phthalate (*204.23*), KHP, given below required the listed volumes of NaOH solutions to reach the phenolphthalein end point. Calculate the formality of each of the NaOH solutions.

	Grams, KHP taken	ml, NaOH solution required
(a)	1.0000	45.64
(b)	1.1256	39.25
(c)	0.8567	28.75
(d)	0.8169	40.00
(e)	0.0679	8.96
(f)	1.1212	35.62

. The monoprotic weak base 2-amino-2-(hydroxymethyl)-1,3-propanediol (*121.14*), $(HOCH_2)_3 \cdot NH_2$, sometimes termed *tris*, is used as a standard for acid solutions (see Section 14.2). Calculate the formalities of the HCl solutions that are standardized by titration of the amounts of tris given below.

	Grams, tris	ml, HCl solution required
(a)	0.4592	32.67
(b)	0.5115	40.11
(c)	0.4179	38.67
(d)	0.3992	45.95

. A 1.0000-g sample of Na_2CO_3 (*105.99*) is reacted with 65.00 ml of HCl solution and the excess of HCl is back-titrated with a NaOH solution, of which 24.62 ml is required. In a separate titration, 25.00 ml of the HCl solution required 27.85 ml of the NaOH solution. Calculate the formalities of both the HCl and NaOH solutions.

15-15. An amount of 0.1523 g of calcium carbonate (*100.09*) of 99.50% w/w purity is treated with 50.00 ml of a nitric acid solution and the excess acid is back-titrated, 18.03 ml of a barium hydroxide solution being required. A volume of 24.34 ml of the nitric acid is equivalent to 29.94 ml of the base. Calculate the formalities of the acid and the base.

15-16. What volume of 0.1000 *F* HCl will be required to titrate 0.2000 g of a material that contains 34.2% w/w $Ba(OH)_2$ (*171.35*) and 65.8% w/w NaOH (*40.00*)?

15-17. What volume of 0.2000 *F* HCl is required to titrate 0.4000 g of a mixture containing 30.0% w/w NaOH (*40.00*) and 53.0% w/w Na_2CO_3 (*106.0*) to the phenolphthalein end point?

15-18. A volume of 20.00 ml of a solution containing 1.600 g of NaOH (*40.00*) per 100 ml and a volume of 50.00 ml of a solution containing 0.530 g of Na_2CO_3 (*106.0*) per 100 ml are mixed in a 250-ml volumetric flask and diluted to mark with pure water. A one-fifth aliquot of the solution is titrated to the methyl orange end point with 0.1000 *F* HCl. What volume of HCl solution is required?

15-19. What volume of 0.1000 *F* HCl will be required to titrate 200.0 mg of a material, containing 60.0% w/w $NaHCO_3$ (*84.01*) and 40.0% w/w Na_2CO_3 (*106.0*), to the phenolphthalein end point?

15-20. A material contains 35.0% w/w $NaHCO_3$ (*84.01*), 45.0% w/w Na_2CO_3 (*106.0*), and inert material. What volume of 0.1500 *F* HCl will be required to titrate a 1.000-g sample to the methyl orange end point?

15-21. A mixture contains 60.0% w/w $NaHCO_3$ (*84.01*) and 40.0% w/w Na_2CO_3 (*106.0*). What volume of 0.1200 *F* HCl will be required to titrate a 1.000-g sample to the methyl orange end point?

15-22. A material contains only NaOH (*40.00*) and Na_2CO_3 (*106.0*). An amount of 0.4240 g of the material requires 93.00 ml of 0.1000 *F* HCl to reach the methyl red end point. Calculate the % w/w composition of the material.

15-23. A mixture is known to contain only NaOH (*40.00*), Na_2CO_3 (*106.0*), and inert material. An amount of 0.5000 g is titrated with 0.1230 *F* HCl. A volume of 28.32 ml is required to reach the phenolphthalein end point and an additional volume of 22.04 ml to reach the methyl orange end point. What is the % w/w composition of the mixture?

15-24. A mixture contains only $CaCO_3$ (*100.09*) and CaO (*56.08*). A 0.5000-g sample is dissolved in 118.52 ml of 0.1000 *F* HCl and the excess of acid back-titrated to the methyl orange end point, 4.26 ml of 0.2000 *F* NaOH being required. Calculate the % w/w composition of the mixture.

15-25. An amount of 0.5000 g of a material containing only $CaCO_3$ (*100.1*), $MgCO_3$ (*84.32*), and some acid-insoluble material is dissolved in 80.00 ml of 0.1000 *F* HCl. The excess of the acid is back-titrated, 7.50 ml of 0.2000 F NaOH being required. The insoluble material is separated by filtration, washed, dried, and found to weigh 195.0 mg. Calculate the % w/w composition of the material.

6. An amount of 1.000 g of a mixture containing Na_2CO_3 (*106.0*), $NaHCO_3$ (*84.01*), and inert material requires 74.12 ml of 0.1500 F HCl to reach the methyl orange end point. A second 1.000-g sample is ignited to 550°C, and a loss of 110.7 mg in weight is observed. Calculate the % w/w composition of the mixture.

7. A solution contains both H_3PO_4 (*98.00*) and H_2SO_4 (*98.08*). When a 25.00-ml portion of the solution is titrated to the methyl orange end point, a volume of 38.32 ml of 0.1000 F NaOH is required. A 10.00-ml portion requires 12.21 ml of the same NaOH solution to reach the phenolphthalein end point. Calculate the grams per liter of each of the two acids present in the solution.

8. In forensic analysis the carbon monoxide content of blood can be determined as follows: In a dry test tube are mixed 2.00 ml of blood (having a few crystals of sodium fluoride added to prevent coagulation) and 1.00 ml of water. Exactly 2 ml of this mixture is pipetted into a Conway chamber. This small device has two separate compartments and when it is closed a gaseous constituent can diffuse from one compartment to the other. The blood preparation and 0.25 ml of 2 F sulfuric acid are placed in one compartment; in the other is placed 1.00 ml of 0.01 F palladium chloride. The chamber is closed and allowed to stand in the dark for 3 hours. The carbon monoxide diffuses into the latter solution and the following reaction takes place:

$$Pd^{2+} + H_2O + CO \rightarrow Pd + 2H^+ + CO_2$$

The chamber is then opened and the acid in the palladium solution is titrated with 0.0200 F sodium hydroxide. A blank is performed under identical conditions but with 2.00 ml of water placed into the Conway chamber instead of the blood preparation. The following formula is used to calculate the result:

$$(\text{ml}_{\text{titrant in titration}} - \text{ml}_{\text{titrant in blank}}) \times (\text{factor})$$
$$= \text{vol. \% CO} \qquad \text{(gas volume at standard conditions)}$$

Calculate the numerical value of this *factor* if the determination is performed exactly as described above. Carbon monoxide has a formula weight of 28.0 and is assumed to behave as an ideal gas with a mole occupying 22.4 liters at exactly 0°C and 1 atm pressure.

Answer: 16.8

16 ACID–BASE TITRATIONS IN NONAQUEOUS MEDIA

Introduction 16.1

So far the discussion of acid–base phenomena has been essentially restricted to aqueous media. However, acid–base reactions, that is, proton-transfer reactions, take place in other solvents as well. Liquid ammonia is an amphiprotic solvent, that is, it can function as either an acid or a base, and undergoes the following autoprotolytic reaction:

$$\underset{acid_1}{NH_3} + \underset{base_2}{NH_3} \rightleftharpoons \underset{base_1}{NH_2^-} + \underset{acid_2}{NH_4^+} \qquad (16\text{-}1)$$

This scheme is completely analogous to the autoprotolysis of water. The ions formed are the amide ion and the ammonium ion. If an acid, HA, is dissolved in liquid ammonia, the protolysis reaction taking place is

$$HA + NH_3 \rightleftharpoons A^- + NH_4^+ \qquad (16\text{-}2)$$

Ammonium ion is the strongest acid that can exist in liquid ammonia, as the hydronium ion, that is, the hydrated hydrogen ion, is in water. Because of the experimental difficulties of working in liquid ammonia, the system has not found application in analytical practice. But this fact does not detract from the importance of the analogy with water. Organic derivatives of ammonia, that is, amines, such as butylamine, $C_4H_9NH_2$, do find such application.

Another system, which is of considerable significance, is water-free acetic acid. Its autoprotolysis proceeds according to

$$CH_3COOH + CH_3COOH \rightleftharpoons CH_3COO^- + CH_3COOH_2^+ \qquad (16\text{-}3)$$

The acetate and acetonium ions formed are fully analogous to the hydroxide and the hydronium ions in water. If an acid, for example, perchloric acid, is dissolved in water-free acetic acid, the following protolytic reaction occurs:

$$HClO_4 + CH_3COOH \rightleftharpoons ClO_4^- + CH_3COOH_2^+ \qquad (16\text{-}4)$$

Note that in each of the three amphiprotic solvents considered above, the "active" acid species is the solvated proton.

The "dissociation" of an acid does not necessarily require an amphiprotic solvent. Such a reaction can take place in so-called aprotic or nonprotonic solvents that offer a pair of nonbonding electrons. 1,4-Dioxane is such a solvent:

$$HClO_4 + :O\begin{matrix} CH_2\!-\!CH_2 \\ CH_2\!-\!CH_2 \end{matrix}O: \rightleftharpoons \left[:O\begin{matrix} CH_2\!-\!CH_2 \\ CH_2\!-\!CH_2 \end{matrix}O:H^+, ClO_4^-\right] \qquad (16\text{-}5)$$

Dioxane has no proton to donate, but a pair of nonbonding electrons can accept a proton. Dioxane has a quite low dielectric constant and therefore does not support the separation of opposite charges. As a consequence, the dioxonium cation and the perchlorate anion by electrostatic attraction remain close together as an entity known as an ion pair. It should be appreciated that in the ion pair the transfer of the proton is complete and that the cation and anion with their separate charges persist as such. The formation of an ion pair also can occur in amphiprotic solvents, such as acetic acid, but to a lesser degree.

16.2 Leveling and Differentiating Effects

As was pointed out in Section 8.1, two factors govern the extent to which the protolysis of an acid proceeds: First, the "intrinsic" strength of the acid, that is, its inherent tendency to donate a proton. Second, the basicity of the solvent, that is, its tendency to accept a proton. If the basicity of the solvent is high, then acids with even relatively low intrinsic strengths are able to donate their protons fully to the solvent and consequently to behave as strong acids in that solvent. Water as a solvent has relatively high basicity, and perchloric acid, sulfuric acid (first step), hydrochloric acid, and nitric acid all have sufficient intrinsic strength to donate their protons completely to the water. Consequently, these four acids behave as strong acids in water and their strength is leveled to that of the hydronium ion, which is the strongest acid that can exist in water. This is known as the leveling effect, and water may be described as a leveling solvent for these acids.

Water-free acetic acid is a far less basic solvent than water; that is, its tendency to accept a proton is smaller. Nitric acid has considerably less intrinsic strength than perchloric acid, and its protolysis is not complete in acetic acid. Consequently, nitric acid in this solvent does not behave as a strong acid, but perchloric acid does. In acetic acid the strength of the acids listed above decreases in the order in which they are named. Because acetic acid shows this differentiating effect, it may be described as a differentiating solvent for these acids.

Analogous reasoning holds for other solvents and for bases. It is the leveling effect of water that prevents reaction (16-3) from taking place in aqueous media. Acetonium ion is a very strong acid, but it is leveled in water to the strength of the hydronium ion. Analogously amide ion is a very strong base, which in water is leveled to the strength of the hydroxide ion.

Nonaqueous Acid–Base Titrimetry 16.3

From the preceding discussion it can be appreciated that proton-transfer reactions in nonaqueous media can be made the basis of titrimetric processes and that such an expansion of acid–base titrimetry not only should supplement titrations in aqueous media but offers distinctive advantages in some cases, including the following:

1. Acids or bases too weak to be titrable in water may be stronger in an appropriate nonaqueous solvent and may become titrable in that medium. For example, acetate has too low an intrinsic basicity for its protolysis with hydronium ion to go to completion. Therefore, in water acetate is too weak a base to be readily titrated with acid. In contrast, in the more acidic solvent, water-free acetic acid, acetate accepts protons more completely from the much stronger acid acetonium ion, consequently acts as a stronger base, and can be titrated to a sharp end point, for example, with a solution of perchloric acid in acetic acid. Many other bases (per se or as the salts of weak acids) that cannot be titrated in water can be determined in acetic acid or other acidic solvents.
2. Acids or bases identical in strength in water can often be differentiated in a nonaqueous solvent of appropriate acidity or basicity. For example, perchloric acid and hydrochloric acid, due to the leveling effect of water, are both strong acids in that medium; thus the titration curve for a mixture of these two acids shows only a single break. However, in water-free acetic acid, which acts as a differentiating solvent, the mixture can be titrated with two breaks, using a

strong base, for example, acetate ion (as sodium acetate) as the titrant.

3. Substances that are insoluble in water may be titrated in a solvent in which they are soluble. For example, many water-insoluble organic compounds can thereby be subjected to simple titrimetric procedures.

Nonaqueous titrimetry has found special interest for the determination of pharmaceuticals. With the selection of suitable differentiating solvents, mixtures of several compounds can be often analyzed titrimetrically, avoiding involved and tedious separations.

Common solvents, in nonaqueous acid–base titrimetry, in addition to acetic acid and dioxane, include acetonitrile, dimethylformamide, and the so-called "G-H" solvent mixtures. These mixtures contain a glycol (usually ethylene glycol) and a hydrocarbon solvent (such as a hydrocarbon, halocarbon, or an alcohol), commonly in a 1:1 proportion.

The titration process proper in nonaqueous media is not significantly different from that in aqueous media. The end point can be detected visually in exactly the same manner—by employing organic dyes that show acid–base properties and that change color when transformed from the acid form to the conjugate base form. The dyes methyl violet and crystal violet find frequent use as indicators in nonaqueous media. Potentiometric end-point detection with a glass electrode is possible in solvent systems having a sufficiently high dielectric constant (see Section 27.3).

The exclusion of water is in many cases a critical factor for the success of a nonaqueous titration. Water, depending on the solvent, may act as an acid or base and then, of course, can upset the situation. The removal of water from acetic acid media is readily achieved by the addition of acetic anhydride, which undergoes the following reaction:

$$(CH_3CO)_2O + H_2O \rightarrow 2CH_3COOH \qquad (16\text{-}6)$$

In many cases, special precautions are necessary to avoid errors due to the high volatility and the large thermal expansion coefficient of the organic solvent. The standard solution may increase significantly in its concentration if losses of solvent by evaporation are not prevented. Temperature changes may cause significant changes in weight–volume concentration.

Lewis Acid–Base Concepts 16-4.

Many organic reactions that are catalyzed by Brønsted acids are also catalyzed by species that have no protons that can be donated. Thus it seems that acid–base properties are not restricted to the transfer of protons. Lewis, in 1923, proposed a concept of acids and bases that is more general than that of Brønsted. Consider as an example the base ammonia. Its acceptance of a proton in terms of electronic structure may be written

$$\begin{matrix} & \text{H} & \\ \text{H}: & \ddot{\text{N}}: \\ & \ddot{\text{H}} & \end{matrix} + \text{H}^+ \rightarrow \left[\begin{matrix} & \text{H} & \\ \text{H}: & \ddot{\text{N}} & :\text{H} \\ & \ddot{\text{H}} & \end{matrix}\right]^+ \qquad (16\text{-}7)$$

The essential fact is that ammonia, or any Brønsted base, shares its nonbonding pair of electrons with the proton it accepts [compare the analogous situation with dioxane, equation (16-5)]. The extension of the concept of a base effected by Lewis consists in removing the restriction that the sharing of a pair of electrons be with a proton. The sharing may be with a proton but equally well with another species. An analog to the above reaction may be written with the copper(II) ion replacing the proton:

$$\begin{matrix} & \text{H} & \\ \text{H}: & \ddot{\text{N}}: \\ & \ddot{\text{H}} & \end{matrix} + \text{Cu}^{2+} \rightarrow \left[\begin{matrix} & \text{H} & \\ \text{H}: & \ddot{\text{N}} & :\text{Cu} \\ & \ddot{\text{H}} & \end{matrix}\right]^{2+} \qquad (16\text{-}8)$$

This reaction is commonly termed complex formation but, by the Lewis theory, is viewed only as one type of acid–base reaction. According to this theory the formation of any coordinate covalent bond is an acid–base reaction, a neutralization. Since those species that have an unshared pair of electrons can combine with a proton as well as with other electron acceptors, the number of bases according to Lewis is not greatly altered. However, the number of substances that are considered to be acids is considerably increased. Any substance that can accept a pair of electrons is termed a Lewis acid. Thus, for example, boron trifluoride is a Lewis acid that reacts with ammonia as follows:

$$\begin{matrix} & \text{F} \\ & | \\ \text{F}- & \text{B} \\ & | \\ & \text{F} \end{matrix} + \begin{matrix} & \text{H} & \\ & | & \\ : & \text{N} & -\text{H} \\ & | & \\ & \text{H} & \end{matrix} \rightarrow \begin{matrix} & \text{F} & & \text{H} & \\ & | & & | & \\ \text{F}- & \text{B} & : & \text{N} & -\text{H} \\ & | & & | & \\ & \text{F} & & \text{H} & \end{matrix} \qquad (16\text{-}9)$$

Lewis concepts are rarely employed in theoretical considerations of inorganic analysis because almost all relevant phenomena can be treated satisfactorily either in terms of the Brønsted concepts or as complex formation (Chapter 20). Of the few exceptions, acid–base phenomena in fused media may be mentioned; these are briefly considered in Chapter 51.

16.5 Questions

16-1. Speculate on the reasons perchloric acid in a suitable solvent is the most widely applied titrant solution in nonaqueous titrimetry.

16-2. Differentiating solvents are also called nonleveling solvents. Is this term appropriate? Explain.

16-3. What is meant by the statement "water levels $HClO_4$ and HCl"?

16-4. What are the properties required by a solvent to be designated as "differentiating"?

16-5. Why is it necessary with some solvents to exclude water strictly when titrations are performed?

16-6. What is the strongest acid that can exist in liquid ammonia?

16-7. What is a "G-H" solvent?

16-8. Explain why many organic substances that are excellent indicators in aqueous media are unsuitable as indicators for titrations in organic solvent?

16-9. Substances A, B, and C are assumed to behave in water and the nonaqueous solvents 1 and 2 as indicated by the table below.

	Water	Solvent 1	Solvent 2
A	$K_b = 10^{-12}$	Medium strong base	Very weak base
B	$K_b = 10^{-4}$	Very weak base (B and C)	Medium strong base
C	$K_b = 10^{-10}$		Strong base

Elaborate on a scheme to determine these substances in their admixture by sole application of titrimetry.

16-10. To titrate a weak base successfully, the solvent should, at most, have only weakly basic properties. Explain.

17 REVIEW QUESTIONS AND PROBLEMS ON ACID–BASE TITRIMETRY

Questions 17.1

Is an aqueous solution "neutral" to methyl orange, actually neutral in the sense of having equal concentrations of hydrogen ion and hydroxide ion?

Elaborate on the statement: The equivalent weight of a substance is independent of the titration reaction.

How could you standardize a hydrochloric acid solution gravimetrically?

What would be the effect on the result if pH 7.0 were taken as the end point in the titration of a weak base with a strong acid?

What anionic species will predominate in a solution obtained by dissolving 5 millimoles of Na_2HPO_4 and 3 millimoles of Na_3PO_4 in water?

What points on an acid–base titration curve should be calculated using approximate formulas to permit a rough, rapid judgment of the feasibility of the titration?

Could the sodium salt of a very weak acid ($K_a = 10^{-11}$) be used reasonably well as a titrant in the titration of a strong acid?

What is stated by the electroneutrality rule and what are the possible applications of the "charge balance"?

A very weak monoprotic organic acid ($K_a = 1 \times 10^{-12}$) forms a water-insoluble silver salt. How might an acid–base titration of this acid be achieved?

Discuss the interrelation of formality, molarity, and normality of a titrant. Elaborate on the meaning of equivalence and equivalent weight in acid–base titrimetry.

Differentiate between the concepts of the equivalence point and the end point in broad terms and in relation to acid–base titrimetry. What is the most desirable relation between these two points?

Why are weak acids and bases hardly ever used as titrants in acid base titrimetry?

On the basis of your knowledge of the color transition interval of an acid–base indicator, explain why a factor of at least 10^3 must separate the dissociation constants of a diprotic acid so that its visual titration to two distinct end points is possible.

How many useful visual end points can be secured in the titration of a triprotic acid with the successive acidity constants 1×10^{-4},

1×10^{-6}, and 1×10^{-12}? In practice to what pH should the titration be performed in an aqueous medium? *Answer:* pH 9.0. Explain.

17-15. Give reasons why calcium carbonate, although readily obtainable in a highly pure form, is not a very suitable standard for the standardization of an acid.

17-16. What is temporary water hardness? How is it determined? What is its significance for industry?

17-17. Explain why perchloric acid is preferred as the titrant in many acid–base titrations in nonaqueous media.

17-18. List some errors that may be encountered in an acid–base titration procedure. Which errors may be related to the selection of the indicator and its functioning?

17-19. Discuss the functioning of an acid–base indicator.

17-20. A carbonate-containing sodium hydroxide solution was standardized using phenolphthalein as the indicator. Explain why incorrect results are obtained if this solution is then used in a titration in which methyl orange serves as the indicator. Would the application of an indicator blank remedy the situation?

17-21. In a titration curve of an acid or a base, what can be concluded from the presence of (a) a nearly horizontal portion and (b) a nearly vertical portion?

17-22. Chloride ion is a base. Why does this base, for example, added in the form of NaCl, not interfere with an acid–base titration?

17-23. A weighed sample containing only two components is given. Compare the indirect analysis of the mixture via gravimetry and by acid–base titrimetry. (For example, what parameters must be considered and what are the optimum conditions?)

17-24. Discuss the statement: The titration of substance A can never yield a correct result since the pH at the end point does not coincide with that at the equivalence point.

17-25. Predict how an aqueous solution of sodium carbonate will react to phenolphthalein. When this solution is treated with an excess of barium chloride and the precipitate separated by filtration, how will the filtrate react toward the same indicator? What will be the result if a solution of sodium carbonate is treated with excess barium chloride and then titrated with hydrochloric acid to the methyl orange end point (a) with and (b) without removal of the precipitate? Write equations for all the reactions taking place.

17-26. The equivalence point in an otherwise satisfactory titration of an acid falls within the range of the high buffering action of another system which is present in the aqueous solution. Consider the feasibility of the titration under this condition.

17-27. In the determination of the carbonate content of calcium carbonate by back-titration of an added excess amount of standard acid, explain why it is advisable to boil the solution to expel carbon dioxide even when methyl orange is used as the indicator. Could phenolphthalein be substituted as the indicator in this back-titration? State how the titration results with each indicator would be influenced if the carbon dioxide were not expelled fully.

In an aqueous solution of H_2SO_4 also containing $NaNO_3$ and KCl, the sulfate content is determined both gravimetrically and by an acid–base titration. Compare and discuss the two methods with regard to (a) speed, (b) the possible influence of the neutral salts, and (c) how the use of a balance enters each determination. State what reagents are required for each of the methods and suggest how they might be prepared. Which set of reagents can be prepared more rapidly? List the laboratory equipment required for the two methods.

Problems 17.2

The monoequivalent base 2-amino-2-(hydroxymethyl)-1,3-propanediol (i.e., tris) (*121.14*) is a primary standard for acids. In a standardization 306.2 mg of this base required 25.32 ml of a H_2SO_4 solution. What is the formality of this solution?

Answer: 0.04991 *F*

In the titration of 50.00 ml of a solution of H_3PO_4 (*98.00*) with 0.1026 *F* NaOH to the "second" end point (thymol blue), a volume of 48.30 ml of the base is required. (a) If the titration of the same volume of the H_3PO_4 solution were performed to the "first" end point, what volume of base would be required? (b) What is the formality of the H_3PO_4 solution? (c) How many grams of H_3PO_4 is present in 50.00 ml of the solution? (d) How many grams of silver ion (*107.87*) is equivalent to the phosphate present in 50.00 ml of the solution if precipitated as Ag_3PO_4?

Answers: (a) 24.15 ml; (b) 0.04956 *F*; (c) 0.2428 g; (d) 0.8018 g

In water the pK_a of HCl has a value of about −3. Calculate K_a for HCl and K_b for its conjugate base, chloride ion.

Answer: $K_a \simeq 10^3$, $K_b \simeq 10^{-17}$

A volume of 50.00 ml of an NH_4NO_3 (*80.04*) solution is treated with concentrated NaOH solution, and the ammonia is distilled into 20.00 ml of an HCl solution. The excess of HCl is back-titrated with 0.1010 *F* NaOH, of which a volume of 6.22 ml is required. In a separate titration a volume of 28.66 ml of the HCl solution is found to be equivalent to 24.50 ml of the standard base. What weight of NH_4NO_3 is present in the sample taken?

Answer: 87.9 mg

In Problem 17-4 what volume of the acid would be required to exactly neutralize the total ammonia distilled if the nitrate ion in another 50.00-ml portion of the solution were reduced to ammonia by the action of Devarda's alloy?

Answer: 25.4 ml

17.2 PROBLEMS

17-6. A 1.000-g sample of a uniform mixture known to contain only Na_2CO_3 (*106.0*) and $NaHCO_3$ (*84.0*) is titrated with 0.2000 F HCl to the methyl orange end point, requiring 73.45 ml of the acid. Calculate the percentage composition of the mixture. (*Hint:* When formulating the simultaneous equations remember that two HCO_3^- ions are equivalent to one CO_3^{2-} ion!)

Answer: 40% Na_2CO_3, 60% $NaHCO_3$

17-7. To a 0.1232-g sample of MgO (*40.32*) known to be of 95.66% purity (the remainder being inert material) is added 50.00 ml of a HNO_3 solution. The excess of the acid is back-titrated with 0.0987 F NaOH, of which a volume of 14.40 ml is required. Calculate the formality of the HNO_3 solution.

Answer: 0.1453 F

17-8. A volume of 50.00 ml of a HCl solution is treated with $AgNO_3$ in excess and the precipitate is separated by filtration and dried. The weight of the dried AgCl (*143.32*) obtained is 0.7225 g. Calculate the formality of the HCl solution.

Answer: 0.1008 F

17-9. A material contains 65.0% K_2HPO_4 (*174.2*) and 35.0% Na_3PO_4 (*164.0*). How many milliliters of 0.1000 F HCl will be required to titrate 0.7000 g of this material to the methyl orange end point?

Answer: 56.0 ml

17-10. A volume of 100.0 ml of a 0.1000 F solution of a triprotic acid ($K_{a,1} = 1.0 \times 10^{-4}$, $K_{a,2} = 1.0 \times 10^{-6}$, $K_{a,3} = 3.0 \times 10^{-8}$) is titrated with 0.1000 F NaOH. (a) Calculate the pH value of the three equivalence points (use the proper solution volume in calculating C_{salt}!). (b) Which of these equivalence points would be used in practice and what volume of base would be required to reach that point? (c) When the solution is titrated as in (b) and then 150 ml of 0.100 F HCl is added, what will be the pH of the resulting solution?

Answers: (a) pH 5.00, 6.76, 9.96; (b) the last one, 300 ml; (c) pH 6.00

17-11. A solution is obtained by mixing 20.00 ml of 0.1000 F NaH_2PO_4 and 30.00 ml of 0.0500 F Na_2HPO_4. (a) How many milliliters of 0.1000 F NaOH will be required to titrate this solution to the second equivalence point of H_3PO_4? (b) How many milliliters of 0.1000 F HCl will be required to bring the solution resulting from this titration to the pH of the first equivalence point of H_3PO_4?

Answers: (a) 20.00 ml; (b) 35.00 ml

17-12. How many milliliters of 0.10 F $Ba(OH)_2$ must be added to 50 ml of 0.10 F formic acid ($K_a = 1.8 \times 10^{-4}$) to attain pH 3.30? Would the resulting solution exert a reasonable buffering action?

Answer: 6.6 ml; yes

17-13. A volume of 20.00 ml of a solution containing 1.600 g of NaOH (*40.00*) per 100 ml and a volume of 50.00 ml of a solution containing 0.530 g of Na_2CO_3 (*106.0*) per 100 ml are mixed in a 250-ml volumetric flask and diluted to mark with pure water. A one-fifth aliquot of the solution is titrated to the methyl orange end point with 0.1000 *F* HCl. What volume of HCl solution is required?

An amount of 0.1523 g of calcium carbonate (*100.09*) of 99.50% w/w purity is treated with 50.00 ml of a nitric acid solution and the excess of acid is back-titrated, 18.03 ml of a barium hydroxide solution being required. A volume of 24.34 ml of the nitric acid is equivalent to 29.94 ml of the base. Calculate the formalities of the acid and the base.

When Rochelle salt, $KNaC_4H_4O_6 \cdot 4H_2O$, is ignited, $KNaCO_3$ is formed. A 0.8050-g sample of Rochelle salt is ignited, the residue dissolved in pure water, and titrated with 0.0635 *F* H_2SO_4 to the methyl orange end point. What volume of titrant will be required if the original material was 98.55% w/w pure?

A solution is known to have been prepared by dissolving H_3PO_4 (*98.00*) and Na_3PO_4 (*163.94*) in pure water. This solution requires 40.00 ml of 0.100 *F* HCl in the titration to the methyl orange end point. Then a neutral solution of $CaCl_2$ is added and the solution titrated to the phenolphthalein end point, 100.0 ml of 0.100 *F* NaOH being required. Calculate the milligrams of H_3PO_4 and Na_3PO_4 present in the original solution.

A solution contains both H_3PO_4 (*98.00*) and H_2SO_4 (*98.08*). When a 25.00-ml portion of the solution is titrated to the methyl orange end point, a volume of 38.32 ml of 0.1000 *F* NaOH is required. A 10.00-ml portion requires 12.21 ml of the same NaOH solution to reach the phenolphthalein end point. Calculate the grams per liter of each of the two acids present in the solution.

The saponification value of an oil or fat is the number of milligrams of KOH (*56.1*) consumed per gram of sample in its saponification. A 2.600-g sample of a fat is refluxed with 50.00 ml of a 0.320 *F* alcoholic KOH solution. After the reaction is complete, the excess of base in the cooled solution is titrated with 0.4200 *F* HCl, 15.20 ml being required. Calculate the saponification value of the fat.

An amount of 0.0821 g of a material is dissolved, the solution made slightly acidic, and potassium precipitated as the tetraphenylborate salt. The precipitate is separated by filtration and dissolved on the filter by acetone, and the resulting solution collected in a platinum crucible. After evaporation of the acetone, the residue is ignited to potassium metaborate, KBO_2. The KBO_2 is dissolved in water and titrated with 0.01122 *F* HCl (HBO_2 is a very weak acid!), of which a volume of 24.07 ml is required to reach the methyl red end point. Calculate the % w/w of K_2O (*94.20*) in the sample.

18 PRECIPITATION TITRATIONS

General 18.1

In a precipitation titration, the species to be determined forms a sparingly soluble compound with the titrant. Although it might appear that countless precipitation reactions could be made the basis of a titration, requirements must be met that seriously limit the number. The reaction must be sufficiently rapid and complete, lead to a product of reproducible composition and of low solubility, and a method must exist to locate the end point. Some difficulties in meeting these requirements may be noted. Precipitation frequently proceeds slowly or starts only after some time. Many precipitates show a tendency to adsorb and thereby coprecipitate the species being titrated or the titrant. The indicator may also be adsorbed on the precipitate formed during the titration and thereby be unable to function appropriately at the end point. If a precipitate is highly colored, visual detection of a color change at the end point may be impossible, but even where a white precipitate is formed, the milky solution may still make location of the end point troublesome.

Because of these difficulties, precipitation titrations receive only limited attention in present-day routine analysis. The major areas of application include the determination of silver or a halide (or pseudohalide) by the precipitation of the silver salt (that is, so-called argentometric titrations) and the determination of sulfate by precipitation as barium sulfate. The considerations here are limited largely to these applications, but the principles are basic to all precipitation titrations.

Titration Curve for Precipitation Titration 18.2

The titration curve for a precipitation is similar to that for the titration of a strong acid with a strong base. As can be seen from the following example, the reasoning involved and the calculations necessary are analogous and straightforward, pH and K_w being replaced by pX and K_{sp}, respectively, where X is the ion titrated and K_{sp} is the solubility product of the compound precipitated.

18.2 TITRATION CURVE, PRECIPITATION TITRATION

Assume a 100.0-ml portion of a 0.100 F sodium chloride solution is titrated with 0.100 F silver nitrate solution. During the titration, silver chloride is precipitated, the solubility product of which for simplicity is taken as $K_{sp} = [Ag^-][Cl^-] = 1.0 \times 10^{-10}$. The titration curve, defined in Section 10.1, is obtained by plotting the negative logarithm of the molar concentration of the species being titrated, Cl^-, versus the milliliters of the titrant solution added.

The value of pCl at the start of the titration is readily calculated: $pCl = -\log[Cl^-] = -\log(1.0 \times 10^{-1}) = 1.00$. After the addition of, for example, 5.00 ml of the silver nitrate solution, the calculation of the pCl value takes the following form. Initially the chloride ion in solution amounted to $100 \times 0.100 = 10.0$ millimoles. The 5.00-ml portion of the titrant introduces $5.00 \times 0.100 = 0.5$ millimole of silver ion and removes an equivalent amount of chloride ion from the solution as insoluble silver chloride. Consequently, the number of millimoles of chloride ion remaining in solution is $10.0 - 0.5 = 9.5$. The total solution volume is $100 + 5 = 105$ ml. Thus, the molar concentration of chloride ion is 9.5/105, and $pCl = -\log(9.5/105) = \log 105 - \log 9.5 = 2.02 - 0.98 = 1.04$.

In this same manner, further points of the titration curve up to the vicinity of the equivalence point may be calculated. At the equivalence point this mode of calculation fails since it yields the value $[Cl^-] = 0$, which is impossible. However, at this point silver and chloride ions are present in exactly equivalent amounts and the titration corresponds to a saturated solution of silver chloride. From the solubility product of silver chloride it follows that

$$[Cl^-] = [Ag^+] = \sqrt{1.0 \times 10^{-10}} = 1.0 \times 10^{-5}$$

and

$$pCl = 5.00.$$

Beyond the vicinity of the equivalence point, the following reasoning is applied. The silver ion contributed by the saturated solution of silver chloride is neglected and the concentration of silver ion is assumed to depend solely on the amount of silver nitrate added in excess. Then the chloride concentration is obtained via the solubility-product expression. Assume that a total of 110 ml of silver nitrate solution is added. A volume of 100 ml is required to reach the equivalence point. The excess of 10 ml corresponds to $10 \times 0.100 = 1.0$ millimole of silver, and this is present in a volume of 210 ml. Hence, the molar concentration of silver ion, $[Ag^+]$, is 1.00/210. The molar concentration of chloride is then calculated from the solubility-product relationship. $[Cl^-] = 1.0 \times 10^{-10}/[Ag^+] = 1.0 \times 10^{-10}/(1.00/210) = 2.10 \times 10^{-8}$ *M*, and $pCl = -\log(2.1 \times 10^{-8}) = 7.68$. The curve, with further additions of silver nitrate, approaches asymptotically a pCl value of 9.00, corre-

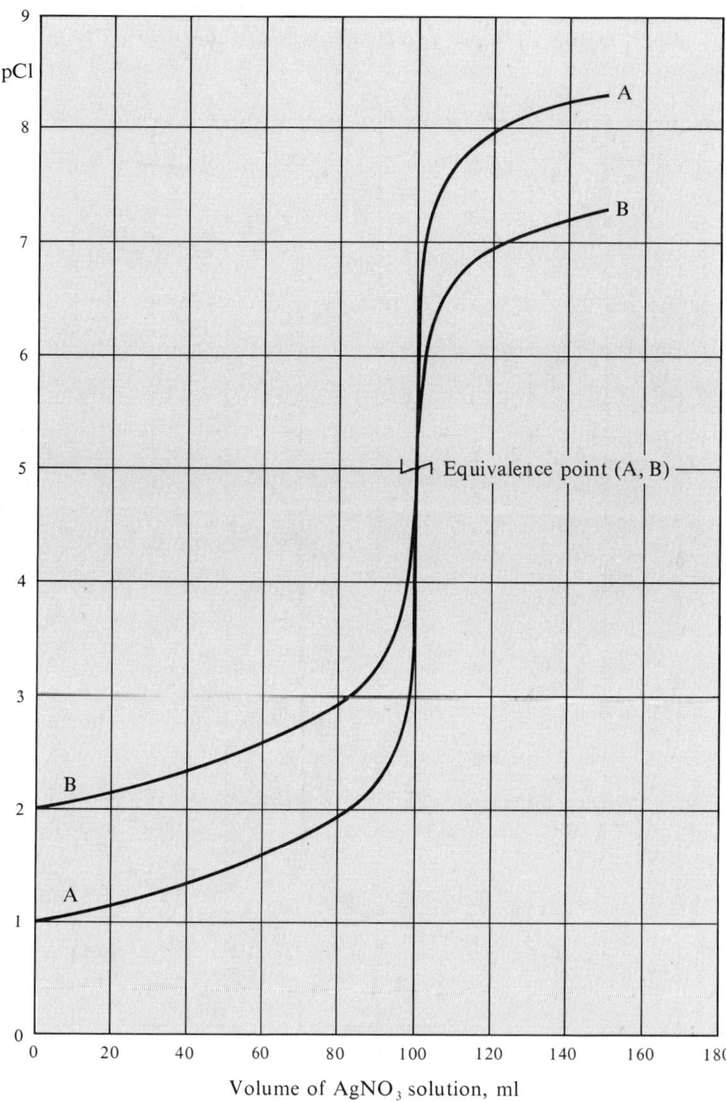

Figure 18-1. Theoretical titration curves for the titration of NaCl with $AgNO_3$. Curve A: 100 ml of 0.100 *F* NaCl titrated with 0.100 *F* $AgNO_3$. Curve B: 100 ml of 0.0100 *F* NaCl titrated with 0.0100 *F* $AgNO_3$.

sponding to saturating the 0.1 F titrant solution with silver chloride.

The complete titration curve, thus calculated, is shown in Figure 18-1 as curve A. The figure also shows (curve B) the curve for the titration of 100 ml of a 0.010 F sodium chloride solution with a 0.010 F silver nitrate solution. This curve starts at a pCl value of 2.00 and approaches asymptotically a pCl

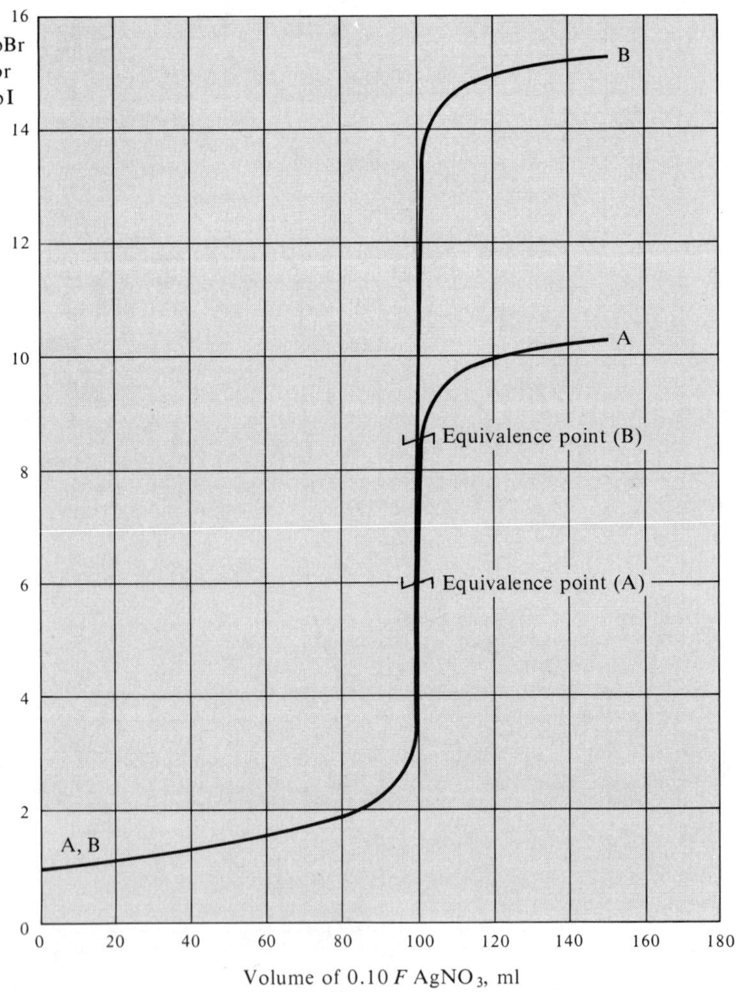

Figure 18-2. Theoretical titration curves for the titration of NaBr and NaI with $AgNO_3$. Curve A: 100 ml of 0.10 F NaBr titrated with 0.10 F $AgNO_3$. Curve B: 100 ml of 0.10 F NaI titrated with 0.10 F $AgNO_3$.

value of 8.00. This curve, as well as those for titrations involving the chloride–silver system at any other concentrations, passes at the equivalence point through the value pCl = 5.00. The reverse titration, that is, the titration of a silver ion solution with chloride ion as the titrant, can also be performed; the plot will be identical except that pAg is plotted on the ordinate. The shape of the titration curve is not peculiar to the example selected but is characteristic of all precipitation titrations in which the reactants combine in a 1:1 ratio. The extent of the "break" around the equivalence point increases as the concentrations of the titrated species and of the titrant increase. Further, the break is the greater, the lower the solubility of the compound precipitated. Thus, in the respective titrations of bromide and iodide ion with silver nitrate to form silver bromide ($K_{sp} = 1 \times 10^{-12}$) and silver iodide ($K_{sp} = 1 \times 10^{-16}$), the equivalence points are at pBr = 6.0 and pI = 8.0. If the titrant is 0.1 *F*, the asymptote approached is at pBr = 11.0 and pI = 15.0, respectively. Curves for the two titrations are shown in Figure 18-2. The student's attention is again called to the comparison of a precipitation titration with the titration of a strong acid with a strong base. In the latter titration, a single titration reaction, $H^+ + OH^- \rightleftharpoons H_2O$, and a single constant, K_w, are involved. Hence, for given formal concentrations, a single titration curve is obtained regardless of the acid titrated. In contrast, in precipitation titrations for given formal concentrations, different titration curves are obtained, depending on the particular titration equation and the solubility product of the substance precipitated.

Visual End-Point Detection in Argentometry 18.3

In neutralization titrations, the visual indication functions as follows. The indicator is a weak acid (or base) and undergoes an acid–base reaction with the titrant. This reaction is accompanied by a color change and takes place at a pH close to that of the equivalence point. By analogy it may be surmised that in a precipitation titration the indicator would be a substance that gives with the titrant a precipitation reaction that is accompanied by a color change (that is, a colored precipitate is formed) and that occurs at a pX value close to that of the equivalence point. Indeed, this approach to indication is used in the so-called Mohr titration (Section 18.4). However, there exist several additional possibilities for indicating a precipitation titration, and the more commonly used ones will be considered in subsequent sections.

18.4 Method of Mohr

Mohr in 1856 introduced chromate ion as the indicator in the titration of chloride with silver. If some potassium chromate is added to the neutral chloride solution and the titration with silver ion is started, a reddish color appears due to the precipitation of silver chromate. However, the color disappears rapidly as the solution is stirred because the silver chromate converts to the less soluble silver chloride. As the titration proceeds the red disappears more and more slowly. Finally, when all the chloride has been precipitated as silver chloride, the next drop of titrant solution imparts a permanent red-brown color to the liquid, thereby signaling the end point. Ideally the chromate concentration should be such that the amount of silver chromate precipitated is just sufficient to yield a visible color change exactly at the equivalence point. However, the chromate concentration necessary to attain this would impart to the solution a yellow color so intense as to obscure the end point. In practice, a chromate concentration of about 0.005 *M* is used; this results in an end point that is somewhat late. Consequently, the use of an indicator blank is advisable and is mandatory when very dilute solutions are involved.

On the basis of their solubility products, bromide, iodide, and thiocyanate should be titrable by this method; however, in practice the titration can only be extended to bromide because silver iodide and silver thiocyanate strongly adsorb the chromate ion. The reverse titration of silver ion with a standard chloride solution is not feasible when chromate is used as the indicator. Upon addition of chromate to the silver salt solution a precipitate of silver chromate would form; this precipitate would react only extremely slowly at the end point.

Since the solubility of silver chromate increases rapidly with temperature, the titration must be performed at room temperature. Silver chromate is readily soluble in acidic solutions due to the conversion of chromate to dichromate ion:

$$2CrO_4^{2-} + 2H \rightleftharpoons Cr_2O_7^{2-} + H_2O \qquad (18\text{-}1)$$

As a consequence, the titration cannot be performed below about pH 6.5. On the other hand, the pH cannot be higher than about 10 because in too alkaline solutions brown hydrous silver oxide precipitates. In practice the appropriate pH is commonly maintained by buffering with hydrogen carbonate or with borate.

Interferences in the Mohr titration of chloride are many. Any anion that forms a silver salt more insoluble than the chloride (e.g., I^-, Br^-, SCN^-) will be precipitated first. Cations which form insoluble chromates (e.g., Ba^{2+}) also interfere by rendering the indicator inactive. The relatively high pH at which the Mohr titration must be performed imposes a serious limitation because difficulties occur if ions are present that form insoluble hydroxides or oxides, including iron and aluminum. Reducing substances may interfere by converting chromate ion to chromium(III). Finally, no substance can be tolerated that forms strong complexes with chloride ion (e.g., Hg^{2+}) or with silver ion (e.g., CN^-, NH_3 above pH 7).

Method of Fajans 18.5

For the precipitation titration of chloride and certain other ions, Fajans and his co-workers introduced indicators that function via adsorption on the precipitate. For an understanding of the action of adsorption indicators the adsorptive behavior of a precipitate must be appreciated. When silver nitrate is added to a dilute sodium chloride solution, turbidity develops but, if other electrolytes are absent, coagulation is not immediate. The silver chloride particles adsorb chloride ions and these attract counterions. The situation may be represented by $AgCl \cdot Cl^-{:}Na^+$. Consequently, these colloidal particles are electrically charged, repel each other, and coagulation is prevented. As the titration proceeds, the chloride ion concentration decreases and hence also the extent of surface charge. This charge is so diminished shortly before the equivalence point that coagulation is initiated and may visually signal that the equivalence point is close. However, abrupt coagulation occurs only in more concentrated solutions; in dilute solution, the turbidity persists. On the addition of an excess of silver nitrate, silver ions are adsorbed on the precipitate and attract counterions. The situation may now be represented by $AgCl \cdot Ag^+{:}NO_3^-$. The point at which equal quantities of positive and negative charge are adsorbed is called the isoelectric point. This point may not coincide with the equivalence point.

The adsorption indicators of Fajans interact with this adsorption process. Such indicators are colored, weakly acidic or basic organic compounds and their action can be exemplified by fluorescein. This compound is a weak acid, denoted by HFl, and its aqueous solution is yellow-green. When a small amount of this indicator is present during the titration, shortly after the isoelectric point is past, the indicator anion, Fl^-, is attracted to the positively charged particles, replacing nitrate ion as the counterion, and imparts a pink color to the particles. The process may be represented by

$$\underset{\text{white}}{AgCl \cdot Ag^+{:}NO_3^-} + \underset{\text{yellow}}{HFl} \rightleftharpoons \underset{\text{pink}}{AgCl \cdot Ag^+{:}Fl^-} + NO_3^- + H^+ \tag{18-2}$$

The titration of chloride using fluorescein as the adsorption indicator takes the following course. The chloride sample solution containing a small amount of fluorescein is initially clear and yellow-green in color. On addition of titrant the white silver chloride precipitate forms, imparting a milky look to the colored solution. Shortly before the end point, coagulation occurs and the end point is signaled by rapid adsorption of the dye on the precipitate, imparting a pink color to it. The solution remains yellow-green since only a small portion of the indicator is adsorbed. On addition of more titrant the color of the particles does not deepen. It is therefore difficult to judge whether or not, and to what extent, the end point has been exceeded.

In the use of fluorescein, if coagulation is avoided, the color change to pink is not confined to the bulky and rapidly settling precipitate but rather is viewed throughout the entire volume. This advantageous situation can be achieved by keeping the concentration of inert electrolytes low, by operating in an adequate dilute solution, and by addition of a "protective" colloid, such as dextrin.

Since the indicator is adsorbed only in its anionic form, the acidity constant of the dye defines the lowest pH value at which a satisfactory end point can be secured. With fluorescein, the titration of chloride can be performed in the pH range 6.5 to 10.3. The upper pH limit for the titration is again controlled by the precipitation of hydrous silver oxide. Fluorescein can also serve in the titration of bromide and iodide. Dichlorofluorescein is a stronger acid and can be used in the titration of chloride down to pH 4. Eosin, a tetrabromofluorescein, can be employed in the titration of bromide, iodide, or thiocyanate down to pH 2, but not in the titration of chloride since this dye is adsorbed on the silver chloride precipitate well before the equivalence point.

The dyes sensitize silver halides toward photochemical reduction, manifested by blackening of the precipitate. Much of the particle surface is blocked by the silver metal formed and the end point becomes less distinct. Consequently, the titration employing an adsorption indicator should not be conducted in direct sunlight but in diffuse daylight or under subdued illumination from an incandescent (not fluorescent) light source.

In the titration of chloride with fluorescein as the indicator, the required pH range is usually maintained by buffering with hydrogen carbonate or with borate. As in the titration after Mohr (Section 18.4), among others, cations that precipitate under the prevailing pH conditions are potential interferences.

Method of Volhard and Modifications 18.6

Silver thiocyanate, AgSCN, is only slightly soluble in water ($K_{sp} = 10^{-12}$) and hence a very accurate precipitation titration of silver can be effected with thiocyanate ion as the titrant. Iron(III), added as iron(III) nitrate or iron(III) ammonium sulfate, serves as the indicator; the first excess of thiocyanate ion imparts a reddish color in the supernatant liquid due to the formation of the soluble iron(III)–thiocyanate complex, $FeSCN^{2+}$. Since the reverse titration of thiocyanate ion with silver ion is also possible, the Volhard method can be applied to the determination of halides by means of a back-titration procedure. The halide solution is treated with an excess of standard silver ion solution, and the excess of silver is back-titrated with a standard thiocyanate solution. In this way chloride, bromide, and iodide can be determined. With iodide, the iron(III) must be added *after* the silver iodide has been precipitated; otherwise iodide would be oxidized to iodine by the iron(III):

$$Fe^{3+} + I^- \rightleftharpoons \tfrac{1}{2}I_2 + Fe^{2+} \qquad (18\text{-}3)$$

A major advantage of the Volhard method is that the titrations can be performed in fairly acidic media; indeed the method fails at higher pH values due to the precipitation of hydrous iron(III) oxide. At the low pH employed, interference from hydrolyzable ions is excluded; the Volhard method, therefore, has a much wider range of application than the methods of Mohr and Fajans.

In the determination of chloride, the end point fades after a few seconds and more titrant must be added to obtain a persistent red coloration. This has been explained by the fact that silver thiocyanate is less soluble than silver chloride, causing the metathetical reaction $AgCl + SCN^- \rightarrow AgSCN + Cl^-$ to proceed.

Division of the solubility-constant expressions of the two precipitates involved yields

$$\frac{K_{\text{sp, AgCl}}}{K_{\text{sp, AgSCN}}} = \frac{[Ag^+][Cl^-]}{[Ag^+][SCN^-]} = \frac{[Cl^-]}{[SCN^-]} = \frac{2 \times 10^{-10}}{1 \times 10^{-12}} = 2 \times 10^2$$

It can be seen that the molar concentration of chloride ion must be about 200 times that of thiocyanate ion, or else the metathetical reaction occurs. A fading end point can be prevented by removing the silver chloride by filtration and performing the back-titration in the filtrate. The results are then satisfactory, but the filtration is tedious. A second approach is to inactivate the silver chloride precipitate by coating it with an immiscible organic solvent, commonly nitrobenzene. When the suspension of silver chloride is shaken vigorously with the nitrobenzene, the latter is adsorbed on the surface of the precipitate, and then in the back-titration the metathesis proceeds so slowly that a stable end point is secured.

Schulek, however, has suggested that metathesis is not the major cause of the fading end point. He claims that silver ions are adsorbed on the surface of the precipitate and that the adsorbed silver reacts with the titrant only slowly. He has proposed that, following the precipitation of silver chloride, potassium nitrate be added and the solution be boiled, thereby promoting coagulation and desorption of the silver ion. The end point in the subsequent back-titration with thiocyanate is then sharp and stable.

18.7 Method of Swift

If a high concentration of iron(III) is provided in the chloride solution and only a very small amount of thiocyanate is added, the titration of chloride can be performed directly. Upon addition of silver nitrate, silver chloride is precipitated, *not* the less soluble silver thiocyanate. At an iron(III) concentration of about 0.2 M, the equilibrium

$$Fe^{3+} + SCN^- \rightleftharpoons FeSCN^{2+} \tag{18-4}$$

is shifted far to the right, and the concentration of "free" thiocyanate ion becomes very small. The value of the ion product $[Ag^+][SCN^-]$ does not reach that of the solubility product of silver thiocyanate and consequently the latter does not precipitate. Only when all the chloride is precipitated does the silver ion concentration become sufficiently high for silver thiocyanate to precipitate. Then the above equilibrium is shifted to the left and the end point is indicated by the disappearance of the red solution color.

Precipitation Titration of Sulfate 18.8

Barium sulfate has about the same solubility product as silver chloride, and hence it might be expected that the precipitation titration of sulfate with barium would be as feasible as that of chloride. Unfortunately, barium sulfate shows such pronounced tendencies to coprecipitation phenomena that accurate results are not readily attained. However, this titration finds application in rapid control analysis where an error of up to 3% can be tolerated. Various indicators have been employed, probably the most explored being tetrahydroxy-*p*-benzoquinone. The titration can sometimes be improved by making the titration medium 30 to 50% in ethanol.

Other Precipitation Titrations 18.9

Of other precipitation titrations receiving attention in routine analysis, two may be singled out for mention. Fluoride may be titrated with thorium ion to form sparingly soluble thorium fluoride, ThF_4. The end point may be detected by the formation of the insoluble red-colored precipitate of thorium with the dye Alizarin Red S. This titration is extremely sensitive to foreign ions and hence is most frequently applied after the separation of fluoride, commonly by distillation as hexafluorosilicic acid from solutions containing nonvolatile mineral acids in the presence of silica.

Zinc can be titrated with hexacyanoferrate(II) ion according to the reaction

$$3Zn^{2+} + 2K^{+} + 2[Fe(CN)_6]^{4-} \rightleftharpoons \underline{K_2Zn_3[Fe(CN)_6]_2} \qquad (18\text{-}5)$$

The end point may be detected in various ways, including the use of redox indicators (Section 24.9).

Questions 18.10

What are the requirements for a reaction in order that it may be made the basis for a precipitation titration?

What methods of end-point detection may be used in precipitimetry?

What are the pH limits in the titration of chloride according to the methods of Mohr, Fajans, Volhard, and Swift?

Discuss the advantages and disadvantages of the four titration methods named in Question 18-3.

18.10 QUESTIONS

18-5. For the four titration methods named in Question 18-3, name the indicators used and describe the end-point color changes and the indicator mechanism.

18-6. Compare the interferences in the four methods named in Question 18-3.

18-7. Explain why the titration of iodide according to the method of Mohr fails.

18-8. What is to be understood by the terms coagulation and isoelectric point?

18-9. What is the principle of the Swift method?

18-10. Explain why it is impossible to titrate silver ion with chloride ion using chromate ion as the indicator.

18-11. Explain why a pH of about 10 is the upper limit for argentometric titrations.

18-12. How will strong light affect argentometric titrations? Which of the modifications is most affected? Explain.

18-13. Explain in what way the acidity constant of the adsorption indicator is related to the lower pH limit at which the halide titration after Fajans can be performed.

18-14. Would barium ion interfere in the titration of chloride after Mohr? Explain your answer.

18-15. Explain why the results of the titration of sulfate with barium ion are limited in accuracy.

18-16. Contrast the titration curve for a precipitation titration with that of an acid–base titration. Compare the shape of the two curves. What is plotted on the ordinate and abscissa? What factors govern the position with respect to the ordinate of the equivalence point of the two titrations?

18-17. At the concentrations of sample and titrant solutions commonly employed, the pH changes by about 10 to 12 units during an acid–base titration in aqueous medium. (a) Within what approximate range might the pX change during precipitation titrations? (b) What would be the range in acid–base titrations if nonaqueous solvents were included in the considerations? Elaborate on your answers.

18-18. Discuss the requirements for the precipitate in a precipitation titration and in a gravimetric determination. Which requirements are identical and which are different?

18-19. Suggest an example where a solution to be used as a titrant in a precipitation titration might be standardized (a) by an acid–base titration and (b) by a gravimetric determination.

Problems 18.11

The bromide present in 80.0 ml of 0.050 F KBr is titrated with 0.080 F $AgNO_3$. For simplicity, take the solubility product of AgBr to be 1.0×10^{-12}. Calculate the molar concentration of silver ion and bromide ion at the following points of the titration curve:

(a) At the start of the titration.
Answer: $[Ag^+] = 0, [Br^-] = 5 \times 10^{-2}\ M$

(b) After the addition of 20.0 ml of silver solution.
Answer: $[Ag^+] = 4.2 \times 10^{-11}\ M, [Br^+] = 2.4 \times 10^{-2}\ M$

(c) At the equivalence point.
Answer: $[Ag^+] = [Br^-] = 1.0 \times 10^{-6}\ M$

(d) After the addition of 10.00 ml of silver solution in excess.
Answer: $[Ag^+] = 5.7 \times 10^{-3}\ M, [Br^-] = 1.8 \times 10^{-10}\ M$

Calculate the pAg values at the starting point, halfway to the equivalence point, 10.0% before the equivalence point, at the equivalence point, and 10.0% beyond the equivalence point for the following titrations of silver ion. The solubility-product constants for AgCl, AgBr, AgI, and AgSCN are 1.8×10^{-10}, 5.2×10^{-13}, 8.3×10^{-17}, and 1.0×10^{-12}, respectively.

	$AgNO_3$, ml	$AgNO_3$, F	Titrant
(a)	25.00	0.108	0.150 F KCl
(b)	50.00	0.162	0.110 F NaCl
(c)	20.00	0.120	0.100 F $BaCl_2$
(d)	30.00	0.110	0.100 F KBr
(e)	25.00	0.101	0.132 F KBr
(f)	10.00	0.098	0.109 F KI
(g)	20.00	0.132	0.092 F KI
(h)	25.00	0.119	0.096 F KSCN

Calculate the pBa values after the addition of 0.0, 5.0, 10.0, 15.0, 19.0, 20.0, 21.0, and 25.0 ml of 0.150 F Na_2SO_4 to 100.0 ml of a solution containing 0.6248 g of $BaCl_2$ (*208.25*). The K_{sp} value of $BaSO_4$ is 1.3×10^{-10}.

A solution obtained by mixing 10.0 ml of 0.100 F KCl, 10.0 ml of 0.200 F KBr, 10.0 ml of 0.150 F KI, and 70.0 ml of pure water is titrated with 0.200 F $AgNO_3$. The solubility-product constants of AgCl, AgBr, and AgI are 1.8×10^{-10}, 5.2×10^{-13}, and 8.3×10^{-17}, respectively. Calculate the pAg values after the addition of 1.0, 4.0, 7.5, 10.0, 14.0, 17.5, 20.0, and 22.5 ml of titrant.

19 CALCULATION OF RESULTS IN PRECIPITATION TITRATIONS

General Considerations 19.1

The result for a precipitation titration can be calculated in a manner which, in principle, is analogous to that employed for acid–base titrations (Chapter 15). There is no need to repeat the reasoning and derivations leading to the final formulas, and they may be used as given in Chapter 15. However, slight modifications will become necessary in some cases for the following reasons. In an acid–base titration, regardless of the acid and base involved, the titration process proper is a transfer of protons and, at least in aqueous media, a single principal titration reactions exists, $H^+ + OH^- \rightarrow H_2O$. In addition, the relevant equivalence numbers are simply the number of moles of hydrogen ion or hydroxide ion, respectively, provided per mole of acid or base and actually reacting according to this titration equation. In precipitation titrations, however, different reacting species are provided by the various titrants and substances titrated and in different molar amounts. In addition, these reacting species may not combine in a 1:1 molar ratio, as do hydrogen and hydroxide ions. This fact requires in some cases the additional consideration of the combining ratio; in other words, it then becomes necessary to base the calculation on the number of moles of sought-for substance corresponding to one mole of the titrant.

The best way to introduce the novice to this approach is to illustrate the calculations by examples, starting with simple cases and passing gradually to more involved ones.

Example 19-1. The chloride in a NaCl solution is titrated with 0.1000 F $AgNO_3$, of which a volume of 25.30 ml is required. (a) What amount of Cl^- (*35.45*) in milligrams is present? (b) What amount of NaCl (*58.44*) in milligrams is present? The titration reaction is

$$Cl^- + Ag^+ \rightarrow \underline{AgCl}$$

The equation previously derived and extensively used in Chapter 15 may be applied:

$$\text{ml}_t \times F_t \times f_t \times \frac{\text{FW}_s}{f_s} = \text{mg}_s$$

Since one mole of $AgNO_3$ provides one mole of Ag^+, that is, of reacting species, the equivalence number of the titrant, f_t, is unity.

(a) The sought-for substance in this part of the problem is Cl^-, which is also the reacting species; consequently, f_s is also unity. Substitution of the numerical values gives

$$25.30 \times 0.1000 \times 1 \times \frac{35.45}{1} = 89.69 \text{ mg of } Cl^-$$

(b) To obtain the answer to this portion of the problem requires merely the application of a chemical factor that converts the result for Cl^- to that for NaCl; this factor is FW_{NaCl}/FW_{Cl}. Then

$$ml_t \times F_t \times 1 \times \frac{FW_{Cl}}{1} \times \frac{FW_{NaCl}}{FW_{Cl}} = \text{mg of NaCl}$$

As can be seen, the term FW_{Cl} cancels and the following is obtained:

$$ml_t \times F_t \times 1 \times \frac{FW_{NaCl}}{1} = \text{mg of NaCl}$$

This equation, of course, could have been obtained directly by the following reasoning. One mole of Cl^- corresponds to one mole of NaCl, and consequently the formula weight of NaCl and its equivalence number of one could have been substituted in the equation from the very beginning.

Substitution of the numerical values gives

$$25.30 \times 0.1000 \times 1 \times \frac{58.44}{1} = 147.9 \text{ mg of NaCl}$$

Example 19-2. The chloride in a solution of $BaCl_2$ is titrated with 0.0800 *F* $AgNO_3$, of which a volume of 18.34 ml is required. (a) What amount of Cl^- (*35.45*) in milligrams is present? (b) What amount of $BaCl_2$ (*208.25*) in milligrams is present?

The titration reaction is the same as in Example 19-1. For part (a) of the problem the reasoning is the same as in Example 19-1 and the calculation takes the form

$$18.34 \times 0.0800 \times 1 \times \frac{35.45}{1} = 52.0 \text{ mg of } Cl^-$$

The answer to part (b) is obtained in a fashion analogous to that of Example 19-1; but here the conversion factor is $FW_{BaCl_2}/2FW_{Cl}$.

Substitution yields

$$ml_t \times F_t \times 1 \times \frac{FW_{Cl}}{1} \times \frac{FW_{BaCl_2}}{2FW_{Cl}} = \text{mg of } BaCl_2$$

Again FW_{Cl} cancels.

Also here the formula can be obtained directly according to the following reasoning. Two moles of reacting species, that is, of Cl^-, are provided by one mole of the sought-for substance, that is, $BaCl_2$, and consequently the equivalence number of $BaCl_2$ is 2.

$$ml_t \times F_t \times 1 \times \frac{FW_{BaCl_2}}{2} = \text{mg of } BaCl_2$$

Substitution of the numerical values gives

$$18.34 \times 0.0800 \times 1 \times \frac{208.25}{2} = 152.8 \text{ mg of } BaCl_2$$

Example 19-3. What is the formality of a solution of $BaCl_2$, of which a volume of 25.00 ml requires 16.00 ml of 0.07500 F $AgNO_3$?

The relevant formula previously derived applies

$$\text{ml}_t \times F_t \times f_t = \text{ml}_s \times F_s \times f_s$$

According to the reasoning in the two preceding examples the equivalence number of the titrant and the substance titrated are 1 and 2, respectively. Substituting the numerical data yields

$$16.00 \times 0.07500 \times 1 = 25.00 \times F_s \times 2$$

And solving,

$$F_s = \frac{16.00 \times 0.07500 \times 1}{25.00 \times 2} = 0.02400\ F$$

Example 19-4. The Ag^+ in a solution of Ag_3PO_4 is titrated with 0.0400 F $BaCl_2$, of which a volume of 27.20 ml is required. (a) What amount of Ag (*107.87*) in mg is present? (b) What amount of Ag_3PO_4 (*418.58*) in mg is present?

The titration reaction is the same as in Example 19-1. With $BaCl_2$ the titrant, two moles of reacting species, that is, Cl^-, are provided per one mole. Consequently, the equivalence number of the titrant $f_t = 2$.

In answering part (a) Ag^+ is the reacting species and also the sought-for substance; consequently, $f_s = 1$. Substituting the numerical values yields

$$27.20 \times 0.0400 \times 2 \times \frac{107.87}{1} = 235 \text{ mg of Ag}$$

In solving part (b), it must be recognized that three moles of the reacting species, that is, of Ag^+, are furnished by one mole of the sought-for substance, that is, by Ag_3PO_4; consequently, $f_s = 3$. Hence

$$\text{ml}_t \times F_t \times 2 \times \frac{\text{FW}_{Ag_3PO_4}}{3} = \text{mg of } Ag_3PO_4$$

The student should derive this formula by applying factors as shown in part (b) of both Examples 19-1 and 19-2.

Substituting the numerical values gives

$$27.20 \times 0.0400 \times 2 \times \frac{418.58}{3} = 304 \text{ mg of } Ag_3PO_4$$

Example 19-5. The Zn^{2+} in a $ZnSO_4$ solution is titrated with 0.1020 F $K_4[Fe(CN)_6]$, of which a volume of 20.40 ml is required. (a) What amount of Zn (*65.37*) in milligrams is present? (b) What amount of $ZnSO_4$ (*161.43*) in milligrams is present?

The titration reaction may be written as follows:

$$3Zn^{2+} + 2K^{+} + 2[Fe(CN)_4)]^{4+} \rightarrow \underline{Zn_3K_2[Fe(CN)_6]_2}$$

For part (a) of the example, since Zn^{2+} is both the reacting species and the sought-for substance, its equivalence number is unity. So is that of $[Fe(CN)_6]^{4+}$. However, the molar combining ratio is not 1:1 as it was in the previous examples, but rather $[Fe(CN)_6]^{4+}$:$[Zn^{2+}]$ = 2:3 or, referred to one mole of zinc ion, 1:$\frac{3}{2}$. Consequently, this fraction $\frac{3}{2}$ must be incorporated into the formula

$$ml_t \times F_t \times 1 \times \frac{FW_{Zn}}{1} \times \frac{3}{2} = \text{mg of Zn}$$

Substitution of the numerical data yields

$$20.40 \times 0.1020 \times 1 \times \frac{65.37}{1} \times \frac{3}{2} = 204.0 \text{ mg of Zn}$$

Part (b) is simply solved. Since one mole of Zn corresponds to one mole of $ZnSO_4$, the equivalence number of the latter is also unity. The calculation takes the form

$$20.40 \times 0.1020 \times 1 \times \frac{161.43}{1} \times \frac{3}{2} = 503.9 \text{ mg of } ZnSO_4$$

Example 19-6. A solution of $K_4[Fe(CN)_6]$ is standardized by employing it in the titration of 625.5 mg of $Zn_2P_2O_7$ (*277.68*) dissolved in acid. What is the formality of this titrant solution if a volume of 28.32 ml is required? The titration reaction is the same as in Example 19-4. The equivalence number of the titrant is unity. The equivalence number of the substance titrated, however, is f_t = 2, since one mole of $Zn_2P_2O_7$ provides two moles of reacting species, that is, of Zn^{2+}. Again the combining ratio must be taken care of by incorporating the factor $\frac{3}{2}$ as in the preceding example.

The calculation takes the form

$$28.32 \times F_t \times 1 \times \frac{277.68}{2} \times \frac{3}{2} = 625.5$$

Solving yields

$$F_t = \frac{625.5 \times 2 \times 2}{28.32 \times 277.68 \times 3} = 0.1061\,F$$

Example 19-7. A 0.5324-g sample of $BaCl_2$ is dissolved in water and the chloride (*35.45*) titrated with 0.0924 *F* $AgNO_3$, of which a volume of 27.67 ml is required. Calculate the per cent Cl in $BaCl_2$.

The calculation, in consolidated form, is

$$\frac{27.67 \times 0.0924 \times 1 \times 35.45 \times 100}{532.4 \times 1} = 17.02\% \text{ Cl}$$

It should be pointed out that to the novice the problem may appear similar to that of part (b) of Example 19-2, and he may be tempted to use the formula weight of $BaCl_2$ and its equivalence number of 2. However, the situation is different. The problem asks for Cl^- to be determined and be expressed as its percentage in $BaCl_2$.

Use of Titer in Precipitation Titrations 19.2

The concept of a titer was briefly considered in Chapter 3. The student, having encountered both acid–base and precipitation titrimetry, is now better prepared to understand the use and significance of titer values. The concentration of a titrant may be directly expressed in terms of the milligrams of the sought-for substance corresponding to 1 ml of the titrant. Such an expression is called a titer.

The use of a titer is of special practical value where a given titration must be performed many times with the same type of sample. Indeed, the concentration of the titrant may be adjusted in such a way that 1 ml corresponds to a certain number of milligrams of the sought-for substance or, for a specified sample weight, to 0.1%, 1%, etc., as desired. The calculation of results is thereby simplified and the introduction of arithmetic errors greatly reduced. Where the titration reaction proceeds stoichiometrically, the concentration of the titrant required for such simplification can be readily calculated.

A titer value is of special interest if the titration reaction does not proceed stoichiometrically, does not go to completion, or if other difficulties are encountered. For example, in a precipitation titration, the end point signaled by the indicator used may be too early or too late, the precipitate formed during the titration may show an unduly high solubility, the titrant may be absorbed on the precipitate, and the presence of impurities may exert an unfavorable influence. Even in such cases a practical titration may be possible and a correct value obtained, provided the titration is performed under reproducible conditions. The titrant may then be standardized under these conditions against a known amount of the substance to be determined. The concentration is then best expressed as a titer.

A titer value expressed for one substance can be used analogously to a formality value in the calculation of the results for the titration of other substances with this titrant. All that is needed is to combine the established titer value with appropriate chemical factors.

The use of titer values and the calculations encountered can be appreciated from illustrative examples.

Example 19-8. If a volume of 35.34 ml of a $AgNO_3$ solution is required for the titration of 0.2461 g of NaCl, what is the NaCl titer of the $AgNO_3$ solution?

$$\text{NaCl titer} = \frac{246.1}{35.34} = 6.964 \text{ mg of NaCl per milliliter}$$

Example 19-9. How much KBr (*119.01*) in milligrams is present in a sample if the titration of its bromide content requires 28.30 ml of a $AgNO_3$ solution having a NaCl (*58.44*) titer of 1.205 mg of NaCl per milliliter?

The result may be first expressed in terms of NaCl by application of the given titer value.

$$28.30 \times 1.205 = 34.10 \text{ mg of NaCl}$$

The result may then be converted to terms of KBr by application of the relevant chemical factors.

$$34.10 \times \frac{KBr}{NaCl} = 34.10 \times \frac{119.01}{58.44} = 69.44 \text{ mg of KBr in sample}$$

Example 19-10. Zinc (*65.37*) is to be determined routinely in an alloy by titration with potassium hexacyanoferrate(II). What must be the formality of the $K_4[Fe(CN)_6]$ titrant solution so that its zinc titer will be exactly 0.1% Zn per milliliter if a sample weight of 0.5000 g is always taken?

Extension of the reasoning of Example 19-5 leads to

$$\frac{\text{ml}_t \times F_t \times 1 \times \dfrac{\text{FW}_{\text{Zn}}}{1} \times \dfrac{3}{2} \times 100}{\text{sample wt. in mg}} = \%\ \text{Zn}$$

Introduction of the numerical data yields

$$\frac{1.000 \times F_t \times 65.37 \times 3 \times 100}{500.0 \times 2} = 0.1\%$$

$$F_t = 0.00509\ F\ K_4[Fe(CN)_6]$$

19.3 Indirect Analysis and Precipitation Titrations

Indirect analysis, discussed in Section 7.4 for gravimetric methods and in Section 15.3 for acid–base titrations, is also possible with precipitation titrations. The calculation procedure is similar to that previously outlined and a single example will serve to illustrate the approach.

Example 19-11. A mixture is known to contain only NaCl (*58.44*) and $BaCl_2$ (*208.25*). A 0.5000-g sample of this mixture is dissolved and the chloride titrated with 0.1000 F $AgNO_3$, of which a volume of 63.00 ml is required. Calculate the % NaCl and % $BaCl_2$ in the mixture.

Let x and y represent the % NaCl and % $BaCl_2$, respectively, in the mixture. From the description of the mixture it follows that

$$x + y = 100.00$$

This is the first equation of the required pair. The result for the titration of chloride may be expressed either as % NaCl or % $BaCl_2$.

Selection of the former yields

$$\frac{63.00 \times 0.1000 \times 58.44 \times 100}{500} = 73.63\% \text{ total chloride expressed as NaCl}$$

The second equation of the required pair is obtained as follows:

$$x + y \times \frac{2 \times NaCl}{BaCl_2} = \% \text{ total chloride expressed as NaCl}$$

$$x + y \times \frac{2 \times 58.44}{208.25} = \% \text{ total chloride expressed as NaCl}$$

$$x + 0.56125y = 73.63$$

Solving the simultaneous equations for y yields $y = 60\%$ $BaCl_2$ and $x = 100.00 - y = 40\%$ NaCl.

Problems 19.4

1. A silver nitrate solution is standardized against KCl (*74.56*) of 99.60% purity. A 0.3254-g sample of this salt is titrated according to Mohr and 44.22 ml of the $AgNO_3$ solution is required. Calculate the formality of this solution.

   Answer: 0.09830 F

2. A coin weighing 12.52 g is analyzed for silver (*107.87*). The coin is dissolved in nitric acid and the solution diluted with water to exactly 250 ml in a volumetric flask. A 25.00-ml portion of the solution is titrated with 0.1050 F KCl of which a volume of 44.22 ml is required. What is the percentage of silver in the coin?

   Answer: 40.00%

3. A mixture of $BaCl_2$ and Na_2CO_3 is to be analyzed for its barium chloride content by the Volhard method. The following data are given. The $AgNO_3$ titrant solution is prepared by dissolving 4.983 g of silver metal of 99.10% purity in nitric acid and diluting with water to exactly 500 ml in a volumetric flask. In the standardization of the KSCN solution, it is found that a volume of 25.00 ml of the $AgNO_3$ solution requires 22.00 ml of the KSCN solution. A 0.5000-g sample is dissolved, acidified, and a volume of 40.30 ml of $AgNO_3$ solution is added; the excess of silver is then back-titrated, requiring 6.22 ml of the KSCN solution.

   (a) Calculate the formality of the silver (*107.87*) solution.

   Answer: 0.0915_6 F

   (b) Calculate the formality of the KSCN solution.

   Answer: 0.1040_4 F

   (c) Calculate the percentage of $BaCl_2$ (*208.25*) in the mixture.

   Answer: $63.3_7\%$

4. Sulfur (*32.1*) in coal is determined by dry combustion of the sample, absorption of the SO_3 formed in water, and titration of the sulfate, thus formed, with a $BaCl_2$ solution. The titer of the titrant solution is established by titrating a freshly dried sample of Na_2SO_4 (*142.0*). In

a particular determination, a 0.2500-g sample of coal is subjected to combustion. Titration of the sulfate obtained requires 4.80 ml of a $BaCl_2$ solution, of which a volume of 17.22 ml is needed in the titration of 0.0188 g of Na_2SO_4. (a) What is the titer of the $BaCl_2$ solution in milligrams of sulfur per milliliter? (b) What is the percentage of sulfur in the coal?

Answers: (a) 0.2468 mg of S per milliliter; (b) 0.474% S

19-5. A mixture is known to contain only NaCl (*58.44*) and KBr (*119.0*). In the argentimetric titration of the total halide in a 0.2500-g sample of this mixture, a volume of 27.64 ml of 0.1000 *F* $AgNO_3$ is required. Calculate the composition of the mixture.

Answer: 30.5% NaCl, 69.5% KBr

19-6. One approach to the determination of fluoride proceeds as follows. Fluoride is precipitated quantitatively as PbClF. The isolated precipitate is either dried and weighed or is dissolved in acid and the chloride in the solution is titrated by the Volhard method. A 0.8400-g sample of a material was dissolved and fluoride precipitated as PbClF. The precipitate was separated and dissolved. To the resulting solution a volume of 35.00 ml of 0.09530 *F* $AgNO_3$ was added and the titration with 0.1025 *F* KSCN required 2.16 ml. Calculate the percent of fluorine (*19.0*) present in the material.

19-7. The production of a preparation containing barium chloride (*208.25*) as the active ingredient is to be checked routinely by a Mohr titration. How many grams of silver (*107.87*) must be dissolved to 2 liters in order to obtain a solution of which each milliliter is equivalent to 1.00% $BaCl_2$ if a 1.00-g sample of the preparation is always to be taken?

20 COMPLEXATION EQUILIBRIA AND COMPLEXIMETRIC TITRATIONS

Compleximetric titrations are based on the formation of a soluble complex upon the reaction of the species titrated with the titrant. Many reactions lead to a soluble complex but only a few can be employed for the purpose of a titration. Besides fulfillment of the general requirements for titration reactions, such as reasonable rate, stoichiometry, etc. (see Chapter 10), special requirements must be met with respect to the stability of the complex formed and the number of steps involved in its formation. Insoluble complexes may be formed and titrations can be based on such reactions. However, such titrations are advantageously considered as precipitation titrations. Formally, protonation reactions may be described as complexation reactions. Where this point of view is adopted an acid is considered as the "proton complex" of its conjugate base. However, titrations based on protonation reactions are usually better classified as acid–base rather than as compleximetric titrations. Nevertheless, the reader should compare the calculation of the titration curve and the treatment of metal indicators given below with the relevant sections on acid–base titrations and note the analogies.

Almost all compleximetric titrations of practical significance entail the formation of a complex involving the cation of a metal; consequently, it is necessary to discuss first the concept of metal complexes.

Nature of Metal Complexes 20.1

A metal complex involves the coordination of a metal ion with a species that has one or more pairs of nonbonding electrons available for sharing. The electron-donating species is known as a ligand (Latin, *ligare* = to bind) or as a complexing agent. The metal ion, in the complex, is known as the central ion. The ligand may have one or more electron-donating groups (or atoms) in each molecule and is then said to uni-, bi-, tri-, quadri-, . . . , multidentate (Latin, *dentatus* = toothed). The greatest number of unidentate ligands known to coordinate with a given central ion is termed the (maximum) coordination number of that ion. The charge of a complex is the algebraic sum of the charges of the central ion and of the ligand species. Thus a metal complex may be neutral, or either positively or negatively charged.

When a ligand is multidentate, that is, has two or more donor groups as part of a single molecule, there is a possibility of forming a ring structure containing the central ion. Such rings are known as chelate rings. The term is apt, as the central ion is held in each ring as in the claws of a crab (Greek, *chele* = crab's claw). The entire multidentate complex is known as a chelate complex or simply as a chelate. The stability of a chelate ring is commonly greatest if it is five-membered or six-membered.

It should be appreciated that all metal ions in aqueous solution are present in the form of complexes, at least as aquo complexes. The oxygen atom of the water molecule serves as an electron donor, $H:\ddot{\underset{..}{O}}:H$. For example, the copper(II) ion in water persists as $Cu(H_2O)_4^{2+}$, but the unsolvated ion Cu^{2+} is usually written for simplicity, unless the solvation must be explicitly shown.

20.2 Nomenclature of Metal Complexes

Complexes are named systematically in the following manner. First, the number of ligand species bonded to a central ion is indicated by a Greek prefix: mono-, di-, tri-, tetra-, penta-, hexa-, etc. Next, the ligand is named, and if anionic is usually terminated by an -o. Water and ammonia, as ligands, are named aquo and ammine. Finally, the name of the central ion is given. In some cases, the root of the Latin name of the element is preferred, for example, *argent-* for silver, *plumb-* for lead, and *ferr-* for iron. If the complex is anionic, the name is terminated by -ate. The oxidation state of the central ion is indicated, when desirable, by a Roman numeral within parentheses. (For a discussion of the determination of the oxidation state, see Section 21.5.) The application of these rules is best appreciated by some simple examples: $Cu(NH_3)_4^{2+}$, tetraamminecopper(II) ion; $Ag(NH_3)_2^+$, diamminesilver ion; HgI_4^{2-}, tetraiodomercurate(II); $Fe(CN)_6^{4-}$, hexacyanoferrate(II); $Fe(CN)_6^{3-}$, hexacyanoferrate(III); $[CrCl_4(H_2O)_2]^-$, tetrachlorodiaquochromate(III); $Ag(CN)_2^-$, dicyanoargentate(I).

It is often convenient to employ the following simpler description of a complex: The name of the metal is separated from that of the ligand (and its prefix, if any) by a hyphen; for example, copper(II)-tetraammine complex, mercury(II)-tetraiodo complex, metal-chloro complexes, etc. Some complex species have traditional names that are in common use, for example, ferrocyanide and ferricyanide for hexacyanoferrate(II) and hexacyanoferrate(III), respectively.

Complexation Equilibria 20.3

The formation of a soluble metal complex is a reversible process. The equilibrium is dynamic and the law of mass action applies. For the formation of the copper(II)-tetraammine complex the reaction and the expression for the equilibrium constant may be written

$$Cu^{2+} + 4NH_3 \rightleftharpoons Cu(NH_3)_4^{2+} \qquad (20\text{-}1)$$

$$K_{st} = \frac{[Cu(NH_3)_4^{2+}]}{[Cu^{2+}][NH_3]^4} \qquad (20\text{-}2)$$

The equilibrium constant corresponding to the reaction leading to the formation of the complex is known as the stability constant or the formation constant of the complex. The equilibrium constant for the reverse reaction is known as the instability constant (or dissociation constant) for the complex. Obviously the two constants are reciprocals:

$$K_{st} = \frac{1}{K_{inst}}$$

Complexes involving more than one ligand molecule or ion form in a stepwise fashion and an equilibrium expression may be written for each step. For example, silver has a coordination number of 2 and hence the silver ion combines in two steps with ammonia to form the diamminesilver(I) ion:

$$Ag^+ + NH_3 \rightleftharpoons Ag(NH_3)^+ \qquad (20\text{-}3)$$

$$k_{st,1} = \frac{[Ag(NH_3)^+]}{[Ag^+][NH_3]} \qquad (20\text{-}4)$$

$$Ag(NH_3)^+ + NH_3 \rightleftharpoons Ag(NH_3)_2^+ \qquad (20\text{-}5)$$

$$k_{st,2} = \frac{[Ag(NH_3)_2^+]}{[Ag(NH_3)^+][NH_3]} \qquad (20\text{-}6)$$

The overall stability constant, K_{st}, is the product of the two stepwise stability constants:

$$K_{st} = k_{st,1} \times k_{st,2} = \frac{\cancel{[Ag(NH_3)^+]}}{[Ag^+][NH_3]} \times \frac{[Ag(NH_3)_2^+]}{\cancel{[Ag(NH_3)^+]}[NH_3]}$$

$$= \frac{[Ag(NH_3)_2^+]}{[Ag^+][NH_3]^2} \qquad (20\text{-}7)$$

The use of stability constants rather than dissociation constants has the advantage that positive exponents are encountered. The reasoning and mode of calculation, however, are fully analogous to those encountered with dissociation constants as used, for example, in the consideration of acid–base equilibria. A few examples will illustrate the point.

Example 20-1. The complex MY, formed from a metal ion M and a complexing agent Y (charges are omitted as they are not relevant to the point) has a stability constant of 4.0×10^8. What is the concentration of free, that is, uncomplexed metal ion, in a solution 1.0×10^{-2} F in the complex?

The expression for the stability constant of the complex is

$$K_{\mathrm{st}} = \frac{[\mathrm{MY}]}{[\mathrm{M}][\mathrm{Y}]}$$

According to the material balance the total concentration of the metal ion, C_{M}, which equals 1.0×10^{-2} F, is given by

$$C_{\mathrm{M}} = [\mathrm{MY}] + [\mathrm{M}]$$

Since upon dissociation of the complex one free ion of the complexing agent is formed for each free metal ion, the relation

$$[\mathrm{M}] = [\mathrm{Y}]$$

holds. Substitution of the last two equations into the expression for the stability constant yields

$$K_{\mathrm{st}} = \frac{C_{\mathrm{M}} - [\mathrm{M}]}{[\mathrm{M}]^2}$$

Since the complex is rather stable, neglecting [M] with respect to C_{M} is a reasonable approximation leading to

$$K_{\mathrm{st}} \simeq \frac{C_{\mathrm{M}}}{[\mathrm{M}]^2}$$

Solution for [M] and substitution of the numerical data gives

$$[\mathrm{M}] = \sqrt{\frac{C_{\mathrm{MY}}}{K_{\mathrm{st}}}} = \sqrt{\frac{1.0 \times 10^{-2}}{4.0 \times 10^{8}}} = 5.0 \times 10^{-6}\ M$$

The concentration of free metal ion is, in fact, considerably smaller than the concentration of the complex and consequently the approximation was allowed.

Example 20-2. A complex MY has a stability constant of 2.0×10^3. What is the concentration of free metal ion in a solution 2.0×10^{-3} F in the complex? Solution in a manner analogous to that for Example 20-1 leads to

$$[\mathrm{M}] = \sqrt{\frac{2.0 \times 10^{-3}}{2.0 \times 10^{3}}} = 1.0 \times 10^{-3}\ M$$

A check of the approximation reveals that the metal ion concentration is half that of the complex concentration; the 5% rule adopted previously is therefore violated, and the approximation is not permitted. Consequently, the solution must involve the full quadratic equation, as follows:

$$2.0 \times 10^{3} = \frac{2.0 \times 10^{-3} - [\mathrm{M}]}{[\mathrm{M}]^{2}}$$

$$2.0 \times 10^{3}[\mathrm{M}]^{2} + [\mathrm{M}] = 2.0 \times 10^{3} = 0$$

$$[\mathrm{M}] = \frac{-1.0 \pm \sqrt{(1.0)^{2} + 4 \times 2.0 \times 10^{3} \times 2.0 \times 10^{-3}}}{2 \times 2.0 \times 10^{3}}$$

$$= \frac{-1.0 \pm \sqrt{1.0 + 16}}{4.0 \times 10^{3}}$$

Only the positive root has physical meaning; hence

$$[\mathrm{M}] = \frac{-1.0 + \sqrt{17}}{4.0 \times 10^{3}} = 8 \times 10^{-4}\,M$$

Example 20-3. A complex MY has a stability constant of 5.0×10^{6}. What is the concentration of free metal ion in a solution $2.0 \times 10^{-3}\,F$ in the complex and made additionally $2.0 \times 10^{-1}\,M$ in the free complexing agent? The material balance for the metal, as in Example 20-1, is

$$C_{\mathrm{MY}} = [\mathrm{MY}] + [\mathrm{M}] \simeq [\mathrm{MY}]$$

However, the equality $[\mathrm{M}] = [\mathrm{Y}]$ no longer applies. Free complexing agent stems from two sources: the dissociation of the complex and from the addition. The complex is quite stable and the contribution from the dissociation will be small, especially since the high concentration of the added agent will repress the dissociation. Consequently, the concentration of the free complexing agent, [Y], may be taken as equal to $2.0 \times 10^{-1}\,M$. Substitution of the numerical values into the expression for the stability constant yields

$$5.0 \times 10^{6} = \frac{2.0 \times 10^{-3}}{[\mathrm{M}] \times 2.0 \times 10^{-1}}$$

and

$$[\mathrm{M}] = 2.0 \times 10^{-9}\,M$$

Checking the validity of the approximation expressed in the material balance is left to the student.

Example 20-4. Into 20.0 ml of a $5.0 \times 10^{-2}\,F$ solution of ammonia is pipetted 5.0 ml of a $1.0 \times 10^{-3}\,F$ solution of copper sulfate. Calculate the concentration of free copper(II) ion.

The solution contains the various ammine complexes of copper(II) ion, and an exact treatment would require the consideration of this fact. The calculation would then involve equations of higher order and the mathematical effort would be great. The treatment, however, can be simplified. The concentration of ammonia is considerably larger than that of the copper ion. As an approximation it may therefore be assumed that the complex $Cu(NH_3)_4^{2+}$ is the only one present; that is, complexes with less than four ammonia molecules coordinated to copper are assumed to be present in negligible quantities. Then the overall stability constant (2.0×10^{12}) can be employed. Applying this and the approximations made in the preceding examples leads to

$$2.0 \times 10^{12} = \frac{\dfrac{5.0 \times 1.0 \times 10^{-3}}{25}}{[Cu^{2+}] \times \left(\dfrac{20.0 \times 5.0 \times 10^{-2}}{25}\right)^4}$$

and $$[Cu^{2+}] = 3.9 \times 10^{-11}\ M$$

20.4 EDTA Titrations, General Considerations

The preponderance of compleximetric titrations involves use of the titrant *e*thylen*d*iamine*t*etra*a*cetic acid, also known as (*e*thylene*d*initrilo)*t*etra*a*cetic acid. The acronym EDTA applies to either name (as indicated here by the italic letters in the name), and also to the anionic form of the compound (that is, the A may represent *a*cetate).

EDTA, introduced as a complex-forming titrant by G. Schwarzenbach shortly after World War II, forms stable, water-soluble, 1:1 chelates with a large number of polyvalent metal ions. It can be obtained in primary standard purity, as the free acid and as the disodium salt dihydrate. Either compound has a favorably high equivalent weight. The salt can be dissolved directly in water; the acid requires addition of sodium hydroxide or another base for dissolution. A titrant solution prepared in either way is stable indefinitely.

In Figure 20-1, the structural formula of EDTA, as the free acid, is shown as well as a representation of the "normal"

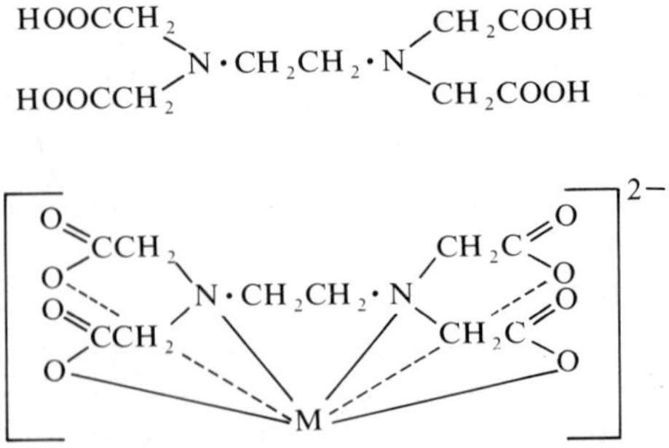

$$(HOOCCH_2)_2N \cdot CH_2CH_2 \cdot N(CH_2COOH)_2$$

$$\left[(O{=}C(O)CH_2)_2N \cdot CH_2CH_2 \cdot N(CH_2C(O){=}O)_2\ M\right]^{2-}$$

Figure 20-1. Structural formulas of EDTA, free acid, and of a 1:1 chelate of a dipositive metal ion and EDTA.

1:1 chelate of a dispositive metal ion, M^{2+}. A count of atoms will reveal that all of the chelate rings are five-membered and that potentially six rings may form. The high stability of EDTA complexes can thereby be appreciated.

EDTA is a tetraprotic acid. The values of the four acid dissociation constants are 1×10^{-2}, 2.1×10^{-3}, 6.9×10^{-7}, and 7.4×10^{-11}. From these values it is evident that two of the protons are strongly acidic and two only weakly acidic. From the values of these constants it can be concluded that in a solution of about pH 5 (closely attained in a solution of the disodium salt of EDTA) the species predominantly present is H_2Y^{2-}, if EDTA is formally represented by H_4Y. At that pH the formation of a complex between EDTA and a metal ion, M^{n+}, may be presented as

$$M^{n+} + H_2Y^{2-} \rightleftharpoons MY^{n-4} + 2H^+ \qquad (20\text{-}8)$$

As can be seen, two hydrogen ions are freed. This reaction is reversible and, consequently, the complex is more extensively dissociated the higher the acidity of the solution. In other words, the proton competes with the metal ion for complexation with the EDTA. This is an important fact because it delineates the necessity of performing any EDTA titration above a certain limiting pH value; otherwise, the complex dissociates to too great an extent and the titration fails. The limiting pH value depends on the stability of the complex formed during the titration. The lower the stability of the relevant metal–EDTA complex, the higher must be the pH value maintained during the titration.

Consequently, evaluation of the feasibility of a complexímetric titration cannot be based simply on the value of the stability constant as such, but among other factors the pH must be taken into account. The influence of the pH can be evaluated by an α factor in a manner analogous to that employed in the treatment of coexistent solubility and acid–base equilibria (see Section 8.13). Consideration of the α factor leads to the conditional stability constant, which is briefly described in Section 20.5.

Conditional Stability Constants 20.5

The expression of the stability constant of a complex formed between a metal ion M^{n+} and EDTA is based on the reaction

$$M^{n+} + Y^{4-} \rightleftharpoons MY^{n-4} \qquad (20\text{-}9)$$

and is then written as

$$K_{st} = \frac{[MY^{n-4}]}{[M^{n+}][Y^{4-}]} \tag{20-10}$$

It should be noted that in this expression only the fully dissociated form of EDTA, that is, the anion Y^{4-}, occurs. By M^{n+} is implied the free metal ion, that is, the aquo complex.

However, only at high pH values (above about pH 10) is the portion of the EDTA not bound to the metal ion present in the form Y^{4-}. The lower the pH, the larger the portion of the uncomplexed EDTA present in various protonated forms. It is possible to write the expression for the conditional stability constant, K_{st}^{H}, that takes care of the influence of the pH as follows:

$$K_{st}^{H} = \frac{[MY^{n-4}]}{[M^{n+}][Y]^*} \tag{20-11}$$

Here $[Y]^*$ is the molar concentration of the EDTA uncombined with the metal ion and in whatever stage of protonation it may be present. The following relationship obviously exists:

$$[Y]^* = [Y^{4-}] + [HY^{3-}] + [H_2Y^{2-}] + [H_3Y^{-}] + [H_4Y] \tag{20-12}$$

α_H is defined as the ratio of the concentration of total uncomplexed EDTA to that of the EDTA present in the form Y^{4-}. Hence

$$[Y]^* = [Y^{4-}]\alpha_H \tag{20-13}$$

Combination of equations (20-11) and (20-13) yields

$$K_{st}^{H} = \frac{[MY^{n-4}]}{[M^{n+}][Y^{4-}]\alpha_H} = \frac{K_{st}}{\alpha_H} \tag{20-14}$$

Consequently, if α_H is known, the value of the conditional stability constant can be calculated from that of the stability constant.

With the acid dissociation constants of EDTA known the factor α_H can be calculated for any pH. The derivation of the required formula proceeds via the successive replacement of all concentration terms other than $[Y^{4-}]$ in equation (20-12) by expressions obtained from the acid dissociation constants:

$$[H_3Y^{-}] = \frac{[Y^{4-}][H^+]}{K_{a,4}}$$

$$[H_2Y^{2-}] = \frac{[H_3Y^{-}][H^+]}{K_{a,3}} = \frac{[Y^{4-}][H^+]^2}{K_{a,3}K_{a,4}} \qquad \text{etc.}$$

The result is

$$[Y]^* =$$

$$[Y^{4-}]\underbrace{\left(1 + \frac{[H^+]}{K_{a,4}} + \frac{[H^+]^2}{K_{a,3}K_{a,4}} + \frac{[H^+]^3}{K_{a,2}K_{a,3}K_{a,4}} + \frac{[H^+]^4}{K_{a,1}K_{a,2}K_{a,3}K_{a,4}}\right)}_{\alpha_H} \qquad (20\text{-}15)$$

By substituting into the parenthetical portion of equation (20-15) the relevant numerical data, the values can be secured for plotting the log α_H versus pH curve for EDTA shown in Figure 20-2. From this curve the log α_H, and consequently α_H, values can be obtained at any pH within the limits at which titrations are performed.

It should be realized that an α_H factor is independent of the metal ion involved and solely related to the ligand. For a given ligand the factor depends only on the pH of the solution.

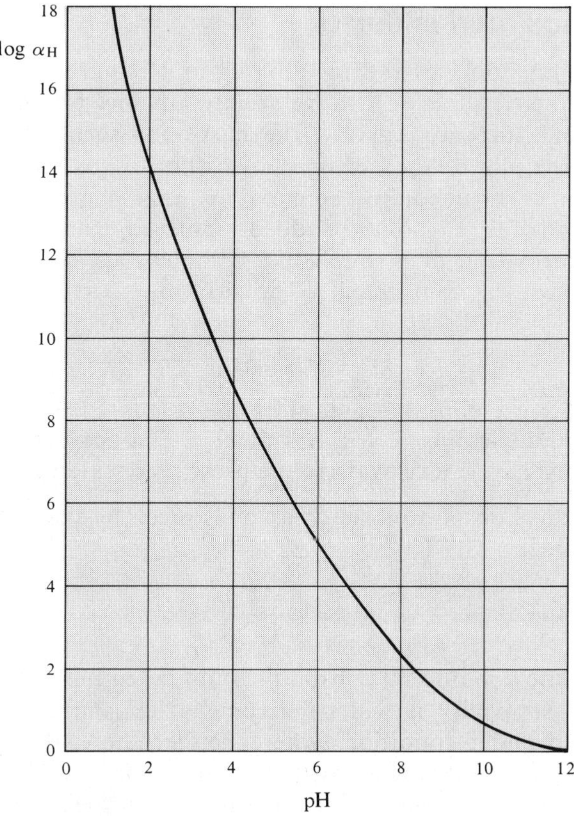

Figure 20-2. Plot of log α_H versus pH for EDTA.

It is, of course, also possible that the metal ion undergoes a side reaction. For example, if some ammonia is added to a solution containing the copper(II)–EDTA complex, ammonia will compete with the EDTA for the copper in a manner analogous to the competition of the proton with the metal for reaction with EDTA. It is possible to derive a factor α_{NH_3} to take care of this influence. The mode of calculation is similar to that for α_H and a formula can be derived that contains the concentration of the additional complexing agent (in the example, ammonia) and the stability constants of the various complexes formed by it with the metal.

The advantage of such factors resides in the fact that the stability constants can be tabulated and the various α factors can be either tabulated or plotted; with these data at hand the value of the conditional constant for any actual solution conditions can be calculated rapidly and then an evaluation of the situation is readily possible.

20.6 Compleximetric Titration Curves

The understanding of the functioning and requirements of a compleximetric titration can be greatly advanced from a discussion of a titration curve. The curve for such a titration involves the plot of the negative logarithm of the molar concentration of the uncomplexed metal ion, that is, of pM, versus the volume of titrant solution added. Only an example involving the formation of a simple 1:1 complex, as in an EDTA titration, will be considered. The titration reaction may be written

$$M + Y \rightleftharpoons MY \tag{20-16}$$

where M represents the metal ion and Y the EDTA (or any other ligand yielding a 1:1 complex). Charges are omitted because they are not relevant to the present consideration.

The stability constant of the complex is given by the following expression:

$$K_{st} = \frac{[MY]}{[M][Y]} \tag{20-17}$$

Actually the conditional constant should be employed. However, for simplicity it will be assumed that the titration is performed under conditions where neither the hydrogen ion nor any other constituent of the solution has any significant influence on the stability of the complex; then the constant as written above can be employed.

Assume that exactly 100 ml of a 0.100 F solution of metal M is titrated with a 0.100 F solution of EDTA, and that the stability constant has a value of 1.0×10^{11}. (The case is close to the titration of calcium at pH 10 or higher.)

At the start of the titration of the pM value is that of the original sample solution. Therefore,

$$\text{pM} = -\log 0.100 = 1.00$$

For the addition of 10.0 ml of titrant solution the calculation of the pM value proceeds as follows. Initially the metal ion present amounted to $100 \times 0.100 = 10.0$ millimoles. The ligand added amounts to $10.0 \times 0.100 = 1.00$ millimole. Since a 1:1 complex is formed, this quantity of titrant reacts with an equimolar amount of the metal ion. As a consequence, $10.0 - 1.0 = 9.0$ millimoles of "free" metal ion remains in a total solution volume of $100 + 10 = 110$ milliliters. Hence

$$[\text{M}] = \frac{9.0}{110}$$

and
$$\text{pM} = -\log(9.0/110) = \log 110 - \log 9.0$$
$$= 2.04 - 0.95 = 1.09$$

Since the complex is rather stable, as revealed by the large value of the stability constant, it can be assumed that essentially all the titrant added combines with the metal ion; in other words, the dissociation of the complex can be neglected. This approximation also applies to further additions of titrant.

For the addition of 20.0 ml of titrant, the calculation proceeds analogously and may be written in consolidated form:

$$[\text{M}] = \frac{100.0 \times 0.100 - 20.0 \times 0.100}{100.0 + 20.0} = \frac{8.0}{120} = \frac{1.0}{15}$$

and
$$\text{pM} = -\log(1.0/15) = \log 15 = 1.18$$

On the addition of 99.0 ml of the complexing agent, the calculation gives

$$[\text{M}] = \frac{100.0 \times 0.100 - 99.0 \times 0.100}{100.0 + 99.0} = \frac{0.10}{199} \simeq \frac{1.0}{2000}$$

and
$$\text{pM} = -\log(1/2000) = 3.30$$

At the equivalence point this mode of calculation results in $[\text{M}] = 0$, an unacceptable result. At this point dissociation of the complex can no longer be neglected. As a matter of fact, dissociation is the only process by which free M is provided. Here the calculation is based on the expression for the stability constant (20-17) and takes the form already discussed in Example 20-1. However, C_M now is the total concentration of metal at the equivalence point. Its numerical value can be calculated from the initial concentration and the dilution that has taken place.

$$C_{M,\text{eq pt}} = \frac{C_{M,\text{initial}} \times 100}{200} = 5.0 \times 10^{-2} F$$

Then

$$[M] = \sqrt{\frac{C_{M,\text{eq pt}}}{K_{st}}} = \sqrt{\frac{5.0 \times 10^{-2}}{1.0 \times 10^{11}}} = \sqrt{5.0 \times 10^{-13}}$$

and $\quad pM = \frac{1}{2}(13 - \log 5.0) = 6.15$

For the calculation of pM values beyond the equivalent point the approach shown in Example 20-3 is applied. On the addition of 110.0 ml of titrant solution, that is, of 10.0 ml beyond the equivalence point, the molar concentration of the free ligand is

$$[Y] = \frac{10.0 \times 0.100}{100 + 110} = \frac{1.00}{210}$$

The molar concentration of the complex is given by

$$[MY] = C_M - [M] \simeq C_M$$

The value of C_M at this point is given by

$$C_M = \frac{100 \times 0.100}{100 + 110} = \frac{10.0}{210}$$

Substitution of the numerical values into the expression for the stability constant,

$$1.00 \times 10^{11} = \frac{10.0/210}{[M] \times 1.00/210}$$

and

$$[M] = \frac{10.0}{1.00 \times 10^{11}} = 1.00 \times 10^{-10} M$$

$$pM = 10.00$$

The value of [M] is very small in comparison with that of C_M, consequently, the above approximation that $[MY] \simeq C_M$ is permissible. Additional points beyond the equivalence point are calculated in the same manner.

The whole curve is presented in Figure 20-3. The shape is again the typical one encountered in acid–base and precipitation titrations; the modes of calculation and the reasonings should be compared. In analogy to these other titration types,

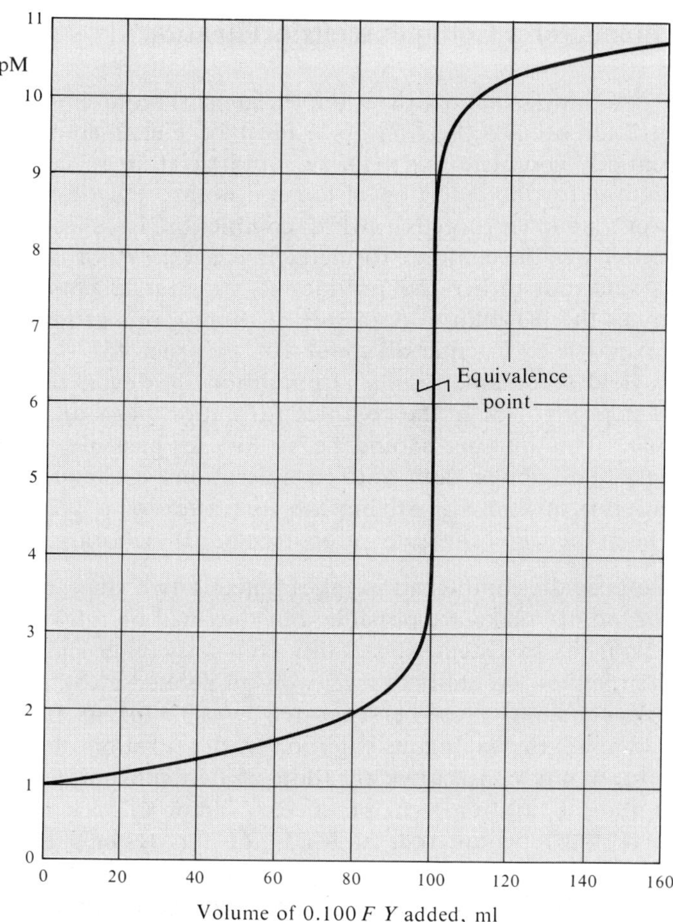

Figure 20-3. Calculated titration curve for titration of 100 ml of a 0.100 F solution of metal ion M with a solution 0.100 F in complexing agent Y to form MY ($K_{st,MY} = 1.0 \times 10^{11}$).

it can be appreciated that the magnitude of the break for a given system will depend on the concentration of the titrant and of the metal ion: The smaller the two, the smaller the break. Further, for different systems at given concentrations of titrant and substance to be titrated, the extent of the break will be the larger the greater the stability of the complex formed.

An increase in the stability of a 1:1 complex by a factor of 10^2 will move the inflection point, that is, the equivalence point, up by 1 pM unit.

20.7 Requirements for a Compleximetric Titration

For a compleximetric titration, the general requirements of any titration process (Section 10.1) must be met, including an essentially complete and relative rapid titration reaction and a method for the detection of the end point. How far the titration reaction proceeds toward completion is reflected in the stability of the complex formed. It is actually not the stability as such but rather the product $K_{st}C_M$ that essentially determines the feasibility. As a rule of thumb, this product should have at least a value of about 10^7 in order for the titration to yield acceptable results. In addition, however, the number of steps involved in the complex formation plays an important role. This number should be as low as possible, preferably one. The above data and considerations consequently hold only for titrations involving the formation of a 1:1 complex. The influence of the number of steps needs elaboration.

Suppose the complexation takes place in two steps. Then two end-point breaks are possible, but they will be fully developed only if the two stepwise stability constants differ sufficiently in their values. Actually to give two well-defined breaks, the stepwise constants must be separated by a factor of 10^4 or 10^5. (Compare the analogous situation in the titration of a diprotic acid, or the consecutive titration of two substances in a precipitation titration.) In practice, a factor of such magnitude is seldom encountered, at least not for systems that fulfill the other requirements for a titration. The two titration steps usually overlap and one poorly defined break results. Even a high value for the overall stability constant will not necessarily assure a satisfactory titration, since this constant is "split" into two stepwise constants. The situation becomes even worse if more than two steps are involved.

Most inorganic ligands are unidentate; therefore, compleximetric titrations involving such ligands are restricted to only a few metal ions, notably silver(I), which has a coordination number of 2, and mercury(II), which, although having a (maximum) coordination number of 4, effectively coordinates with only two particles of many unidentate ligands (see Section 20.11). Compleximetric titrations for a large number of metals are therefore based on the use of organic multidentate ligands that permit the complexation to proceed in a single step. Since chelate complexes are formed, this type of titration may be assigned the more restrictive term "chelatometric titration," where the distinction is useful. EDTA and some of its analogs are the most used representatives of this class of chelate-forming titrants.

So far the discussion has centered on the titration of a metal ion with a complexing agent. Obviously there is a possibility for the reverse titration, that is, the titration of a complexing agent with a suitable metal ion. The treatment is analogous and needs no special consideration.

Indicators in Compleximetric Titrations 20.8

There are a number of possibilities for indicating the end point in compleximetric titrations, including the formation of a precipitate (see Section 20.11) and the use of redox indicators or instrumental methods. However, the technique most often employed is visual indication with a complex-forming indicator. Metallochromic indicators, or simply metal indicators, form colored complexes with a metal ion, and the color of the "metallized" indicator is different from that of the free, "unmetallized" form. The value of the stability constant of the metal-indicator complex must be sufficiently high to prevent extensive dissociation of this complex; such dissociation would cause a sluggish end point. However, the value must be sufficiently smaller than that of the metal-titrant complex to permit complete removal of the metal from the indicator complex at the end point.

The indicator equilibrium and the expression of the stability constant of the metal-indicator complex take the following form, where charges are again omitted since they are not relevant to the discussion:

$$\mathrm{M} + In \rightleftharpoons \mathrm{M}In \qquad (20\text{-}18)$$

$$K_{\mathrm{st,\,M}In} = \frac{[\mathrm{M}In]}{[\mathrm{M}][In]} \qquad (20\text{-}19)$$

Rearrangement of equation (20-19) yields

$$[\mathrm{M}] = \frac{1}{K_{\mathrm{st,\,M}In}} \times \frac{[\mathrm{M}In]}{[In]} \qquad (20\text{-}20)$$

and, written in logarithmic form,

$$\mathrm{pM} = \log K_{\mathrm{st,\,M}In} - \log \frac{[\mathrm{M}In]}{[In]} \qquad (20\text{-}21)$$

The analogy of the last equation with equation (12-4) for acid–base indicators is obvious, and all the considerations advanced in Chapter 12 concerning visibility, transition range, etc., are fully applicable here. The ideal case would be for the pM value of the midpoint of the color transition range to be identical with the pM value of the equivalence point in the compleximetric titration. In practice, however, because of the large breaks usually encountered, it is sufficient to have the transition range located on the steep portion of the titration curve, and then to perform the titration to the disappearance of one limiting color. Metal indicators exist that also form other than 1:1 complexes. The theoretical treatment is then somewhat more involved, but for practical application the considerations are the same as those given above.

Metal indicators complex not only with metal ions but also with protons and thus show acid–base indicator properties. Consequently, the pH of the solution is of importance to their functioning and not only with respect to the necessity of considering the conditional stability constant of the indicator complex.

As an example, Eriochrome Black T may be discussed; this is a typical metal indicator that finds widespread application in EDTA titrations. The compound is a triprotic acid. One proton corresponds to a relatively strong acid and the other two to quite weak ones. Consequently, the indicator can be denoted by H_2D^-. Eriochrome Black T can act as an acid–base indicator and the color changes and perspective pH values involved may be presented as follows:

$$\begin{array}{ccccc} H_2D^- & \underset{+H^+}{\overset{-H^+}{\rightleftharpoons}} & HD^{2-} & \underset{+H^+}{\overset{-H^+}{\rightleftharpoons}} & D^{3-} \\ \text{red} \leftarrow & \text{pH 6–7} & \rightarrow \text{blue} \leftarrow & \text{pH 11–12} & \rightarrow \text{orange} \\ & (pK_{a,1}, 6.3) & & (pK_{a,2}, 11.6) & \text{(decomposition)} \end{array} \tag{20-22}$$

All metal complexes of Eriochrome Black T are red, although with different metals the red differs in shade. In order to employ this dye as a metal indicator that gives a distinctive color change, the pH of the solution must be above 7 and below 11. Only then is it possible to secure a color change from the red of the metal complex to the blue of the free indicator.

Of course, such a pH range is not the only consideration in the selection of the conditions for a titration. The indicator as a weak acid has a high affinity toward the protons. Consequently, at too low a pH, even when within the range 7 to 11, the hydrogen ion concentration may become sufficiently high for the indicator to prefer combination with protons instead of the particular metal; the compound will then not function to indicate the end point. Thus it becomes obvious that the use of metal indicators requires that the appropriate pH be adjusted for a titration and be maintained within reasonable limits during the titration. For this purpose appropriate buffers are commonly employed.

Some metals form more stable complexes with a given indicator than with the titrant. For those metals, the given indicator is inappropriate. Moreover, in the presence of even traces of such metals the indicator will react with them and not function. The indicator is then said to be "blocked" (see Section 20.10).

Masking 20.9

Masking is defined as the exclusion of the action of an interfering substance by addition of an appropriate reagent and without removal of the substance. Masking may be effected by precipitation, oxidation, reduction, complexation, or by combination of more than one of these processes. Masking is widely employed as a powerful tool in many branches of analytical chemistry. However, because masking is quite often applied in EDTA titrations and because the majority of masking techniques involve complex formation, it is appropriate to discuss this topic in the present chapter. Some examples randomly drawn from practical analysis will illustrate various masking approaches.

The EDTA titration of magnesium is performed at pH 10 and, under the conditions prevailing, barium (among various other metals) reacts with the titrant. If, however, sodium sulfate is added, barium precipitates as barium sulfate and then no longer interferes with the magnesium determination. The titration is performed in the presence of the precipitate. If the barium sulfate were removed, separation would be involved rather than masking.

Precipitation masking frequently suffers many of the adverse phenomena discussed for precipitation in connection with gravimetry. Coprecipitation of the substance to be determined or the reagent used in the determination has quite detrimental effects. The presence of the precipitate may obscure other phenomena, for example, the color change at the end point in a titration. Consequently, only a few cases of precipitation masking exist that are of practical value.

In many cases the chemical behavior of a metal ion changes

drastically when the oxidation state is changed. For example, chromium(III) precipitates readily as the hydrous oxide and thereby interferes with the gravimetric determination of, say, iron(III). If the chromium is oxidized to chromium (VI) it is present in anionic form as chromate or dichromate ion and as such does not precipitate together with iron upon the addition of ammonia. As another example of oxidation masking may be cited arsenic(III), which interferes with the color reactions of some other elements; by oxidation to arsenic(V) the interference is often obviated.

Reduction masking is quite effective where iron(III) causes disturbances. Conversion to iron(II), often effected by reduction with ascorbic acid, is in many analytical procedures the choice for eliminating the interference. Another example is the reduction of mercury(II) to mercury(I), which at the same time is precipitated as the insoluble chloride; here a mixed masking process is encountered involving both reduction *and* precipitation. However, the number of metals that can readily change their oxidation state is limited and as a consequence so is the application of reduction masking.

In contrast, the number of complexation reactions is enormous and such reactions are less prone to difficulties than, for example, precipitation reactions. Accordingly, complexation is the most frequently employed process to effect the masking of metal ions.

An example of a most versatile and effective complexing agent used in masking is the cyanide ion. It forms very stable complexes with a variety of metal ions including those of iron, copper, mercury, cobalt, nickel, cadmium, zinc, and the noble metals. The masking of iron(II) and of iron(III) leads to the formation of the respective hexacyanoferrates. The masking is so effective that no chemical reaction known for either iron(II) or iron(III) gives a positive test. This "vanishing" of iron was very puzzling to early chemists. The cyano complexes of zinc and cadmium are relatively weak and the cyanide is able to react with formaldehyde; as a consequence zinc and cadmium can be freed, that is, "demasked." Based on this reaction the selective detection and determination of zinc or cadmium is possible in the presence of the other elements named above, the cyano complexes of which are not affected by formaldehyde.

Iron(III) in chloride-containing solutions produces an intensely yellow color due to the formation of chloro complexes. Although the iron(III) itself may undergo no interfering reaction, the intense color can cause difficulties in the determination (or detection) of other substances. If phosphoric acid is added, a colorless iron(III)-phosphato complex forms and the yellow solution color disappears. Silver forms an ammine complex of sufficient stability to prevent the precipitation of silver chloride, but not that of silver iodide; consequently, ammonia masks silver against chloride and allows the selective detection or determination of iodide. The stability of the complexes formed by many metals with EDTA is the reason for EDTA having become one of the most versatile masking agents. An example of mixed redox-complexation masking may be cited. Cobalt(II) upon addition of ammonia and hydrogen peroxide is oxidized to cobalt(III) and strongly bound in an ammine complex. The ammine complexes of other metals are sufficiently weak to permit reaction of these metals with other reagents. This masking of cobalt is especially valuable when nickel is to be determined.

Additional examples of masking will be found below.

Application of EDTA Titrations 20.10

An EDTA titration is most frequently employed in the determination of the *total* hardness of water. (The concept of water hardness has already been discussed in Section 14.8.) Since both calcium and magnesium complex strongly with EDTA, the sum of these metals is readily determined. The determination can be effected by adding pH 10 ammoniacal buffer and some Eriochrome Black T to the water sample, and then titrating with a standard EDTA solution to an end point marked by a color change of red to blue. Usually some potassium cyanide is also added to mask traces of heavy nickel that would block the indicator. EDTA titrations also permit the rapid determination of both the calcium and magnesium hardness separately. In one aliquot of the water sample, the *sum* of calcium and magnesium is titrated as described above. To a second aliquot sodium hydroxide is added to attain a pH above 12. Magnesium is precipitated as the hydroxide and is thereby masked in the titration of calcium with EDTA. Murexide is an indicator often employed in this titration. The magnesium hardness is found by difference.

The principles for determining calcium and magnesium in water described above are employed for the determination of these two elements in a large variety of materials, ranging from blood serum and pharmaceuticals to rocks and minerals.

The analysis of alloys can frequently be accomplished rapidly and simply via EDTA titrations. For example, a bismuth–lead alloy can be analyzed as follows. A sample of the material is dissolved in nitric acid and then the bismuth is titrated with EDTA at a pH 1 to 1.5, employing Pyrocatechol Violet as the indicator. When the end point is reached (blue to yellow), the volume of EDTA required is recorded and the pH is increased to about 5. Then lead is titrated with EDTA in the same solution using the same indicator (again a blue to yellow color change).

A mixture of zinc and lead may be analyzed in the following manner. To the weakly acidic sample solution some tartaric acid is added; the tartrate masks the lead against precipitation as the hydroxide (but does not impair its reaction with EDTA) when ammoniacal buffer of pH 10 is added in the next step. (The ammonia complexes with the zinc and thus masks it against hydroxide precipitation.) To the buffered solution potassium cyanide is added next. This agent masks the zinc against EDTA but does not affect the lead, which is then titrated with EDTA using Eriochrome Black T as the indicator. The volume of EDTA is recorded and formaldehyde is added to demask the zinc (see Section 20.9), which is then titrated using the same titrant and indicator.

20.11 Other Compleximetric Titrations

Of the limited examples of complexation titrations not involving a chelating titrant, two will be selected that are of practical interest.

The titration of cyanide with silver was developed as early as 1851 by Liebig. The titration reaction is

$$2CN^- + Ag^+ \rightleftharpoons Ag(CN)_2^- \qquad (20\text{-}23)$$

This actually is a two-step reaction, but the formation constant of the complex has an exceptionally high value (1.1×10^{21}). The titration is self-indicating and no added indicator is necessary. Near the equivalence point, all of the cyanide ion is transferred into the dicyanoargentate(I) ion. On addition of more titrant solution (that is, silver nitrate) insoluble silver dicyanoargentate(I) is formed. The "indicator reaction equation" is

$$Ag(CN)_2^- + Ag^+ \rightleftharpoons \underline{Ag[Ag(CN)_2]} \qquad (22\text{-}24)$$

The appearance of a permanent turbidity signals the end point. Good stirring and slow addition of titrant solution are essential in this titration. Because of local overconcentration of titrant a transient turbidity may occur in the course of the titration. In addition, the permanent turbidity indicating the end point develops slowly. The cyanide titration is routinely applied to the analysis of cyanide-containing plating baths.

The mercurimetric determination of chloride offers another example of compleximetric titration of practical significance. The titration is simply performed by adding a titrant solution of mercury(II) nitrate or perchlorate to the acidic sample solution until an appropriate indicator signals the end point. The combining ratio between mercury(II) and chloride ion in this titration is 1:2.

Although mercury(II) ion has a coordination number of 4, it often acts with an "effective" coordination number of 2. This is evident, for example, from the logarithmic values of the stability constants of the chloro complexes of mercury(II)—$HgCl^+$, $HgCl_2$, $HgCl_3^-$, and $HgCl_4^{2-}$: 6.75, 6.48, 0.95, and 1.05, respectively. The constants for the first and second steps are close in their value. In a simple way it may be reasoned that the coordination of the two chloride ions takes place simultaneously in a quasi-single step. In addition, the overall stability constant for the two steps is high (1.7×10^{13}). Mercury(II) chloride, as a complex, is a weak electrolyte and is little dissociated in aqueous solution. In contrast, mercury nitrate and perchlorate are strong electrolytes and are commonly used as titrants. A stable, mercury(II) standard solution may be prepared by dissolving a weighed amount of mercury(II) oxide in either nitric or perchloric acid and diluting to known volume with water; mercury(II) oxide is an excellent primary standard due to its high equivalent weight, exact stoichiometric composition, and ready availability in high purity.

In mercurimetric titrations *sym*-diphenylcarbazide [i.e., 1,5-diphenylcarbohydrazide, $(C_6H_5NHNH)_2CO$] or its oxidation product *sym*-diphenylcarbazone is often employed as a metallochromic indicator. These compounds are yellow in solution and form blue-violet complexes with mercury(II). With the first indicator the optimum range for the titration is pH 1.5 to 2.0, and with the second, pH 3.0 to 3.5.

The compleximetric titration of chloride with mercury(II) ion, since it is operative in acidic solution, is often employed in place of the Mohr titration with silver ion, especially in the determination of chloride in blood serum and other biological fluids.

20.12 Calculation of Results in Compleximetric Titrations

The calculations are based on the reaction describing the complex formation. This reaction may be written in the general form

$$nS + mT \rightleftharpoons S_nT_m \tag{20-25}$$

where S and T are the species titrated and the titrant, respectively. From this equation it follows that one mole of the titrant is equivalent to n/m moles of the species titrated. If the species titrated and the sought-for substance are not identical, an appropriate conversion factor may be additionally applied. The calculations are especially simple in the case of EDTA titrations, because here 1:1 complexes are formed almost exclusively, yielding $n = m = 1$. (Only molybdenum and tungsten form 2:1 complexes.) Some examples will illustrate the approach.

Example 20-5. How much $CaCl_2$ (*110.99*) in milligrams is present in a sample which on titration requires 25.22 ml of 0.01350 *F* EDTA?

Since calcium and EDTA react in a 1:1 ratio, the calculation takes the form

$$\text{ml}_t \times F_t \times \text{FW}_s \times 1 = \text{mg}_s$$
$$25.22 \times 0.01350 \times 110.99 \times 1 = 37.79 \text{ mg of } CaCl_2 \text{ in sample}$$

Example 20-6. What amount of KCN (65.12) in milligrams is present in a sample if 18.35 ml of 0.1050 *F* $AgNO_3$ is required for its compleximetric titration?

From the titration equation (20-15), silver and cyanide ions react in a 1:2 ratio; hence

$$\text{ml}_t \times F_t \times \text{FW}_s \times 2 = \text{mg}_s$$

Substitution of the numerical data yields

$$18.35 \times 0.1050 \times 65.12 \times 2 = 250.9 \text{ mg of KCN in sample}$$

Example 20-7. A solution containing nickel ion is titrated in an ammoniacal medium with 0.1160 *F* KCN, of which a volume of 20.15 ml is required. Express the titration result in milligrams of nickel pyrophosphate, $Ni_2P_2O_7$ (*291.36*).

The titration reaction is

$$Ni^{2+} + 4CN^- \rightarrow Ni(CN)_4^{2-}$$

Thus, one-fourth mole of nickel is equivalent to one mole of cyanide ion or of potassium cyanide; however, two moles of nickel ion are furnished by one mole of nickel pyrophosphate. Consequently, one-eighth mole of nickel pyrophosphate is equivalent to one mole of potassium cyanide and the required formula becomes

$$\text{ml}_t \times F_t \times \text{FW}_s \times \tfrac{1}{8} = \text{mg}_s$$

Here the factor $\frac{1}{8}$ contains both the combining ratio according to the complexation reaction $(\frac{1}{4})$ and the equivalence factor for the conversion of nickel to nickel pyrophophate $(\frac{1}{2})$.

Alternatively, the overall stoichiometry may be shown by writing the reaction of interest as

$$Ni_2P_2O_7 + 8KCN \rightarrow 2K_2[Ni(CN)_4] + K_4P_2O_7$$

From the equation it is immediately evident that one mole of potassium cyanide is equivalent to one-eighth mole of nickel pyrophosphate.

Substitution of the numerical data yields

$$20.15 \times 0.1160 \times 291.36 \times \tfrac{1}{8} = 85.13 \text{ mg of } Ni_2P_4O_7$$

Questions 20.13

What is the relationship between the dissociation constant and the stability constant of a complex?

Define in your own words the terms chelate ring, chelate complex, central ion, coordination number, masking, and blocking of an indicator.

Explain why a low number of steps is essential if a complexation process is to be made the basis of a titration.

Compare the titration curves and parameters influencing their shape for acid–base, precipitation, and compleximetric titrations. What influences the magnitude of the "break" and the position of the equivalence point? What changes take place in the shape of the titration curves and in the location of the equivalent point if changes are made in concentrations of the species titrated and of the titrant?

When zinc is titrated with EDTA using Eriochrome Black T, an ammoniacal buffer solution is added. What function does this solution perform in addition to buffering the solution? Explain.

Explain how a metallochromic indicator functions. Suggest some requirements for a substance which is to serve satisfactorily as such an indicator.

Discuss the possibility of the stepwise titration, at a single pH value and without resort to masking, of two metal ions with a single titrant that complexes with both ions. Compare your reasoning with that for the stepwise titration of a diprotic acid with a base.

20-8. In the EDTA titration of magnesium any iron(III) present in the original sample solution precipitates as hydrous oxide upon addition of ammoniacal buffer of pH 10. The hydrous oxide does not react with the indicator and the titrant. Critically discuss the possibility of using these facts for a masking of iron.

20-9. Would the total hardness of a water sample change if it were heated before analysis?

20-10. Could ordinary tap water be used in the preparation of a standard EDTA solution? Elaborate on your answer.

20-11. Explain why adequate buffering is important in EDTA titrations.

20-12. Give several examples for each type of masking technique. Draw the examples from your own knowledge and experience and do not cite those discussed in Section 20.9.

20-13. What is the meaning of the statement "EDTA can function as a hexadentate ligand"?

20-14. What do you understand by the expression: "free" metal ion concentration in a solution?

20.14 Problems

20-1. A 1:1 complex, MZ, having a stability constant of 1.00×10^{10}, is present as a 2.50×10^{-3} F solution. Calculate the free metal ion concentration.

Answer: 5.00×10^{-7} M

20-2. The solution of the complex MZ of Problem 20-1 is made 0.100 M in Z. What is the free metal ion concentration?

Answer: 2.5×10^{-12} M

20-3. Exactly 10 ml of a 0.0100 F solution of M is titrated complexi-metrically with 0.00500 F Z. The reactants form a 1:1 complex MZ ($K_{st} = 1.00 \times 10^{11}$). Calculate the pM value at (a) the start of the titration, (b) after the addition of exactly 5 ml of titrant, (c) after the addition of exactly 20 ml of titrant, and (d) after the addition of exactly 30 ml of titrant.

Answers: (a) 2.00; (b) 2.30; (c) 6.74; (d) 10.70

20-4. What is the concentration of free metal ion in a 0.0010 F solution of the complex MZ ($K_{st} = 5.0 \times 10^{2}$)? (*Hint:* Since the complex is a weak one, C_{MZ} cannot be taken as equal to [MZ].)

Answer: 7.3×10^{-4} M

20-5. A metal ion forms a complex MZ ($K_{st} = 1.0 \times 10^{8}$) and a slightly soluble compound MP_2 ($K_{sp} = 1.0 \times 10^{-10}$). Can a 0.100 F solution of MZ be made 0.010 M in P without precipitation of MP_2 occurring? Calculate the free metal ion concentration according to the

relevant solubility and complexation equilibria and compare the figures in making your decision.
Answer: No; $[M]_{complex} = 3.2 \times 10^{-5}\ M$, $[M]_{solubility} = 1.0 \times 10^{-6}\ M$

. A 0.200 *F* EDTA solution is used to titrate exactly 100 ml of a solution of $ZnCl_2$ (*136.3*). A volume of 50.0 ml of EDTA was required to attain the Eriochrome Black T end point. What is the % w/v concentration of the zinc salt in the solution?

Answer: 1.36% w/v $ZnCl_2$

. What volume, in milliliters, of 0.0500 *F* $MgSO_4$ is required for the determination of the substances listed below by way of a back-titration? In each case a volume of 30.00 ml of 0.0500 *F* EDTA has been added to a solution containing 100.0 mg of the substance indicated.

	Substance titrated	
(a)	$CaCO_3$	(*100.1*)
(b)	$MgSO_4$	(*120.4*)
(c)	$Pb(NO_3)_2$	(*331.2*)
(d)	$Al_2(SO_4)_3$	(*342.1*)

. A mixture contains KCl (*74.56*), LiCl (*42.39*), and $MgSO_4 \cdot 7H_2O$ (*246.48*). An amount of 1.8650 g of the mixture is dissolved in pure water and diluted to mark with water in a 250-ml volumetric flask. A one-fifth aliquot of the solution is treated with 100.0 ml of 0.1000 *F* $AgNO_3$. The excess silver is back-titrated with 0.1885 *F* KSCN, 20.00 ml being required. Another one-fifth aliquot of the sample solution requires 11.50 ml of 0.0100 *F* EDTA. Calculate the % w/w composition of the sample.

. Calculate the molar concentrations of uncomplexed metal ion, M, and ligand, Z, in a solution 1.0×10^{-3} *F* in complex MZ, from the assumed values of the stability constant.

	Stability constant of MZ
(a)	1.0×10^{9}
(b)	5.0×10^{10}
(c)	7.5×10^{8}
(d)	6.2×10^{6}
(e)	8.4×10^{9}
(f)	4.0×10^{5}
(g)	3.2×10^{14}

. Calculate the molar concentrations of uncomplexed metal ion, M, and ligand, Z, when the concentrations of the complex, MZ ($K_{st} = 2.0 \times 10^{8}$) and ligand, Z, in the solution have the specified values.

	Formality in MZ	Molarity in excess Z
(a)	1.0×10^{-2}	1.0
(b)	1.0×10^{-3}	5.0×10^{-1}

(c) 1.0×10^{-1} 1.0×10^{-3}
(d) 5.0×10^{-2} 2.0×10^{-2}
(e) 5.0×10^{-1} 1.0×10^{-3}
(f) 5.0×10^{-3} 8.0×10^{-1}

20-11. A standard mercury(II) solution is prepared by dissolving exactly 1 g of mercury (*200.59*) of 99.71% purity in nitric acid and diluting to exactly 1 liter with pure water. A sample solution is titrated with the mercury(II) solution, 24.85 ml being required. Calculate the amount of NaCl, in milligrams, in the sample.

20-12. A solution of $FeCl_3$ is titrated with 0.04930 *F* mercury(II) nitrate, 35.21 ml being required. What volume, in milliliters, of 0.0832 *F* EDTA will be required to titrate the iron in an identical aliquot of that solution?

20-13. The chloride content of blood serum is often determined by a mercurimetric titration. In a centrifuge tube are mixed 2.00 ml of water, 0.50 ml of blood serum, and 0.50 ml of trichloroacetic acid. Protein is thereby precipitated and is removed by centrifugation. Exactly 2 ml of the clear supernatant liquid is transferred to a vessel and titrated with 0.0100 *F* mercury(II) nitrate using *sym*-diphenylcarbazide as the indicator. A blank is performed in exactly the same manner but substituting an equal volume of water for the serum. The following formula is used to calculate the result:

$$(\text{ml}_{\text{titrant in titration}} - \text{ml}_{\text{titrant in blank}}) \times (\text{factor}) = \text{mg }\%\text{ Cl}$$

By mg % Cl in the jargon of clinical analysis is meant the milligrams of chloride in 100 ml of the serum. If the determination is performed as described, the *factor* has a certain numerical value. Calculate this value. Chloride has a formula weight of 35.45. Take into account that one mercury(II) ion in the titration complexes with two chloride ions.

Answer: 213

20-14. The normal concentration of calcium (*40.1*) in blood serum is 9 to 11 mg % (that is, 9 to 11 mg of calcium in 100 ml of serum). A value of 7 mg % or lower or of 13 mg % or higher is considered a sign of an abnormal condition. A 1-ml sample of serum in a complexi-metric titration of calcium was found to require 2.85 ml of 1.560×10^{-3} *F* EDTA.

(a) Express the result of the titration in mg % Ca.
(b) Does the result suggest an abnormal condition?
(c) The concentration is often given in milliequivalents of calcium per liter of serum, with one equivalent here taken as one-half mole. Express the titration result in this way.

Answers: (a) 17.8 mg % Ca; (b) yes; (c) 8.9 meq of Ca per liter

20-15. The normal level of magnesium (*24.3*) in blood serum is 2 to 3 mg % (that is, 2 to 3 mg of magnesium in 100 ml of serum).

(a) How many grams of EDTA as the disodium salt dihydrate (*372.2*) must be dissolved in water and be diluted to a volume of 100.0 ml

to obtain a solution of which 0.100 ml is equivalent to 0.20 mg % Mg when a 1-ml sample of serum is taken and titrated after calcium is removed by an oxalate precipitation?
(b) What is the formality of this EDTA solution?
(c) What mg % Ca of calcium (*40.1*) is equivalent to 0.100 ml of this EDTA solution when 1 ml of serum is taken for the titration?
(d) Describe the conditions, including the indicator used, for the titration of the magnesium. What should be the titration result of the magnesium titration if calcium were not removed before this titration? Explain.

Answers: (a) 0.0306_3 g; (b) 8.23×10^{-4} *F*; (c) 0.330 mg % Ca; (d) the result would be high

21 OXIDATION-REDUCTION EQUATIONS (REDOX EQUATIONS)

Introduction 21.1

Originally, the term "oxidation" was used to designate a reaction in which a substance is reacted or combined with oxygen, and "reduction" a reaction in which oxygen is removed from a substance. These concepts have undergone generalization and, in terms of the electron concepts of matter, are applied to reactions in which electrons are lost or gained. If a substance loses one or more electrons, it is said to undergo oxidation; if it gains one or more electrons, it is said to undergo reduction. The overall *red*uction-*ox*idation process, abbreviated *redox* process, involves the transfer of one or more electrons from the reducing agent (also known as the reductant or reducer) to the oxidizing agent (the oxidant or oxidizer). Since free electrons do not accumulate in chemical reactions, the number of electrons lost must exactly equal the number of electrons gained.

A redox reaction, involving the transfer of one or more electrons, is formally analogous to an acid–base reaction in which one or more protons are transferred (see Chapter 16).

Because electrons do not appear as reactants or products in an overall redox equation, it is often profitable to split the reaction into two partial reactions, known as half-reactions: one related to the oxidation and having one or more electrons appearing as a product, the other related to the reduction and having one or more electrons appearing as a reactant. The correctly balanced half-reaction equations may be multiplied by appropriate factors so that on their addition the electrons cancel out and the balanced overall redox equation is obtained. The details of this balancing process for redox equations are considered below.

The concept of the half-reaction has an experimental basis since many such reactions can occur in electrochemical half-cells (Chapter 22). When two such half-cells are suitably connected, the redox process proceeds without physical mixing of the components of the two half-cells. It should be appreciated, however, that half-reaction equations—as most chemical equations—do not delineate the actual mechanism of the process. The considerations in this textbook are confined to half-reactions occurring in aqueous medium and, consequently, where necessary, water and hydrogen ion or hydroxide ion can be considered as participants in half-reactions in order to achieve a balance of material in the equation.

A half-reaction equation must be balanced with regard to material (that is, the same number of atoms of each element must appear on both sides of the equation) and also with regard to electrical charge. The preferred approach, which is delineated below, is to balance material first, with the addition of water, hydrogen ion, or hydroxide ion as a reactant or product where necessary. Charge can then be balanced by simply adding the arithmetically necessary number of electrons as either a product or reactant. An alternative approach, considered in Section 21.4, involves establishing initially the number of electrons involved in a half-reaction equation; this requires the assignment of oxidation states to the atoms involved in the reduction and oxidation processes. Such assignments must often be arbitrary in the case of composite ions and covalent compounds, and may then be a source of confusion. Of course, either method, correctly followed, leads to the same redox equation.

21.2 Balancing Redox Equations: Electron-Change Method

The first step in balancing a redox equation is obviously to write down the reactants and products, which must either be established experimentally or be inferred from chemical knowledge. Once the reactants and products are written, the balancing of the redox equation merely involves balancing of material and charge.

Here redox equations and half-reaction equations are written with equal signs; where the reactions are known to be fully reversible, the double arrow, $\rightleftharpoons$, might be substituted appropriately. The balancing process, however, is independent of the reversibility or nonreversibility of the redox process.

Example 21-1. Write the redox equation for the reaction of iron(III) with tin(II) to yield iron(II) and tin(IV).

The half-reaction equations are

$$Sn^{2+} = Sn^{4+} + 2e \qquad \text{(oxidation, balanced)}$$
$$Fe^{3+} + e = Fe^{2+} \qquad \text{(reduction, balanced)}$$

The second half-reaction equation is then multiplied by 2 and the two equations added:

$$\begin{array}{rcl} Sn^{2+} & = & Sn^{4+} + 2e \\ 2Fe^{3+} + 2e & = & 2Fe^{2+} \\ \hline 2Fe^{3+} + Sn^{2+} & = & 2Fe^{2+} + Sn^{4+} \end{array} \qquad \text{(material and charge balanced)}$$

In the above example, the number of electrons involved in each half-reaction equation is obvious since simple ions are involved. However, with composite ions or covalent compounds, the determination of the number of electrons becomes more difficult.

Example 21-2. Iron(II) is oxidized in *acidic* medium by permanganate ion, MnO_4^-, to yield iron(III) and manganese(II). Balance the redox equation.

$$Fe^{2+} + MnO_4^- = Fe^{3+} + Mn^{2+} \qquad \text{(unbalanced)}$$

The half-reactions are

$$Fe^{2+} = Fe^{3+} \qquad \text{(unbalanced)}$$
$$MnO_4^- = Mn^{2+} \qquad \text{(unbalanced)}$$

Notice that at this point it is unnecessary to recognize explicitly which half-reaction represents oxidation and which reduction.

The first half-reaction equation is already balanced with regard to material, and to achieve a balance of charge one electron is added to the right side:

$$Fe^{2+} = Fe^{3+} + e \qquad \text{(balanced)}$$

In the second half-reaction equation, manganese is already balanced, since one atom appears on both sides. To balance oxygen, four molecules of water are added to the right side, and then to balance the added hydrogen, eight hydrogen ions are added to the left side. Hydrogen ion and water are used, rather than hydroxide ion and water because the reaction occurs in acidic medium. The result thus obtained is

$$MnO_4^- + 8H^+ = Mn^{2+} + 4H_2O \qquad \text{(material balanced, charge unbalanced)}$$

To balance charge, it is now only necessary to note the net electric charge for each side of the equation—+7 on the left and +2 on the right—and to add electrons so that the net charges become equal. Consequently, five electrons are added to the left side. The fully balanced half-reaction equation is therefore

$$MnO_4^- + 8H^+ + 5e = Mn^{2+} + 4H_2O \qquad \text{(balanced)}$$

To obtain the overall redox equation, the first half-reaction equation, which involves a single electron, is multiplied by 5 and the two half-reaction equations are added. The final result is

$$5Fe^{2+} + MnO_4^- + 8H^+ = 5Fe^{3+} + Mn^{2+} + 4H_2O$$

21.2 ELECTRON-CHANGE METHOD

Example 21-3. The following oxidation of the hexanitritocobaltate(III) ion occurs in acidic solution:

$$Co(NO_2)_6^{3-} + MnO_4^- = Mn^{2+} + Co^{2+} + NO_3^- \qquad \text{(unbalanced)}$$

The half-reactions may be written

$$MnO_4^- = Mn^{2+} \qquad \text{(unbalanced)}$$
$$Co(NO_2)_6^{3-} = Co^{2+} + NO_3^- \qquad \text{(unbalanced)}$$

The first half-reaction equation has already been considered in Example 21-2:

$$MnO_4^- + 8H^+ + 5e = Mn^{2+} + 4H_2O \qquad \text{(balanced)}$$

In the second half-reaction equation, six nitrate ions must appear on the right side to balance the six atoms of nitrogen on the left:

$$Co(NO_2)_6^{3-} = Co^{2+} + 6NO_3^- \qquad \text{(unbalanced)}$$

Twelve oxygen atoms appear on the left side and eighteen on the right; therefore to balance oxygen, six molecules of water are added to the left side:

$$Co(NO_2)_6^{3-} + 6H_2O = Co^{2+} + 6NO_3^- \qquad \text{(unbalanced)}$$

To balance hydrogen, twelve hydrogen ions must be added to the right side. (Water and hydrogen ion are added because the reaction occurs in acidic medium.) The balancing of material is thereby completed:

$$Co(NO_2)_6^{3-} + 6H_2O = Co^{2+} + 6NO_3^- + 12H^+$$
$$\text{(material balanced, charge unbalanced)}$$

The net charge on the left side is -3 and on the right side $+8$; to balance charge, eleven electrons must be added to the right side to yield the fully balanced half-reaction equation:

$$Co(NO_2)_6^{3-} + 6H_2O = Co^{2+} + 6NO_3^- + 12H^+ + 11e$$
$$\text{(balanced)}$$

The overall redox equation is obtained by multiplying the first half-reaction equation by eleven and the second by five and adding:

$$11MnO_4^- + 88H^+ + 55e = 11Mn^{2+} + 44H_2O$$
$$5Co(NO_2)_6^{3-} + 30H_2O = 5Co^{2+} + 30NO_3^- + 60H^+ + 55$$
$$\overline{5Co(NO_2)_6^{3-} + 11MnO_4^- + 28H^+ = 5Co^{2+} + 30NO_3^- + 11Mn^{2+} +}$$

Example 21-4. In an alkaline aqueous medium, the citrate anion, $C_6H_5O_7^{3-}$, is completely oxidized by permanganate ion according to the equation

$$C_6H_5O_7^{3-} + MnO_4^- = CO_2 + MnO_2 \qquad \text{(unbalanced)}$$

Notice that in alkaline media the reduction product of the permanganate ion is insoluble manganese(IV) oxide, rather than manganese(II), which is obtained in acidic media. Carbon dioxide is written here as a product. Depending on the solution conditions, this may persist in the solution as hydrogen carbonate ion or carbonate ion. Where the exact conditions are known, one of these ions could be written as the product and the redox equation be balanced accordingly. The half-reactions are

$$MnO_4^- = MnO_2 \qquad \text{(unbalanced)}$$
$$C_6H_5O_7^{3-} = CO_2 \qquad \text{(unbalanced)}$$

The first half-reaction equation is already balanced with regard to manganese; to achieve a material balance, oxygen must be considered. As an alkaline medium is involved, it is desirable to operate with water and hydroxide ion. In such a case, add the difference in oxygen in the form of water to the side with the excess in oxygen and then add twice that amount of hydroxide ion to the other side. In the reaction of immediate interest, this results in

$$MnO_4^- + 2H_2O = MnO_2 + 4OH^- \qquad \text{(material balanced, charge unbalanced)}$$

Since the net charges on the left and right sides are -1 and -4, respectively, charge is balanced by the addition of three electrons to the left side to yield the fully balanced half-reaction equation:

$$MnO_4^- + 2H_2O + 3e = MnO_2 + 4OH^- \qquad \text{(balanced)}$$

In the second half-reaction equation, citrate contains both hydrogen and oxygen. Whenever such a situation exists, the measures suggested above in the balancing of the permanganate half-reaction in alkaline medium may be inadequate. However, the material balance can be readily achieved as follows:

First, balance all elements except hydrogen and oxygen. For the second half-reaction this gives

$$C_6H_5O_7^{3-} = 6CO_2 \qquad \text{(unbalanced)}$$

Then balance oxygen by adding water where needed. Thus

$$C_6H_5O_7^{3-} + 5H_2O = 6CO_2 \qquad \text{(unbalanced)}$$

Next balance hydrogen. For an alkaline medium, as in this example, add water to the side deficient in hydrogen and an equal number of hydroxide ions to the other side. (For an acidic medium, add the appropriate number of hydrogen ions to the side deficient in hydrogen.) Thus,

$$C_6H_5O_7^{3-} + 5H_2O + 15OH^- = 6CO_2 + 15H_2O \qquad \text{(material balanced; charge unbalanced)}$$

Often, as in this example, water molecules will appear on both sides of the equation and some can be canceled.

$$C_6H_5O_7^{3-} + 15OH^- = 6CO_2 + 10H_2O \qquad \text{(material balanced; charge unbalanced)}$$

Because the net charges on the left and right sides are -18 and zero, respectively, charge is balanced by the addition of eighteen electrons to the right side. The fully balanced half-reaction equation is

$$C_6H_5O_7^{3-} + 15OH^- = 6CO_2 + 10H_2O + 18e \qquad \text{(balanced)}$$

Since the first half-reaction involves three electrons and the second eighteen, the first equation is multiplied by six and the two half-reaction equations are added to yield the overall redox equation:

$$\begin{array}{rcl} 6MnO_4^- + 12H_2O + 18e & = & 6MnO_2 + 24OH^- \\ C_6H_5O_7^{3-} + 15OH^- & = & 10H_2O + 6CO_2 + 18e \\ \hline C_6H_5O_7^{3-} + 2H_2O + 6MnO_4^- & = & 6CO_2 + 6MnO_2 + 9OH^- \end{array}$$

It is always advisable to inspect the final equation to assure that a balance of both material and charge has been attained, in other words, to rule out any arithmetic error. The coefficients of the reactants and products in the final equation should be examined to see if any common denominator is present, which should be divided through so that the smallest integers possible appear as coefficients.

Redox reactions may also take place in a slightly acidic, slightly alkaline, or neutral medium. The question then arises as to whether to use hydrogen ion or hydroxide ion in securing a material balance. It should be recalled that both hydrogen ion and hydroxide ion are present in an aqueous solution and the designation "acid" or "alkaline" refers only to the predominance of one of these species, not to the absence of the other species. As noted in Section 21.1, neither the overall redox equation nor the half-reaction equations represent the actual reaction mechanism; the writing of hydrogen ion rather than hydroxide ion or vice versa is therefore to some degree arbitrary. For example, the reduction of permanganate to manganese(IV) oxide can also occur in slightly acidic medium; since this process is the general one occurring in alkaline solution, the redox equation may be appropriately written with hydroxide ion as a product as in Example 21-4.

21.3 Determination of Oxidation State

In the application of the electron-change method for the balancing of redox equations, as presented in Section 21.2, it is unnecessary to determine the oxidation state of the various atoms involved, since the number of electrons in each half-reaction equation is obtained, in one sense, automatically. It is sometimes of interest to determine the oxidation state and the changes in oxidation states involved in a reduction or oxidation process without the necessity of balancing an equation. Further, the calculation of an oxidation state is necessary in writing the systematic names for those substances in which the oxidation state is given in parenthetical roman numerals (a positive sign being understood). It is also possible to balance a half-reaction via the change in oxidation states (Section 21.4).

The determination of the oxidation state is accomplished by application of the following rules:

1. The oxidation state of a simple (monoatomic) ion is equal to its ionic charge.
2. The sum of the oxidation states of each atom in a molecule or ion must equal the total charge of that molecule or ion.
3. The oxidation state of oxygen is -2 except in peroxy compounds, oxygen gas, and substances in which oxygen is bonded to fluorine, where it is -1, 0, and $+2$, respectively.
4. The oxidation state of hydrogen is $+1$ except in hydrogen gas and hydrides, where it is 0 and -1, respectively.

These rules may be exemplified:

In the ion Fe^{2+} by rule 1, the oxidation state is $+2$, and hence it is termed the iron(II) ion.

In the permanganate ion, MnO_4^-, the oxidation state of manganese is determined by rules 2 and 3. The sum of the oxidation states of manganese and oxygen must equal -1; four oxygens equal a total of $-2 \times 4 = -8$; hence manganese is present in the oxidation state of $+7$. Or, in other terms, manganese(VII) is present in the permanganate ion. In the reduction of this ion in acidic solution to manganese(II), it follows that $7 - 2 = 5$ electrons are involved.

The oxidation state of manganese in MnO_2, by rules 2 and 3, is $+4$, and the compound can be named as manganese(IV) oxide. In the reduction of permanganate in alkaline medium to this compound, it follows that $7 - 4 = 3$ electrons are involved.

The oxidation states of cobalt and nitrogen in the ion $Co(NO_2)_6^{3-}$ are calculated as follows. In the nitrite ion, NO_2^-, nitrogen has an oxidation state of $+3$, since the charge on the ion is -1. Six nitrite ions are present in the complex ion

$Co(NO_2)_6^{3-}$, and consequently the oxidation state of cobalt must be +3. The systematic name of this ion is hexanitritocobaltate(III); the ion was known classically as the cobaltinitrite ion. If this ion is oxidized to the products Co^{2+} and NO_3^- (Example 21-3), the oxidation state of cobalt is +2 and indeed cobalt is reduced. The oxidation state of nitrogen in the nitrate ion, NO_3^-, is +5, and hence in the oxidation each nitrogen has lost two electrons. Hence in the overall oxidation of the $Co(NO_2)_6^{3-}$ ion, $2 \times 6 - 1 = 11$ electrons are involved.

21.4 Balancing Redox Equations: Oxidation-State Method

A half-reaction can also be balanced by consideration of the oxidation states involved. In this procedure, the number of electrons involved in the half-reaction is determined from the difference in the oxidation state between the reactant and its corresponding product. For example, consider the half-reaction for the reduction of permanganate ion in acidic medium:

$$MnO_4^- = Mn^{2+} \qquad \text{(unbalanced)}$$

The oxidation states of manganese, calculated as considered above, are +7 and +2 in the reactant and product, respectively. The half-reaction therefore involves $7 - 2 = 5$ electrons:

$$MnO_4^- + 5e = Mn^{2+} \qquad \text{(unbalanced)}$$

Material is then balanced in the same manner as in the previously described method: The four atoms of oxygen of the reactant will combine with eight hydrogen ions to yield four molecules of water:

$$MnO_4^- + 8H^+ + 5e = Mn^{2+} + 4H_2O \qquad \text{(balanced)}$$

As a check, charge balance is calculated for both sides of the equation: The net charge on each side is found to be +2.

21.5 Problems

21-1. The reaction of iodate ion with iodide ion in acidic solution yields elemental iodine. Write the balanced redox equation.

Answer: $IO_3^- + 5I^- + 6H^+ = 3I_2 + 3H_2O$

21-2. The complete oxidation of methanol, CH_4O, to carbon dioxide in alkaline solution by the action of permanganate ion yields manganese(IV) oxide. Write the balanced redox equation.

Answer: $CH_4O + 2MnO_4^- = CO_2 + 2MnO_2 + H_2O + 2OH^-$

In the oxidation of iron(II) ion in acidic solution, dichromate ion is reduced to chromium(III) ion. Write the balanced overall equation.

Answer: $6Fe^{2+} + Cr_2O_7^{2-} + 14H^+ = 6Fe^{3+} + Cr^{3+} + 7H_2O$

Complete and balance the following redox equations (check your results to assure that material and charge are fully balanced).

(a) $IO_4^- + I^- = I_2$ (acidic solution)
(b) $IO_4^- + Mn^{2+} = MnO_4^- + IO_3^-$ (acidic solution)
(c) $MnO_4^- + Mn^{2+} = \underline{MnO_2}$ (alkaline solution)
(d) $MnO_4^- + \underline{Mn} = Mn^{2+}$ (acidic solution)
(e) $\underline{PbO_2} + Cr^{3+} = Pb^{2+} + Cr_2O_7^{2-}$ (acidic solution)
(f) $AsO_3^{3-} + I_2 = AsO_4^{3-} + I^-$ (alkaline solution)
(g) $\underline{ZnS} + NO_3^- = Zn^{2+} + \underline{S} + NO\uparrow$ (acidic solution)
(h) $BrO_3^- + Br^- = Br_2$ (acidic solution)
(i) $Hg^{2+} + Cl^- + Sn^{2+} = \underline{Hg_2Cl_2} + Sn^{4+}$ (acidic solution)
(j) $Bi(OH)_3 + Na_2SnO_2 = \underline{Bi} + Na_2SnO_3$ (alkaline solution)
(k) $FeCl_3 + H_2S = FeCl_2 + HCl + \underline{S}$ (acidic solution)
(l) $As_2O_3 + Br_2 = H_3AsO_4 + Br^-$ (acidic solution)
(m) $S_2O_3^{2-} + I_2 = SO_4^{2-} + I^-$ (alkaline solution)
(n) $Ag + ClO_3^- = Cl^- + \underline{AgCl}$ (acidic solution)
(o) $H_3PO_3 + Ce^{4+} = H_3PO_4 + Ce^{3+}$ (acidic solution)
(p) $Cu(NH_3)_4^{2+} + CN^-$
$= Cu(CN)_3^{2-} + NCO^- + NH_3$ (alkaline solution)
(q) $\underline{CuSCN} + IO_3^- + Cl^-$
$= Cu^{2+} + HCN + SO_4^{2-} + ICl_2^-$ (acidic solution)
(r) $Ce(IO_3)_4 + H_2C_2O_4$
$= I_3^- + Ce_2(C_2O_4)_3 + CO_2\uparrow$ (acidic solution)
(s) $NH_3 + H_2O_2 = N_2\uparrow$ (alkaline solution)
(t) $\underline{PbS} + H_2O_2 = \underline{PbSO_4}$ (acidic solution)

22 ELECTROCHEMICAL PRINCIPLES

Electrical conductance is the process by which electrical charges pass through a conductor. With respect to the phenomena to be discussed in this chapter, two types of electrical conductors may be distinguished: metallic and electrolytic. The former are materials, usually metals, that conduct electricity by the passage of electrons; the latter conduct electricity by the movement of ions (cations and anions). If a current flows across the junction between a metallic conductor and an electrolytic conductor, an electrochemical reaction has to take place. The nature of this reaction is the central topic of electrochemistry.

Electrochemical phenomena underlie some of the most important instrumental methods of quantitative analysis as well as redox titrimetry. Before these methods are considered, some basic electrochemical principles will be reviewed.

Definition of Electrical Units 22.1

From his study of physics the student is familiar with electrical phenomena and may recall that various systems of units are in use. The international, practical system of units, which is employed here in the consideration of electrochemical phenomena, is briefly summarized below. For additional background information, including the absolute system of units, the student should consult a textbook on physics.

The *ampere* is the unit of strength of a current. One ampere (A) is defined as the unvarying, direct current that will deposit under specified conditions 1.11800 mg of silver from a solution of silver nitrate in 1 second.

The *ohm* is the unit of electrical resistance. One ohm (Ω) is defined as the resistance offered on passage of an unvarying current at 0°C through a column of mercury of uniform cross sections 106.300 cm in length and 14.4521 g in weight. Such a column at 0°C has a cross-sectional area of almost 1 mm^2.

The *volt* is the unit of electromotive force. One volt (V) is the difference in electrical potential, defined by Ohm's law (see below), required to maintain the flow of a current of 1 ampere (A) through a resistance of 1 Ω.

Ohm's law describes the relation between the electromotive force, E, the current, I, and the resistance, R, and may be expressed as $E = IR$. In the present work, the practical units, defined above, are used as the parameters of this law; thus, volts = amperes × ohms.

The *coulomb* is the unit quantity of electricity. One coulomb (C) is defined as the quantity of electricity passed in 1 sec by the flow of an unvarying current of 1 A. By the definition of the ampere, 1 C will deposit 1.11800 mg of silver from a solution of silver nitrate.

22.2 Electrochemical Cells

If a piece of zinc metal is placed in an aqueous solution containing copper(II) ion, the zinc is observed to dissolve forming zinc ion and copper plates on the surface of the zinc. The redox reaction may be written†

$$Zn + Cu^{2+} \rightarrow Zn^{2+} + Cu \qquad (22\text{-}1)$$

To examine the electrical phenomena involved and also to make possible practical use of the changes in electrical energy associated with the reaction, it is necessary to physically separate the two redox systems but still have them connected by an electric conductor. Such an arrangement is shown in Figure 22-1 and is called a cell. It consists of the two corresponding half-cells, which are connected by a U-tube filled with a salt solution. This solution is "solidified" by the addition of gelatin or agar-agar. This salt bridge effects a conducting connection without allowing the two solutions to mix. An electrolytically conducting connection must be employed instead of a simple wire loop for the following reason. In a metallic conductor an electrical current is the flow of electrons; in an electrolytic conductor a current is sustained by the translatory movement of ions. In order that the current pass the phase boundary between a metallic and electrolytic conductor, or vice versa, it is necessary that an electrochemical reaction take place at this phase boundary. If a metal wire were used to connect the two half-cells, an additional half-cell would be present in each of

†Zinc and copper metals, as solids, might be underlined in equation (22-1), however, this practice will usually be omitted in the consideration of electrochemistry as an unnecessary complication.

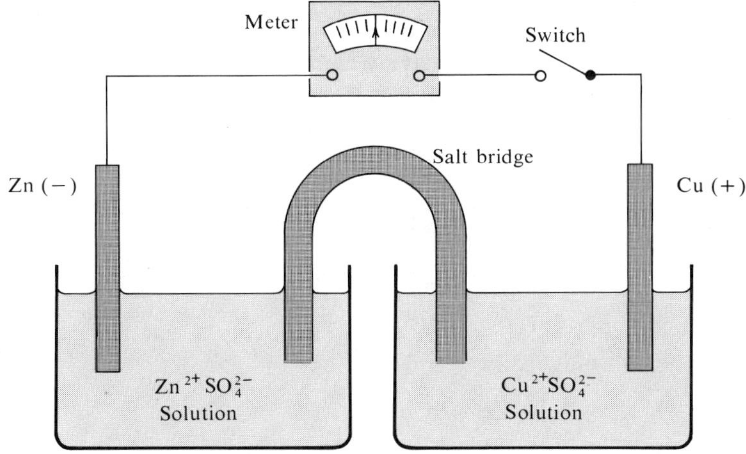

Figure 22-1. Zinc–copper voltaic cell.

the two vessels and, instead of the desired single cell, two cells connected by the wire would exist.

Instead of providing a full diagram or a detailed description of a cell under discussion, it is easier to represent it schematically. The zinc–copper cell of Figure 22-1 may be represented as follows:

$$Zn \mid ZnSO_4 \parallel CuSO_4 \mid Cu$$

A single vertical line here denotes a phase boundary between a metallic conductor and an electrolyte solution (often a solid–liquid interface), and a double line, a liquid–liquid junction, that is, an electrolytically conducting connection (for example, a salt bridge or a porous, sintered-glass disk) between two solutions. Unless otherwise stated, the electrolytes [zinc(II) sulfate and copper(II) sulfate in the example] are assumed to be present as aqueous solutions. Where the state of a substance must be stated explicitly, *s* and *g* may be used parenthetically to designate solid and gas, respectively, and concentrations may also be given parenthetically, for example, $AgCl(s)$, $H_2(g)$, and $ZnSO_4\,(0.1\ F)$. The use of these conventions for the representation of a cell will become clearer as further examples are encountered.

Voltaic Cells and Electrolysis Cells 22.3

When the pieces of zinc and copper metal of the zinc–copper cell are connected by a metallic wire, the zinc dissolves progressively and copper plates on the copper piece. In addition, electrons flow through the wire from the zinc to the copper, and the zinc piece is negatively charged compared to the copper piece. From these findings it can be inferred that in the right half-cell the reaction occurring is

$$Cu^{2+} + 2e \rightarrow Cu$$

which is a reduction process. In the left half-cell, the reaction occurring is

$$Zn \rightarrow Zn^{2+} + 2e$$

which is an oxidation process. The overall reaction equation of the complete cell is

$$Zn + Cu^{2+} \rightleftharpoons Zn^{2+} + Cu \qquad (22\text{-}2)$$

The double arrow is used since, as will be considered below, the reaction is reversible.

In this cell, a chemical reaction is used to generate electrical energy. Such a cell is known as a voltaic cell or a galvanic cell.

If an outside source of electrical energy is used to force electrons on to the zinc metal and off at the piece of copper, under appropriate conditions zinc ion will be reduced to zinc metal and copper metal will be oxidized to copper(II) ion; that is, the direction of the cell reaction will be reversed. Here, then, electrical energy is used to cause a chemical reaction to proceed. This process is known as electrolysis and the cell is called an electrolysis cell or electrolytic cell.

It is important to appreciate that the piece of zinc is negative and the copper is positive, regardless of whether the cell acts as a voltaic cell or as an electrolysis cell. In the voltaic cell, oxidation proceeds at the negative electrode; in the electrolytic cell, reduction occurs at this electrode. Instead of the terms positive and negative electrodes, the terms cathode and anode are often used. Unfortunately there is no uniform agreement as to the definition of these terms. Some workers define the anode and cathode as the positive and negative electrodes, respectively. Others define the anode as the electrode at which oxidation occurs and the cathode as the electrode at which reduction occurs. From the above considerations, the student should realize that these two definitions are equivalent for an electrolysis cell but are directly opposed for a voltaic cell. To avoid this confusion, it is best to limit the use of the terms cathode and anode to the consideration of electrolysis cells and never to use them in connection with voltaic cells. This practice is followed in the present textbook.

Faraday's Laws 22.4

In many instances it is possible to treat the electrons in a balanced half-reaction equation as a reactant or product just as any other participating species. Then the principle of Le Châtelier may be applied. For example, for the equilibrium

$$Cu^{2+} + 2e \rightleftharpoons Cu \qquad (22\text{-}3)$$

it can be concluded that adding electrons will shift the position of the equilibrium toward the right. However, it must be stressed that this treatment is strictly formal and that electrons, of course, cannot be added from a reagent bottle in the form of a solution. In a solution electrons never occur as free, isolated entities but rather are provided or consumed by another redox system or, and this is of basic interest in electrochemistry, by an electrode.

The above equation implies that two electrons are required to convert one copper(II) ion into one copper atom. Consequently, for the conversion of one mole of copper(II) ion, two moles of electrons are necessary. As has been established by careful measurements, one mole of electrons corresponds to 9.649×10^4 C. Since 1 C equals 1 A-sec, it can be established that the conversion of one mole of copper(II) ion to one mole of copper metal requires $2 \times 9.649 \times 10^4$ A-sec. Consequently, the quantity of material reacting at an electrode is related both to the strength of the direct current flowing and to the time that it flows.

A half-cell can never operate alone; there must always be two electrodes in one cell, even when the two half-cells forming the cell are physically separated but joined by an electrolytic conductor. When the circuit is completed the electrode reactions proceed and, if no branches exist, the same number of amperes flow in any element of the circuit. Consequently, the number of electrons involved in the reaction at one electrode must exactly equal the number of electrons involved in the reaction taking place at the other electrode, or, in chemical terms, the quantity of substance reacting at one electrode must be *equivalent* to the quantity of substance reacting at the other electrode. Thus, if one mole of copper(II) reacts at one electrode in the cell under consideration, one mole of zinc metal reacts at the other electrode. If a cell consists of a copper(II)–copper half-cell and a silver(I)–silver half-cell, then for one mole of copper(II) reacting two moles of silver should and will react, since only one electron is involved in the silver(I)–silver half-reaction ($Ag^+ + e \rightarrow Ag$). It is therefore possible to define the equivalent weight of a substance in an electrochemical reaction as the formula weight divided by the number of electrons released or consumed per reacting molecule, ion, or atom of that substance. The quantity of 9.649×10^4 C is the electrochemical equivalent; it is termed one faraday (F). One faraday corresponds to one mole of electrons, to as many electrons as there are atoms in exactly 12 g of carbon-12, and to the number of electrons given by Avogadro's number, 6.02291×10^{23}.

The considerations of the preceding paragraphs are implicit in discoveries by Michael Faraday about 1833. These findings may be expressed in the form of two laws:

1. The quantity of a substance reacting at an electrode of a cell is directly proportional to the quantity of electricity passed. Since the quantity of electricity is given by current multiplied by time (ampere $\times$ second = coulombs), the quantity of a substance reacting at the electrode in unit time is directly proportional to the current.
2. The quantities of different substances reacting during the passage of equal quantities of electricity are proportional to the chemical equivalent weights of the substances. In other words, the same quantity of electricity causes the same number of chemical equivalents to react at an electrode.

With this information at hand, it is possible to derive an equation for the relationship among time, current, and the quantity of a reacting substance.

The chemical equivalents of reacting substance are also given by g/EW, where g is the grams of the substance. The equivalent weight, EW, is given by FW/f, where f is the equivalence number, that is, here, the number of electrons reacting per ion, atom, or molecule. Thus the equivalents are g $\times$ f/FW. Since 1 F is required for the reaction of one equivalent of substance, the number of coulombs required for g grams of substance will be F $\times$ g $\times$ f/FW. This number of coulombs is the number of ampere-seconds necessary:

$$\frac{\mathrm{F} \times \mathrm{g} \times f}{\mathrm{FW}} = It \qquad (22\text{-}4)$$

where I is the current in amperes and t the time in seconds.

Rearrangement and substitution of the numerical value for the faraday yields

$$\frac{I \times t \times \mathrm{FW}}{9.649 \times 10^4 \times f} = \mathrm{g} \qquad (22\text{-}5)$$

A few examples will illustrate the application of the facts discussed and the use of these equations.

Example 22-1. How many grams of copper (*63.5*) will be deposited on the cathode if a constant current of 0.50 A is passed for 5.0 min?

Application of equation (22-5) yields

$$\frac{\underset{\text{amp}}{0.50} \times \overbrace{5.0 \times 60}^{\text{sec}} \times \underset{\text{FW}}{63.5}}{9.65 \times 10^4 \times \underset{f}{2}} = 0.049 \text{ g or } 49 \text{ mg of Cu}$$

Example 22-2. Permanganate can be cathodically reduced according to the reaction

$$MnO_4 + 8H^+ + 5e \rightleftharpoons Mn^{2+} + 4H_2O$$

For what time must a constant current of 0.0250 A be applied to reduce 5.2 mg of $KMnO_4$ (*158.0*)?

$$\frac{\overbrace{2.50 \times 10^{-2}}^{\text{amp}} \times \underset{\text{sec}}{t} \times \underset{\text{FW}}{158.0}}{9.65 \times 10^4 \times \underset{f}{5}} = 5.20 \times 10^{-3} \text{ g}$$

Solving for time,

$$t = \frac{5.20 \times 5 \times 9.65 \times 10^4}{25 \times 158} = 0.0635 \times 10^4 = 635 \text{ seconds}$$

or 10 minutes 35 seconds

Example 22-3. A cell in which silver is deposited and a cell in which dichromate is reduced ($Cr_2O_7^{2-} + 14H^+ + 6e \rightleftharpoons 2Cr^{3+} + 7H_2O$) are connected in series and an unvarying current is passed for a certain time. After that time it is found that 125.0 mg of Ag (*107.9*) has been plated. How many grams of Cr^{3+} (*52.0*) has formed?

According to Faraday's second law the quantities of Ag deposited and of Cr^{3+} formed must be equivalent. The deposition of one atom of Ag involves one electron and the formation of two ions of Cr^{3+} involves six electrons; that is, for each Cr^{3+} ion formed, three electrons have reacted, or in turn, three Ag atoms have plated. Consequently the relation

$$(\text{moles of Ag deposited}) = (\text{moles of } Cr^{3+} \text{ formed}) \times 3$$

exists. Then

$$\frac{125 \times 10^{-3}}{3 \times 107.9} = \frac{x}{52.0}$$

and

$$x = \frac{52.0 \times 125 \times 10^{-3}}{3 \times 107.9} = 0.0201 \text{ g of } Cr^{3+}$$

22.5 Half-Cell Potential; Standard Hydrogen Electrode

It is impossible to obtain the *absolute* potential of a single half-cell from the measurement of cell voltages. This difficulty is a result of the fact that if a certain amount of one substance is reduced, an exactly equivalent amount of another substance in the system must be oxidized and vice versa. Although it is possible to separate the oxidation and reduction half-reactions so that they take place in two separate half-cells, only the combined voltage can be measured, since the half-cell reactions occur simultaneously and interdependently. To make quantitative comparisons between various half-cells, it is necessary to assign arbitrarily a potential value to one selected, closely defined half-cell and to relate the potentials of all other half-cells to it.

By international agreement, the standard hydrogen electrode† is assigned the exact value of zero volts at all temperatures. This electrode consists of a piece of platinum metal (coated with finely divided platinum) dipping into an aqueous solution which has hydrogen ion at unit activity and which is saturated with hydrogen gas at a pressure of 1 atm. The approximation is often made that molar concentration equals activity and this practice is followed in this text (see Section 4.3 for the definition of activity). Thus, the hydrogen electrode may be represented by

$$\text{Pt, H}_2\,(\text{g, 1 atm}) \mid \text{H}^+\,(1\,M)$$

As a consequence of the assignment of zero volts to the standard hydrogen electrode, the total voltage of any cell in which one half-cell is this standard hydrogen electrode can be ascrib-

†The term electrode is used in two senses in the discussion of electrochemical cells. On one hand, the piece of metal that dips into the electrolyte solution is often called the electrode. This metal may participate in the cell reaction as in the zinc–copper cell considered above. Alternatively, the metal may simply provide a means of electric connection or a site for an electron transfer to occur. In these cases the metal does not participate in the cell reaction and is called an inert electrode (or indicator electrode). Platinum often serves for this purpose. On the other hand, the term electrode is applied to an entire half-cell. The two usages may be seen in a single context: The hydrogen electrode (i.e., half-cell) consists of an inert platinum electrode (that is, a rod, wire, or sheet of platinum metal) dipped into an electrolyte solution and at which the reaction $H^+ + e \rightleftharpoons \frac{1}{2}H_2$ occurs. When an inert metal is used, the meaning of the term electrode is usually obvious from the context. When the metal participates in the cell reaction, however, the term is ambiguous.

able to the other half-cell. Thus, the cell

$$Pt, H_2 (g, 1 atm) \mid H^+ (1\ M) \parallel Zn^{2+} (1\ M) \mid Zn$$

has experimentally a total voltage of 0.76 V and the zinc electrode is found to be negative. The (relative) electrode potential of the zinc(II)–zinc electrode is then stated to be −0.76 V. Because in the above cell zinc metal and zinc ion (1 $M \simeq$ unit activity) are at "standard state," the value of −0.76 V represents the "standard potential," designated by the superscript zero. Thus, we may write $E^0_{Zn^{2+}/Zn} = -0.76$ V.

Under the conventions adopted, the term potential (that is, electrode potential or half-cell potential), without any further qualification, should be used to describe the electrical potential (including sign) of the half-cell with respect to the standard hydrogen electrode only under the condition that the reaction equation of the half-cell to be specified is written as a reduction.† In reference to the electromotive force of whole cells, batteries, circuitry, etc., the term *voltage* will be used rather than potential.

A cell consisting of a standard hydrogen electrode and the standard copper(II)–copper electrode

$$Pt, H_2 (g, 1 atm) \mid H^+ (1\ M) \parallel Cu^{2+} (1\ M) \mid Cu$$

has experimentally a voltage of 0.34 V and the copper electrode is positive. The standard electrode potential of the copper(II)–copper half-cell, $E^0_{Cu^{2+}/Cu}$, therefore has a value of +0.34 V.

Nernst Equation 22.6

In the zinc–copper voltaic cell, if both copper(II) and zinc ions are present at a concentration of one mole per liter, the cell at 25°C will be found experimentally to have a voltage of 1.10 V, which corresponds to the difference in the electrode potentials, 0.34 − (−0.76). If the electrodes are connected externally and the cell is allowed to discharge, progressively the copper(II) ion concentration decreases, the zinc ion concentration increases, and the cell voltage decreases. Thus, it is evident that

†In practice, the hydrogen electrode has some disadvantages and other electrodes of constant and well-defined electrode potential are substituted (see Section 22.12). In such a case, the electrode potential of a half-cell may be numerically expressed in reference to the substitute electrode. The expression of an electrode potential in this way requires the further qualification that the exact nature of the substitute (reference) electrode be stated.

the voltage of a voltaic cell and the potential of a half-cell will depend on the concentration of the species involved in the cell reaction.

By the convention expressed above, a half-cell reaction is written as a reduction and in general terms takes the form

$$\text{Ox} + ne \rightleftharpoons \text{Red} \qquad (22\text{-}6)$$

where Ox and Red denote the oxidized and reduced species forming the so-called redox pair or redox couple. The quantitative relation of the potential of the half-cell to concentration, termed the Nernst equation, after the electrochemist who derived it, is given by

$$E = E^0 + \frac{RT}{n\text{F}} \ln \frac{[\text{Ox}]}{[\text{Red}]} \qquad (22\text{-}7)$$

In the Nernst equation, which is applicable only if the half-cell reaction is reversible,† the parameters have the following meaning: E is the potential of the half-cell relative to the standard hydrogen electrode. E^0, the standard electrode potential, is the potential of the half-cell relevant to the standard hydrogen half-cell with each of the species involved in the cell reaction in its standard state: the pure liquid, pure solid, gas at a pressure of 1 atm, and dissolved species at unit concentration (strictly unit activity, but 1 M in the approximation here employed). R is the gas constant of the ideal gas law and has the value 8.316 V-C/degree. T is the absolute centigrade temperature (that is, degrees Kelvin), F is the value of the faraday, 9.649×10^4 C, and n is the number of electrons transferred in the half-cell reaction as written. The terms [Ox] and [Red] stand for the products of the concentrations of all species occurring on the oxidation and reduction sides of the half-reaction, respectively. Concentrations are molarities except for gases; their concentrations are expressed as partial pressures in atmospheres. Solids and water, however, are at constant concentration and do not occur in the expression; or, in other words, they are at standard state and thus their activity is unity. As in the case of expressions of the law of mass action, a coefficient in the half-cell reaction equation appears as the power of the concentration of the relevant species in the Nernst equation.

---

†Some irreversible reactions, however, follow the Nernst equation reasonably well, as, for example, the reduction of permanganate. This and other reactions (e.g., reduction of dichromate) will be used in the exercises in some of the examples and problems given.

On the conversion from natural logarithms to Briggsian logarithms (i.e., to base 10) and insertion of the values of the constants with T equal to 298.1°K (i.e., 25.0°C), the result is†

$$E = E^0 + \frac{0.059}{n} \log \frac{[\mathrm{Ox}]}{[\mathrm{Red}]} \quad \text{(at 25°C)} \qquad (22\text{-}8)$$

Example 22-4. What is the potential of the electrode Pt | Fe^{2+} (1.0×10^{-1} M), Fe^{3+} (1.0×10^{-2} M)? $E^0_{Fe^{3+}/Fe^{2+}}$ has the value +0.771 V at 25°C.

The half-cell reaction is

$$Fe^{3+} + e \rightleftharpoons Fe^{2+}$$

and the potential after equation (22-8) is given by

$$E = E^0 + \frac{0.059}{1} \log \frac{[Fe^{3+}]}{[Fe^{2+}]}$$

Substitution of the numerical data yields

$$E = +0.771 + 0.059 \log \frac{1.0 \times 10^{-2}}{1.0 \times 10^{-1}}$$

$$= +0.771 + 0.059(-1.00) = +0.712 \text{ V}$$

This implies that a cell consisting of the standard hydrogen electrode and this particular iron(III)–iron(II) electrode would have an experimental voltage of 0.712 V and that the latter electrode would be positive.

Often the reduced species is the metal of the electrode (as in the zinc or copper electrode) and since, as mentioned above, the metal is in its standard state, its concentration term is not written. In such cases the Nernst equation simplifies to

$$E = E^0 + \frac{0.059}{n} \log [\mathrm{Ox}] \quad \text{(at 25°C)} \qquad (22\text{-}9)$$

Example 22-5. What is the potential of the half-cell Ni | Ni^{2+}(1.0×10^{-3} M) at 25°C? $E^0_{Ni^{2+}/Ni} = -0.25$ V.

The potential after equation (22-9) is given by

$$E = E^0 + \frac{0.059}{2} \log [Ni^{2+}]$$

†The value of the factor in the Nernst equation varies, as follows, with the temperature:

Temperature, °C	10	15	20	25	30	35	40
Value of factor	0.05618	0.05717	0.05816	0.05916	0.06014	0.06114	0.06213

In many calculations, especially where formal or molar concentrations are used rather than activities, the value 0.059 is sufficiently precise for the temperature range 20 to 30°C.

Hence

$$E = -0.25 + \frac{0.059}{2}\log(1.0 \times 10^{-3}) = -0.25 - 0.09 = -0.34 \text{ V}$$

This finding implies that a cell consisting of the standard hydrogen half-cell and the given nickel electrode would have a voltage of 0.34 V with the nickel electrode being negative.

Example 22-6. What is the potential of the following half-cell?

$Pt \mid Cr_2O_7^{2-}(1.0 \times 10^{-2}\,M), Cr^{3+}(2.0 \times 10^{-3}\,M), H^+(1.0 \times 10^{-2}\,M)$

The standard potential for the reaction $Cr_2O_7^{2-} + 14H^+ + 6e \rightleftharpoons 2Cr^{2+} + 7H_2O$ is $E^0_{Cr_2O_7^{2-}/Cr^{3+}} = +1.33$ V. The Nernst equation for the given half-cell is

$$E = E^0 + \frac{0.059}{6}\log\frac{[Cr_2O_7^{2-}][H^+]^{14}}{[Cr^{3+}]^2}$$

Inserting the relevant numerical data and solving yields

$$E = +1.33 + \frac{0.059}{6}\log\frac{(1.0 \times 10^{-2})(1.0 \times 10^{-2})^{14}}{(2.0 \times 10^{-3})^2} = +1.09 \text{ V}$$

22.7 Cell Voltage and Electrode Polarity

If the necessary data are available, the Nernst equation can be used to calculate half-cell potentials. From these potentials cell voltages can be computed, and a scheme can be developed that allows the polarity of the electrodes to be established and the direction in which the cell reaction proceeds spontaneously. Several different procedures are possible, all of which are satisfactory if followed consistently. The procedure to be described is consistent with the conventions recommended by the International Union of Pure and Applied Chemistry, most of which have already been introduced in the preceding paragraphs. The procedure is best explained through several examples.

Example 22-7. For the following cell calculate the cell voltage, indicate the polarity of the electrodes, write the cell reaction, and indicate in which direction it will proceed spontaneously.

$$Zn \mid Zn^{2+}(1.0 \times 10^{-2}\,M) \parallel Cu^{2+}(1.0 \times 10^{-1}\,M) \mid Cu$$

At 25°C $\quad E^0_{Cu^{2+}/Cu} = +0.34$ V $\quad$ and $\quad E^0_{Zn^{2+}/Zn} = -0.76$ V

First, write the reaction equations as reductions with that for the right electrode above that for the left electrode. Proceed similarly with the Nernst equations for the two electrodes.

$$Cu^{2+} + 2e \rightleftharpoons Cu$$

$$E_r = E^0 + \frac{0.059}{n}\log[Cu^{2+}] = +0.34 + \frac{0.059}{2}\log(1.0 \times 10^{-1})$$

$$Zn^{2+} + 2e \rightleftharpoons Zn$$

$$E_l = E^0 + \frac{0.059}{n}\log[Zn^{2+}] = -0.76 + \frac{0.059}{2}\log(1.0 \times 10^{-2})$$

Then subtract the reaction equation for the left electrode from that of the right electrode, and subtract the corresponding Nernst equations similarly.

$$Cu^{2+} + Zn \rightleftharpoons Cu + Zn^{2+}$$

$$E_{cell} = E_r - E_l = +0.34 - (-0.76) + \frac{0.059}{2}\log\frac{1.0 \times 10^{-1}}{1.0 \times 10^{-2}}$$

$$= +0.34 + 0.76 + \frac{0.059}{2}\log(1.0 \times 10) = (+)1.13\ \text{V}$$

The voltage of the cell is thus 1.13 V. The positive sign indicates that the right electrode is positive and that the cell reaction, as written, proceeds spontaneously from left to right; that is, copper(II) ion is reduced and zinc metal oxidized.

It should be stressed that a cell has a certain voltage and that this voltage has no sign. The plus or minus sign for the cell voltage is an algebraic consequence and is placed in parentheses. This sign is useful, however, since it indicates the polarity of the right half-cell and the direction of the spontaneous reaction. The following example will show how to proceed if species other than those involved directly in the redox process participate in the half-reaction.

Example 22-8. What is the voltage and polarity of the cell given below? What is the cell reaction and in which direction will it proceed spontaneously?

$$Zn \mid Zn^{2+}(1.0 \times 10^{-4}\,M) \parallel H^+(1.0 \times 10^{-2}\,M),$$
$$V^{3+}(1.0 \times 10^{-1}\,M),\quad VO^{2+}(1.0 \times 10^{-3}\,M) \mid Pt$$

$$E^0_{Zn^{2+}/Zn} = -0.763\ \text{V} \quad \text{and} \quad E^0_{VO^{2+}/V^{3+}} = +0.337\ \text{V}$$

First, write the reaction equations and Nernst equations in the manner described in Example 22-7.

$$VO^{2+} + 2H^+ + e \rightleftharpoons V^{3+} + H_2O$$

$$E_r = +0.337 + 0.059\log\frac{[VO^{2+}][H^+]^2}{[V^{3+}]}$$

$$= +0.337 + 0.059\log\frac{(1.0 \times 10^{-3})(1.0 \times 10^{-2})^2}{1.0 \times 10^{-1}}$$

$$Zn^{2+} + 2e \rightleftharpoons Zn$$

$$E_l = -0.763 + \frac{0.059}{2}\log[Zn^{2+}]$$

$$= -0.763 + \frac{0.059}{2}\log(1.0 \times 10^{-4})$$

Before the reaction equations can be subtracted, that for the right electrode must be multiplied by 2 so that the number of electrons involved will cancel on subtraction. The question arises whether the terms in the Nernst equation for this electrode should also be multiplied by 2. The answer is no! In analogy to a water-distribution system, the electrode potential compares to the water pressure, and this is independent of the volume of the system. It is true that when the reaction equation is multiplied throughout by 2, in the Nernst equation, the n will be multiplied by that factor. However, the concentration terms will also be raised to a power equal to this factor, so that the increase in n is nullified. Thus in the example

$$E_r = E_r^0 + \frac{0.059}{1} \log \frac{[VO^{2+}][H^+]^2}{[V^{3+}]}$$

$$= E_r^0 + \frac{0.059}{2} \log \frac{[VO^{2+}]^2[H^+]^4}{[V^{3+}]^2}$$

Multiplication by 2, as indicated above, and subtraction yields the balanced overall reaction equation for the cell, and mere subtraction of the Nernst equations yields the expression for the cell voltage.

$$2VO^{2+} + 4H^+ + Zn \rightleftharpoons 2V^{3+} + Zn^{2+} + 2H_2O$$

$$E_{cell} = E_r - E_l$$

$$= +0.337 - (-0.763) + 0.059 \times (-6.00) - \frac{0.059}{2} \times (-4.00)$$

$$= (+)0.864 \text{ V}$$

The voltage of the cell is therefore 0.864 V, the vanadyl–vanadium(III) electrode is positive, and the reaction proceeds spontaneously from left to right, as the equation is written.

The procedure for the calculation of the voltage and polarity of a voltaic cell may be summarized:

1. Write each of the electrode reactions as a reduction and each of the Nernst equations for the potential of the electrodes, with those for the right electrode above those for the left one.†
2. Multiply the reaction equations by such factors as are necessary to make the number of electrons involved in each identical. (Do not multiply the Nernst equations by these factors!)
3. Subtract the equations for the left electrode from the corre-

†In Chapter 21, in the balancing of redox reactions, one half-reaction equation is written as an oxidation and one as a reduction. This can be done since by the statement of the problem the species that are oxidized and reduced are known. In the present considerations, this information is in general to be ascertained by following a definite scheme, and this scheme requires all half-reactions to be written as reductions. Furthermore, electrode potentials are involved and these by definition are related to half-reactions written as reductions.

sponding equations for the right one, thus securing the balanced equation for the cell reaction and an expression for the voltage of the cell, which is solved.

4. The sign associated with the cell voltage is the polarity of the right electrode. A positive sign indicates that the cell reaction will proceed spontaneously from left to right as written in step 3. A negative sign indicates that this reaction proceeds spontaneously from right to left. If the cell voltage is zero, the system is at equilibrium and no reaction will occur.

Formal Potentials 22.8

In analytical chemistry, solutions are seldom encountered in which the species involved in a cell or half-cell reaction are present alone. Foreign substances often influence the voltage either by changing the activities or through the formation of soluble complexes. The magnitude of these effects is difficult to evaluate and, even where possible, the calculations are tedious. For this reason, electrode potentials are often determined and tabulated for solutions in which the given redox couple may be frequently encountered in practice. These are called formal electrode potentials and may be designated as E^f. When a formal electrode potential is used in the Nernst equation, the concentration terms are expressed in the formalities of the relevant species in the solution of interest. Thus, in general terms

$$E = E^f + \frac{0.059}{n} \log \frac{C_{\text{Ox}}}{C_{\text{Red}}} \qquad \text{(at 25°C)} \qquad (22\text{-}9)$$

For example, the standard potential of the iron(III)–iron(II) redox couple is +0.771 V. In 0.1 F, 1 F, and 5 F hydrochloric acid, the formal potential, E^f, is +0.73, +0.70, and +0.64 V, respectively. The change in potential with increasing concentration of hydrochloric acid is due to the formation of chloro complexes such as $FeCl_4^-$. The concentration, C_{Ox}, used in this Nernst equation should be the formal concentration of iron(III) rather than the molar concentration of the uncomplexed iron(III).

Measurement of Cell Voltage: The Potentiometer 22.9

The measurement of the voltage of a voltaic cell is complicated by the necessity of avoiding, as much as possible, the passage of current through the cell. If current flows, as will be discussed subsequently, the voltage will not correspond to the "spontaneous" cell voltage sought. Therefore, a simple voltmeter cannot be used, since it requires the passage of a nonnegligible current. Usually a "potentiometer" is employed, which allows the measurement of a voltage with no current flowing.† Frequently a technique for the determination of a cell voltage is employed known as the Poggendorff compensation method.

Potentiometer Circuitry. The essential parts of a potentiometer circuit are shown schematically in Figure 22-2. The working battery, WB, is connected across a uniform slide-wire

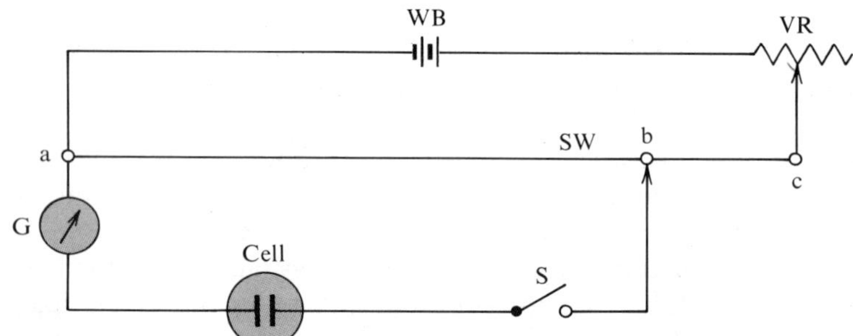

Figure 22-2. Potentiometer circuit.

resistance, SW, in series with a variable resistance, VR. The total voltage drop, E_{ac}, across the slide wire is controlled by the setting of VR. The voltage drop, E_{ab}, from point *a* to the movable slide-wire contact, *b*, is related to the total voltage drop by $E_{ab} = E_{ac} \times (\overline{ab}/\overline{ac})$, where the symbols $\overline{ab}$ and $\overline{ac}$ are the length of the slide wire between points *a* and *b* and between points *a* and *c*, respectively. Thus, if $\overline{ac}$ is 100 cm and $\overline{ab}$ is 75 cm and the total voltage drop across the slide wire is 1.00 V, the voltage drop between points *a* and *b*, E_{ab}, is $1.00 \times (75/100) = 0.75$ V.

The remainder of the circuit is a means of momentarily applying the voltage between points *a* and *b* to a cell and observing whether the applied voltage is greater or smaller than the cell voltage. The cell is momentarily connected to the potentiometer circuit by taping the key, *S*, and if a current flows its size and direction are revealed by the displacement of the needle of the galvonometer, G. The slide-wire contact is

†The potentiometer here referred to should not be confused with a device used in electric and electronic circuitry that is simply a variable resistor with three terminals.

moved until a point is found at which no current flows when the key is closed momentarily. At this point the cell voltage is then equal to the voltage, E_{ab}.

Determination of Half-Cell Potentials 22.10

In the determination of the potential of a half-cell it is not mandatory to couple it experimentally with a standard hydrogen electrode as the reference. Any other half-cell may be used provided its potential remains constant during the measurement and is exactly known relative to the standard hydrogen electrode. The voltage of the cell thus formed is the difference of the two half-cell potentials. Consequently, the unknown potential is obtained by adding the potential of the reference half-cell to the measured cell voltage. Full recognition of signs, including the formal sign of the cell voltage, is mandatory. The procedure is as follows. Let E_c, E_x, and E_{ref} stand for the voltage of the cell, the unknown potential, and the reference potential, respectively. Then

$$E_c = E_x - E_{\text{ref}} \tag{22-10}$$

$$E_x = E_c + E_{\text{ref}} \tag{22-11}$$

and for the formal computation E_c must have a sign attached. The equation is analogous to $E_c = E_r - E_l$ used in the previously established computation scheme in which the sign of E_c indicated the polarity of the right half-cell. Some examples will illustrate the procedure and reference to the accompanying "scale" will aid the understanding.

Volts:	−0.76	−0.52	−0.32	0.00	+0.34	+0.49
Electrode:	Zn^{2+}/Zn	X	Z	H^+/H_2	Cu^{2+}/Cu	Y

Example 22-9. A cell consists of a half-cell of unknown potential X and a zinc(II)–zinc electrode having a potential of −0.76 V. The cell voltage is found to be 0.24 V with the unknown electrode the positive electrode of the cell. What is the electrode potential of the unknown electrode?

$$X = (+)0.24 + (-0.76) = -0.52 \text{ V}$$

Example 22-10. A cell consists of an electrode of unknown potential Y and a copper(II)–copper electrode having a potential of +0.34 V. The cell voltage is found to be 0.15 V and the unknown electrode to be the positive electrode. What is the electrode potential of the unknown electrode?

$$Y = (+)0.15 + 0.34 = +0.49 \text{ V}$$

Example 22-11. A cell consists of an electrode of unknown potential Z and a copper(II)–copper electrode having a potential of +0.34 V. The cell voltage is found to be 0.66 V and the unknown electrode is negative. What is the electrode potential of the unknown electrode?

$$Z = (-)0.66 + 0.34 = -0.32 \text{ V}$$

22.11 Secondary Reference Electrodes

The standard hydrogen electrode is troublesome when used in practice. This electrode is easily "poisoned" by impurities in either the solution or the hydrogen gas. The electrode reaches equilibrium only slowly, and consequently a stable voltage is not attained promptly. In addition, the adjustment of the appropriate partial pressure of hydrogen gas, although not overly critical (see Problem 22-5), is tedious and the excess of gas must safely be carried away to avoid the formation of an explosive hydrogen–air mixture. For these reasons, secondary reference electrodes that are free of most of these shortcomings are usually employed, most frequently calomel electrodes.

One form of the calomel electrode is shown in Figure 22-3. A pool of mercury is covered with a thin layer of solid calomel,

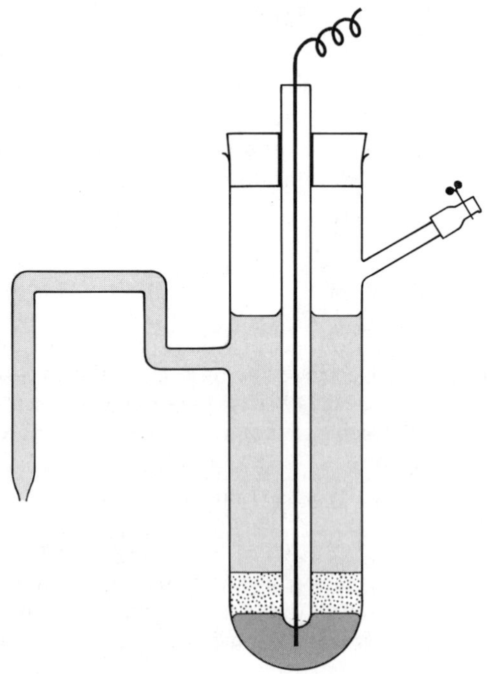

Figure 22-3. Calomel reference electrode.

Hg_2Cl_2. The solution is saturated with respect to calomel and also contains potassium chloride. A platinum wire sealed into glass tubing makes electrical contact with the mercury. The side arm is filled with gelatin containing potassium chloride and functions as the salt bridge. The potential of a calomel electrode depends on the chloride concentration and is given at 25°C by the Nernst equation $E = E^0 - 0.059 \log [Cl^-]$. The saturated calomel electrode (abbreviated S.C.E.), containing a solution saturated with potassium chloride, is preferred for routine analytical applications, since on any evaporation of the solution the potential remains unchanged. This particular calomel electrode has a potential of +0.241 V at 25°C.

Another reference electrode often used in practice is the silver chloride–silver electrode. It consists of a silver wire coated with silver chloride and dipping into a solution 1 *M* in chloride ion (usually as potassium chloride) saturated with silver chloride. This electrode at 25°C has a potential of +0.222 V.

Example 22-12. The potential of a half-cell *A* is found to be −0.362 V versus S.C.E. (a) What is the electrode potential (i.e., versus the standard hydrogen electrode) of *A*? (b) What is the potential of *A* versus the silver chloride–silver (1 *F* in KCl) electrode? The electrode potential of the S.C.E. is +0.241 V; that of the silver chloride–silver (1 *F* in KCl) electrode, +0.222 V.

(a) Since *A* is the negative electrode of the cell, the calculation takes the form $E = -0.362 + 0.241 = -0.121$ V (versus standard hydrogen electrode).

(b) Using the value calculated in part (a), the calculation takes the form $E = -0.121 - 0.222 = -0.343$ V versus the silver chloride–silver (1 *F* in KCl) electrode.

The understanding of the student will be aided by examination of the scale given below:

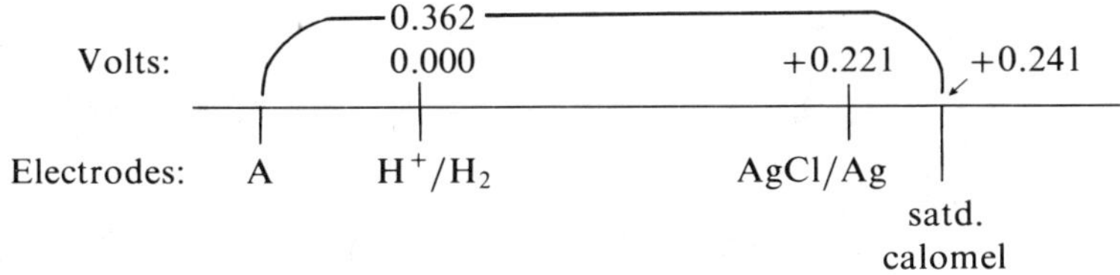

Questions 22.12

What is the basic distinction between voltaic and electrolytic cells?

What is the function of a salt bridge?

In your own words, explain the impossibility of measuring the *absolute* potential of a single half-cell.

22-4. At what temperature is the standard hydrogen electrode defined to have a potential of zero volts?

22-5. Discuss the two meanings of the term electrode in the consideration of electrochemical cells.

22-6. What is the significance of the sign of a calculated cell voltage?

22-7. What does a cell voltage of zero imply with regard to the cell reaction?

22-8. Explain why the term for the metal concentration is omitted from the Nernst equation when the electrode material is the metal itself.

22-9. Elaborate on the distinction between the terms voltage and potential.

22.13 Problems

For each of the following voltaic cells, calculate the cell voltage at 25°C, give the polarity of the right-hand electrode, write a balanced equation for the overall cell reaction, and indicate the direction of the spontaneous reaction by use of a single arrow.

22-1. Pt | VO_2^+ (1.0×10^{-2} *M*), VO^{2+} (1.0×10^{-3} *M*), H^+ (1.0×10^{-4} *M*) ‖ Fe^{2+} (1.0×10^{-1} *M*), Fe^{3+} (1.0×10^{-3} *M*) | Pt. $E^0_{VO_2^+/VO^{2+}} = +1.00$ V, $E^0_{Fe^{3+}/Fe^{2+}} = +0.77$ V.

Answer: 0.07 V, positive,

$$Fe^{3+} + VO^{2+} + H_2O \rightarrow Fe^{2+} + VO_2^+ + 2H^+$$

22-2. Pt | $Cr_2O_7^{2-}$ (1.0×10^{-1} *M*), Cr^{3+} (1.0×10^{-3} *M*), H^+ (1.0×10^{-2} *M*) ‖ Fe^{2+} (1.0×10^{-2} *M*), Fe^{3+} (1.0×10^{-3} *M*) | Pt. $E^0_{Cr_2O_7^{2-}/Cr^{3+}} = +1.33$ V, $E^0_{Fe^{3+}/Fe^{2+}} = +0.77$ V.

Answer: 0.39 V, negative,

$$Cr_2O_7^{2-} + 6Fe^{2+} + 14H^+ \rightarrow 2Cr^{3+} + 6Fe^{3+} + 7H_2O$$

22-3. Pt | Ti^{3+} (1.0×10^{-1} *M*), TiO^{2+} (1.0×10^{-3} *M*), H^+ (1.0×10^{-1} *M*) ‖ V^{2+} (1.0×10^{-2} *M*), V^{3+} (1.0×10^{-1} *M*) | Pt. $E^0_{V^{3+}/V^{2+}} = -0.26$ V, $E^0_{TiO^{2+}/Ti^{3+}} = +0.10$ V.

Answer: 0.06 V, negative,

$$V^{2+} + TiO^{2+} + 2H^+ \rightarrow V^{3+} + Ti^{3+} + H_2O$$

22-4. The standard electrode potential, E^0, of the copper(II)–copper half-cell has the value +0.337 V. What is the potential of this half-cell (a) versus a silver chloride–silver (1 *F* in KCl) reference electrode ($E^0 = +0.222$ V), (b) versus the saturated calomel electrode ($E^0 = +0.241$ V)?

Answers: (a) +0.115 V versus AgCl/Ag (1 *F* in KCl) electrode;
(b) +0.096 V versus saturated calomel electrode

22-5. What will be the change in the potential of a standard hydrogen electrode if the partial pressure of hydrogen is allowed to drop from the required values of 1.00 atm (= 760 torr) to 730 mm? Remember: For $H^+ + e \rightleftharpoons \frac{1}{2}H_2$, the Nernst equation takes the form

$$E = 0.059 \log \frac{[H^+]}{\sqrt{p_{H_2}}} \quad \text{(with } p_{H_2} \text{ in atmospheres)}$$

Answer: $\Delta E = 0.00052$ V

. How much time, in hours, is required to deposit all the copper from 100.0 ml of 0.10 *F* $CuSO_4$ when the current is 0.10 A? Assume 100% efficiency.

. A solution containing Cu^{2+} and Ag^+ is electrolyzed for 32 minutes and 10 seconds with a current of 200.0 mA, thereby depositing copper and silver completely. The total weight of deposited copper (*63.54*) and silver (*107.87*) was 279.3 mg. Calculate the milligrams of copper and of silver present in the solution assuming that only their ions reacted at the cathode.

. Pure crystals of $CuSO_4 \cdot 5H_2O$ are dissolved in 100.0 ml of pure water and the solution is electrolyzed. The cathode gains 0.4280 g. What is the pH of the resulting solution?

. A volume of 200.0 ml of 0.10 *F* $NiSO_4$ is electrolyzed with a current of 6.0×10^{-2} A for 1 hour and 20 minutes. Oxygen is evolved at the anode. Calculate the pH of the solution after electrolysis if the initial pH was 4.00 due to the presence of a small amount of H_2SO_4. Assume 100% efficiency for the electrolysis.

. Calculate the cell voltage, indicate the polarity of the right electrode, and write a balanced equation for the cell reaction, indicating the direction of the spontaneous reaction with a single arrow, for each of the following cells at 25°C.

(a) $Pt \mid Tl^{3+}(1.0 \times 10^{-2}\ M),$
$Tl^{+}(0.10\ M) \parallel Cu^{2+}(1.0 \times 10^{-3}\ M) \mid Cu$
$E^0_{Tl^{3+}/Tl^+} = +0.13$ V, $E^0_{Cu^{2+}/Cu} = +0.34$ V

(b) $Sb \mid SbO^{+}(5.0 \times 10^{-2}\ M),$
$H^{+}(1.0 \times 10^{-2}\ M) \parallel Pb^{2+}(1.0 \times 10^{-4}\ M) \mid Pb$
$E^0_{SbO^+/Sb} = +0.2$ V, $E_{Pb^{2+}/Pb} = -0.13$ V

(c) $Ag \mid Ag^{+}(6.0 \times 10^{-4}\ M) \parallel Fe^{3+}(1.0 \times 10^{-3}\ M),$
$Fe^{2+}(5.0 \times 10^{-1}\ M) \mid Pt$
$E^0_{Ag^+/Ag} = +0.80$ V, $E^0_{Fe^{3+}/Fe^{2+}} = +0.77$ V

. How many milliliters of a solution 0.10 *F* in iron(II) and 1.0 *F* in HCl must be added to 50.0 ml of a solution which is 1.0×10^{-2} *F* in iron(III), 1.0×10^{-3} *F* in iron(II), and 1.0 *F* in HCl, so that a platinum indicator electrode in this solution will have a potential of +0.440 V with reference to a saturated calomel electrode? $E_{Fe^{3+}/Fe^{2+}} =$ 0.700 V, $E_{S.C.E.} = +0.241$ V.

. What is the potential of a half-cell obtained by mixing 30.0 ml of 1.0×10^{-1} *M* Fe^{2+} with 60.0 ml of 1.0×10^{-3} *M* Fe^{3+}? $E^0_{Fe^{3+}/Fe^{2+}} =$ +0.77 V.

23 ELECTROMETRIC DETERMINATION OF pH

In many cases it is possible to compute (via the Nernst equation) the concentration of an ion from the measured potential of a half-cell involving this ion, and the known standard potential of the half-cell. The most common analytical application of this technique is the determination of the hydrogen ion concentration of solutions and consequently the pH. This determination, under certain circumstances, can be effected with a simple potentiometer circuit of the type described in Section 22.10. But commonly a more elaborate device, known as a pH meter, is employed. The purpose of this chapter is to give the student an insight into the facts underlying the electrometric determination of pH.

Measurement of pH by Means of Voltaic Cell 23.1

Consider the following cell:

$$\text{Pt, H}_2\text{(1 atm)} \mid \text{H}^+ (x\ M) \parallel \text{KCl, Hg}_2\text{Cl}_2\text{(satd. soln.)} \mid \text{Hg}_2\text{Cl}_2(s)\text{, Hg}$$

The right half-cell is a saturated calomel electrode, which at 25°C has a potential of +0.241 V. The left half-cell is a hydrogen electrode, the potential of which depends on the concentration of the hydrogen ion, x, and at 25°C follows the relation

$$E = 0.0592 \log[\text{H}^+]$$

The voltage of the cell is given by

$$E_{\text{cell}} = E_r - E_l = +0.241 - 0.0592 \log[\text{H}^+]$$

Rearrangement yields

$$\frac{E_{\text{cell}} - 0.241}{0.0592} = -\log[\text{H}^+] = \text{pH} \qquad (23\text{-}1)$$

Thus the pH of the solution in the left half-cell can, at least theoretically, be computed from the measured cell voltage.

23.2 Operational Definition of pH

In an actual attempt to obtain the concentration of hydrogen ion in the manner just described, problems, both theoretical and practical in nature, are encountered. First, the Nernst equation and the definition of pH, on a rigorous basis, involve activities; the use of molar concentrations is an approximation. Second, the measured value of the cell voltage contains, in addition to the potentials of the two half-cells, a small, unknown "junction potential" of the order of a few millivolts. Such a potential occurs at liquid–liquid junctions, for example, at the ends of the salt bridge, and has as its source the difference in the rates of diffusion of anions and cations across the junction.

To eliminate the difficulties in practical pH measurements an operational pH scale is established, in the following manner. A standard buffer solution of assigned pH value, pH_s, is made the electrolyte in the left half-cell and the cell voltage, E_s, is measured. Then the buffer is removed and replaced by the solution of unknown pH value, pH_x, and, under otherwise identical conditions, the cell voltage, E_x, is obtained. For the pH standard and the sample solution, respectively, the following analogs to equation (23-1) may be written:

$$\frac{E_s - k}{0.0592} = pH_s \qquad (23\text{-}2)$$

$$\frac{E_x - k}{0.0592} = pH_x \qquad (23\text{-}3)$$

The term k includes the electrode potential of the reference half-cell and any junction potentials, and its value is assumed to remain constant if the measurements of E_s and E_x are made consecutively and with the same experimental arrangement. Then k may be eliminated by combining equations (23-2) and (23-3) to yield the following equation, which allows calculation of the pH of the sample solution:

$$pH_x = pH_s + \frac{E_x - E_s}{0.0592} \qquad (23\text{-}4)$$

It should be emphasized that the operational pH scale does not solve the theoretical problems mentioned above. It merely allows the pH measurements to be made on a useful basis and in a comparable manner.

In the United States, the National Bureau of Standards has assigned pH_s values to five solutions as primary pH standards and to two further solutions as secondary standards. The pH values of these seven reference solutions for the temperature range 10 to 40°C are given in Table E in the Appendix.

The Glass Electrode 23.3

In Section 22.12 various experimental difficulties encountered with the hydrogen electrode are mentioned. Owing to these shortcomings, considerable effort has been expended in the development of other pH-sensitive electrodes. Most of these electrodes, including the antimony(III)–antimony and the quinhydrone electrodes, have been largely replaced by the glass electrode in association with the required electronic circuitry of the "pH meter."

One of the many existing forms of a glass electrode is shown schematically in Figure 23-1. A very thin membrane of a

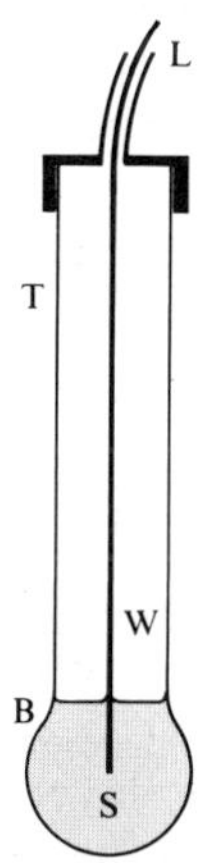

Figure 23-1. Glass electrode.

special glass in the form of a small bulb, B, is sealed onto a tube of thicker ordinary glass, T. Within this glass envelope is placed a solution, S, of definite pH. This internal solution usually has a substantial buffer capacity, so its hydrogen ion concentration remains constant. Into this internal solution dips a metal wire, W. The interior of the glass electrode is thereby made a half-cell, in practice frequently a silver chloride–silver electrode. The internal electrode has a shielded wire lead, L, that allows connection to a voltage-measuring device.

To determine the pH, the glass electrode and a salt bridge leading to an external reference electrode are immersed in the sample solution and the voltage developed across the cell thus formed is measured. To simplify the discussion it will be assumed that the internal and external reference electrodes are identical silver chloride–silver electrodes, although in practice the saturated calomel electrode commonly serves as the external reference. The complete cell may then be represented as follows:

$$\underbrace{\overbrace{\mathrm{Ag, AgCl}(s) \mid \mathrm{Cl}^-(a\ M)}^{\text{internal electrode}},\ \mathrm{H}^+(s\ M)}_{\text{glass electrode}} \overset{\text{glass membrane}}{\vdots}\ \mathrm{H}^+(x\ M) \parallel$$

$$\overbrace{\mathrm{Cl}^-(a\ M) \mid \mathrm{AgCl}(s),\ \mathrm{Ag}}^{\text{external (reference) electrode}}$$

Since the chloride concentration is the same in the two electrodes, no net cell voltage results from these electrodes. If the concentration of hydrogen ion in the sample solution, x, differs from that in the internal solution, s, a potential develops across the glass membrane and therefore across the cell. The origin of this potential is not the same as that for the voltaic cells previously discussed. The phenomena underlying this so-called membrane potential are not fully elucidated, but the following nonrigorous description will provide the student with at least some insight into the functioning of the glass electrode.

The unique feature of the special glass used in the glass bulb is its pronounced selectivity for protons. Protons, unlike most other ions, are capable of accumulating on the membrane surface. This accumulation is due to adsorption, or, according to another interpretation, proceeds via exchange with other cations present in the outer layer of the silicate structure of the membrane. Whatever the mechanism, the number of protons accumulated depends on the hydrogen ion concentration of the solution. When the concentrations of hydrogen ion in the internal and sample solution differ, more protons accumulate at one side of the membrane than at the other and a membrane potential develops.

The relation between this potential and the hydrogen ion concentration of the external solution follows the equation†

$$E = k + 0.0592 \log[\mathrm{H}^+] \qquad (23\text{-}5)$$

†The equation is, in fact, an approximation, because the factor of the logarithmic term is not 0.0592 but, depending on the glass, a number close to it. However, this has no bearing on actual pH measurements, because in this connection the formula is not used as such, but the pH meter is calibrated employing a standard solution.

or, rewritten in terms of pH,

$$E = k - 0.0592 \text{ pH} \qquad (23\text{-}6)$$

Thus the potential is a linear function of the pH of the test solution. The value, k, depends on the composition of the glass, the temperature, and the hydrogen ion concentration of the internal solution.

Equations (23-5) and (23-6) apply even if the internal and external electrodes are not identical. The potential difference between these two electrodes is constant, can be incorporated into k, and consequently merely changes the numerical value of this term. For practical determinations k is not evaluated as such but a calibration with a standard solution is employed and the reasoning given in Section 23.2 applies.

Even when the internal and external electrodes and the hydrogen ion concentrations of the internal and the sample solutions are identical, a small voltage usually develops across the cell. This potential, known as the asymmetry potential, is the result of strain in the glass membrane, unequal conditions of the inner and outer membrane surfaces, and other causes not fully understood. This potential is also of no practical consequence because, during the calibration, it is incorporated into k. However, because the value of the asymmetry potential undergoes changes with time, frequent recalibration of a glass electrode is warranted.

The pH Meter 23.4

The voltage of a cell containing a glass electrode cannot be measured by a potentiometer with a galvanometer as the null detector (Section 22.10). The resistance of a glass electrode is of the order of 100 megohms ($1 \text{ M}\Omega = 10^6 \ \Omega$). With so great a resistance in the circuit, the current flowing is far too small to move the needle of the galvanometer. Consequently it is necessary to resort to more elaborate vacuum-tube or transistor circuits. The theory and operation of such circuits is beyond the scope of this textbook and only the method of their application in pH meters will be summarized.

One type of pH meter uses a conventional potentiometer circuit (Figure 22-2) in which the galvanometer is replaced by a high resistance. Any imbalance between the potentiometer and cell circuits will result in a voltage (IR) drop across the high resistance. This voltage drop is amplified by an electronic circuit so that it can be presented by even a relatively insensitive meter. When the potentiometer and cell voltages are exactly balanced, no current flows, no voltage drop occurs across the resistance, and the meter reads zero. Here the amplifier circuit aids only in the detection of the balance point of the potentiometer (null detector). Consequently, such pH meters are described as the potentiometer or null type. The slide-wire dial of such instruments is usually calibrated both in pH units and volts.

The second type of pH meter does not use a potentiometer circuit, but rather is a vacuum-tube voltmeter (or a transistor analog). The voltage across the cell is applied through a high resistance to the input of the instrument and the pH (or voltage) is read directly from a meter which is driven by the output of the amplifier circuit. The electronic requirements for such a pH meter are more rigorous than those for the potentiometer type because a strictly linear relationship between input voltage and the deflection of the meter needle is mandatory. This type of instrument is usually called a direct-reading pH meter. It has an additional practical advantage over the potentiometer type in that the electrical output can be fed directly to a chart recorder, permitting a graphical presentation of pH changes as a function of time or of any parameter proportional to time. For example, such an instrument coupled with a recorder can be used to plot the titration curve during an acid–base titration if the titrant is added from a buret of constant flow rate. The electrical output of a direct-reading pH meter can also be fed to a servomechanism to control an industrial process.

23.5 "Sodium Error"

With common glass electrodes, a sizable error in the pH measurement is encountered if the sample solution contains sodium ion and has a pH value greater than about 9.5. This error imposes a serious limitation, since sodium is often a major constituent of alkaline solutions. For solutions 1 *M* in sodium ion the sodium error is negligible below pH 9, but may reach −0.2 pH unit at pH 10 and −1 pH unit at pH 12. Some manufacturers provide a nomograph that allows a correction to be read for the sodium error. The magnitude of the correction depends on the temperature, the pH value, and the sodium ion concentration. Other alkali metal cations cause an analogous, but smaller, error. Thus the error is sometimes called the alkali-metal error or simply the alkali error.

In recent years, special glasses have been developed that yield a glass electrode showing so small an alkali error that it can be neglected for most practical purposes.

Comparison of Glass and Hydrogen Electrodes 23.6

The glass electrode has the following advantages over the hydrogen electrode: (1) It is not readily poisoned; (2) it does not change the composition of the sample solution; (3) it functions properly in solutions containing oxidants or reductants; and (4) it comes rapidly to equilibrium. The glass electrode has the following shortcomings: (1) it is fragile and cannot readily be improvised; (2) its surface can become fouled with large organic molecules such as proteins; (3) its high electrical resistance requires the use of special circuitry; (4) it is subject to an alkali error unless fashioned from special glass; and (5) it must be stored in an aqueous solution or, if dried, be reconditioned by prolonged soaking in water before use.

Ion-Selective Electrodes 23.7

With respect to pH measurements, the alkali error is quite detrimental, and much research has been expended to find glasses that do not show this error. Studies have also been directed in the opposite direction, toward the goal of finding glasses that are extremely sensitive to changes in the concentration of some metal ions; indeed, so sensitive that pH changes cause no, or only a negligible, response. Electrodes fashioned from such glasses can be used to measure a metal ion concentration (or rather activity) in the same simple manner as pH. Such cation-selective electrodes are available and exhibit a quite satisfactory degree of selectivity. Electrodes, for example, that respond to sodium, potassium, or calcium are commercially available.

The action of a glass membrane, cation-selective electrode is related to the presence of anionic sites attractive to a cation of appropriate charge-to-size ratio. Similarly if a membrane with cationic sites is prepared, an anion-selective electrode might be realized. Such membrane electrodes have been developed based on impregnating a rubber with an insoluble salt (precipitate membrane electrode), use of an inorganic single crystal doped with impurities (solid-state electrode), or a liquid ion exchanger in contact with the sample solution (liquid–liquid electrode). For example, a solid-state fluoride-selective electrode is available that responds to fluoride activities in the range 10^0 to 10^{-6} *M*, with hydroxide ion apparently the only major interference.

Cation- and anion-selective electrodes are becoming important analytical tools and can be used analogously to a pH glass electrode, and indeed with some pH meters. Ion-selective electrodes serve both to establish directly the activity of an ion in an aqueous solution and for indication of titrations involving that ion.

23.8 Questions

23-1. What is the purpose of adopting an operational definition of pH?

23-2. Explain why frequent standardization of a pH meter is necessary.

23-3. Explain why the potential of a glass electrode cannot be measured with an ordinary potentiometer.

23-4. Describe the difference between and discuss the advantages and disadvantages of the two types of pH meters mentioned.

23-5. What difficulty may be encountered in the use of a glass electrode to measure the pH of a sodium hydroxide solution?

23-6. Reasoning from a knowledge of the use of a hydrogen-ion-sensitive electrode and pH meter, indicate the steps that might be involved in establishing the chloride activity of a biological fluid using a chloride-selective electrode. If the chloride content of the fluid were determined by titration with silver nitrate, how might the chloride-selective electrode be used?

23.9 Problems

23-1. A hydrogen electrode in combination with a saturated calomel electrode is used at 25°C to determine the pH of a solution. The measured cell voltage is 0.321 V with the hydrogen electrode negative in polarity. Assuming the pressure of hydrogen gas to be 1 atom, calculate the pH of the solution.

Answer: pH 1.3_6

23-2. A hydrogen electrode–calomel electrode pair is used to measure the pH of a solution. Neither the pressure of the hydrogen gas nor the chloride ion concentration is known exactly. However, the measured cell voltage is 0.561 V, with the hydrogen electrode having negative polarity, when the cell contains a phthalate solution of pH 4.01. When the cell is filled with the unknown solution, other conditions remaining constant, the voltage is found to be 0.828 V. Calculate the pH of the solution.

Answer: pH 8.54

A hydrogen electrode–saturated calomel electrode pair is used to measure the pH of solutions. Calculate the voltage of this cell for each of the pH values given below assuming the hydrogen pressure is exactly 1 atm, and indicate the polarity of the S.C.E. $E_{S.C.E.} = +0.241$ V.

(a) 1.00 *Answer:* 0.300 V with S.C.E. positive
(b) 3.75
(c) 8.26
(d) 9.07
(e) 5.22
(f) 6.98

The voltage of the following cell is found to be 0.118 V. Calculate the dissociation constant of the weak acid, HA.

$$^{(+)}Pt \mid H_2(1\ atm), H^+(0.20\ M) \parallel HA(0.040\ F), H_2(1\ atm) \mid Pt^{(-)}$$

Answer: $K_a = 1.0_5 \times 10^{-4}$

The voltage of the following cell is found to be 0.477 V. Calculate the dissociation constant of the weak acid. $E_{S.C.E.} = +0.241$ V.

$$^{(+)}Hg \mid Hg_2Cl_2(s), KCl(satd.) \parallel HA(1.0 \times 10^{-2}\ F), H_2(1\ atm) \mid Pt^{(-)}$$

Calculate the voltage of the following cell.

$$Hg \mid Hg_2Cl_2(s), KCl(satd.) \parallel NH_4Cl(0.18\ F), H_2(1\ atm) \mid Pt$$

$$K_{b,NH_3} = 1.8 \times 10^{-5},\ E_{S.C.E.} = +0.241\ V$$

24 THE THEORY OF REDOX TITRATIONS

To be made the basis of a titration, a redox reaction must fulfill the general requirements for any titration reaction (Chapter 10); that is, the reaction should proceed rapidly and to completion, a definite, known equivalence should exist between the oxidant and reductant, and a suitable technique should be available for the detection of the end point. Since many of the elements can exist in more than one oxidation state, applications for redox titrations are numerous (Chapter 25). In this chapter the feasibility of such titrations will be examined by the use of the Nernst equation to calculate titration curves.

Titration Curves 24.1

For redox titrations, as for titrations previously considered, a titration curve can be obtained by plotting the negative logarithm of the concentration of the species being titrated as a function of the volume of titrant added. However, the plotting of half-cell potentials is preferable for a number of reasons. First, from the Nernst equation, such potentials are a logarithmic function of the concentration of the reacting species. Second, the usage facilitates the choice of visual redox indicators, since these are redox couples for which electrode potentials are available from tables (Section 24.9). Finally, the potentiometric method of end-point detection, widely used in redox titrimetry, is directly related to the half-cell potential (Chapter 27).

The model used for the calculation of a redox titration curve is a cell consisting of the half-cell containing the solution being titrated and a standard hydrogen electrode, connected via a salt bridge. The cell potential is calculated after the assumed addition of a known volume of the titrant. As will become evident in the subsequent consideration of methods of end-point detection, such a physical arrangement may or may not be used in the performance of the actual redox titration. Such a model, however, aids in understanding the course of the titration.

Although the same calculational principles apply to all redox titrations, it is convenient to distinguish several cases: (1) titrations in which the same number of electrons appear in both balanced half-reaction equations (Section 24.2); (2) titrations in which different numbers of electrons appear in the two balanced half-reaction equations (Section 24.3); (3) titrations in which a half-reaction equation involves the solvent and its dissociation products explicitly (Section 24.4); and (4) titrations in which one molecule or ion of a reactant yields two or more molecules or ions of product, or vice versa (Section 24.5). The titration curve for these various cases will be considered individually.

24.2 Titrations with Same Number of Electrons in Both Half-Reaction Equations

Consider the titration of exactly 50 ml of a solution 0.10 *F* in iron(II) sulfate with 0.10 *F* cerium(IV) sulfate solution. The titration equation is

$$Fe^{2+} + Ce^{4+} \rightarrow Fe^{3+} + Ce^{3+} \qquad (24\text{-}1)$$

The half-reaction equations and Nernst equation expressions take the form†

$$Fe^{3+} + e \rightarrow Fe^{2+} \qquad E = +0.68 + 0.059 \log \frac{[Fe^{3+}]}{[Fe^{2+}]} \qquad (24\text{-}2)$$

$$Ce^{4+} + e \rightarrow Ce^{3+} \qquad E = +1.44 + 0.059 \log \frac{[Ce^{4+}]}{[Ce^{3+}]} \qquad (24\text{-}3)$$

At the start of the titration, the solution as a first approximation might be considered to contain only Fe^{2+}. This assumption implies that the concentration of Fe^{3+} is zero and, hence,

†This titration is performed in a sulfuric acid solution and cerium(IV) is largely present as an anionic sulfato complex, $Ce(SO_4)_3^{2-}$. In order to simplify the treatment of this example it will be assumed that the cerium(IV) ion is present as the simple ion and that its concentration can be shown as $[Ce^{4+}]$. However, the formal potential of the couple, $E^f_{Ce^{4+}/Ce^{3+}} = +1.44$ V in 1 *F* H_2SO_4, will be used rather than the standard potential. The formal potential of the iron couple, $E^f_{Fe^{3+}/Fe^{2+}} = +0.68$ in 1 *F* H_2SO_4, will also be used. The concept of formal potential was introduced in Section 22.8 and is further treated in Section 28.8. Throughout Chapter 24 wherever formal potentials are used with the Nernst equation, concentrations will be shown as molarities, although formalities should properly be used for the relevant species.

from the Nernst equation that the potential is minus infinity. This conclusion is unreasonable and in conflict with experimental observation. Actually there is an extremely small, but unknown, concentration of Fe^{3+} present; the potential, although finite, cannot be calculated for this point.

On the addition of a small volume of the titrant, the situation changes. Since the reaction goes essentially to completion, it can be seen from the titration equation, (24-1), that the number of millimoles of Fe^{3+} formed is equal to the number of millimoles of Ce^{4+} added. The number of millimoles of Fe^{2+} remaining unreacted equals the original number of millimoles of Fe^{2+} minus the number of millimoles of Fe^{3+} formed.

After the addition of a volume of 1.0 ml of the 0.10 M Ce^{4+} solution, the amount of Fe^{3+} formed equals $1.0 \times 0.1 = 0.10$ millimole. The amount of Fe^{2+} remaining is $50 \times 0.10 - 0.10 = 4.9$ millimoles. The half-cell potential can be calculated from the Nernst equation for the iron couple:

$$\begin{aligned} E &= +0.68 + 0.059 \log \frac{0.10}{4.9} \\ &= +0.68 - 0.059 \times 1.69 = +0.58 \text{ V} \end{aligned}$$

In this calculation millimoles may be substituted for molarities since the solution volume cancels. (The student will recall that a similar simplification was possible in the calculation of the titration curve for a weak acid in Section 11.4.) This possibility implies that for this and analogous examples the potential is independent of the dilution of the solution.

In principle, the potential could also be calculated from the Nernst equation for the cerium couple. However, the reduction goes essentially to completion and only an extremely small concentration of Ce^{4+} remains. This Ce^{4+} concentration could be calculated from the value of the equilibrium constant if this were known. It is obviously easier, however, to calculate the half-cell potential from the expression for the iron couple. (The value of an equilibrium constant can be calculated from potential data; see Section 24.7.)

When a total volume of 10 ml of the Ce^{4+} solution has been added, the amount of Fe^{3+} formed is 10×0.10 millimole and the amount of Fe^{2+} unreacted corresponds to $5.0 - 1.0 = 4.0$ millimoles. At this point the half-cell potential is given by

$$\begin{aligned} E &= +0.68 + 0.059 \log \frac{1.0}{4.0} \\ &= +0.68 - 0.059 \times 0.60 = +0.65 \text{ V} \end{aligned}$$

At the midpoint of the titration, that is, when a total volume of exactly 25 ml of the Ce^{4+} solution has been added, the amount of Fe^{3+} formed as $25 \times 0.10 = 2.5$ millimoles and the amount of Fe^{2+} remaining is $5.0 - 2.5 = 2.5$ millimoles. Hence, the half-cell potential is given by

$$E = +0.68 + 0.059 \log \frac{2.5}{2.5} = +0.68 \text{ V}$$

Note that at the midpoint of this titration the potential is equal to the formal potential (or standard electrode potential) of the system being titrated, $E^{f}_{Fe^{3+}/Fe^{2+}}$. This equality offers an analogy to the titration of a weak acid with a strong base, where at the midpoint $pH = pK_a$; the analogy is more obvious when the Nernst equation is contrasted with the "buffer" equation, $pH = pK_a + \log([\text{salt}]/[\text{acid}])$ (see Sections 11.4 and 13.3).

The equivalence point is reached when a total volume of exactly 50 ml of the Ce^{4+} solution has been added. At this point, following the same mode of calculation, the amount of Fe^{3+} formed is $50 = 0.10 = 5.0$ millimoles, and the amount of Fe^{2+} unreacted is $5.0 - 5.0 = 0.0$ millimole. Substitution of these values into the Nernst equation yields an electrode potential of infinity, which is obviously incorrect. (An analogous situation exists in the calculation of the equivalence point for the titration of a strong acid with a strong base; see Section 11.2.) Actually at equilibrium small amounts of Fe^{2+} and Ce^{4+} must be present. To obtain the potential at this point one proceeds as follows. The equivalence-point potential is written via the Nernst equation for each of the two couples involved:

$$E_{\text{eq pt}} = +0.68 + 0.059 \log \frac{[Fe^{3+}]}{[Fe^{2+}]}$$

$$E_{\text{eq pt}} = +1.44 + 0.059 \log \frac{[Ce^{4+}]}{[Ce^{3+}]}$$

Addition of these two equations yields

$$2E_{\text{eq pt}} = +0.068 + 1.44 + 0.059 \log \frac{[Fe^{3+}][Ce^{4+}]}{[Fe^{2+}][Ce^{3+}]}$$

At the equivalence point, the total millimoles of the iron species exactly equals the total millimoles of the cerium species. Inspection of the titration equation reveals that the change of one ion of Fe^{2+} to Fe^{3+} requires the conversion of one ion of Ce^{4+} to Ce^{3+}. In other words, at the equivalence point for each Fe^{+3} present one Ce^{+3} has formed, and for each Fe^{+2} ion remaining unoxidized one Ce^{+4} ion remains unreduced. Consequently,

$$[Fe^{3+}] = [Ce^{3+}]$$
$$[Fe^{2+}] = [Ce^{4+}]$$

Division of one of these equations by the other yields

$$\frac{[Fe^{3+}]}{[Fe^{2+}]} = \frac{[Ce^{3+}]}{[Ce^{4+}]}$$

Substitution in the above equation yields

$$2E_{\text{eq pt}} = +0.68 + 1.44 + 0.059 \log \frac{\cancel{[Ce^{3+}]}\cancel{[Ce^{4+}]}}{\cancel{[Ce^{4+}]}\cancel{[Ce^{3+}]}}$$

Since log 1 = 0,

$$2E_{\text{eq pt}} = +0.68 + 1.44$$

Hence,

$$E_{\text{eq pt}} = \frac{+0.68 + 1.44}{2} = +1.06 \text{ V}$$

Thus, in this example, the potential of the half-cell at the equivalence point is the average of the formal electrode potentials of the two redox couples. (Where standard electrode potentials are applicable, the equivalence-point potential will be the average of these two values.)

Beyond the equivalence point the half-cell potential can be calculated from the Nernst equation for the cerium couple. In principle, the potential could be calculated from the Nernst equation for the iron couple; however, the concentration of the Fe^{2+} ion is not readily available. The reasoning is similar to that mentioned above for the nonuse of the cerium couple before the equivalence point.

When a total volume of 60 ml of the Ce^{4+} solution has been added, 5 millimoles of Ce^{3+} and 1 millimole of Ce^{4+} are present in the solution. Here

$$E = +1.44 + 0.059 \log \frac{1.0}{5.0} = +1.40 \text{ V}$$

After addition of a total of 100 ml of the Ce^{4+} solution,

$$E = +1.44 + 0.059 \log \frac{5.0}{5.0} = +1.44 \text{ V}$$

Thus, after addition of twice the volume of titrant required to reach the equivalence point, the half-cell potential equals the formal potential (or standard electrode potential) of the redox system of the titrant, $E^f_{Ce^{4+}/Ce^{3+}}$. The complete titration curve, calculated in this way, is given in Figure 24-1.

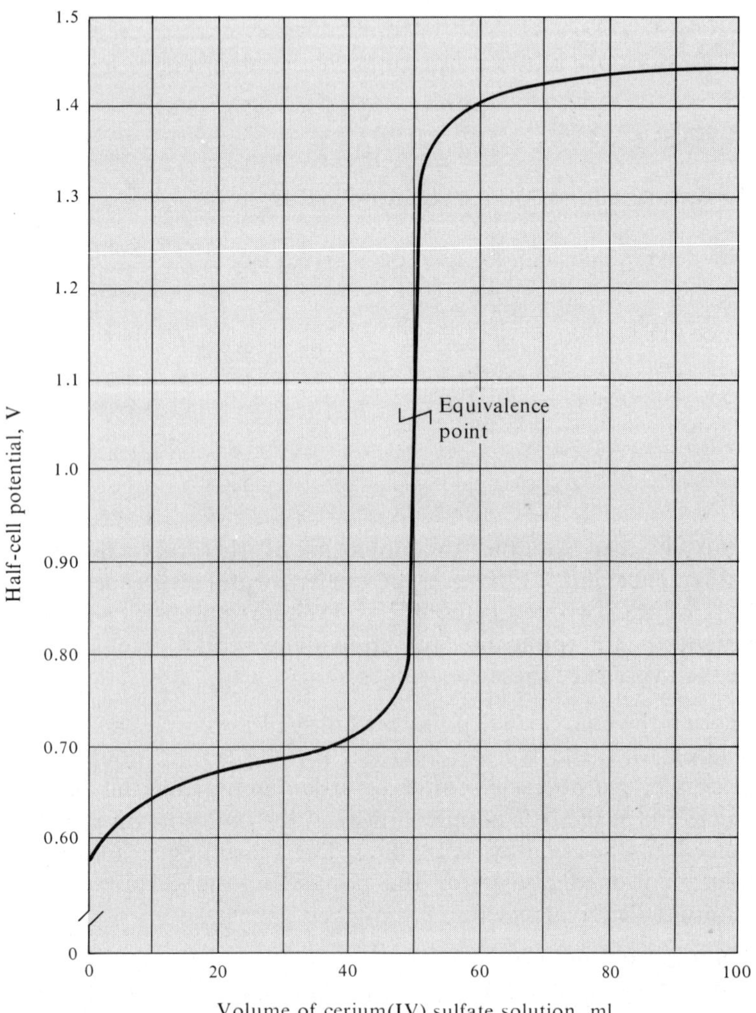

Figure 24-1. Calculated titration curve for the titration of 50 ml of 0.10 *F* $FeSO_4$ with 0.10 *F* cerium(IV) solution (in a titration medium 1 *F* in H_2SO_4).

24.3 Titrations with Different Numbers of Electrons in the Half-Reaction Equations

When the redox titration involves a different number of electrons in the two half-reaction equations, the shape of the titration curve differs from that where the same number of electrons is involved.

Consider the titration of exactly 50 ml of a solution 0.050 *F* in tin(IV) chloride with a solution 0.10 *F* in chromium(II) chloride. Since tin(IV) is present as a chloro complex, the formal potential in 1 *F* hydrochloric acid will be used for the tin couple, but to simplify the notation used, the symbol $[Sn^{4+}]$ will be allowed to denote the formal, rather than the molar, concentration of tin(IV). Similar considerations apply to the chromium couple, for which the formal potential in 1 *F* hydrochloric acid will also be used.

The titration reaction is

$$Sn^{4+} + 2Cr^{2+} \rightleftharpoons Sn^{2+} + 2Cr^{3+} \qquad (24\text{-}4)$$

The half-reaction equations and Nernst equation expressions take the form

$$Sn^{4+} + 2e \rightleftharpoons Sn^{2+} \qquad E = +0.14 + \frac{0.059}{2} \log \frac{[Sn^{4+}]}{[Sn^{2+}]} \qquad (24\text{-}5)$$

$$Cr^{3+} + e \rightleftharpoons Cr^{2+} \qquad E = -0.38 + 0.059 \log \frac{[Cr^{3+}]}{[Cr^{2+}]} \qquad (24\text{-}6)$$

At the start of the titration, the potential cannot be calculated, for reasons noted in the case considered in Section 24.2.

Upon addition of a volume of 10 ml of the Cr^{2+} solution, the amount of Cr^{3+} formed is $10 \times 0.10 = 1.0$ millimole. Initially, the amount of Sn^{4+} present was $50 \times 0.050 = 2.5$ millimoles. Since from the titration equation two ions of Cr^{3+} are produced by the reaction of one Sn^{4+} ion, the amount of Sn^{4+} reacted is $1.0/2 = 0.50$ millimole. The amount of Sn^{4+} remaining is therefore $2.5 - 0.5 = 2.0$ millimoles. The half-cell potential can be calculated from the Nernst equation for the tin couple, (24-5):

$$E = +0.14 + \frac{0.059}{2} \log \frac{2.0}{0.5} = +0.16 \text{ V}$$

In this calculation, millimoles can be substituted for molarities since the solution volume cancels; this implies that, for this and analogous examples, the potential is independent of the dilution of the solution.

When a total volume of 25 ml of the Cr^{2+} solution has been added, the amount of Cr^{3+} formed is $25 \times 0.10 = 2.5$ millimoles. The amount of Sn^{4+} reacted is $2.5/2 = 1.25$ millimoles.

The amount of Sn^{4+} remaining is 2.5 − 1.25 = 1.25 millimoles. At this point the potential is given by

$$E = +0.14 + \frac{0.059}{2} \log \frac{1.25}{1.25} = +0.14 \text{ V} = E^{f}_{Sn^{4+}/Sn^{2+}}$$

Here, as in the case considered in Section 24.2, the potential at the midpoint of the titration is equal to the formal potential (or standard electrode potential) of the system being titrated.

The equivalence point is reached upon addition of exactly 50 ml of the Cr^{2+} solution. Here the total number of millimoles of the chromium species equals twice the total number of millimoles of the tin species present. At equilibrium not all the chromium is present as Cr^{3+} and not all the tin is present as Sn^{2+}. Inspection of the titration equation reveals that the change of one ion of Sn^{4+} to Sn^{2+} requires the conversion of two ions of Cr^{2+} to Cr^{3+}. As a consequence, at the equivalence point

$$2[Sn^{4+}] = [Cr^{2+}] \qquad \text{and} \qquad 2[Sn^{2+}] = [Cr^{3+}]$$

Division of the first of these equations by the second yields

$$\frac{[Sn^{4+}]}{[Sn^{2+}]} = \frac{[Cr^{2+}]}{[Cr^{3+}]}$$

Substitution of this equation into the Nernst equation for the tin couple, (24-5), yields for the half-cell potential at the equivalence point

$$E_{\text{eq pt}} = +0.14 + \frac{0.059}{2} \log \frac{[Cr^{2+}]}{[Cr^{3+}]} \tag{24-7}$$

To eliminate the logarithmic term on addition of this equation to the Nernst equation for the chromium couple, (24-6), it is necessary to multiply (24-7) by 2:

$$2E_{\text{eq pt}} = 2 \times 0.14 + 0.059 \log \frac{[Cr^{2+}]}{[Cr^{3+}]} \tag{24-8}$$

For the chromium couple,

$$E_{\text{eq pt}} = -0.38 + 0.059 \log \frac{[Cr^{3+}]}{[Cr^{2+}]} \tag{24-6}$$

Addition of equations (24-6) and (24-8) yields

$$3E_{\text{eq pt}} = 2 \times 0.14 - 0.38$$

and $$E_{\text{eq pt}} = \frac{2 \times 0.14 - 0.38}{3} = -0.03 \text{ V}$$

It is seen that the value of the potential of the equivalence point does not fall halfway between the values of the formal potentials (or standard electrode potentials) of the two redox couples, as in the case considered in Section 24.2. Rather the value is two thirds of the way and closer to the potential involving the two-electron change.

Beyond the equivalence point the potential is calculated from the expression for the chromium couple. Thus, the potential value corresponding to the addition of a total of 60 ml of the Cr^{2+} solution is given by

$$E = -0.38 + 0.059 \log \frac{5.0}{1.0} = -0.34 \text{ V}$$

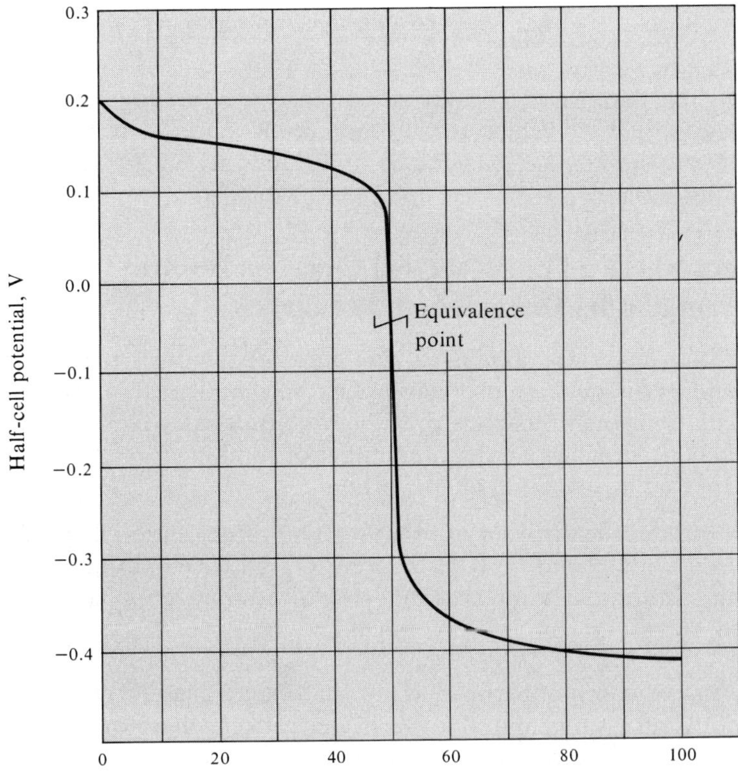

Figure 24-2. Calculated titration curve for the titration of 50 ml of 0.050 F $SnCl_4$ with 0.10 F $CrCl_2$ solution (in a titration medium 1 F in HCl).

After addition of a total volume of 100 ml of the Cr^{2+} solution, the potential is given by

$$E = -0.38 + 0.059 \log \frac{5.0}{5.0} = -0.38 \text{ V} = E^f_{Cr^{3+}/Cr^{2+}}$$

Here, as in the case previously considered, the potential after addition of twice the volume of titrant required to reach the equivalence point corresponds to the formal potential (or standard electrode potential) represented by the titrant. The complete titration curve for this example is given in Figure 24-2.

In general, where the titration equation can be written in a form $n_1\text{Ox}_1 + n_2\text{Red}_2 \rightleftharpoons n_1\text{Red}_1 + n_2\text{Ox}_2$, the half-cell potential at the equivalence point is given by

$$E_{\text{eq pt}} = \frac{n_1 E^f_{\text{Ox}_1/\text{Red}_1} + n_2 E^f_{\text{Ox}_2/\text{Red}_2}}{n_1 + n_2} \qquad (24\text{-}9)$$

In the two cases considered so far, $n_1 = n_2$ and $2n_1 = n_2$, respectively. While formal potentials appear in equation (24-9), standard electrode potentials may be substituted when applicable.

24.4 Titrations where a Half-Reaction Equation Involves the Solvent and Its Dissociation Products

Titrations where a half-reaction equation involves the solvent and its dissociation products are more complicated in their calculations than the cases previously considered. This difficulty may occur when a different number of combined oxygen atoms appears in a reactant and its product.

Consider the titration of exactly 50 ml of a solution 0.10 *M* in Fe^{2+} with a solution 0.02 *M* in MnO_4^- under conditions where the titration solution is acidic. The titration reaction is

$$5Fe^{2+} + MnO_4^- + 8H^+ \rightleftharpoons 5Fe^{3+} + Mn^{2+} + 4H_2O \qquad (24\text{-}10)$$

The reduction of permanganate ion to manganese(II) ion is not fully reversible and the Nernst equation, therefore, is not strictly applicable. However, the agreement between calculated and experimental titration curves is close enough to make this a worthwhile example, especially because of the practical importance of permanganate titrations.

Since the half-cell potential prior to the equivalence point is determined by the ratio $[Fe^{3+}]/[Fe^{2+}]$ and the concentration ratios are identical to those in the example considered in Section 24.3, the titration curve in both examples will be identical up to the vicinity of the equivalence point. At the equivalence point, the following equalities exist:

$$[Fe^{3+}] = 5[Mn^{2+}] \qquad \text{and} \qquad [Fe^{2+}] = 5[MnO_4^-]$$

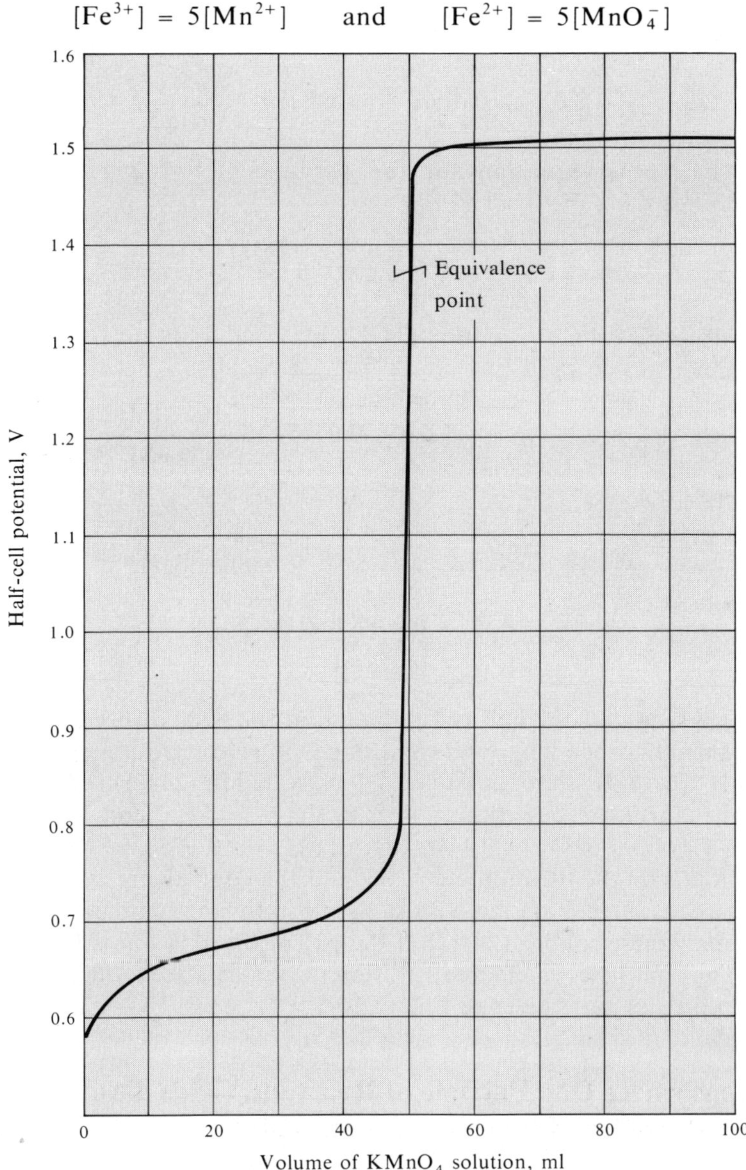

Figure 24-3. Calculated titration curve for the titration of 50 ml of 0.10 *F* $FeSO_4$ with 0.02 *F* $KMnO_4$ (in a titration medium 1 *M* in hydrogen ion).

Hence,

$$\frac{[Fe^{3+}]}{[Fe^{2+}]} = \frac{[Mn^{2+}]}{[MnO_4^-]}$$

Substitution of this equation into the Nernst equation for the iron couple, (24-2), yields

$$E_{\text{eq pt}} = +0.68 + 0.059 \log \frac{[Mn^{2+}]}{[MnO_4^-]} \tag{24-11}$$

The Nernst equation for the permanganate–manganese(II) couple takes the form

$$E_{\text{eq pt}} = +1.51 + \frac{0.059}{5} \log \frac{[MnO_4^-][H^+]^8}{[Mn^{2+}]} \tag{24-12}$$

Multiplication of equation (24-12) by 5 and addition to (24-11) yields

$$6E_{\text{eq pt}} = +0.68 + 5 \times 1.51 + 0.059 \log \frac{\cancel{[Mn^{2+}]}\cancel{[MnO_4^-]}[H^+]^8}{\cancel{[MnO_4^-]}\cancel{[Mn^{2+}]}}$$

On rearrangement,

$$\begin{aligned} E_{\text{eq pt}} &= \frac{+0.68 + 5 \times 1.51}{6} + 0.0098 \log[H^+]^8 \\ &= +1.37 + 0.0098 \log[H^+]^8 \\ &= +1.37 - 0.078 \text{ pH} \end{aligned} \tag{24-13}$$

It is seen that the half-cell potential at the equivalence point is strongly dependent on the hydrogen ion concentration. Only if $[H^+] = 1\ M$, that is, pH $= 0$, does equation (24-13) reduce to the form of equation (24-9) with $n_1 = 5n_2$, and $E_{\text{eq pt}} = +1.37$ V. If $[H^+] = 0.1\ M$, that is, pH $= 1.0$, $E_{\text{eq pt}} = +1.29$ V. Points on the titration curve beyond the equivalence point can be calculated from the Nernst equation for the permanganate–manganese couple, (24-12), if the pH of the solution is known. The complete, calculated titration curve for a medium 1 *M* in hydrogen ion is given in Figure 24-3.

24.5 Titrations where One Particle of Reactant Yields Two or More Particles of Product or Vice Versa

In the preceding cases, the titration curve is independent of the total concentration of the redox couple. That is, the half-cell potential is essentially unchanged by dilution (or concentration) of the solution, except as this changes the pH in the systems considered in Section 24.4. Cases exist where the species actually reduced (or oxidized) written on one side of the balanced equation occurs with a coefficient different from that of the species formed and written on the other side of the equation. Typical examples include

$$S_4O_6^{2-} + 2e \rightleftharpoons 2S_2O_3^{2-}$$
$$14H^+ + Cr_2O_7^{2-} + 6e \rightleftharpoons 2Cr^{3+} + 7H_2O$$

Here one mole of tetrathionate or dichromate is equivalent to two moles of thiosulfate or chromium(III), respectively. An analogous situation occurs, for example, in all titrations involving iodine, bromine, and oxalate. (Write the relevant half-reaction equations!)

In such cases the concentration terms of the relevant species have different exponents in the Nernst equation. As a result the solution volume will not cancel in the ratio of concentrations (if expressed as millimoles per milliliter), and the potential will vary with the dilution of the solution. In addition, the logarithmic term will not vanish from the equation for the potential of the equivalence point and, consequently, the potential at this point will also be a function of the initial concentration of the species titrated and of the titrant.

Redox "Buffers" 24.6

Analogies between redox and acid–base titrations have already been stressed. One more is of interest. The shape of the curve for a redox titration may be compared with that of the titration of a weak acid with a strong base. Both curves show the following features: a sharp rise at the beginning, a section of low slope around the halfway point, a "jump" about the equivalence point, and a horizontal portion beyond that jump. Of special interest is the situation around the halfway point. The situation for an acid–base titration has been discussed in Section 13.3. The portion of low slope indicates that addition of oxidant or reductant to a solution having a composition corresponding to that region will not cause an appreciable change in potential. In this region the redox system is said to be well "poised." The term "poised" is analogous to "buffered."

Calculation of Equilibrium Constants from Potentials 24.7

The equilibrium constant for the reaction equation in redox titrations (as well as for many other ionic reactions) can be calculated from potential data. At any point in a redox titration, when equilibrium has been established, the potential calculated from the Nernst equation of one of the redox couples present must equal the potential calculated from the Nernst equation of the second couple. By equating the two Nernst equations, a relationship can be obtained between formal potential (or standard electrode potential) data and the equilibrium constant that allows calculation of its value.

Example 24-1. Estimate the value of the equilibrium constant in 1 *F* HCl for the reaction $Cr^{2+} + Fe^{3+} \rightleftharpoons Cr^{3+} + Fe^{2+}$, if $E^f_{Cr^{3+}/Cr^{2+}} = -0.38$ V and $E^f_{Fe^{3+}/Fe^{2+}} = +0.70$ V.

The expression for the equilibrium constant takes the form

$$K = \frac{[Cr^{3+}][Fe^{2+}]}{[Cr^{2+}][Fe^{3+}]}$$

The Nernst equations for the two redox couples, with the formal potentials in 1 *F* HCl inserted, are

$$E = +0.70 + 0.059 \log \frac{[Fe^{3+}]}{[Fe^{2+}]}$$

$$E = -0.38 + 0.059 \log \frac{[Cr^{3+}]}{[Cr^{2+}]}$$

At equilibrium, the potentials of the two couples have the same value; hence

$$+0.70 + 0.059 \log \frac{[Fe^{3+}]}{[Fe^{2+}]} = -0.38 + 0.059 \frac{[Cr^{3+}]}{[Cr^{2+}]}$$

Rearrangement yields

$$0.70 + 0.38 = 0.059 \log \frac{[Cr^{3+}]}{[Cr^{2+}]} + 0.059 \frac{[Fe^{2+}]}{[Fe^{3+}]}$$

or

$$1.08 = 0.059 \log \frac{[Cr^{3+}][Fe^{2+}]}{[Cr^{2+}][Fe^{3+}]} = 0.059 \log K$$

Consequently,

$$\log K = \frac{1.08}{0.059} = 18.3$$

or

$$K = \sim 10^{18} \qquad \text{(in 1 } F \text{ HCl)}$$

The large value of the equilibrium constant indicates that the reaction goes essentially to completion. As far as the criterion of the completeness of the reaction is concerned, the titration of chromium(II) with iron(III) or vice versa is feasible.

Example 24-2. Estimate the value of the equilibrium constant for the reaction $5Fe^{2+} + MnO_4^- + 8H^+ \rightleftharpoons 5Fe^{3+} + Mn^{2+} + 4H_2O$ in acidic solution, if $E^0_{Fe^{3+}/Fe^{2+}} = +0.77$ V and $E^0_{MnO_4^-/Mn^{2+}} = +1.51$ V.

The expression for the equilibrium constant takes the following form:

$$K_{eq} = \frac{[Fe^{3+}]^5[Mn^{2+}]}{[Fe^{2+}]^5[MnO_4^-][H^+]^8}$$

Equating the two Nernst equations yields

$$E = +0.77 + 0.059 \log \frac{[Fe^{3+}]}{[Fe^{2+}]}$$

$$= +1.51 + \frac{0.059}{5} \log \frac{[MnO_4^-][H^+]^8}{[Mn^{2+}]}$$

Next the concentration terms of the iron species are raised to the fifth power and at the same time the factor in front of the logarithm is divided by 5. This does not change the numerical value of the term but makes possible the combination of the two logarithmic terms to give

$$1.51 - 0.77 = \frac{0.059}{5} \log \frac{[Fe^{3+}]^5[Mn^{2+}]}{[Fe^{2+}]^5[MnO_4^-][H^+]^8}$$

Hence

$$\log K = \frac{5 \times 0.74}{0.059} = 62.7$$

$$K = \sim 10^{63}$$

A solubility product is a special form of an equilibrium constant and can be calculated from suitable potential data, notably from the known standard electrode potentials for the reduction of one of the ions involved from the simple (aquo) ion and from the slightly soluble strong electrolyte.

Example 24-3. Calculate the solubility product of mercury(I) bromide, Hg_2Br_2, given that

$$Hg_2Br_2 + 2e \rightleftharpoons 2Hg + 2Br^- \qquad E^0 = +0.14 \text{ V}$$

$$Hg_2^{2+} + 2e \rightleftharpoons 2Hg \qquad E^0 = +0.79 \text{ V}$$

If the second equation is subtracted from the first, the desired net reaction is obtained:

$$Hg_2Br_2 \rightleftharpoons Hg_2^{2+} + 2Br^-$$

If the two redox couples are in equilibrium, their potentials must be equal, and the two Nernst equations can be equated.

$$E = +0.79 + \frac{0.059}{2} \log[Hg_2^{2+}] = +0.14 + \frac{0.059}{2} \log \frac{1}{[Br^-]^2}$$

$$0.14 - 0.79 = \frac{0.059}{2} \log[Hg_2^{2+}][Br^-]^2$$

$$-0.65 = \frac{0.059}{2} \log K_{sp}$$

$$\log K_{sp} = \frac{2 \times (-0.65)}{0.059} = -22.0$$

$$K_{sp} = \sim 10^{-22}$$

24.8 Detection of End Points in Redox Titrations

The end point of a redox titration can be detected by measurement of the half-cell potential by the potentiometer method described in Section 22.10. This may be described as a redox titration to a potentiometric end point or simply as a potentiometric redox titration. Consideration of this technique is deferred until Chapter 27.

The end point in a redox titration can also be detected visually if the titration system is self-indicating or if a suitable indicator is available. Permanganate ion is probably the most common example of a self-indicating oxidizing titrant; the end point corresponds to the appearance of a pink solution color on the addition of the first drop of permanganate solution in excess (Sections 25.5 and 25.6).

Redox indicators are organic substances that can be reduced or oxidized and exhibit different colors in their oxidized and reduced forms. Obviously the color of at least one of the forms should be so intense that only a small amount of the indicator need be added and hence the amount of titrant consumed in its oxidation or reduction can be neglected.

The color transition interval of a reversible, two-color redox indicator is centered around a definite potential, which is analogous to the transition interval of a two-color acid–base indicator. In Table H in the Appendix some common redox indicators are listed with their formal potentials and color changes.

In a redox titration the transition interval of the indicator should include the equivalence point, and ideally the titration should be stopped when the color is reached that corresponds exactly to the potential of the titration half-cell at that point. In practice, a titration is considered satisfactory when the color change occurs within the potential range corresponding to the "jump" in the titration curve. The end point is then sufficiently close to the equivalence point and any difference between these two points becomes negligible.

The treatment of a simple, reversible two-color redox indicator parallels that given for a two-color acid–base indicator in Chapter 12. The equation for the reaction is $In_{ox} + ne \rightleftharpoons In_{red}$. The corresponding Nernst equation is

$$E_{In} = E_{In}^{f} + \frac{0.059}{n} \log \frac{[In_{ox}]}{[In_{red}]} \qquad (24\text{-}14)$$

where E^f_{In} is the formal potential of the indicator under the existing solution conditions.

For a one-color redox indicator, consider the case where the oxidized form is colored. The formal (total) concentration of the indicator is C_{In}; hence, the Nernst equation takes the form

$$E_{In} = E^f_{In} + \frac{0.059}{n} \log \frac{[In_{ox}]}{C_{In} - [In_{ox}]} \tag{24-15}$$

Consequently for a one-color indicator the end-point color (more correctly the color intensity) to which the titration should be performed depends on the concentration of the indicator. This is analogous to the behavior of a one-color, acid–base indicator (see Chapter 12).

Complexing Agents in Redox Titrations 24.9

Complexing agents are used in redox titrations for a number of purposes. If the product of a titration reaction is complexed more strongly than a corresponding reactant, the equilibrium position will be shifted toward the products and the reaction will proceed more completely. Thus, addition of a suitable complexing agent may make a titration feasible that otherwise would not be so. For example, the break around the equivalence point may be increased, or the curve may be shifted so that it passes through the transition interval of an available indicator. For example, phosphoric acid is often added in the titration of iron(II) ion with dichromate ion in acidic media. Iron(III) forms a stable, almost colorless, soluble complex with the orthophosphate ion in acidic solution. As a consequence, the formal potential of the iron couple is lowered and the titration reaction proceeds further to completion. Diphenylaminesulfonic acid, that is, *N*-phenylanthranilic acid, as the sodium or barium salt is usually employed as the redox indicator. This indicator has a formal potential of about +0.84 V (see Table H in the Appendix). The dichromate–chromium(III) and iron(III)–iron(II) couples have standard electrode potentials of +1.36 and +0.77 V, respectively. With this indicator, the end point would therefore occur before the equivalence point. By formation of the iron(III)–phosphate complex, the electrode potential of the iron couple is lowered by about 0.12 V. Consequently, the formal potential of the indicator more nearly coincides with the potential of the equivalence point. The colorless nature of the iron(III) phosphate complex also makes the end-point color change more discernable. If permanganate is used in the analogous, self-indicating titration of iron(II), the decolorizing action of phosphate is even more important (Section 25.7).

24.10 Questions

24-1. At every point in a redox titration the numerical values of the half-cell potentials of the two redox systems involved are identical. Explain.

24-2. What is the difficulty in calculating the half-cell potential at the start of a redox titration?

24-3. Is a redox titration possible if the formal potentials of the two redox couples involved are close in value? Discuss in terms of the shape of the titration curve.

24-4. Explain why a more precise result can be expected in a redox titration when the standard potentials of the two redox couples involved are widely separated in value.

24-5. Elaborate on a possible classification of redox titrants into groups wherein the titration curves are (a) independent of concentration and pH, (b) pH-dependent, (c) concentration-dependent, and (d) dependent on both concentration and pH.

24-6. Is a large difference in formal potentials sufficient to ensure a practical redox titration? What other requirements must be fulfilled?

24-7. Compare the action of a redox indicator to that of an acid–base indicator.

24-8. Compare a redox titration curve (where the same number of electrons appears in each half-reaction) equation with the titration curves for (a) a strong monoprotic acid with a strong base and (b) a weak monoprotic acid with a weak base. Compare the methods of calculating the starting points, halfway points, equivalence points, and points beyond the equivalence point.

24-9. Elaborate on the possibility of titrating consecutively two or more redox couples in one solution. (*Hint:* The criteria can be deduced by analogy from the titration curve of a dibasic acid with a strong base.)

24.11 Problems

24-1. The titration of 50 ml of a 0.10 M Fe^{2+} solution with 0.02 F MnO_4^- solution is considered in Section 24.4. Calculate the points on the titration curve for the addition of 50, 60, 80, and 100 ml of the titrant, assuming that the hydrogen ion concentration is maintained (a) at 1.0 M and (b) at 1.0×10^{-2} M.

Answers: (a) +1.37, +1.50, +1.51, +1.51 V; (b) +1.21, +1.31, +1.32, +1.32 V

Calculate the solubility product of ZnS, given that

$$Zn^{2+} + 2e \rightleftharpoons Zn \qquad E^0 = -0.76 \text{ V}$$
$$\underline{ZnS} + 2e \rightleftharpoons Zn + S^{2-} \qquad E^0 = -1.44 \text{ V}$$

Answer: $\sim 10^{-23}$

Uranium(VI) can be reduced to uranium(IV) by chromium(II). Derive the formula for the calculation of the potential at the equivalence point for the titration of uranyl ion, UO_2^{2+}, with a standard Cr^{2+} solution. Calculate the value of this potential if the pH of the solution is 2.00.

$$UO_2^{2+} + 4H^+ + 2e \rightleftharpoons U^{4+} + 2H_2O \qquad E^f = +0.31 \text{ V}$$
$$Cr^{3+} + e \rightleftharpoons Cr^{2+} \qquad E^f = -0.38 \text{ V}$$

Answer: $E_{\text{eq pt}} = 0.08 - 0.08\,\text{pH}$, $E_{\text{eq pt}}$ at pH 2.0 $= -0.08$ V

A volume of exactly 25 ml of a solution 1.0×10^{-2} F in Cr^{2+} is titrated with a solution 5.0×10^{-3} F cerium(IV) sulfate solution.

$$Cr^{3+} + e \rightleftharpoons Cr^{2+} \qquad E^f = -0.41 \text{ V}$$
$$Ce^{4+} + e \rightleftharpoons Ce^{3+} \qquad E^f = +1.44 \text{ V}$$

(a) Calculate the values of the half-cell potential after the addition of 10, 20, 30, 40, 49, 50, and 55 ml of the cerium(IV) solution.
(b) Estimate the value of the equilibrium constant for the titration reaction.

Answers: (a) −0.45, −0.42, −0.40, −0.37, −0.31, +0.52, +1.38 V;
(b) $\sim 10^{31}$

A volume of 50.0 ml of 0.0500 F iron(II) nitrate is titrated with 0.100 F cerium(IV) solution with both solutions 1 F in HNO_3. Calculate the potential of a platinum indicator electrode in the titration solution after the addition of 5.0, 10.0, 12.5, 15.0, 20.0, 24.0, 25.0, 26.0, 30.0, 40.0, and 50.0 ml of titrant.

$$E^f_{Ce^{4+}/Ce^{3+}} = +1.60 \text{ V}, E^f_{Fe^{3+}/Fe^{2+}} = +0.68 \text{ V}$$

A volume of 10.0 ml of 0.0700 F iron(II) perchlorate is titrated with 0.100 F cerium(IV) solution with both solutions 1 F in $HClO_4$. Calculate the potential of a platinum indicator electrode in the titration solution after the addition of 1.0, 2.0, 3.5, 5.0, 6.0, 6.9, 7.0, 7.1, 8.0, 10.0, 12.0, and 14.0 ml of titrant.

$$E^f_{Ce^{4+}/Ce^{3+}} = +1.70 \text{ V}, E^f_{Fe^{3+}/Fe^{2+}} = +0.75 \text{ V}$$

A volume of 30.0 ml of 0.100 F iron(III) chloride is titrated with 0.0800 F chromium(II) chloride with both solutions 1 F in HCl. Calculate the potential of a platinum indicator electrode in the titration solution after the addition of 10.0, 20.0, 30.0, 37.0, 37.5, 38.0, 45.0, 60.0, and 75.0 ml of titrant.

$$E^f_{Fe^{3+}/Fe^{2+}} = +0.70 \text{ V}, E^f_{Cr^{3+}/Cr^{2+}} = -0.38 \text{ V}$$

A volume of 20.0 ml of 0.0500 F iron(III) chloride is titrated with 0.0200 F tin(II) solution with both solutions 1 F in HCl. Calculate the potential of a platinum indicator electrode in the titration solution after the addition of 5.0, 10.0, 15.0, 20.0, 24.0, 25.0, 26.0, 30.0, 40.0, and 50.0 ml of titrant.

$$E^f_{Fe^{3+}/Fe^{2+}} = 0.70\ V,\ E^f_{Sn^{4+}/Sn^{2+}} = +0.14\ V$$

24-9. A volume of 25.00 ml of 0.020 *F* uranium(IV) sulfate is titrated with 0.050 *F* vanadium(V) solution. The titration solution is maintained 0.50 *F* in H_2SO_4 throughout the titration. Calculate the potential of a platinum indicator electrode in the solution after the addition of 5.0, 10.0, 15.0, 19.0, 20.0, 21.0, 25.0, 30.0, and 40.0 ml of titrant. The relevant half-reaction equations and potentials are

$$VO_2^+ + 2H^+ + e \rightleftharpoons VO^{2+} + H_2O \qquad E^f = +1.02\ V$$
$$UO_2^{2+} + 4H^+ + 2e \rightleftharpoons U^{4+} + 2H_2O \qquad E^f = +0.41\ V$$

24-10. A volume of 30.0 ml of 0.0500 *F* uranium(IV) sulfate is titrated with 0.0200 *F* $KMnO_4$. The titration solution is kept 1 *F* in H_2SO_4 throughout the titration. Calculate the potential of a platinum indicator electrode in the solution after the addition of 10.0, 15.0, 20.0, 29.0, 30.0, 31.0, 40.0, and 50.0 ml of titrant. The relevant half-reaction equations and potentials are

$$UO_2^{2+} + 4H^+ + 2e \rightleftharpoons U^{4+} + 2H_2O \qquad E^f = +0.41\ V$$
$$MnO_4^- + 8H^+ + 5e \rightleftharpoons Mn^{2+} + 4H_2O \qquad E^0 = +1.51\ V$$

24-11. Calculate the potential at the equivalence point in the titration of uranium(IV) with cerium(IV) in 0.5 *F* H_2SO_4 solution.

$$UO_2^{2+} + 2H^+ + 2e \rightleftharpoons UO^{2+} + H_2O \qquad E^f = +0.41\ V$$
$$Ce^{4+} + e \rightleftharpoons Ce^{3+} \qquad E^f = +1.44\ V$$

24-12. Calculate the change in potential from 0.10% before the equivalence point to 0.10% after it in the titration of iron(III) with tin(II) in 1.0 *F* HCl solution.

24-13. Calculate the change in electrode potential from 0.10% before the equivalence point to 0.10% after it in the titration of iron(II) with cerium(IV) in 1 *F* $HClO_4$ solution.

24-14. A volume of 100.0 ml of a 0.100 *F* solution of iron(III) is titrated with a 0.100 *F* solution of titanium(III). Assuming the solution is 2.00 *F* in H_2SO_4, calculate the formal concentration of iron(III) remaining (a) at the equivalence point and (b) after the addition of one drop (0.050 ml) of titrant past the equivalence point.

$$Fe^{3+} + e \rightleftharpoons Fe^{2+} \qquad E^f = +0.68\ V$$
$$TiO^{2+} + 2H^+ + e \rightleftharpoons Ti^{3+} + H_2O \qquad E^f = +0.12\ V$$

24-15. Calculate the value of the solubility product of silver chloride from the following data.

$$Ag^+ + e \rightleftharpoons Ag \qquad E^0 = +0.80\ V$$
$$\underline{AgCl} + e \rightleftharpoons Ag + Cl^- \qquad E^0 = +0.22\ V$$

Calculate the value of the solubility product of silver bromide from the following data.

$$Ag^+ + e \rightleftharpoons Ag \qquad E^0 = +0.80\ V$$
$$\underline{AgBr} + e \rightleftharpoons Ag + Br^- \qquad E^0 = +0.07\ V$$

Calculate the value of the solubility product of lead(II) bromide from the following data.

$$Pb^{2+} + 2e \rightleftharpoons Pb \qquad E^0 = -0.13\ V$$
$$\underline{PbBr_2} + 2e \rightleftharpoons Pb + 2Br^- \qquad E^0 = -0.28\ V$$

Calculate the value of the solubility product of mercury(I) chloride from the following data.

$$Hg_2^{2+} + 2e \rightleftharpoons 2Hg \qquad E^0 = +0.79\ V$$
$$\underline{Hg_2Cl_2} + 2e \rightleftharpoons 2Hg + 2Cl^- \qquad E^0 = +0.27\ V$$

Calculate the value of the equilibrium constant for the reaction $Cu^{2+} + Zn \rightleftharpoons Cu + Zn^{2+}$ from the following data.

$$Cu^{2+} + 2e \rightleftharpoons Cu \qquad E^0 = +0.34\ V$$
$$Zn^{2+} + 2e \rightleftharpoons Zn \qquad E^0 = -0.76\ V$$

Calculate the value of the equilibrium constant for the reaction $Sn^{4+} + 2Cr^{3+} \rightleftharpoons Sn^{2+} + 2Cr^{2+}$ in 1.0 F HCl.

$$Sn^{4+} + 2e \rightleftharpoons Sn^{2+} \qquad E^f = +0.14\ V$$
$$Cr^{3+} + e \rightleftharpoons Cr^{2+} \qquad E^f = -0.41\ V$$

Calculate the value of the equilibrium constant for the reaction $Tl^{3+} + 2Fe^{2+} \rightleftharpoons Tl^+ + 2Fe^{3+}$ in 1.0 F HCl.

$$Tl^{3+} + 2e \rightleftharpoons Tl^+ \qquad E^f = +0.78\ V$$
$$Fe^{3+} + e \rightleftharpoons Fe^{2+} \qquad E^f = +0.70\ V$$

25 APPLIED REDOX TITRIMETRY

Oxidizing or reducing titrants are widely applied to either the direct or indirect titration of numerous substances. Some substances can be determined by either an oxidation or a reduction titration, depending upon the oxidation state of the substance present in the solution after preliminary treatment. The selection of the best method must be based on the consideration of various factors, including the stability of the oxidation states, the nature of accompanying substances, the presence of interferences, convenience and rapidity of the method, the accuracy required, the number of samples to be analyzed, and the frequency of the determination. Selected examples of the use of the most common titrants will be discussed to demonstrate the principles and some of the possible practical approaches.

Preliminary Treatment 25.1

For the success of a redox titration, it is essential that the substance to be titrated be present in a single oxidation state, a situation that is sometimes not attained by simple dissolution of the sample. In addition, it is frequently more convenient to titrate the substance in another oxidation state than that existing after such dissolution. In these circumstances, a preliminary oxidation or reduction is necessary before the actual titration. As an example, consider the determination of iron, which is usually present as iron(III). This ion can be titrated with a strong reductant such as tin(II) or titanium(III), but the solutions of these agents require special storage to prevent their air oxidation (Section 25.11). An oxidimetric titration is therefore often preferred; for this purpose, iron(III) is first reduced to iron(II), which is somewhat more stable to air oxidation. The iron(II) is then titrated with an oxidizing titrant.

It is essential in any such pretreatment that the oxidation or reduction proceed completely to a single oxidation state and that the excess of the oxidant or reductant can be removed or destroyed without affecting the species to be titrated. Furthermore, it is often necessary that the pretreatment be such that only the substance to be determined reacts. The preliminary treatments can be divided into reductions and oxidations.

25.2 Preliminary Reduction

Metals and Reductor Columns. Many metals are strong reductants and offer the advantage that their excess is readily separated. The metal, in the form of a sheet or wire coil, may be immersed in the solution and after the reaction is complete, be simply removed with rinsing. Since the redox reaction proceeds at the metal surface, the use of metal granules, shot, turnings, or powder is advantageous because they offer a larger surface area and therefore afford a more rapid reaction.

An especially convenient manner for the use of a metal is to pack it into a vertically mounted tube to form a reductor column (known simply as a reductor). Thus a zinc reductor, also known as Jones reductor, contains zinc granules. The tube is filled, washed with dilute acid, and then the solution containing the species to be reduced is passed through the column. The liquid level must not be allowed to fall below the top of the zinc layer, to avoid the entrance of air bubbles. Such bubbles would decrease the available surface area and, in addition, oxygen might be reduced to hydrogen peroxide, which interferes in some titrations. The zinc(II) ion formed in the redox reaction does not interfere in the subsequent titration. Often the reduced species emerging from the column is susceptible to air oxidation. In such cases, the effluent should be collected under a protective atmosphere of nitrogen or carbon dioxide. Sometimes it is convenient to collect the effluent in a measured volume of an oxidizing solution, and to back-titrate the excess of the oxidant. In another variation, the effluent is passed into an iron(III) solution and the iron(II) produced is titrated with a standard solution of an oxidant.

Zinc is such a strong reductant ($E^0_{Zn^{2+}/Zn} = -0.76$ V) that hydrogen gas is formed when an acidic solution is passed through the column. Not only is zinc wasted by this undesired reaction, but the gas bubbles block some of the available surface area. The zinc granules are therefore usually shaken with a mercury(II) chloride solution before filling the reductor. The thin coating of zinc amalgam formed on the metal surface reduces or even prevents the formation of hydrogen because of the high kinetic overpotential of hydrogen on mercury (Section 28.7).

Some reductions of analytical importance that can be accomplished with a Jones reductor include

$$\text{Fe(III)} \rightarrow \text{Fe(II)}, \quad \text{Cr(III)} \rightarrow \text{Cr(II)}, \quad \text{Cr(VI)} \rightarrow \text{Cr(II)},$$
$$\text{Ti(IV)} \rightarrow \text{Ti(III)}, \quad \text{Ce(IV)} \rightarrow \text{Ce(III)}, \quad \text{V(V)} \rightarrow \text{V(II)},$$
$$\text{Mo(VI)} \rightarrow \text{Mo(III)}, \quad \text{and} \quad \text{Cu(II)} \rightarrow \text{Cu}$$

A reductor column filled with silver granules is also frequently used. The sample solution must contain hydrochloric acid. Then the oxidation of silver involves the half-reaction

$$\text{Ag} + \text{Cl}^- \rightarrow \underline{\text{AgCl}} + e \qquad (E^0 = +0.22\ \text{V})\dagger \qquad (25\text{-}1)$$

The silver reductor, sometimes called a Walden reductor, is milder than the zinc reductor and offers the advantage that no hydrogen gas is evolved. In addition, the insoluble oxidation product, silver chloride, is retained in the reductor and the effluent remains uncontaminated. The packing of the Walden reductor can be regenerated by treatment with zinc metal and sulfuric acid. (The student should write the reaction equation.)

Some reductions of analytical value that can be accomplished in a silver reductor, somewhat dependent on the hydrochloric acid concentration established, are

$$\text{Fe(III)} \rightarrow \text{Fe(II)}, \quad \text{Cr(VI)} \rightarrow \text{Cr(III)}, \quad \text{Ce(IV)} \rightarrow \text{Ce(III)},$$
$$\text{V(V)} \rightarrow \text{V(IV)}, \quad \text{Mo(VI)} \rightarrow \text{Mo(V), Mo(III)},$$
$$\text{and} \quad \text{Cu(II)} \rightarrow \text{Cu(I)}$$

Comparison of these reductions with those occurring in the zinc reductor reveals that, in the silver reductor, chromium(VI) is reduced to chromium(III) rather than to chromium(II), titanium is not reduced at all, vanadate is reduced only to vanadium(IV) [the further reduction to vanadium(III) is extremely slow], and copper(II) is reduced to copper(I), which at a sufficient hydrochloric acid concentration remains in solution as soluble chloro complexes. Molybdenum(VI) is reduced only to molybdenum(V) in 2 *F* hydrochloric acid; however, reduction to molybdenum(III) is possible in 4 *F* acid.

Schemes for the analysis of mixtures can be devised using the difference in the reducing power of metal reductors. For example, one aliquot of a sample solution containing iron and

---

†Throughout this chapter a half-reaction is written in the direction in which it actually proceeds. The E^0 value given parenthetically, however, corresponds to the particular half-reaction written as a reduction equation.

titanium is passed through a zinc reductor and a second aliquot through a silver reductor. The effluent from each column is titrated with an oxidizing titrant; the results of the titrations correspond to the total of iron and titanium, and to the iron alone, respectively.

A liquid amalgam offers a further convenient form for a reducing metal since it can be readily separated from an aqueous solution once the reduction has been effected. The reducing power of an amalgam depends on the particular metal dissolved in the mercury.

Tin(II) Salts. Tin(II) salts are frequently used to reduce iron(III) in hydrochloric acid solution. The disappearance of the yellow solution color indicates that the reduction is complete. The excess of tin(II) is readily oxidized with mercury(II) chloride, which does not affect iron(II):

$$2HgCl_2 + Sn^{2+} \rightarrow \underline{Hg_2Cl_2} + Sn^{4+} + 2Cl^- \qquad (25\text{-}2)$$

The calomel formed is insoluble and under appropriate conditions does not react in the subsequent titration of iron(II). The presence of chloride ion is essential; in its absence mercury(I) is reduced further to elemental mercury. The metal appears in a finely divided, black form, is reoxidized in the subsequent titration, and thereby interferes.

Sulfurous Acid. Sulfurous acid acts as a mild reductant in acidic solution,

$$H_2SO_3 + H_2O \rightarrow SO_4^{2-} + 4H^+ + 2e \qquad (E^0 = +0.17\text{ V}) \qquad (25\text{-}3)$$

Any excess of this reductant can be readily removed by boiling the solution, thereby volatilizing sulfur dioxide.

Other Reductants. Other reductants employed prior to redox titrations include hydrochloric acid [especially for oxidants such as manganese(IV) oxide and lead(IV) oxide] and hypophosphorous acid, H_3PO_2, and its salts.

25.3 Preliminary Oxidation

Peroxydisulfate Ion. Peroxydisulfate ion, $S_2O_8^{2-}$, commonly termed persulfate ion, employed as either the potassium or ammonium salt, is a strong oxidizing agent in acidic solution.

$$S_2O_8^{2-} + 2e \rightarrow 2SO_4^{2-} \qquad (E^0 = +2.0\text{ V}) \qquad (25\text{-}4)$$

Any excess of this oxidant can be decomposed by boiling the solution; the following reactions then occur:

$$S_2O_8^{2-} + 2H_2O \rightarrow 2SO_4^{2-} + 2H^+ + H_2O_2 \qquad (25\text{-}5)$$

$$2H_2O_2 \rightarrow 2H_2O + O_2\uparrow \qquad (25\text{-}6)$$

The important uses of this oxidant, often with a silver salt added as a catalyst, include the conversion of manganese to permanganate, chromium to dichromate, and vanadium to vanadate ion.

Perchloric Acid. Hot, concentrated perchloric acid (50 to 70%) is a strong oxidant:

$$2ClO_4^- + 16H^+ + 14e \rightarrow Cl_2 + 8H_2O \qquad (E^0 \sim +2\text{ V}) \qquad (25\text{-}7)$$

The acid is often used to oxidize chromium and vanadium in ores and alloys to dichromate and vanadate, respectively. It is unnecessary to remove the excess of perchloric acid since cooling and dilution of the solution effectively checks the oxidizing power of the perchlorate ion. However, care must be taken to exclude organic substances from the system because they may react violently, even explosively, with the hot, concentrated acid.

Sodium Bismuthate and Lead(IV) Oxide. Sodium bismuthate, $NaBiO_3$, is a fine, water-insoluble, brown powder used principally to oxidize manganese(II) to permanganate in nitric acid solution, even at room temperature.

$$\underline{NaBiO_3} + 6H^+ + 2e \rightarrow Na^+ + Bi^{3+} + 3H_2O \qquad (E^0 \sim +1.6\text{ V}) \qquad (25\text{-}8)$$

Chloride ion must be excluded since it would be oxidized to chlorine gas. The excess of sodium bismuthate is removed by filtration through an inert filter material such as sintered glass or porcelain. Lead(IV) oxide, PbO_2, is used in a similar manner.

$$\underline{PbO_2} + 4H^+ + 2e \rightarrow Pb^{2+} + 2H_2O \qquad (E^0 \sim +1.5\text{ V}) \qquad (25\text{-}9)$$

Hydrogen Peroxide. Hydrogen peroxide is a strong oxidant, even in alkaline solution.

$$H_2O_2 + 2H^+ + 2e \rightarrow 2H_2O \qquad (E^0 = +1.77\text{ V}) \qquad (25\text{-}10)$$

This compound can also function as a reductant according to the half-reaction

$$H_2O_2 \rightarrow O_2 + 2H^+ + 2e \qquad (E^0 = +0.69\text{ V}) \qquad (25\text{-}11)$$

Hydrogen peroxide can disproportionate according to the reaction

$$2H_2O_2 \rightarrow 2H_2O + O_2 \qquad (25\text{-}12)$$

which is a combination of the two preceding reactions. This reaction is used to decompose any excess of the reagent. The solution is simply boiled, often either with a trace of a nickel or iodide salt added or with a piece of platinum foil placed temporarily in the solution to catalyze the reaction. Hydrogen peroxide is frequently used to assure complete oxidation of iron after its ores and alloys have been dissolved.

Nitric Acid. Nitric acid is chiefly used to attack materials, such as alloys, that are inert to hydrochloric acid, and also to ensure that metals such as iron and copper are present in the solution in their higher oxidation state.

25.4 Permanganate as an Oxidizing Titrant

Permanganate ion is one of the most widely used oxidizing titrants because of its high oxidizing power and because it allows a self-indicating titration. The end point can be indicated visually by the pink color imparted to the solution by the first fraction of a drop of titrant in excess. A redox indicator, however, may be used to advantage where very dilute solutions are involved.

In strongly acidic medium, permanganate undergoes a five-electron reduction to manganese(II):

$$MnO_4^- + 8H^+ + 5e \rightarrow Mn^{2+} + 4H_2O \qquad (E^0 = +1.51\ V) \qquad (25\text{-}13)$$

In weakly acidic, neutral, or alkaline solutions, the reduction of permanganate involves its three-electron reduction to manganese(IV), which precipitates as hydrous manganese(IV) oxide, usually written simply as MnO_2. The half-reaction may be formulated as

$$MnO_4^- + 4H^+ + 3e \rightarrow \underline{MnO_2} + 2H_2O \qquad (E^0 = +1.69\ V) \qquad (25\text{-}14)$$

or in alkaline medium as

$$MnO_4^- + 2H_2O + 3e \rightarrow \underline{MnO_2} + 4OH^- \qquad (E^0 = +0.57\ V) \qquad (25\text{-}15)$$

Permanganate has several disadvantages that are prompting a shift to other oxidizing titrants. Even though potassium permanganate is available in high purity, it cannot serve as a primary standard, since freshly prepared solutions are unstable. In such solutions permanganate decomposes by the reaction

$$4MnO_4^- + 2H_2O \rightarrow \underline{4MnO_2} + 4OH^- + 3O_2 \quad (25\text{-}16)$$

This reaction, which proceeds slowly in neutral and more rapidly in acidic solution, is "autocatalytic," since the manganese(IV) oxide formed hastens the decomposition. Even when reagent-grade potassium permanganate is used, a trace of this oxide is present and, furthermore, organic dust particles in the water or the vessel reduce some of the permanganate. It is essential to remove this manganese(IV) oxide to make the solution stable. The initial solution is either boiled or allowed to stand at room temperature for about 1 week, and is then filtered. After this treatment, a neutral potassium permanganate solution (0.02 F or greater), stored in the dark and protected from the entry of dust, maintains its strength for a protracted period of time. Storage in the dark is necessary since the decomposition is enhanced by strong light. In any event, occasional restandardization of the solution is necessary. If titrations are performed only infrequently, it is usually more expedient to prepare a potassium permanganate solution, to standardize it, and then to use it at once.

Standardization of Potassium Permanganate Solutions. A number of suitable standards for permanganate solutions are available; prominent ones are briefly considered below.

Oxalate. Permanganate solutions may be standardized by the titration of an acidic solution containing a known amount of oxalate ion. Suitable sources of oxalate include oxalic acid dihydrate, $H_2C_2O_4 \cdot 2H_2O$, sodium oxalate, $Na_2C_2O_4$, and potassium tetroxalate, $KHC_2O_4 \cdot H_2C_2O_4 \cdot 2H_2O$. The solution titrated must be acidic and is usually made at least 0.5 F in sulfuric acid. At insufficient acidity, the reaction does not proceed to the formation of manganese(II) ion, but stops with the formation of manganese(IV) oxide. The titration should be performed promptly without allowing the acidified oxalate solution to stand long.

At the start of the titration, the pink color of the added permanganate persists for some time, indicating the slowness of the reaction. As further increments of titrant are added, the decolorization proceeds more rapidly, since the manganese(II) formed catalyzes the reaction. Heating of the solution also hastens the reaction, but at too high a temperature as well as too great an acidity, oxalic acid decomposes according to the reaction

$$H_2C_2O_4 \rightarrow CO_2\uparrow + CO\uparrow + H_2O \qquad (25\text{-}17)$$

Thus precautions must be taken in the permanganate titration of oxalate.

Probably the best method of effecting the titration is to add about 90% of the required permanganate at room temperature, allow to stand until decolorization is complete, then to heat to 55 to 60°C, and to finish the titration. It is important to perform a blank determination to correct for any oxidizable substances present in the water and reagents used and for the amount of permanganate solution required to impart a pink color under the titration conditions. (The establishment of a blank is also necessary with other standards and indeed in most permanganate titrations.) The importance of a blank increases when titrant solutions of low concentrations are involved.

Arsenic(III) Oxide. Arsenic(III) oxide is an excellent primary standard. It is nonhygroscopic, contains no water of crystallization, and is readily obtainable in a purity of 99.95% or higher. The relevant half-reaction in its use in the standardization of oxidants is

$$H_3AsO_3 + H_2O \rightarrow H_3AsO_4 + 2H^+ + 2e \qquad (E^0 = +0.56\ V) \qquad (25\text{-}18)$$

In the direct titration of arsenic(III) with permanganate in acidic solution, usually a trace of an iodine compound (e.g., KI or KIO_3) is added to catalyze the reaction and thereby to speed the titration.

Iron. Iron, in the form of wire of 99.90% or higher purity, is available for standardization. It must be dissolved in sulfuric acid with exclusion of air to avoid oxidation to iron(III), or a prereduction before the titration becomes necessary.

Other Standards for Permanganate. Other standards for permanganate occasionally used include iron(II) ammonium sulfate hexahydrate (Mohr's salt), $FeSO_4 \cdot (NH_4)_2SO_4 \cdot 6H_2O$, and potassium hexacyanoferrate(II) trihydrate, $K_4Fe(CN)_6 \cdot 3H_2O$. The Mohr's salt used should be free of iron(III). The hexacyanoferrate(II) salt, which is oxidized to hexacyanoferrate(III) in a one-electron step, has an especially favorable equivalent weight.

Determinations with Permanganate 25.5

In view of the prominence of permanganate as an oxidizing titrant, its application in inorganic analysis will be treated in somewhat greater detail than those titrants to be considered subsequently. Many of the considerations, however, apply to the use of other oxidants.

Determination of Iron (Zimmermann–Reinhardt Method). The most prominent application of permanganate in titrimetry is to the determination of iron in a large variety of materials. After the sample is dissolved, all the iron is reduced to iron(II), which is then titrated. No special difficulties arise in the direct permanganate titration of iron(II) in acidic solution if chloride is absent. However, hydrochloric acid is frequently used in the dissolution of samples and its presence is required in some reduction techniques. Chloride ion causes difficulty because permanganate ion can oxidize it to elemental chlorine and hypochlorous acid. These reactions as such are slow but are catalyzed by iron.

To obviate the interference of chloride the sample solution is diluted and manganese(II) is added to it. Both the dilution of the solution (via the decrease in acidity) and the addition of manganese(II) decrease the oxidizing power of permanganate, as can be concluded from an inspection of the Nernst equation for the permanganate half-reaction equation (Section 24.5). In addition, the dilution decreases the chloride concentration. The titration is performed in the cold to keep the reaction with chloride at a low rate. Further, the titrant solution is added slowly and with good stirring to avoid a high local concentration of the titrant.

During the titration, iron(III) forms, which in the presence of chloride yields chloro complexes intensely yellow in color. This color makes detection of the visual end point difficult. Phosphoric acid is added because it forms a colorless, soluble complex with iron(III). The complexation with phosphoric acid also lowers the value of the formal potential of the iron redox couple (Section 24.10) and thus improves the titration. However, the presence of phosphoric acid also makes air oxidation of iron(II) easier! In practice, manganese(II) sulfate, phosphoric acid, and sulfuric acid are combined in a single solution, known as the "Zimmermann–Reinhardt solution," which is added to the sample solution after the preliminary reduction.

Determination of Manganese (Volhard's Method). In neutral or slightly alkaline solution, manganese(II) is oxidized to manganese(IV) oxide by permanganate. The course of the reaction is more complicated than is suggested by the simple overall equation

$$3Mn^{2+} + 2MnO_4^- + 2H_2O \rightarrow \underline{5MnO_2} + 4H^+ \quad (25\text{-}19)$$

The manganese(IV) precipitates as a hydrous oxide, which shows acidic properties. It is capable of forming an insoluble manganese(II) salt or, according to other explanations, to adsorb manganese(II) ion. Whatever the cause, some manganese(II) is coprecipitated and, thereby, escapes being titrated; low results are the consequence. Volhard found that this difficulty can be circumvented by the addition of a zinc salt. Zinc ion then interacts preferentially with the hydrous manganese(IV) oxide and the manganese(II) ion remains in solution. In practice the zinc ion is generated by adding a slurry of zinc oxide in water to the slightly acidic sample solution. Because of the relatively high pH established by the zinc oxide, iron(III), titanium, aluminum, and other metals that form hydrous oxides will also be precipitated. On boiling, a rapidly settling precipitate is obtained, which also carries down the manganese(IV) oxide. The visual end point is then readily discernible in the clear, colorless supernatant liquid.

Metal Determinations via Oxalate Titrations. The titration of oxalate ion has been discussed in the consideration of the standardization of permanganate solutions (Section 25.5). This titration may be used for the determination of oxalate and, in inorganic analysis, receives attention for the indirect determination of metals forming insoluble oxalate salts. This approach was a standard method for the determination of calcium in a variety of samples until the advent of EDTA titration methods (Section 20.10). Calcium oxalate is precipitated, the precipitate is separated by filtration, washed, and dissolved in acid. The oxalate is then titrated with permanganate.

Hydrogen Peroxide and Peroxides. Hydrogen peroxide is oxidized by permanganate in acidic solution. The relevant half-reaction is

$$H_2O_2 \rightarrow O_2 + 2H^+ + 2e \qquad (E^0 = +0.69 \text{ V}) \quad (25\text{-}20)$$

The determination of hydrogen peroxide involves dilution of the sample, addition of sulfuric acid, and titration with permanganate to the first permanent pink color. Also here an induction period is encountered as is the case in the titration of oxalate. Commercial hydrogen peroxide often contains a small amount of an organic compound as a stabilizer. This compound may also be oxidized by the permanganate, and then a slightly high value for the peroxide content will result. Metal peroxides, such as sodium peroxide, can be assayed by a permanganate titration via dissolution in dilute acid and titration of the hydrogen peroxide formed.

Miscellaneous Determinations with Permanganate. Nitrite is often determined by a permanganate titration. The half-reaction involved is

$$HNO_2 + H_2O \rightarrow NO_3^- + 3H^+ + 2e \qquad (E^0 = +0.94\ V) \qquad (25\text{-}21)$$

Arsenic(III) can be titrated as described in the discussion of the standardization of permanganate solutions (Section 25.5). Arsenic(V) can be reduced, for example, with sulfurous acid, and then be titrated similarly. Antimony(III) and antimony(V) are determined analogously.

Titanium(IV) can be reduced to titanium(III) in a Jones reductor and then titrated to the tetravalent state. Since titanium(III) is quite sensitive to air oxidation, it is expedient to catch the effluent from the reductor in an iron(III) sulfate solution, to add Zimmermann–Reinhardt solution if chloride is present, and to titrate the iron(II) formed with permanganate. The method is also applicable to chromium and vanadium after reduction to chromium(II) and vanadium(II) in a Jones reductor.

Permanganate can be employed in the indirect titration of oxidants. The assay of pyrolusite may serve as an example. This mineral contains manganese(IV) oxide as the active ingredient. A powdered sample of the mineral is treated with a measured volume of an acidified solution of oxalate or iron(II) and, after dissolution is complete, the excess of oxalate or iron(II) is titrated with permanganate.

Iodine and Thiosulfate In Redox Titrimetry 25.6

The iodine–iodide redox couple

$$I_2 + 2e \rightarrow 2I^- \qquad (E^0 = +0.54\ V) \qquad (25\text{-}22)$$

is intermediate between strong oxidants and strong reductants. Thus, in some applications iodine is reduced to iodide ion and in others the reverse reaction occurs. Reducing substances can be directly titrated with iodine; oxidizing substances are reacted with iodide ion and the iodine formed is titrated with a reducing titrant, commonly sodium thiosulfate. The two approaches are often distinguished from each other by calling them io*di*metry and io*do*metry, respectively.

Although iodine and thiosulfate solutions do not show good storage stability and thus require frequent standardization, iodine methods receive widespread application because of the large number of substances that can be determined and because of the sharpness of the end point, which allows an accurate titration even where the titrant is only 0.001 *F*.

The iodine–iodide redox potential is independent of acidity over a wide range. About pH 8, however, iodine disproportionates to form iodate and iodide. The reaction may be represented as proceeding in two steps, hypoiodite ion being formed first:

$$I_2 + 2OH^- \rightarrow H_2O + I^- + IO^- \qquad (25\text{-}23)$$

$$3IO^- \rightarrow 2I^- + IO_3^- \qquad (25\text{-}24)$$

In acidic solution, iodide ion is oxidized to iodine by the oxygen of the air:

$$O_2 + 4H^+ + 4I^- \rightarrow 2I_2 + 2H_2O \qquad (25\text{-}25)$$

Sunlight and various ions, notably copper(II) and nitrite, catalyze this reaction.

Iodine is somewhat volatile and its solubility in water is small (0.0013 *F* in I_2 at 20°C). A more concentrated aqueous solution of improved stability can be obtained by adding an excess of iodide ion which combines with iodine to form the soluble triiodide ion, I_3^-. For this ion, the half-reaction can be written

$$I_3^- + 2e \rightleftharpoons 3I^- \qquad (E^f = +0.545 \text{ V in } 0.5\ F\ H_2SO_4) \qquad (25\text{-}26)$$

The formal potential is displaced from the standard potential of the iodine–iodide redox couple by only 9 mV [compare equation (25-22)]. For simplicity and to make the stoichiometry more obvious, reactions for iodine methods are usually written showing iodine as a product or reactant, rather than the triiodide ion.

In neutral or weakly acidic solution, thiosulfate ion is oxidized by iodine to tetrathionate ion, which is the desired reaction in the titration. The half-reaction

$$2S_2O_3^{2-} \rightarrow S_4O_6^{2-} + 2e \qquad (E^0 = +0.09 \text{ V}) \qquad (25\text{-}27)$$

involves one electron per molecule of thiosulfate ion; consequently, in this reaction the equivalent weight of this ion equals its formula weight.

If the pH is too high, thiosulfate ion is converted to sulfate. The half-reaction involved is

$$S_2O_3^{2-} + 10OH^- \rightarrow 2SO_4^{2-} + 5H_2O + 8e \quad (25\text{-}28)$$

In this reaction, the equivalent weight of thiosulfate ion equals one eighth of its formula weight. However, usually an unpredictable combination of reactions (25-27) and (25-28) occurs, making operation at high pH values impractical.

Thiosulfate ion is not stable in acidic solution since thiosulfuric acid decomposes:

$$H_2S_2O_3 \rightarrow H_2SO_3 + S \quad (25\text{-}29)$$

Sulfurous acid also reacts with iodine. The relevant half-reaction is

$$H_2SO_3 + H_2O \rightarrow SO_4^{2-} + 4H^+ + 2e \qquad (E^0 = +0.17\text{ V}) \quad (25\text{-}30)$$

This half-reaction involves two electrons, while the oxidation of thiosulfate to tetrathionate, after (25-27), involves only one electron per molecule of substance oxidized.

The decomposition of thiosulfate occurs already at moderate acidities but causes no difficulties during a titration because the reaction is slow. Thus, if the thiosulfate solution is added to the sample solution slowly and with good stirring, no decomposition occurs and the reaction with iodine is stoichiometric even in strongly acidic solution. However, the decomposition leads to an interesting phenomenon. If portions of a freshly prepared sodium thiosulfate solution are titrated with an iodine solution during a period of a few days, the following will be found. The reducing power of the thiosulfate solution at first increases somewhat; then it decreases and finally reaches a stable value that is below the initial one. The increase is associated with the decomposition of thiosulfate to sulfite ion, which per molecule donates twice as many electrons as does thiosulfate. After some time, the sulfite is oxidized by oxygen of the air to unreactive sulfate and the formality decreases. Consequently, it is necessary to age a sodium thiosulfate solution before its final standardization. After several days of aging, a 0.1 *F* solution will maintain its strength for a protracted period of time. More dilute solutions are less stable and are best prepared as needed from a stock solution, but should be standardized before use.

Indication of the End Point in Iodine Methods. One drop of a 0.05 F iodine solution imparts a just visible yellow color to about 100 ml of water; hence, the appearance or disappearance of this yellow color can afford self-indication. However, the color is far less intense than that of permanganate, and even only slightly colored substances present in the sample solution interfere. Consequently, self-indication is seldom used in practice. Iodine is far more soluble in organic solvents, such as chloroform and carbon tetrachloride, than in water and shows an intense violet color in them. Since these nonpolar solvents are almost immiscible with water, even small amounts of iodine can be extracted from a large volume of a dilute aqueous solution by shaking with 1 or 2 ml of the solvent. The appearance or disappearance of the iodine color in the organic layer affords a sensitive end point, but the technique is tedious since extensive shaking is necessary in the vicinity of the end point to assure equilibration.

Starch is the indicator commonly used in iodine methods. "Soluble" starch and iodine form a complex entity of intensely blue color. The color develops only in the presence of iodide and in cold solution. It fades upon heating but returns on cooling. With old starch solutions, a red-purple color is obtained rather than a blue. The starch-iodine color reaction is quite sensitive and a distinct blue is discernable in solutions about 10^{-5} F in iodine. When titrating solutions containing much iodine the starch must not be added until the end point is near, that is, until the iodine concentration is decreased to a low value; otherwise, the starch-iodine entity reacts very slowly and the end point is readily exceeded. The usual technique is to perform the titration until the yellow color of the iodine starts to fade, then to add the starch, and to continue the titration until the blue disappears.

When the titration of iodine is performed in strongly acidic media or in the presence of substances that catalyze the oxidation of iodide by oxygen of the air, after reaction (25-25), the colorless solution obtained when the end point is reached becomes blue again after some time. The exclusion or at least delay of this "after-blueing" requires special precautions in certain determinations.

Standardization of Iodine and Thiosulfate Solutions. Elemental iodine can be highly purified by sublimation and may serve as a primary standard. A standard iodine solution may be prepared from such material by weighing and dissolving the necessary amount in a solution of potassium iodide. However, the volatility of iodine makes this method impractical. Therefore it is preferable to prepare an iodine solution of approximately the desired concentration and to standardize it against arsenic(III) oxide, which is an excellent primary standard. The oxide is dissolved, as described for its use in the standardization of permanganate; the solution is buffered with hydrogen carbonate and titrated with the iodine solution.

A thiosulfate solution may be standardized with an iodine solution, of reliably known formality, but it is usually preferred to standardize it independently. Elemental iodine may be employed for this purpose, but, as mentioned above, its volatility causes difficulties. Better results are obtained by formation of a known amount of iodine in solution. Commonly, potassium iodate, which is readily obtained in high purity and is stable in aqueous solution, is reacted in acidic solution with an excess of potassium iodide. The reaction leading to the formation of iodine proceeds both rapidly and stoichiometrically:

$$IO_3^- + 5I^- + 6H^+ \rightarrow 3I_2 + 3H_2O \qquad (25\text{-}31)$$

A standard solution of potassium iodate may be prepared and when needed be used to obtain an iodine solution of exactly known iodine content.

Standard solutions of strong oxidizing agents, such as potassium permanganate or dichromate, may also be used to oxidize iodide to iodine. The iodine formed is then titrated with thiosulfate.

Applications of Iodine Methods 25.7

The discussion of iodine methods can be divided into those in which iodine is the titrant (iod*i*metric methods) and those in which thiosulfate is used (iod*o*metric methods).

Arsenic(III) is readily oxidized to arsenic(V) by iodine. The titration with iodine is satisfactory in the pH range 4 to 9 and a medium buffered with hydrogen carbonate is often used. The reaction equation may be written in the form

$$AsO_3^{3-} + 2H_2O + I_2 \rightarrow AsO_4^{3-} + 2H^+ + 2I^- \qquad (25\text{-}32)$$

Antimony(III) can be determined analogously.

25.7 APPLICATIONS OF IODINE METHODS

Hydrogen sulfide reacts readily with iodine according to the reaction

$$H_2S + I_2 \rightarrow \underline{S} + 2H^+ + 2I^- \qquad (25\text{-}33)$$

and thus can be titrated directly. However, because of the volatility of hydrogen sulfide, preferably an excess of iodine is added to the sample and the unreacted iodine is back-titrated with thiosulfate. Hydrogen sulfide can be determined in gaseous or liquid samples by precipitation as cadmium sulfide from an alkaline medium, dissolution of the isolated precipitate in acid, and reaction with iodine.

Sulfite (and sulfur dioxide) can be titrated with iodine in acidic medium. To avoid erroneous results caused by air oxidation of the sulfite, it is also of advantage here to effect first the reaction with iodine and then to back-titrate its excess with thiosulfate.

Tin(II), in alloy analysis, can be titrated directly with iodine. The susceptibility of tin(II) to air oxidation makes careful exclusion of air a necessity.

Water in organic liquids, hydrated salts, and diverse samples is often determined by the Karl Fischer titration. The overall reaction, somewhat simplified, may be represented by

$$C_5H_5N \cdot I + C_5H_5N \cdot SO_2 + C_5H_5N + CH_3OH + H_2O \rightarrow 2C_5H_5NHI + C_5H_5NHOSO_2OCH_3 \qquad (25\text{-}34)$$

where C_5H_5N and CH_3OH are pyridine and methanol, respectively.

The titrant solution, which consists of iodine, sulfur dioxide, and pyridine in anhydrous methanol, is intensely brown-yellow in color. All the products of the reaction are colorless. Thus a self-indicating titration can be performed to the appearance of a permanent yellow. The titrant is not stable and must be standardized frequently against a known amount of water dissolved in anhydrous methanol. Commonly the titer of the iodine solution is expressed in milligrams of water per milliliter.

Many oxidants (e.g., bromine and chlorine) might be directly titrated with thiosulfate, but usually side reactions occur that do not allow establishment of a strictly stoichiometric relationship. Hence, the oxidant is allowed to react with potassium iodide to form iodine, which is then titrated with thiosulfate. This approach is commonly applied to the determination of the "active chlorine" content of bleaching powders and bleaching solutions.

Arsenic(V) reacts with iodide in highly acidic solution by the reverse of equation (25-32), which for high acidities may be formulated as

$$H_3AsO_4 + 2H^+ + 2I^- \rightleftharpoons H_3AsO_3 + I_2 + H_2O \quad (25\text{-}35)$$

The iodine formed can be titrated with thiosulfate.

Copper(II), in alloys and ores, is determined by reaction with an excess of iodide and titration of the liberated iodine with thiosulfate.

$$2Cu^{2+} + 4I^- \rightarrow \underline{2CuI} + I_2 \quad (25\text{-}36)$$

The titration is performed in the presence of the insoluble copper(I) iodide. Interference by iron, often present in samples, is obviated by complexation of iron(III) with fluoride ion.

Hydrogen peroxide oxidizes iodide ion to iodine and therefore can be determined iodometrically. In contrast to the permanganate method for peroxides (Section 25.5), organic substances often used as stabilizers for hydrogen peroxide do not interfere in the determination.

The determination of iodate has been described in the standardization of thiosulfate solutions (Section 25.6). Bromate and chlorate in a strongly acidic medium react with iodide analogously and thus can also be determined.

Nitrite in acidic solution reacts with iodide according to

$$2HNO_2 + 2H^+ + 2I^- \rightarrow 2NO + 2H_2O + I_2 \quad (25\text{-}37)$$

Titration of the iodine formed allows the determination of nitrite. The nitrogen oxide formed reacts readily with oxygen to form nitrogen dioxide, NO_2, which in turn leads to further liberation of iodine; consequently, air must be excluded.

An interesting method for the determination of extremely small amounts of iodide (e.g., in blood serum) proceeds as follows. The iodide is oxidized to iodate by bromine in a weakly acidic medium.

$$I^- + 3H_2O + 3Br_2 \rightarrow IO_3^- + 6H^+ + 6Br^- \quad (25\text{-}38)$$

The excess of bromine is then removed (e.g., by boiling), potassium iodide is added to the solution to react with the iodate, and the iodine formed is then titrated with thiosulfate. In this way the original amount of iodide in the sample is "multiplied" by 6 and the sensitivity and precision of the determination are greatly increased.

The reaction between iodate and iodide to form iodine (25-31) proceeds stoichiometrically also with respect to hydrogen ion; consequently, an acid may be determined iodometrically or a thiosulfate solution standardized by the use of an exactly standardized solution of a strong acid. The calculation is based on the fact that two hydrogen ions correspond to the liberation of one molecule of iodine (I_2).

25.8 Dichromate Methods

Dichromate ion in acidic solution is a slightly weaker oxidant than permanganate.

$$Cr_2O_7^{2-} + 14H^+ + 6e \rightarrow 2Cr^{3+} + 7H_2O \qquad (E^0 = +1.33 \text{ V}) \qquad (25\text{-}39)$$

Dichromate is used in place of permanganate in many titrations, notably that of iron(II), and in various methods where the iron(III)–iron(II) system is employed as an intermediate redox system because the presence of chloride causes fewer difficulties. Potassium dichromate, $K_2Cr_2O_7$, is readily available in a purity greater than 99.95% and can be used directly to prepare a standard solution, which is indefinitely stable. An indicator is required in dichromate titrations because the color of dichromate is not sufficiently intense to allow self-indication. The redox indicator most frequently used is *N*-phenylsulfanilic acid or its barium salt.

25.9 Cerium(IV) Methods

Broad application of cerium(IV) as an oxidizing titrant followed largely from the discovery of suitable redox indicators.

$$Ce^{4+} + e \rightarrow Ce^{3+} \qquad (E^f = +1.44 \text{ V in } 1\ F\ H_2SO_4) \qquad (25\text{-}40)$$

The yellow color of cerium(IV) compounds is insufficient for a self-indicating titration. The indicator most frequently used is the iron(II) chelate of 1,10-phenanthroline (often termed Ferroin), obtained by mixing solutions of iron(II) sulfate and of the organic compound. The intense red color of the indicator changes to a pale blue upon oxidation to the corresponding iron(III) chelate. At the low indicator concentrations used in a titration, the change is from pink to almost colorless.

Cerium(IV) does not exist in acidic solutions as the simple aquo ion but rather forms complexes with the oxyanions present. Cerium(IV) standard solutions are often prepared in 1 to

8 F sulfuric or nitric acid and such complexes as

$$[Ce(SO_4)_n]^{-2n+4} \quad \text{or} \quad [Ce(NO_3)_n]^{-n+4}$$

are then present, respectively. However, in reaction equations, simply Ce(IV) or Ce^{4+} is often written to facilitate recognition of the stoichiometric relationships involved (see Section 24.3).

Cerium(IV) possesses salient advantages as an oxidizing titrant. Its solution in sulfuric or nitric acid is stable even on boiling and it does not oxidize chloride to chlorine as long as the halide concentration is below about 1 M (contrast with permanganate!). The oxidizing power of cerium(IV) solutions is quite high and cerium can be substituted for permanganate in most determinations. Some reactions with cerium(IV), however, are slow; for example, the direct titration of arsenic(III) is only possible if a minute amount of osmium(VIII) oxide, OsO_4, is added as a catalyst.

Bromate and Bromine Methods 25.10

Bromate ion is a strong oxidizing agent in acidic solution.

$$2BrO_3^- + 12H^+ + 10e \rightarrow Br_2 + 6H_2O \qquad (E^0 = +1.5 \text{ V}) \qquad (25\text{-}41)$$

In a bromatometric titration, as long as some of the reductant being titrated is still present, the bromine is further reduced to bromide ion.

$$Br_2 + 2e \rightarrow 2Br^- \qquad (E^0 = +1.09 \text{ V}) \qquad (25\text{-}42)$$

Beyond the equivalence point, the first drop of bromate solution in excess yields free bromine, according to

$$BrO_3^- + 5Br^- + 6H^+ \rightarrow 3Br_2 + 3H_2O \qquad (25\text{-}43)$$

The end point may be detected by the yellow color of the free bromine, but the use of a suitable indicator is preferable. Potassium bromate is readily obtained in highest purity and a standard solution may be directly prepared. The solution is indefinitely stable.

The bromate titration of antimony(III) in hydrochloric acid solution is routinely applied to the analysis of alloys and ores. The sample is dissolved, the antimony brought to the tervalent state by a suitable treatment (for example, sulfurous acid reduction), and the titration is started. Antimony(III) is progressively oxidized to antimony(V) and bromate is reduced to bromide ion. At the end point, the first excess of bromate oxidizes some of the bromide formed to bromine. The end point is commonly signaled by the irreversible decolorization of a dye by the bromine. Often methyl orange is used; it serves here not as an acid–base indicator, but rather as an *irreversible* redox indicator. Since the indicator reaction is slow, the end point must be approached slowly. Arsenic(III) can be determined similarly.

In many cases bromine is a more suitable oxidant than bromate. But even then, a bromate solution may be used as titrant instead of a bromine solution, which is quite unstable. Potassium bromide is added to the solution to be titrated or is incorporated into the bromate standard solution. On addition of the titrant solution to the acidified sample solution, bromine is formed in situ, according to equation (25-43).

25.11 Methods Based on Strong Reductants

Strong reducing agents such as titanium(III), chromium(II), or tin(II) receive attention for the titration of iron(III), vanadate, molybdate, uranyl, and other ions. Since the reducing agents react readily with oxygen, air must be excluded. Chromium(II) is such a strong reductant that it reduces even hydrogen ion of water. Fortunately, this reaction proceeds so slowly that titrations with chromium(II) are still practical, but frequent restandardization of the titrant solution is necessary. Because of the special storage and buret-filling system required, these titrants are usually employed only where reductometric titrations are performed routinely. It is commonly more convenient to reduce the substance to be determined and then to titrate it with an oxidizing titrant that is unaffected by air.

25.12 Questions

25-1. Explain briefly why strong oxidants are more commonly used as titrants than strong reductants.

25-2. What is the chief advantage in the use of a metal as a prereductant?

25-3. What is a disadvantage in the use of nitric acid as a preoxidant?

25-4. Explain briefly why a standard permanganate solution of reliably known strength cannot be prepared by simple dissolution of a known amount of reagent-grade salt in water and dilution to a known volume.

25-5. Name some standards for the standardization of a permanganate solution and discuss their relative advantages and disadvantages.

State the pH limitations for the use in redox titrimetry of (a) an iodine solution and (b) a thiosulfate solution. Explain what side reaction or other facts are causing the pH restrictions.

What are the advantages of a cerium(IV) solution as an oxidizing titrant?

Elaborate on the possibility of a consecutive redox titration of tin(II) and iron(II).

State the number of electrons involved in the two most common redox couples for permanganate and the conditions under which they are realized.

Which redox couples may be readily used in both oxidimetric and reductimetric titrations?

State some redox titrants, other than permanganate, that have more than one useful redox couple that differ in the number of electrons involved.

What possibilities can you think of for excluding the influence of oxygen of the air on solutions of strong reductants?

Suggest how you might proceed to determine both iron and copper in a solution by the use of redox titrations, and similarly both titanium and iron.

When an alloy containing tin is dissolved in hydrochloric acid with the rigorous exclusion of air, to what oxidation state is the tin converted?

What is an extractive end point and what are its principles of operation, advantages, and disadvantages? Give an example of such an end point.

Which redox titrants permit self-indicating visual titrations?

State which of the following ions, present in equal amounts with iron, might interfere in the permanganate titration of iron by the Zimmermann–Reinhard method: Hg^{2+}, Ag^{+}, Zn^{2+}, Al^{3+}, Cu^{2+}, Cr^{3+}, Ca^{2+}, Mn^{2+}, Sn^{4+}, Cl^{-}, SO_4^{2-}, PO_4^{3-}, I^{-}, NO_3^{-}, and CrO_4^{2-}. Elaborate on the interferences presented and the possibilities for their exclusion.

State some substances used as primary standards for redox titrants and write the equations involved in the standardization procedures.

An acidified solution containing iron(III), copper(II), titanium(IV), mercury(II), manganese(II), zinc(II), calcium(II), sodium(I), and silver(I) is passed through a Jones reductor. Indicate which ions will be reduced and to what oxidation state.

Zinc metal reduces permanganate ion readily. Utilizing this fact, suggest a titrimetric procedure for the determination of metallic zinc in zinc oxide used as a pigment. Write the reaction equation. Would you recommend that the titration be performed in a strongly acidic medium?

26 CALCULATION OF RESULTS IN REDOX TITRATIONS

Basis of Calculations 26.1

For the calculation of the result of a redox titration, it is possible to write the balanced overall redox equation for the titration, establish the combining ratio, and then proceed in the manner recommended for precipitation titrations (Chapter 19). However, a more convenient approach is to base the calculation on the number of electrons transferred in the redox process. Reasoning as in earlier considerations of the calculations of results in titrimetry in Chapters 15 and 19 leads to the formula

$$\text{ml}_t \times F_t \times f_t \times \frac{\text{FW}_s}{f_s} = \text{mg}_s \qquad (26\text{-}1)$$

Here f_t is the equivalence number of the titrant, and its value equals the number of electrons lost or gained by *one molecule* of the titrant. Care should be exercised in assigning a value to this number because the titrant molecule may in some cases require a different number of electrons than the reacting titrant species occurring in the titration reaction, in which commonly only ions are written. With the increased knowledge of the student in titration processes it is possible to go one step further with the designation of the term f_s. Previously this term referred to the species actually participating in the titration reaction; if this species and the sought-for substance were not identical, application of a conversion factor was an additional requirement. Here a more direct approach will be taken and by f_s will be designated the number of electrons corresponding to (that is, lost or gained by) one molecule of the sought-for substance. This more direct approach will become obvious, for example, from Examples 26-3 and 26-4.

The assignment of the equivalence number to the titrant as well as the sought-for substance can be effected by the above considerations in combination with either the change of the oxidation states of the reacting species or the relevant balanced half-reaction equations. The modification of equation (26-1) for the calculation of percentage contents is analogous to cases discussed in other titrations:

$$\frac{\text{ml}_t \times F_t \times f_t \times \text{FW}_s \times 100}{W \times f_s} = \%s \qquad (26\text{-}2)$$

As in other titrations, the chemical equivalency existing at the equivalence point takes the form

$$\text{meq}_t = \text{ml}_t \times F_t \times f_t = \text{ml}_s \times F_s \times f_s = \text{meq}_s \quad (26\text{-}3)$$

The relation $F_t \times f_t = N_t$ holds for redox titrimetry as for acid–base titrimetry.

The normality of an acid (or base) is the number of moles of hydrogen ion (or hydroxide ion) provided per liter of solution under the reaction conditions of interest. Here the normality of an oxidant (or reductant) is the number of moles of electrons actually gained (or lost) per liter of solution under the reaction conditions of interest. Again, normality will not be stressed in this textbook because, contrary to the formality, the normality of a given solution depends on the reaction involved (see Example 26-5).

26.2 Illustrative Examples

Example 26-1. What amount of iron (*55.85*) in mg is present in a sample if a redox titration in acidic medium requires 42.34 ml of 0.02500 F $KMnO_4$?

For permanganate, the electron change is five, since Mn(VII) → Mn(II); and for iron, one, Fe(II) → Fe(III). Therefore, $f_t = 5$ and $f_s = 1$. Substitution of the numerical data into equation (26-1) yields

$$42.34 \times 0.02500 \times 5 \times \frac{55.85}{1} = \text{mg}_s$$

$$\text{mg}_s = 295.6 \text{ mg of Fe in sample}$$

Example 26-2. What volume in ml of 0.04000 F $KMnO_4$ is required to titrate 25.00 ml of a 0.1000 F solution of oxalic acid in acidic medium?

Here $f_t = 5$ as in Example 26-1; $f_s = 2$ since the oxidation of oxalic acid is a two-electron process ($C_2O_4^{2-} \rightarrow 2CO_2 + 2e$).

Substitution of the numerical data into equation (26-2) yields

$$\text{ml}_t \times 0.04000 \times 5 = 25.00 \times 0.1000 \times 2$$

$$\text{ml}_t = 25.00 \text{ ml of } KMnO_4 \text{ solution}$$

Example 26-3. What is the percentage of Fe_2O_3 (*159.69*) present in an ore if a 1.000-g sample is dissolved and iron, after complete conversion to iron(II), is titrated with 0.02000 F $K_2Cr_2O_7$, of which a volume of 32.56 ml is required?

The reduction of dichromate to chromium(III) involves six electrons ($Cr_2O_7^{2-} + 14H^+ + 6e \rightarrow 2Cr^{3-} + 7H_2O$); hence $f_t = 6$. The oxidation of iron(II) involves one electron ($Fe^{2+} \rightarrow Fe^{3+} + e$); however, there are two atoms of iron per molecule of the oxide; hence, $f_s = 2$. Substitution of the numerical data into equation (26-7) yields

$$\frac{32.56 \times 0.02000 \times 6 \times 159.69 \times 100}{1000 \times 2} = 31.20\%\ Fe_2O_3 \text{ in sample}$$

It may be noted that the normality of the dichromate solution used is $0.02000 \times 6 = 0.1200\ N$.

Example 26-4. For the standardization of a $KMnO_4$ solution a 0.2134-g sample of As_2O_3 (*197.84*) is titrated in acidic medium. What is the formality of the permanganate solution if a volume of 39.12 ml is required?

As established in Example 26-1, $f_t = 5$. Arsenic(III) is oxidized to arsenic(V) and one molecule of As_2O_3 contains two arsenic atoms; hence, $f_s = 2 \times 2 = 4$. The relevant formula, (26-1), on substitution of the numerical data, takes the form

$$39.12 \times F_t \times 5 \times \frac{197.84}{4} = 213.4$$

$$F_t = 0.02206\ F\ KMnO_4$$

Example 26-5. The $KMnO_4$ solution of Example 26-4 is used to titrate manganese(II) in nearly neutral medium. What amount of manganese (*54.94*) in milligrams is present in a sample which requires 27.20 ml of the $KMnO_4$ solution?

Under these conditions, permanganate is reduced in a three-electron reaction, to manganese(IV) oxide ($MnO_4^- + 4H^+ + 3e \rightarrow MnO_2 + 2H_2O$); hence, $f_t = 3$. Manganese(II) is oxidized to manganese(IV); hence $f_s = 2$. The calculation, after (26-1), takes the form

$$27.20 \times 0.02206 \times 3 \times \frac{54.94}{2} = mg_s$$

$$mg_s = 49.45 \text{ mg of Mn in sample}$$

It may be noted that the normality of a potassium permanganate solution for a redox titration in acidic medium is $F \times 5$, but in nearly neutral medium $F \times 3$. The formality, of course, is the same regardless of the medium employed.

In some redox titrations, notably iodometric determinations, the sought-for substance does not react directly with the titrant but is first allowed to react with some other substance, and the product of this reaction is subjected to the titration.

Example 26-6. A 0.1480-g sample of KIO_3 (*214.00*) is treated with an excess of KI and acid. The iodine liberated is titrated with a $Na_2S_2O_3$ solution, of which a volume of 30.55 ml is required. What is the formality of this $Na_2S_2O_3$ solution?

The reactions occurring are

$$IO_3^- + 5I^- + 6H^+ \rightarrow 3I_2 + 3H_2O \qquad \text{(liberation of } I_2\text{)}$$
$$I_2 + 2e \rightarrow 2I^- \qquad \text{(titration half-reaction for } I_2\text{)}$$
$$2S_2O_3^{2-} \rightarrow S_4O_6^{2-} + 2e \qquad \text{(titration half-reaction for thiosulfate)}$$

Reduction of one molecule of I_2 to iodide is a two-electron process, but three molecules of I_2 result from the reaction of one molecule of iodate. Consequently, $f_s = 2 \times 3 = 6$. Alternative reasoning would give the same result—one molecule of iodate yields six atoms of iodine (five coming from the KI), and each atom is present as iodide ion at the end of the titration. Hence, as far as the iodate is concerned, the reduction is from iodine(V) to iodide ion, a change of six electrons. Thiosulfate loses two electrons for two thiosulfate ions; that is, a single thiosulfate ion is involved in a one-electron oxidation and $f_t = 1$. The calculation, based on equation (26-1), takes the form

$$30.55 \times F_t \times 1 \times \frac{214.00}{6} = 148.0$$

and

$$F_t = 0.1358\ F\ Na_2S_2O_3$$

Example 26-7. Potassium iodide may be determined in the following way. The solution of the salt is acidified and treated with iodate. The first reaction given in Example 26-6 occurs, and the iodine liberated is distilled, collected under appropriate conditions, and titrated with a thiosulfate solution. What is the equivalent weight of potassium iodide (*166.01*) in this procedure?

Six atoms of iodine are obtained from five molecules of potassium iodide. Each iodine atom involves one electron in the thiosulfate titration. Therefore, six electrons are equivalent to five molecules of potassium iodide, and $f_s = \frac{6}{5}$. The equivalent weight is given by

$$EW_{KI} = \frac{FW_{KI}}{\frac{6}{5}} = \frac{166.01 \times 5}{6} = 138.34$$

Some further examples show the different equivalence numbers for the one and the same substance when used in different titrations.

Example 26-8. Potassium tetroxalate, $KHC_2O_4 \cdot H_2C_2O_4 \cdot 2H_2O$, is a dihydrate double salt of potassium hydrogen oxalate and oxalic acid. A solution of this salt is prepared. When a volume of 20.00 ml of this solution is titrated in acidic medium with 0.02000 F $KMnO_4$, a volume of 25.00 ml of the latter is required. What volume of 0.08000 F NaOH is required to titrate 30.00 ml of the tetroxalate solution in an acid–base titration, using phenolphthalein as the indicator?

In the redox titration of the tetroxalate, four electrons are involved since each oxalate reacts in a two-electron oxidation (Example 26-2) and there are two oxalate ions present in one molecule of the tetroxalate salt; hence, $f_s = 2 \times 2 = 4$. For permanganate, in acidic solution $f_t = 5$.

Substitution of the numerical date into equation (26-2) yields

$$25.00 \times 0.02000 \times 5 = 20.00 \times F_s \times 4$$
$$F_s = 0.03125\ F \text{ tetroxalate}$$

When tetroxalate participates in the acid–base titration, all three available hydrogen ions react; hence, $f_s = 3$. For sodium hydroxide, $f_t = 1$.

$$\text{ml}_t \times 0.08000 \times 1 = 30.00 \times 0.03125 \times 3$$
$$\text{ml}_t = 35.16 \text{ ml of NaOH solution}$$

Example 26-9. A volume of 20.00 ml of a K_2CrO_4 solution is treated with an excess of $BaCl_2$ and the $BaCrO_4$ (*253.33*) separated by filtration, washed, and dried. The dried precipitate is found to weigh 0.1755 g. If this chromate solution is used to titrate iron(II) in acidic medium, 1 ml is equivalent to what amount of iron (*55.85*) in milligrams?

Since one mole of barium chromate is equivalent to one mole of chromate ion, the following may be written, where F_t is the formality of the chromate solution.

$$20.00 \times F_t \times 253.33 = 175.5$$
$$F_t = 0.03464\ F\ K_2CrO_4$$

In the titration of iron(II), the chromate is converted to dichromate, which acts as the oxidizing titrant (see Example 26-3). However, one mole of chromate ion is equivalent to one-half mole of dichromate ion ($2CrO_4^{2-} + 2H^+ \rightarrow Cr_2O_7^{2-} + H_2O$); consequently, $f_t = 3$, when related to the formality of the chromate solution. Thus

$$1.00 \times 0.03464 \times 3 \times 55.85$$
$$= 5.80 \text{ mg of Fe per ml of } K_2CrO_4 \text{ solution}$$

Example 26-10. The silver content of 24.00 ml of 0.1000 F $AgNO_3$ is completely precipitated as Ag_3AsO_4. The isolated precipitate is dissolved in strong acid and the arsenate determined iodometrically, with 20.00 ml of a $Na_2S_2O_3$ solution being required. What is the formality of this thiosulfate solution?

One mole of silver yields one-third mole of silver arsenate; hence, the number of millimoles of the latter formed equals $24.00 \times 0.1000 \times \frac{1}{3} = 0.800$. In the redox titration, arsenic(V) is reduced to arsenic(III) providing, via the liberated iodine, two electrons per molecule; consequently from equation (26-3) 2×0.800 meq is present. For thiosulfate, $f_t = 1$.

$$\text{ml}_t \times F_t \times f_t = \text{meq}_t$$

Substitution of the numerical data yields

$$20.00 \times F_t \times 1 = 2 \times 0.800$$
$$F_t = 0.0800\ F\ Na_2SO_3$$

Instead of expressing concentrations in terms of formalities, it is also possible to employ titer values (see Sections 3.3 and 19.2). The answer to Example 26-9 is in the form of a titer. An additional example will illustrate the interconversion of titer values.

Example 26-11. The "iron titer" of a given $KMnO_4$ solution is 1.00 mg/ml. What is its "manganese titer"?

From Examples 26-1 and 26-5, it is evident that, in the two titrations, the electrons involved are 5 and 3, respectively, for permanganate. The values of f_s for iron and manganese are 1 and 2, respectively. Two equations may be written:

$$1 \times F_t \times 5 \times \frac{FW_{Fe}}{1} = 1.00$$

$$1 \times F_t \times 3 \times \frac{FW_{Mn}}{2} = x$$

where x denotes the mg of manganese equivalent to 1 ml of the $KMnO_4$ solution. Division of the first equation by the second yields

$$\frac{5}{3} \times \frac{FW_{Fe}}{FW_{Mn}} \times 2 = \frac{1.00}{x}$$

Solution for x and insertion of the formula weights for manganese (*54.94*) and iron (*55.85*) gives

$$x = 1.00 \times \frac{3}{10} \times \frac{54.94}{55.85}$$

$$= 0.295 \text{ mg of Mn per ml of } KMnO_4 \text{ solution}$$

26.3 Problems

26-1. A 0.2034-g sample of sodium oxalate (*134.00*) is titrated in acidic medium with a $KMnO_4$ solution, of which a volume of 45.32 ml is required. What is the formality of this titrant?

Answer: 0.01340 F

26-2. What percentage of Mn (*54.94*) is present in a steel if a volume of 12.34 ml of 0.0120 F $KMnO_4$ is required to titrate the manganese(II) in a solution obtained by appropriate treatment of a 1.800-g sample of the steel?

Answer: $0.67_8\%$ Mn

26-3. A cerium(IV) sulfate solution is standardized by titration of a 0.2005-g sample of As_2O_3 (*197.84*), a volume of 42.20 ml being required. A volume of 20.00 ml of this cerium(IV) solution is treated with KI and the I_2 liberated requires 25.68 ml of a $Na_2S_2O_3$ solution. What volume of the $Na_2S_2O_3$ solution will be required to titrate the I_2 liberated by the treatment of 0.1000 g of KIO_3 (*214.00*) with an excess of KI and acid?

Answer: 37.48 ml

A 0.2200-g sample of copper metal (*63.54*) of 99.35% purity is dissolved and the copper is determined iodometrically, a volume of 35.06 ml of a $Na_2S_2O_3$ solution being required. When a 100.0-ml volume of a H_2O_2 (*34.01*) solution is acidified, sufficient KI is added, and the I_2 liberated is titrated with this $Na_2S_2O_3$ solution, a volume of 17.31 ml is required. Calculate the concentration of the H_2O_2 solution in milligrams of H_2O_2 per liter.

Answer: 288.8 mg of H_2O_2 per liter

Pyrolusite, a mineral containing MnO_2 (*86.94*) as the active major constituent, is assayed as follows. A solution is prepared by dissolving 3.2340 g of primary standard sodium oxalate (*134.00*) in water and diluting to a volume of 500.0 ml with water. A volume of 50.00 ml of this solution is acidified and a 0.2500-g sample of the pyrolusite is added to it. After the sample has reacted completely ($MnO_2 \rightarrow Mn^{2+}$), with oxalate being oxidized concurrently to carbon dioxide, the excess of oxalate is back-titrated with a $KMnO_4$ solution, of which a volume of 2.25 ml is required. In a separate titration in acidic medium, a volume of 16.80 ml of this $KMnO_4$ solution is found to be equivalent to 20.00 ml of the oxalate solution. What is the percentage of MnO_2 in the pyrolusite?

Answer: 79.43% MnO_2

What amount of $K_2Cr_2O_7$ (*294.19*) in grams must be dissolved and diluted to a volume of 500.0 ml with water to obtain a solution which has a titer of 1.50 mg of Fe_2O_3 (*159.69*) per milliliter?

Answer: 0.4606 g

An amount of 0.2143 g of iron wire containing 99.52% w/w Fe (*55.85*) is dissolved. All the iron is reduced to iron(II) and titrated with a permanganate solution, of which a volume of 38.22 ml is required in acidic medium. Calculate the formality of the $KMnO_4$ solution.

An amount of 0.1205 g of iron wire containing 95.0% Fe (*55.85*) is dissolved, the solution passed through a reductor, and the Fe(II) titrated with a permanganate solution, of which a volume of 21.70 ml is required. Calculate the formality of the permanganate solution.

An amount of 2.160 g of a sample is dissolved and the calcium precipitated with an excess of oxalate. The calcium oxalate is separated from the solution, washed, and dissolved in acid. A volume of 22.0 ml of 0.0100 *F* $KMnO_4$ is required for the titration of the oxalate in acidic medium. Calculate the % w/w of CaO (*56.08*) in the sample.

A material is known to be either FeO, Fe_2O_3, or Fe_3O_4, in pure form. A 0.1500-g sample is fused with $KHSO_4$, the melt dissolved, and the iron reduced to iron(II) and titrated with 0.00900 *F* $KMnO_4$, 43.15 ml being required. The material consists of what oxide of iron?

A solution contains arsenous acid and arsenic acid. The arsenous acid is titrated with 0.01500 *F* $KBrO_3$, 24.60 ml being required. After the titration the total amount of arsenic acid present is precipitated as $MgNH_4AsO_4$ and the isolated and washed precipitate is ignited to $Mg_2As_2O_4$ (*310.46*). The weight of the magnesium pyroarsenate is 1.2531 g. How many millimoles of *o*-arsenic acid, H_3AsO_4, was initially present in the solution?

26-12. When a strong acid is added to a neutral solution containing KIO_3 and KI the following reaction takes place, $IO_3^- + 5I^- + 6H^+ \rightarrow 3I_2 + 3H_2O$. This reaction proceeds stoichiometrically with respect to H^+. Consequently a standard acid solution may be employed to standardize a thiosulfate solution. Calculate the formality of each of the thiosulfate solutions from the data given.

	Acid solution	Thiosulfate solution
(a)	40.00 ml of 0.1000 *F* HCl	35.21 ml
(b)	18.26 ml of 0.1023 *F* HCl	21.46 ml
(c)	23.15 ml of 0.0829 *F* N_2SO_4	26.44 ml
(d)	26.00 ml of 0.0750 *F* H_2SO_4	46.86 ml
(e)	25.00 ml of 0.1025 *F* $HClO_4$	31.02 ml
(f)	24.22 ml of 0.0993 *F* $HClO_4$	18.95 ml

26-13. An amount of 0.2350 g of a copper ore is dissolved in acid, the pH adjusted appropriately, and an excess of KI added. The liberated iodine requires a volume of 5.26 ml of 0.1500 *F* $Na_2S_2O_3$ to reach the starch end point. Calculate the % w/w of copper (*63.54*) in the ore.

26-14. An amount of 0.01589 g of pure Cu (*63.54*) is dissolved and an excess of KI added. The iodine liberated requires a volume of 25.00 ml of a thiosulfate solution to reach the starch end point. Calculate the formality of the thiosulfate solution.

26-15. The arsenic present in a 0.2000-g sample of a material is volatilized as AsH_3 and the latter trapped in 40.00 ml of a 0.0953 *F* I_2 solution. Titration of the excess of iodine requires 21.50 ml of a 0.1062 *F* thiosulfate solution. Calculate the % w/w As_2O_2 in the material.

26-16. Potassium tetroxalate, $KHC_2O_4 \cdot H_2C_2O_4 \cdot 2H_2O$ (*254.2*), can be titrated via an acid–base as well as a redox reaction. In the following problems, one portion of a tetroxalate solution is titrated with a standard NaOH solution and an identical portion titrated with a $KMnO_4$ solution in acidic medium. For each problem calculate the formality of the $KMnO_4$ solution.

	ml, tetroxalate soln.	ml, NaOH soln.	Formality, NaOH soln.	ml, $KMnO_4$ soln.
(a)	25.00	30.00	0.1000	6.00
(b)	20.00	34.25	0.1100	7.25
(c)	25.00	35.00	0.0975	10.00
(d)	10.00	15.00	0.0900	4.25
(e)	30.00	25.00	0.1200	6.80

What volume in milliliters, of a solution exactly $\frac{1}{60}$ *F* in $K_2Cr_2O_7$ is required for each of the back-titrations indicated below? In each case a volume of 100.0 ml of 0.100 *F* $FeSO_4$ in acid solution is added to exactly 0.1 g of the substance indicated.

	Substance titrated	
(a)	MnO_2	(*86.94*)
(b)	$KMnO_4$	(*158.0*)
(c)	$K_2Cr_2O_7$	(*294.2*)
(d)	K_2CrO_4	(*194.2*)

Salicylic acid, $C_7H_6O_3$ (*138.1*), can be oxidized by permanganate (reduced to MnO_2) in alkaline solution to water and carbonate ion. How many milligrams of salicylic acid will be indicated by exactly 1 ml of a $KMnO_4$ solution which contains 7.900 g of $KMnO_4$ (*158.0*) in 500.0 ml?

27 POTENTIOMETRIC TITRATIONS

Principles of Potentiometric Titrations 27.1

In many cases it is possible to make the solution to be titrated the electrolyte of a half-cell, the potential of which is a function of the concentration of one or more of the species participating in the titration reaction. A wire or foil of an appropriate metal is immersed in the solution to act as the "indicator" electrode. The half-cell thus formed is connected via a salt bridge with a reference electrode. The cell established in this way is connected with a potentiometer circuit (Section 22.10) and during the subsequent titration, the cell voltage is measured as a function of titrant additions. To determine the end point graphically, the titration curve may be obtained by plotting the values of the half-cell potential ($= E_{\text{cell}} - E_{\text{reference}}$) versus the volume of titrant solution added. However, because the potential of the reference electrode remains constant, it is common practice to simply plot the total cell voltage rather than the half-cell potential, thus merely shifting the curve vertically without changing its shape or its position relative to the abscissa. A titration so conducted is described as proceeding to a potentiometric end point or simply as a potentiometric titration. Many types of titrations can be effected potentiometrically.

Potentiometric Redox Titrations 27.2

A redox titration can be conducted potentiometrically as has already been implied in the considerations in Chapter 24. In this case the indicator electrode is frequently a platinum wire or foil. The experimental titration curves will often closely approximate the theoretical curves calculated as described in that chapter. The point of maximum slope in the curve is taken as the end point.

Potentiometric Acid–Base Titrations 27.3

In Chapter 11 the titration curves for acid–base titrations in aqueous solution were calculated. These curves are plots of pH versus the volume of acid or base added. Such curves are realized experimentally in acid–base titrations to a potentiometric end point. An electrode sensitive to hydrogen ion is used as the indicator electrode. The principles underlying the electrometric determination of pH have been described in Chapter 23 and from these considerations it is obvious that a pH meter with a glass electrode can be employed to follow the pH changes during the titration. The measured pH values can be plotted versus the volume of titrant solution added and the point of maximum slope be taken as the end point. Alternatively, if the pH of the equivalence point is accurately known, the titration may be performed without plotting and terminated when this pH value is reached. Potentiometric acid–base titrations may be employed when visual indication becomes difficult due to colorants present in the sample solution. A potentiometric titration is of special value in the determination of acids and bases that are too weak to afford a satisfactory visual titration.

27.4 Other Potentiometric Titrations

The presence of an appropriate complexing agent may markedly change the potential of a redox couple involving a metal ion (Section 24.10); consequently, the end point of some compleximetric titrations (Chapter 20) may be detected potentiometrically. Similarly, the formation of a precipitate by a metal ion that is involved in a redox equilibrium will shift the redox potential. Hence, potentiometric precipitation titrations are also feasible. In the titration of a halide with silver nitrate, a silver wire can serve as the indicator electrode.

Potentiometric titrations often allow the stepwise titration of two or more substances with a single titrant. For example, a mixture of bromide and chloride ion can be titrated with silver nitrate. In the visual titration of this mixture by the Mohr method (Section 18.4), the result corresponds to the total of the two halides. In contrast, in the potentiometric titration, two breaks will be obtained corresponding to bromide and chloride, respectively. The accuracy is not especially high because of the closeness of the values of the respective solubility products. More accurate results are obtained in the stepwise potentiometric titration of iodide and chloride. Under favorable concentration conditions, for example, in the analysis of mineral waters, it is even possible to analyze a mixture of iodide, bromide, and chloride by titration with silver ion. Three breaks in the potentiometric titration curve are obtained.

Diverse Techniques 27.5

Instead of using the inflection point as the end point in a potentiometric titration, more precise results can be gained from the plot of the first or second derivative of the curve. The first-derivative curve is readily obtained by adding the titrant solution in small, equal increments and plotting the difference between two consecutive potential (or voltage) readings versus the volume of titrant solution added. The maximum in this

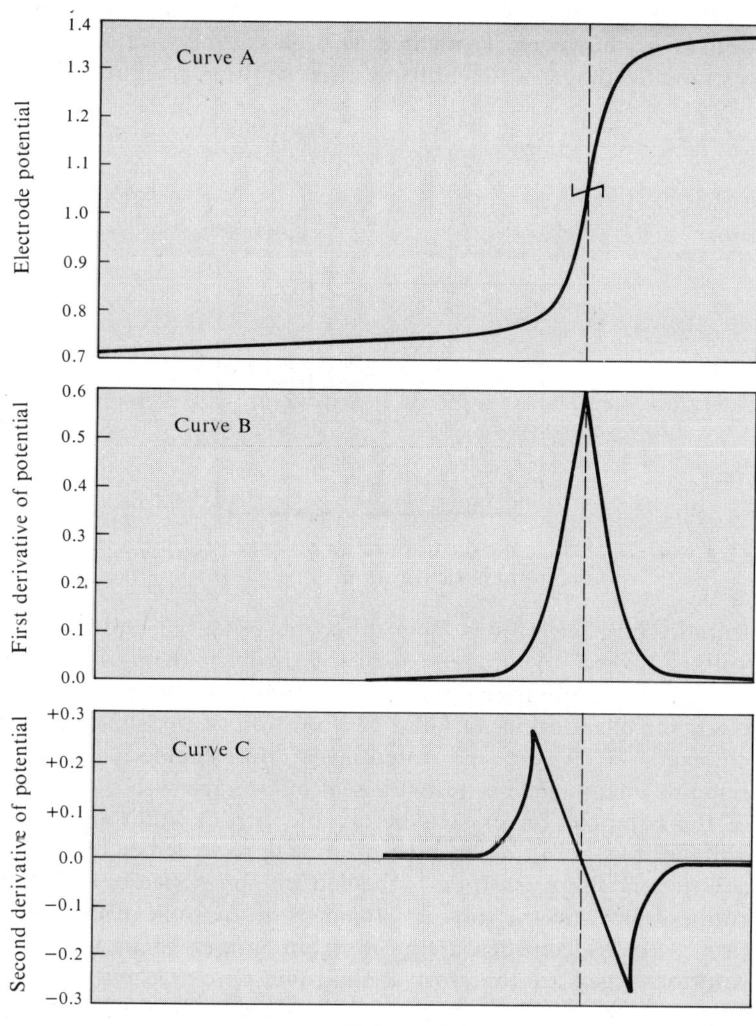

Figure 27-1. Curve A: experimental potentiometric titration curve. Curve B: first-derivative curve of A. Curve C: second-derivative curve of A.

curve is taken as the end point. The curve for the second derivative is obtained by plotting the difference between two consecutive differences versus the volume of titrant solution added. Here the end point corresponds to passage of the curve through the zero line. Representative curves plotted in these three ways are shown in Figure 27-1.

The values of the first derivative may be established directly by a technique which is called a derivative potentiometric titration. Two identical electrodes are placed in the titration solution. One, however, is shielded by a glass tube with a narrow opening leading to the bulk of the solution (Figure 27-2).

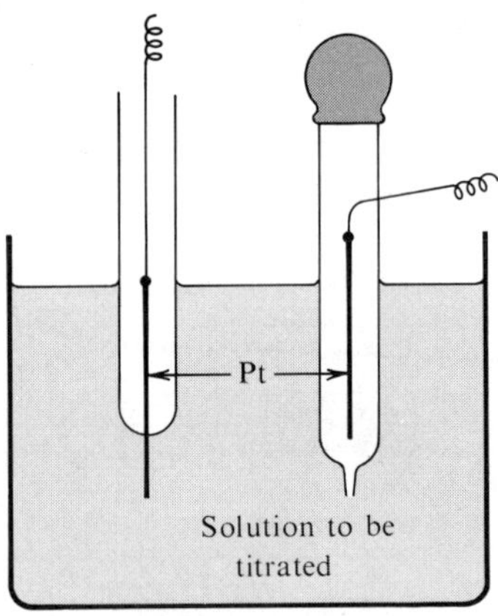

Figure 27-2. Schematic diagram of an arrangement for a derivative potentiometric titration.

Initially both electrodes have the same potential and the cell voltage is zero. An increment of the titrant is then added and the solution is homogenized. The potential of the unprotected electrode changes because the composition of the solution has changed. However, the potential of the shielded electrode remains unchanged because it is still in contact with a solution of the composition present before the titrant addition. The voltage thus resulting is measured and recorded. Then the solution in the glass tube is flushed out by squeezing of the rubber bulb, and the tube is refilled from the bulk of the solution. Both electrodes are now again immersed in identical solutions; they are therefore at the same potential and the cell voltage is zero. This process is repeated for each (equal) increment of titrant. A graph of the cell voltage values, that is, of the potential differences versus the volume of titrant, corresponds to the first derivative of the usual titration curve. For equal increments the potential differences increase as the equivalence point is approached, reach a maximum when passing through this point, and become smaller again when an excess of titrant is added. At the equivalence point addition of the fixed amount of titrant causes the greatest fractional change in concentration and thereby in potential. Consequently, the maximum in the curve is taken as the end point.

Where a potentiometric titration is performed routinely, there is no need to plot the titration curve if the cell voltage corresponding to the end point is well established. The titrant solution need only be added until this voltage is reached. This modification of the potentiometric titration is readily capable of automation. Instruments are available that follow the potential and regulate the addition of titrant from a buret, the delivery of which is controlled electrically and stopped when the preselected cell voltage is reached. In other more sophisticated "autotitrators," a double differentiation is effected by an electronic circuit; when the output signal of the titrator is zero, a relay is triggered by means of which the flow of titrant is terminated. With this latter type of instrument, the potential of the equivalence point need not be established previously.

Questions 27.6

What advantages may be realized by performing an acid–base titration to a potentiometric rather than to a visual end point?

Does a potentiometric titration have any advantage when more than one component of the solution reacts with the titrant? Explain.

Elaborate on the various techniques for the detection of the end point in a potentiometric titration and discuss their relative advantages.

A calomel electrode is commonly used as the reference electrode in potentiometric titrations. However, any electrode may be substituted which exhibits a constant potential during the course of the titration. Is it necessary to know the exact potential of such an electrode? Explain.

Addition of the titrant solution during a potentiometric titration in equal increments, at least in the vicinity of the end point, has the special advantage that the position of the end point can be calculated without plotting the data. Explain how this might be done.

In a potentiometric acid–base titration using a glass electrode and pH meter, it is unnecessary to calibrate the latter. Explain.

What is an inert electrode?

28 THEORY OF ELECTROLYTIC PROCESSES

The purpose of this chapter is to consider the principles underlying electrolytic processes that form the basis of a number of analytical methods. Some of these are discussed in subsequent chapters. Electrolysis occurs when an external source of voltage is applied to an electrochemical cell to force a nonspontaneous electrochemical reaction to proceed. The electrodes connected to the positive and negative terminals of the voltage source are known as the anode and cathode, respectively. Oxidation proceeds at the former, reduction at the latter. (The possible ambiguities existing in the definition of the terms anode and cathode are noted in Section 22.4.) The solution in the cell is known as the electrolyte solution or in brief as the electrolyte.

Decomposition Voltage 28.1

Consider the experimental arrangement illustrated in Figure 28-1. The beaker contains an aqueous solution 0.10 *F* in cadmium iodide, serving as the electrolyte. Two identical

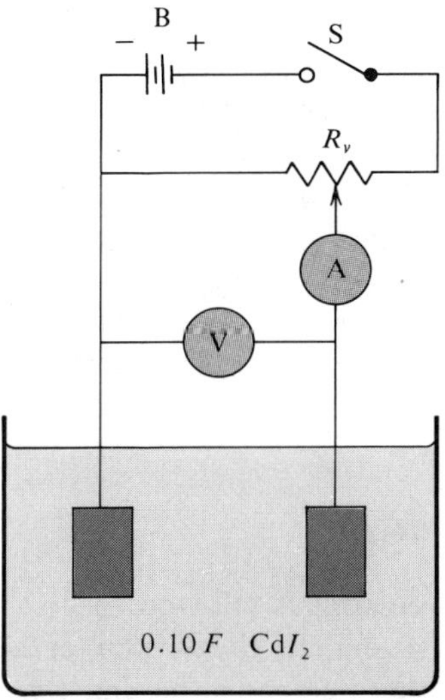

Figure 28-1. Schematic arrangement for electrolysis.

platinum electrodes are dipped into the solution. The left and right electrodes are connected through the external circuit to the negative and positive terminals of the voltage source and hence will be the cathode and anode, respectively. The external circuit consists of the battery, B, the off-on switch, S, the variable resistance, R_v, the ammeter, A, and the voltmeter, V. The resistance, R_v, allows adjustment of the voltage that is applied across the electrolysis cell.

Before the switch is closed the voltmeter shows zero volts because the electrodes are identical and dip into the same homogeneous solution. Then the switch is closed and a small voltage (say a few tenths of a volt) is applied to the cell, by appropriate adjustment of the variable resistor. This voltage is indicated by the voltmeter. The ammeter, however, shows that essentially no current flows and an inspection of the electrodes reveals that no electrode reactions are taking place.

If the voltage applied to the cell is increased progressively, at first only a very feeble rise in current is observed. Then, at a certain voltage the current increases suddenly, and electrode reactions become evident. Cadmium plates at the cathode, and the anode becomes discolored by the deposition of elemental iodine. Iodine dissolves to a slight extent in the solution surrounding the anode but, for simplicity, this fact will be neglected and all the iodine will be assumed to deposit on the anode.

Obviously, when the current starts to flow and electrode reactions become evident, the voltage applied to the cell is sufficient to cause appreciable decomposition of the electrolyte and is then said to have reached the decomposition voltage. This voltage, as any across an electrochemical cell, consists of two portions: the anodic and cathodic decomposition potentials. Since there is some ambiguity in deciding when a current "just starts to flow," it is customary to define the decomposition voltage, E_d, as the zero-current intercept obtained by back-extrapolation of the current versus applied voltage curve, as shown in Figure 28-2.

28.2 Back Electromotive Force

If the electrolysis current is allowed to flow for some time and then the switch is opened, the voltmeter is found to indicate a voltage across the electrodes having the same polarities as during the electrolysis. This voltage arises within the cell, is

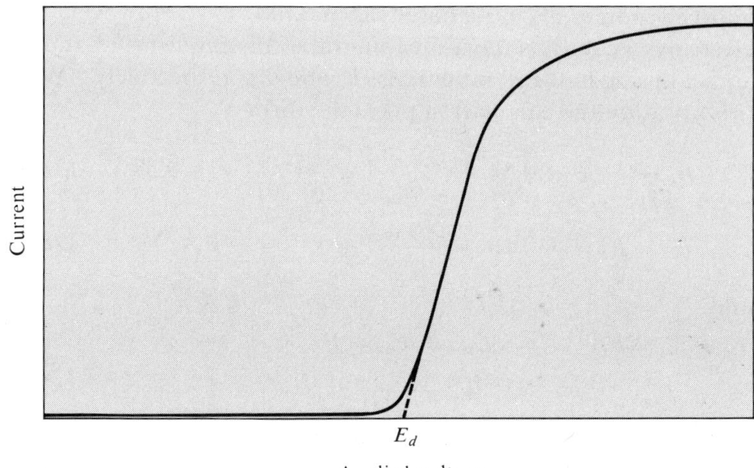

Figure 28-2. Plot of current versus applied voltage (E_d = decomposition voltage).

in opposition to the applied voltage, and is known as the back electromotive force, or in short, the back emf. This back emf consists of two portions: the reversible back emf and the concentration overvoltage.

The reversible back emf arises because a voltaic cell formed when the electrolytic depositions took place. One platinum electrode is now coated with cadmium and the other with iodine; together with their surrounding solutions they form a voltaic cell consisting of a cadmium(II)–cadmium half-cell and an iodine–iodide half-cell. Realizing that a solution 0.10 *F* in CdI_2 is 0.10 *M* in Cd^{2+} and 0.20 *M* in I^-, the newly formed cell may be represented as

$$\mathrm{Cd} \mid \mathrm{Cd}^{2+}(0.10\ M) \parallel \mathrm{I}^-(0.20\ M) \mid \mathrm{I}_2(s)$$

This schematic presentation of the cell can readily be related to the actual electrolysis cell shown in Figure 28-1 by considering the latter to consist formally of three parts. The left one third, which envelopes and includes the cathode, and for which $Cd^{2+} + 2e \rightleftharpoons Cd$ and $E^0 = -0.40$ V; the right one third, which envelopes and includes the anode, and for which $I_2 + 2e \rightleftharpoons 2I^-$ and $E^0 = +0.54$ V; and the central one third, which acts as an electrolytic connection.

The voltage of the newly formed voltaic cell can be calculated by the procedure described in Section 22.8. In that section the half-cells were referred to as "right" and "left" and the half-cell potentials designated as E_r and E_l. However, for electrolysis cells it is more meaningful to distinguish the anodic and cathodic half-cell potentials, E_a and E_c, respectively. With this notation the calculation takes the form

$$E_r = E_a = +0.54 + \frac{0.059}{2} \log \frac{1}{(0.20)^2} = +0.58 \text{ V} \qquad (28\text{-}1)$$

$$E_l = E_c = -0.40 + \frac{0.059}{2} \log 0.10 = -0.43 \text{ V} \qquad (28\text{-}2)$$

and

$$E_{\text{cell}} = E_r - E_l = E_{\text{rev}} = E_a - E_c$$
$$= +0.58 - (-0.43) = (+)1.01 \text{ V} \qquad (28\text{-}3)$$

It may be noted that it is common practice to write $E_{\text{rev}} = E_a - E_c$ in order to obtain a positive value for the reversible emf. The positive sign of the cell voltage indicates the right half-cell, that is, the iodine half-cell, to be the positive pole. Since this electrode is connected to the positive terminal of the voltage source and thus opposes the external voltage, and since the cell reaction $Cd^{2+} + 2I^- \rightleftharpoons Cd + I_2$ is reversible, the term reversible back electromotive force is quite appropriate.

Now it can be appreciated that for an electrolysis the voltage applied to the cell must at least equal the reversible emf of the voltaic cell that forms once electrolysis is initiated. For the cadmium iodide system under consideration, the decomposition voltage is essentially identical to the reversible back emf calculated from the Nernst equation. As can be deduced from an inspection of equations (28-1) through (28-3), the reversible back emf and consequently the decomposition voltage increases as the concentrations of the relevant ions decrease. Since the concentrations of both the cadmium(II) and the iodide ion decrease as the electrolysis proceeds, the reversible back emf increases progressively.

28.3 IR Drop and Applied Voltage

The decomposition voltage, E_d, as defined above, is in essence the applied voltage required to initiate electrolysis. In order to advance the electrolysis, that is, to plate cadmium and iodide at a reasonable rate, an additional voltage is required to overcome the electrical resistance of the cell. With an electrolysis current, I, the voltage required in excess over the decomposition voltage is given according to Ohm's law: $E = IR$. If the cell resistance, R, is known, this voltage, called the IR drop, can be calculated. Consequently, the applied voltage, E_{app}, required to perform an electrolysis with the passage of a certain current may be given as $E_{app} = E_d + IR$. However, as will be shown presently, this formula does not yet fully describe the situation.

Concentration Overvoltage and Overpotential 28.4

From the considerations presented so far, it would seem that once the decomposition voltage is reached the current should increase in a linear fashion with further increase in the applied voltage. Experiments show that this is true for only a moderate voltage rise. At higher voltages, the current increase becomes progressively less than would correspond to a linear relation, and finally a point is reached beyond which an increase in applied voltage does not lead to any appreciable change in the current unless the decomposition of another system, possibly that of the solvent, is reached. The situation is illustrated in Figure 28-2.

To explain this leveling of the current–applied voltage curve to a plateau, closer inspection of the processes taking place at the electrodes is necessary. Consider first the situation at the cathode. Here cadmium metal is deposited and the solution in the vicinity of the cathode becomes depleted of cadmium ions. In the calculation of the half-cell potential in equation (28-2) the bulk concentration of 0.10 M Cd^{2+} was used. However, the Nernst equation only applies to the situation at the electrode surface and here the cadmium ion concentration is lower due to the depletion. Consequently, the value based on the bulk concentration must be suspected of error. Several processes operate to replace the cadmium ions that leave the solution as the electrolysis proceeds. Cadmium ions are transported to the electrode region by diffusion from the more concentrated solution. Other transport mechanisms include electrostatic attraction between the positively charged cadmium ions and the negatively charged electrode, convection, and possibly agitation of the solution. If a current of, say, 0.1 A (which by definition equals 0.1 C/sec) is to be maintained, 0.1/96,500 or approximately 10^{-6} equivalent of cadmium ion must be transported to the electrode each second. As long as the applied voltage, and therefore the current, is moderate, cadmium ions are replenished by these transport processes sufficiently fast. At higher voltages, however, the current cannot follow the voltage rise linearly because the ion transport lags. Finally a point is reached where the current is governed by the rate at which ions are transported to the electrode and a further increase in applied voltage no longer has any pronounced effect on the current.

The situation at the anode is analogous.

To examine the voltage–current relation more closely it is necessary to calculate the "effective" back emf using the concentrations actually prevailing at the electrodes. Assume, for example, that these concentrations are 1×10^{-3} and 2×10^{-3} *M* for cadmium and iodide ions, respectively; a voltage of 1.19 V is obtained (the calculation is left to the student!). Consequently the "effective" back emf is 0.18 V greater than the reversible back emf. This additional voltage can therefore be interpreted as caused by concentration effects and is appropriately termed the concentration overvoltage. It is the composite of two portions: the anodic and cathodic concentration overpotentials. Some workers prefer the term "concentration polarization."

It is now possible to present a more complete formula for the calculation of the voltage to be applied:

$$E_{\text{appl}} = E_d + IR + E_{\text{concn. overv.}} \qquad (28\text{-}4)$$

where, in the case of cadmium iodide, E_d equals the reversible back emf. It is important to realize that the concentration overvoltage is *not* a component of the decomposition voltage since it only arises during the passage of current. Unfortunately, the concentration of the solutions immediately adjacent to the electrodes is not known, and consequently the concentration overpotentials cannot be calculated by use of the Nernst equation. This restriction holds not only for the present example, but generally. In calculations employing equation (28-4), all that can be done is to neglect concentration overvoltage and thereby obtain a minimum value for the applied voltage required that will be approached in practice to the extent that measures can be taken to reduce such overvoltage. Several measures are possible to this end. Raising the temperature increases the diffusion rate and thereby the rate of particle transport to the electrode. Stirring the solution has the same effect. A decrease may further be achieved by decreasing the current density. The current density is defined as the cell current divided by the surface area of the relevant electrode and is usually expressed in amperes per square centimeter (or decimeter). The current density at a given current strength can be decreased by an increase in surface area, for example, by substituting a larger electrode. This electrode will be in contact with more solution from which ions can be drawn.

Kinetic Overvoltage and Kinetic Overpotential 28.5

If the electrolysis experiment, described with cadmium iodide as the electrolyte, is performed with certain other types of electrolytes, a somewhat different situation may result. Consider the same experimental arrangement (Figure 28-1) but with 0.10 *F* sulfuric acid as the electrolyte. As will be established below, this acid has no effect on the overall cell reaction, which is the decomposition of water according to $H_2O \rightarrow H_2 + \frac{1}{2}O_2$.

Before the external voltage is applied, no cell voltage exists since again two identical electrodes dip into the same homogeneous solution. When a certain voltage is applied, current starts to flow and the anode and cathode become coated with small bubbles of oxygen and hydrogen, respectively. Thereby again a voltaic cell is formed, here consisting of an oxygen half-cell and a hydrogen ion–hydrogen half-cell. The electrode reactions and relevant Nernst expression may be written

$$\tfrac{1}{2}O_2 + 2H^+ + 2e \rightleftharpoons H_2O$$

$$E_a = +1.23 + \frac{0.059}{2}\log[H^+]^2\sqrt{p_{O_2}} \tag{28-5}$$

$$2H^+ + 2e \rightleftharpoons H_2$$

$$E_c = 0.00 + \frac{0.059}{2}\log\frac{[H^+]^2}{p_{H_2}} \tag{28-6}$$

where the concentrations of the gaseous components are expressed in terms of their partial pressures. The reversible back emf may now be calculated. The hydrogen ion concentration is 2.0×10^{-1} *M* and, for simplicity, it is assumed that $p_{O_2} = p_{H_2} = 1$ atm. The calculation then takes the form

$$E_a = +1.23 + \frac{0.059}{2}\log(2 \times 10^{-1})^2 = +1.19\text{ V} \tag{28-7}$$

$$E_c = 0.00 + \frac{0.059}{2}\log(2 \times 10^{-1})^2 = -0.04\text{ V} \tag{28-8}$$

and

$$E_{rev} = E_a - E_c = +1.19 - (-0.04) = 1.23\text{ V} \tag{28-9}$$

It is noteworthy that the term for the hydrogen ion concentration vanishes in the subtraction. Consequently, as long as the partial pressures of the two gases are both 1 atm, the reversible back emf will be 1.23 V, regardless of the pH of the solution. Thus the value of 1.23 V holds also for pure water. However, such a medium contains only an extremely small number of ions and consequently shows a very high electrical resistance. To lower the resistance and thus to facilitate the flow of an electric current, sulfuric acid is added. Any other electrolyte that does not undergo electrolytic decomposition (e.g., sodium sulfate, sodium hydroxide, etc.) would serve the purpose equally well.

If, in analogy to the cadmium iodide case, the decomposition voltage were equal to the reversible back emf, an applied voltage of 1.23 V should suffice to just initiate the decomposition of the water. Experimentally, however, the decomposition voltage is found to be about 1.7 V. Thus an overvoltage of 0.5 V (1.7 − 1.23) exists.

The presence of concentration overvoltage might be suspected. However, this assumption is not logical because the decomposition potential was established at essentially zero current condition, and with no current flowing, no concentration changes around the electrode take place. In addition, the overvoltage of 0.5 V persists even on vigorous stirring of the electrolyte. Therefore, the overvoltage must be caused by phenomena other than concentration effects.

This additional type of overvoltage is called kinetic overvoltage or activation overvoltage. Portions of this overvoltage arise at the anode and cathode and are termed anodic and cathodic kinetic overpotentials. The mechanisms involved in the establishment of such overpotentials are complex and not fully understood.

Especially large overpotentials are consistently encountered when diatomic gases are produced in the electrode reaction. Some systems other than those involving diatomic gases also show kinetic overpotential, commonly to a much lesser degree. For example, overpotential is encountered in the electrodeposition of cobalt, chromium, and nickel. Molybdenum exhibits such a high kinetic overpotential that its deposition from aqueous media is impossible under ordinary conditions. Another interesting example is the higher potential required to deposit bismuth or silver on a copper cathode than on a platinum one. Kinetic overpotentials are usually nonexistent where the electrode reaction involves merely a change in oxidation state without a deposition taking place.

The size of kinetic overpotentials depends on various factors. The electrode material and the nature of the electrode surface have a great influence. Commonly soft metals (like lead and tin, and especially mercury) cause higher kinetic overpotentials than hard metals. The overpotential increases with increasing current density and decreases with increasing temperature.

The values for the kinetic overpotential of hydrogen encountered at various electrode materials are listed as a function of current density in Table I in the Appendix. The above-mentioned influence of the nature of the electrode surface manifests itself clearly in the difference between the overpotentials on smooth and on platinized platinum. The latter is platinum metal on which finely divided platinum, known as "platinum black," has been deposited. This difference results largely from the greater effective surface area of the platinized electrode and thereby to the smaller current density it affords. The student is now in the position to appreciate the use of platinized platinum in a hydrogen electrode. The values given in Table I of the Appendix and analogous data are valid under specified conditions and may change if deviations from these conditions are encountered.

Decomposition Voltage and Kinetic Overvoltage 28.6

For cases where no kinetic overpotentials are encountered (as in the example of cadmium iodide) the decomposition voltage equals essentially the reversible back emf, and can be calculated from standard electrode potentials and concentration data via the Nernst equation. This was shown in Section 28.2. Where kinetic overpotentials are encountered, these must be considered in addition to the reversible back emf.

An estimate of the decomposition voltage is of interest when an electrolysis is to be performed. Decomposition potentials are of special importance when a decision has to be made as to which of two or more possible electrode reactions will actually occur under given conditions. The formulas necessary may be derived as follows.

Denoting the kinetic overvoltage by the Greek letter eta, η, the relationship among the decomposition voltage, reversible back emf, and kinetic overvoltage may be formulated as

$$E_d = E_{\text{rev}} + \eta \qquad (28\text{-}10)$$

Both the decomposition voltage and the kinetic overvoltage consist of anodic and cathodic portions. Thus we may write

$$E_d = E_{d,a} - E_{d,c} \tag{28-11}$$

and

$$\eta = \eta_a - \eta_c \tag{28-12}$$

When a combination of these expressions is used in actual calculations, careful attention to signs is necessary, a point often missed by the beginner. The anodic kinetic overpotential is always positive. It increases the magnitude of both the anodic decomposition potential, $E_{d,a}$, and the decomposition voltage. The cathodic overpotential, $E_{d,c}$, is always negative. It acts to make the cathodic decomposition potential more negative, but it also increases the value of the total decomposition voltage (subtraction of a negative number is involved!). Thus for the anode

$$E_{d,a} = E_{\text{rev}} + \eta_a \qquad \text{(where } \eta_a \text{ is a positive number)} \tag{28-13}$$

and for the cathode

$$E_{d,c} = E_c + \eta_c \qquad \text{(where } \eta_c \text{ is a negative number)} \tag{28-14}$$

Upon combination of the appropriate expressions one obtains

$$E_d = \overbrace{E_a - E_c}^{E_{\text{rev}}} + \overbrace{\eta_a - \eta_c}^{\eta} \tag{28-15}$$

Equation (28-15) allows evaluation of the decomposition voltage. Equation (28-13) or (28-14) may be employed when deciding which of several electrode reactions will take place. Of two or more possible anodic reactions, the one with the least positive decomposition potential will occur. Analogously, at the cathode the reaction with the least negative decomposition potential will occur.

Some examples should aid in gaining a better understanding of the concepts.

Example 28-1. Consider a solution 1.0×10^{-1} M in Co^{2+} and 1.0×10^{-1} M in H^+. If this solution is electrolyzed, will hydrogen gas or cobalt metal form if (a) bright platinum and (b) mercury, serve as the cathode? The cathodic kinetic overpotential of hydrogen gas at $p_{H_2} = 1$ atm is -0.09 V and -1.04 V on bright platinum and mercury, respectively.

(a) At the platinum cathode,
for hydrogen:

$$E_{d,c} = E_c + \eta_c$$

$$= \overbrace{0.00 + \frac{0.059}{2}\log(1.0 \times 10^{-1})^2}^{E_c} - \overbrace{0.09}^{\eta_c} = -0.15 \text{ V}$$

for cobalt:

$$E_{d,c} = E_c = -0.28 + \frac{0.059}{2}\log(1.0 \times 10^{-1}) = -0.31 \text{ V}$$

Since the formation of hydrogen has the more positive decomposition potential, hydrogen gas will form.

(b) At the mercury cathode,
for hydrogen:

$$E_{d,c} = E_c + \eta_c$$

$$= 0.00 + \frac{0.059}{2}\log(1.0 \times 10^{-1})^2 - 1.04 = -1.10 \text{ V}$$

for cobalt:

$E_{d,c} = -0.31$ V, as in part (a)

The exceptionally high kinetic overpotential of hydrogen on mercury is of importance for analytical chemistry. The mercury cathode allows the separation of many metal ions from aqueous solution that cannot be plated on a platinum electrode. Further, the wide applicability of the dropping mercury electrode in polarography (Chapter 29) stems in part from the high overpotential of hydrogen on mercury.

It is important to realize that the kinetic overvoltage can never be utilized to supply electrical energy; it acts only as a hindrance to the electrolysis current and does not contribute to the back emf. If a vacuum-tube voltmeter (or a potentiometer) is connected across the electrodes of an electrolysis cell during an electrolysis, the applied voltage will be indicated on the meter. If the external voltage source is now disconnected from the electrodes, the indicated voltage will immediately decrease by the sum of the kinetic overvoltage and *IR* drop. The remaining voltage will be a composite of the reversible back emf and the concentration overvoltage. However, the concentration overvoltage will quickly vanish as the concentration gradients around the electrodes are eliminated by diffusion and only the reversible back emf will be indicated on the meter.

Estimation of Applied Voltage Needed for Electrolysis 28.7

In order to estimate the applied voltage necessary for an electrolysis, the following parameters need chiefly to be considered: the concentrations of all species involved in the electrode reactions, the *IR* drop across the cell, the temperature, concentration overvoltage, kinetic overvoltage, pH, and the presence of complexing agents. The influences of these parameters, with the exception of concentration overvoltage, can be evaluated with an adequate degree of reliability by calculation, or may be taken from tabulations. Estimation of the concentration overvoltage, however, is difficult because of the unknown concentrations of the species at the electrode surfaces. For this reason concentration overvoltage is commonly omitted in the estimation of the applied voltage, hopefully assuming the vigorous stirring during the electrolysis and possibly warming of the solution will decrease it to an insignificant value. The value for the applied voltage thus obtained, although not exact, is still a valuable guide when establishing the actual electrolysis conditions. From the considerations and formulas in the preceding sections the equation employed in the calculation is

$$\begin{aligned} E_{\text{appl}} &= E_d + IR + (E_{\text{concn. overv.}}) \\ &= E_a - E_c + \eta_a - \eta_c + IR + (E_{\text{concn. overv.}}) \end{aligned} \qquad (28\text{-}16)$$

Some examples will illustrate the use of these relationships.

Example 28-2. What applied voltage is needed to electrolyze 5.0×10^{-3} F H_2SO_4 between bright platinum electrodes at a current of 2.0 A? The cell resistance is 0.20 Ω and the electrolyte is well stirred. Assume that η_c for H_2 is -0.44 V, η_a for O_2 is $+0.40$ V, and $p_{H_2} = p_{O_2} = 1$ atm.

Substitution of the data in equation (28-16) yields

$$\begin{aligned} E_{\text{appl}} &= 1.23 + \frac{0.059}{2}\log(1.0 \times 10^{-2})^2 - 0.00 \\ &\quad - \frac{0.059}{2}\log(1.0 \times 10^{-2})^2 + 0.40 - (-0.44) + 2.0 \times 0.20 \\ &= +2.47 \simeq 2.5 \text{ V} \end{aligned}$$

Example 28-3. What applied voltage is needed to plate copper from a solution 1.0×10^{-1} M in Cu^{2+} and 1.0×10^{-1} M in H^+ between bright platinum electrodes at a current of 2.0 A? The cell resistance is 0.30 Ω. Assume that η_a for O_2 is $+0.40$ V and $p_{O_2} = 1$ atm.

Substitution in equation (28-16) yields

$$\begin{aligned} E_{\text{appl}} &= +1.23 + \frac{0.059}{2}\log(1.0 \times 10^{-1})^2 - 0.34 \\ &\quad - \frac{0.059}{2}\log(1.0 \times 10^{-1}) + 0.40 - 0 + 2.0 \times 0.30 \\ &= +1.86 \simeq 1.9 \text{ V} \end{aligned}$$

28.8 Formal Potentials

The concentration term for a metal ion in the Nernst equation relates to the free metal ion, that is, to the aquo complex. However, in many practical cases the ion is not present as such and then erroneous potential data are obtained from calculations employing the unmodified Nernst equation. For example, the metal ion may be partially hydrolyzed, that is, it forms a hydroxo complex, or may form complexes with a complexing agent present in the solution. Similar reasoning holds for nonmetallic species. The changes in potential caused by complex formation can be substantial, as the following example illustrates.

Example 28-4. A solution 1.0×10^{-1} *M* in Cu^{2+} and buffered to pH 5 is electrolyzed. Calculate the cathodic decomposition potential of copper with and without the solution being made 1.0 *F* in EDTA. The conditional stability constant of the Cu–EDTA complex at pH 5 is 2.0×10^{12}.

Since copper shows no kinetic overpotential, the decomposition potential equals the electrode potential and in the absence of EDTA is given by

$$E_{d,c} = E_c = +0.34 + \frac{0.059}{2} \log(1.0 \times 10^{-1}) = +0.31 \text{ V}$$

With EDTA present the calculation takes the following form. The conditional stability constant of the copper(II)–EDTA complex after equation (20-11) is given by

$$K_{\text{st}}^{H} = \frac{[\text{CuY}^{2-}]}{[\text{Cu}^{2+}][\text{Y}]^*} = 2.0 \times 10^{12}$$

where $[Y]^*$ stands for the molar concentration of the EDTA uncombined with the copper and in whatever stage of protonation it persists.

Substitution of the numerical data yields

$$\frac{0.10 - [\text{Cu}^{2+}]}{[\text{Cu}^{2+}](1.0 - 0.1)} = 2.0 \times 10^{12}$$

Neglecting the term $[Cu^{2+}]$ in the numerator yields

$$[\text{Cu}^{2+}] = \frac{0.10}{0.9 \times 2.0 \times 10^{12}} = 5.6 \times 10^{-14}\ M$$

Insertion into the Nernst equation gives the decomposition potential

$$E_{d,c} = +0.34 + \frac{0.059}{2} \log(5.6 \times 10^{-14}) = -0.05 \text{ V}$$

Thus the addition of the complexing agent lowers the cathodic decomposition potential by 0.36 V!

Example 28-4 represents a relatively simple case. All data necessary (that is, stability constant of the complex, pH of the solution, and concentration of the complexing agent) were known, and the processes taking place are simple and known unambigously. So favorable a situation, however, seldom prevails in practice. Even if all data required are available, the calculations can often be quite involved. Frequently rather highly concentrated solutions are encountered for which consideration of activity coefficients would be required if any degree of reliability were expected from the calculated data. Quite often, however, the complex species and equilibria involved are unknown or exact data about them are unavailable. In such cases an escape from the difficulties is possible by using the formal potential, E^f, of the redox couple. Instead of the molar concentrations of the various species involved the formal concentrations are used. Consider, for example, the tin(IV)/tin(II) half-reaction in a hydrochloric acid medium. The tin(IV) is definitely not present as the species Sn^{4+}. It is known that tin(IV) shows a strong tendency to hydrolyze and also forms chloro complexes of the composition $SnCl_n^{4-n}$, where n may be any and all numbers between 1 and 6, depending on the chloride ion concentration. To make matters even worse polynuclear species can also exist in such a solution. Tin(II) is capable of similar behavior, although to a lesser extent. To bypass the difficulties arising from this complicated situation and to allow possibilities for calculations useful to practical applications, the various forms in which tin(IV) and tin(II) may be present are neglected and the formal concentrations of the metal in the two oxidation states are used. The Nernst equation is written

$$E = E^f + \frac{0.059}{2} \log \frac{C_{Sn(IV)}}{C_{Sn(II)}} \qquad (28\text{-}17)$$

It must be realized that the formal potential is strongly dependent on the solution conditions, including the acidity or basicity of the solution, the concentration of the complexing agent, and the temperature. The value of a formal potential must therefore be given together with this relevant information and, if used for any calculations, will only produce sensible results if applied to a case corresponding to identical or at least closely similar conditions. For example, the formal potential for the tin(IV)/tin(I) couple may be given as $E^f = +0.14$ V (25°C, 1 F HCl). Note that the standard potential of this couple is -0.140 V. The student should note the fine difference in writing Sn(IV) and Sn^{4+}, a point he may have either missed so far or not appreciated fully. Calculations involving formal potentials are not different from that with standard potentials and a single example will suffice to make the point.

Example 28-5. What will be the potential of a half-cell obtained by dissolving 7.9 g of tin(IV) chloride pentahydrate (*350.6*) in 150 ml of a 0.010 F solution of tin(II) in 1 F hydrochloric acid?

$$E^{f}_{Sn(IV)/Sn(II)} = +0.14 \text{ V } (25°\text{C}, 1\ F\ \text{HCl})$$

In consolidated form the calculation is

$$E = +0.14 + \frac{0.059}{2} \log \frac{(7.9 \times 10^3)/350.6}{150 \times 0.010}$$

$$= +0.14 + \frac{0.059}{2} \log 15 = 0.19 \text{ V}$$

The advantages of the concept of formal potential are quite obvious; unfortunately many data must be established and tabulated because many values are needed for each system.

Questions 28.9

Does the decomposition voltage depend on concentration overpotential?

In your own words distinguish concentration overpotential and kinetic overpotential.

For an electrolysis cell, define the terms anode, cathode, reversible back emf, IR drop, and applied voltage.

When does the back emf arise if two identical platinum electrodes are placed into the electrolyte of an electrolysis cell?

What is a polarized electrode? Are the electrodes polarized when electrolysis is in progress?

Does any electrolysis occur below the decomposition potential?

In what types of systems should the presence of high kinetic overvoltage be suspected?

Problems 28.10

Two electrolysis cells, A and B, are filled with a solution 1.0×10^{-3} F in $AgNO_3$ and 1.0×10^{-1} F in HNO_3. Cell A has a silver cathode and a platinum anode; cell B has a silver anode and a platinum cathode. The following data are given:

$$\tfrac{1}{2}O_2 + 2H^+ + 2e \rightleftharpoons H_2O \qquad E^0 = +1.23 \text{ V}$$

$$2H^+ + 2e \rightleftharpoons H_2O \qquad E^0 = 0.00 \text{ V}$$

$$Ag^+ + e \rightleftharpoons Ag \qquad E^0 = +0.800 \text{ V}$$

The anodic oxygen overpotential is +0.40 V; the cathodic hydrogen overpotential is −0.09 V on platinum and −0.46 V on silver. Assume $p_{O_2} = p_{H_2} = 1$ atm. (a) Write the equation for the reactions at the anode and cathode of the two cells using a single arrow to indicate the direction. (b) Calculate the decomposition voltage for each cell. (c) If the electrolysis in the two cells is performed at an applied voltage of 1.00 V and the resistance of each cell is 2.0 Ω, estimate the cell current in each cell. Neglect concentration overpotentials.

Answers: (a) Cell A: $H_2O \rightarrow 2H^+ + \frac{1}{2}O_2 + 2e$ at anode, $Ag^+ + e \rightarrow Ag$ at cathode; cell B: $Ag \rightarrow Ag^+ + e$ at anode, $Ag^+ + e \rightarrow Ag$ at cathode. (b) For cell A, $E_d = +0.95$ V; for cell B, $E_d = 0.00$ V. (c) For cell A, $I = 0.025$ A; for cell B, $I = 0.50$ A.

28-2. Portions of a solution 1.0 F in $CuSO_4$ and 0.50 F in H_2SO_4 are placed in each of four different electrolysis cells, A through D. Cell A has two platinum electrodes, cell B has two copper electrodes, cell C has a copper anode and a platinum cathode, and cell D has a platinum anode and a copper cathode. Write the equations for the actual electrochemical reactions at the cathode and anode of each cell using a single arrow to indicate the direction. Calculate the decomposition voltage of each cell. Assume for simplicity that no kinetic overvoltage exists and that $p_{O_2} = p_{H_2} = 1$ atm. E^0 for copper(II) is $+0.34$ V and for oxygen in acidic solution $+1.23$ V.

Answer: E_d for the four cells is $+0.89$, 0.00, 0.00 and $+0.89$ V, respectively

28-3. Calculate the voltage necessary to electrolyze 0.50 F H_2SO_4 using a copper cathode and a platinum anode if the electrolysis cell has a resistance of 1.0 Ω and it is desired to pass a current of 0.10 A through the cell. The anodic overpotential of oxygen on platinum is $+0.40$ V and the cathodic overpotential of hydrogen on copper is -0.82 V. Assume $p_{H_2} = p_{O_2} = 1$ atm. E^0 for oxygen in acidic solution is $+1.23$ V.

Answer: $E_{appl} = +2.55$ V

28-4. Calculate the decomposition potential for the formation of hydrogen from a solution of pH 10.0 when the cathodic kinetic overpotential is -0.40 V.

28-5. Two electrolysis cells are filled with a solution 1.5×10^{-2} F in $AgNO_3$ and 1.0×10^{-2} F in HNO_3. Cell A has a silver cathode and a platinum anode and cell B has a silver anode and a platinum cathode. The overpotential for H_2 is -0.09 V on platinum and -0.46 V on silver. (a) Calculate the decomposition voltage for each cell. (b) Write the electrode reactions in the direction in which they occur. $E^0_{Ag^+/Ag} = +0.799$ V, $E^0_{O_2/H_2O} = 1.229$ V.

28-6. Two electrolysis cells are filled with a solution 1.00 F in $CuSO_4$ and 0.05 F in H_2SO_4. Cell A has a copper anode and a platinum cathode; cell B has a platinum anode and a copper cathode. (a) Calculate the decomposition voltage for each cell if the O_2 overpotential is $+0.40$ V on either metal and the H_2 overpotential is -0.09 V on platinum and -0.60 V on copper. (b) Write the electrode reactions in the direction in which they occur. $E^0_{Cu^{2+}/Cu} = 0.337$ V, $E^0_{O_2/H_2O} = +1.229$ V.

•. Two electrolysis cells are filled with a solution 1.0×10^{-3} F in $CuSO_4$ and 1.0×10^{-1} F in H_2SO_4. Cell A has a copper anode and a platinum cathode; cell B has a platinum anode and a copper cathode. The overpotential for hydrogen is -0.09 V on platinum and -0.60 V on copper. The overpotential of oxygen is $+0.40$ V on either metal. (a) Calculate the decomposition voltage for each cell. (b) Write the electrode reactions in the direction in which they occur. $E^0_{Cu^{2+}/Cu} =$ $+0.337$ V, $E^0_{O_2/H_2O} = +1.229$ V.

29 POLAROGRAPHY AND AMPEROMETRIC TITRATIONS

Introduction 29.1

The relation between the voltage applied to an electrolysis cell and the current flowing through the cell was considered in Chapter 28 and presented graphically in Figure 28-2. As can be seen from that figure, the current increases sharply once the decomposition potential of the substance reacting is reached and later, because concentration overpotential becomes established, the voltage–current curve levels to a plateau. The current at the plateau is called the limiting current and is chiefly controlled by the rate at which the relevant reacting species reach the electrodes. A further increase in the applied voltage does not increase the current significantly until the decomposition potential of some other substance is reached, eventually that of water.

Assume that the electrolysis cell contains a cathode that is quite small in comparison to the anode. In such a case the concentration overpotential at the anode is negligible and the current is controlled by the cathodic concentration overpotential, which in turn is a function of the rate at which reducible species are transported to the cathode. If the small electrode were made the anode, the current would flow in the opposite direction and would be controlled by the transport of oxidizable species.

If in such a cell the solution is not stirred mechanically, the mechanisms that transport ions toward the electrode include diffusion, electromigration (see Section 29.4), and convection.† If migration and convection are made negligibly small, diffusion is the only effective mechanism by which reducible species are transported to the cathode. Under this condition, since the rate of diffusion is proportional to the concentration of the diffusing species, it should be possible to establish a

†As an electrode reaction occurs, the concentration of the solution in the region of the electrode changes and consequently the density of the solution around the electrode. A density gradient becomes established and results in convection, which tends to stir the solution. The term "convection," here and elsewhere, is restricted to mean "natural" convection. Forced convection is termed stirring, that is, agitation of the solution by a glass rod, magnetic stirring bar, shaking, etc.

relationship between limiting current and concentration, and to use this relation as the basis for an analytical determination.

29.2 Polarography

Early attempts to utilize the principles outlined above for analytical purposes failed. With the solid electrodes employed, distorted curves were obtained, which in addition were not reproducible because the shape strongly depended on the history, that is, the pretreatment, of the electrode. In 1922, however, Heyrovský succeeded in securing reproducible results by employing a small electrode consisting of mercury drops formed at the immersed tip of a fine capillary. He named the technique polarography because it is based on "polarization" phenomena.

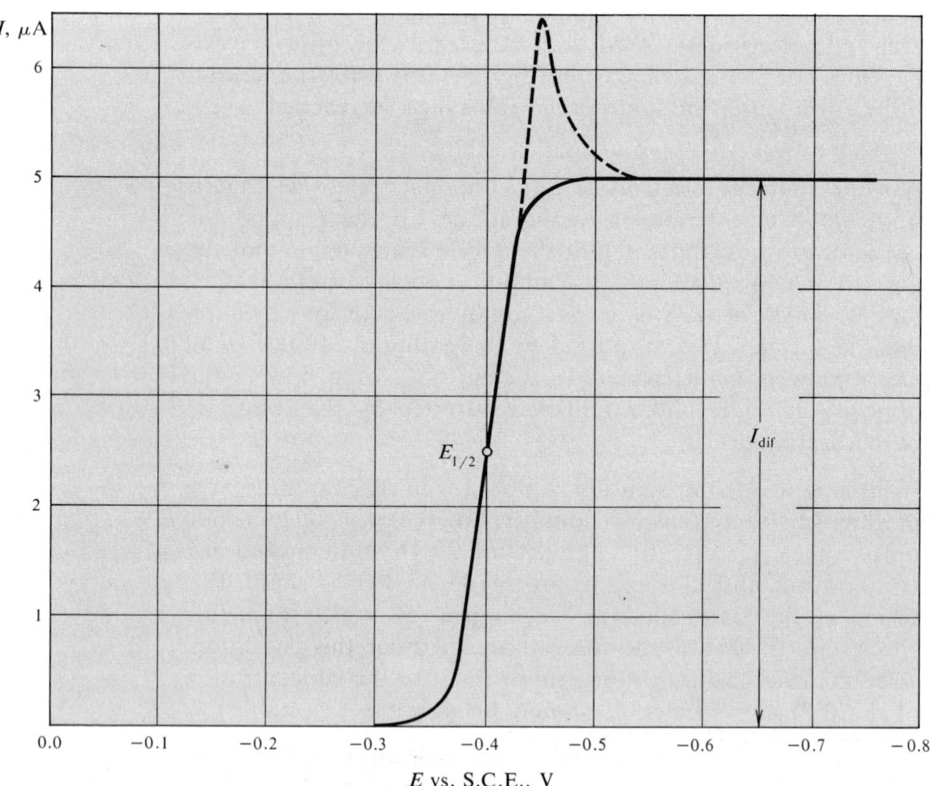

Figure 29-1. Typical plot of current (in microamperes) versus applied potential (in volts versus S.C.E.) for a single polarographic wave. (Polarographic maximum shown as dashed line.)

If to a cell consisting of a dropping mercury electrode and an unpolarizable half-cell, a steadily increasing voltage is applied and the current corresponding to each momentary voltage value is measured and plotted versus the applied voltage, a graph is obtained that is called a polarogram (Figure 29-1). From such a polarogram it is possible to make qualitative deductions as to the nature of the species reacting and from the relationship between the current and concentration, quantitative results can be obtained. Polarography is especially valuable for the determination of small amounts of reducible species and operates best at concentrations below about 10^{-2} *M*.

Polarograph 29.3

The voltage–current measurements resulting in a polarogram are made with an apparatus called a polarograph, which may either be operated manually or be a recording instrument that directly plots the desired curve. Instruments vary greatly in complexity, but the principles on which their operation is based are identical. The arrangements for a manual polarograph are shown in Figure 29-2. The polarographic cell, PC, is a vessel containing the sample solution and is fitted with a stopper containing holes to provide for the insertion of the following items: the inlet, N, for passing nitrogen gas through the solution to remove any dissolved oxygen, which commonly interferes with the measurements (see below); the tube, E, as an exit for the nitrogen; the capillary tube, CT; and the salt bridge, SB, connecting the polarographic cell to the saturated calomel electrode, abbreviated S.C.E., which is the unpolarizable large electrode commonly employed. The capillary tube, CT, is connected with the mercury-containing reservoir, MR, via the flexible tubing, FT, of rubber or plastic. A wire dipping into the mercury makes electrical connection between the mercury electrode and the rest of the circuit. The working battery, WB, is connected via a variable resistance, VR, to the slide-wire resistance, SW. Thus the voltage applied across the slide wire can be regulated by the variable resistor and is measured by the voltmeter, V. If the voltage across the slide wire is known and the wire is calibrated, the voltage taken across the contact points *a* and *b* and applied to the cell assembly is known. The current flowing through the cell assembly is measured by the galvanometer, G, which may be calibrated in microamperes.

29.3 POLAROGRAPH

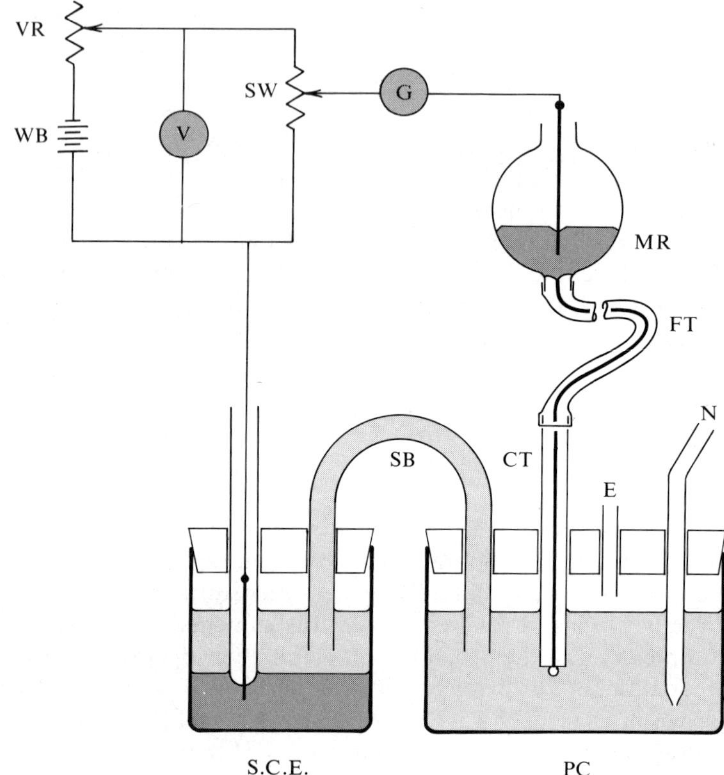

Figure 29-2. Circuit for manually operated polarograph (WB, working battery; VR, variable resistance; V, voltmeter; SW, uniform slide-wire resistance; G, galvanometer; MR, mercury reservoir; FT, flexible tubing; CT, capillary tubing; PC, polarographic cell; S.C.E., saturated calomel electrode; SB, salt bridge; N, nitrogen inlet; E, nitrogen outlet.

The establishment of a polarogram then proceeds as follows. The sample solution is placed into the polarographic vessel, which is then attached to the stopper. Nitrogen is passed through the solution until oxygen is swept out. Then the nitrogen flow is stopped and a certain voltage is applied to the cell assembly by setting the tap on the slide wire at the appropriate position. The voltage is plotted versus the corresponding current reading taken on the galvanometer. Then the potentiometer is set to a different voltage value, and the current is again read and plotted. This process is continued until enough points have been obtained within the desired voltage range to draw the curve.

Polarogram and Polarographic Wave 29.4

The graph resulting from plotting the voltage–current measurements as mentioned above is called a polarogram. Since commonly the S.C.E. is employed as the unpolarizable electrode it is established practice in polarography to express all potential values in volts versus this electrode rather than the standard hydrogen electrode. A typical polarogram is shown in Figure 29-3. It can be seen that increasingly negative potential versus

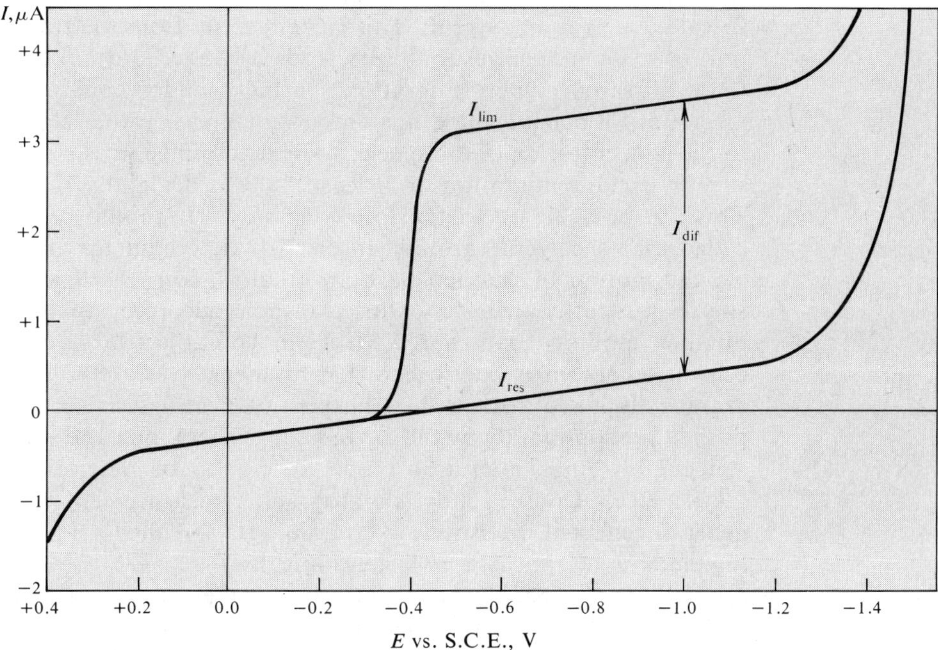

Figure 29-3. Polarographic wave showing typical relationship among residual current, diffusion current, and limiting current.

S.C.E. is plotted to the right on the abscissa. The reason for this practice is that the number of reducible species increases as the cathode potential is made more negative.

The ascent in the curve is called a polarographic wave or step. The height from the base line (zero current) to the plateau corresponds to the limiting current. The limiting current under normal polarographic conditions is the sum of the residual current and the diffusion current. The residual current I_{res} must therefore be subtracted from the limiting current to obtain the desired diffusion current I_{dif}. Either, as shown in Figure 29-3, the line preceding the wave is extrapolated or the residual current is established by "running" a polarogram of the solution without the reducible species present, in other words, by establishing a polarographic "blank." Full explanation of the origin of the residual current is beyond the scope of this textbook. It suffices to mention that it results from the capacitive charging of the growing mercury drop and partially from the presence of polarographically active impurities in the solution.

Where the reducible species is present alone in the solution the limiting current also contains a further component, the so-called migration current. This current arises from the transport of reducible species to the electrode by a mechanism other than diffusion, namely migration, partially under coulombic attraction. Since it is the diffusion current that is to be related to the concentration of the species to be determined, it is necessary to exclude migration or at least make it negligibly small. This can be achieved in the following way. It should be recalled that a current through an electrolytic conductor flows via the motion of charged particles, that is, ions. All ionic species present contribute to this transport according to their number, mobility, and charge. If in the solution a large concentration of ions is established that by themselves are polarographically inactive, then the transport of electricity is accomplished predominantly by the movement of these ions and only a negligibly small migration of the species to be determined takes place. Consequently, polarographic measurements are generally effected in a solution containing a so-called supporting electrolyte, that is, a polarographically inactive salt, often potassium chloride. The concentration of the supporting electrolyte is at least in the order of 0.1 to 1 F, which is large in comparison to the small concentration of the substances to be determined, usually in the order of 10^{-3} to 10^{-4} F. It should be realized that the supporting electrolyte should not undergo an electrode reaction, at least not at a potential that is more positive than that at which the species to be determined reacts.

The supporting electrolyte has the additional function of keeping the resistance of the cell solution low (commonly 100 to 500 Ω) and thereby minimizing the IR drop across the cell. With solutions too high in electrical resistance, drawn-out, distorted polarographic curves are obtained.

The potential–current relation of a polarographic wave is given by

$$E = E_{1/2} + \frac{0.059}{n} \log \frac{I_{\text{dif}} - I}{I} \qquad (29\text{-}1)$$

where I_{dif} is the diffusion current and I the current at any point of the polarographic wave. The half-wave potential, $E_{1/2}$, represents the inflection point of the curve and is the potential where $I = \frac{1}{2}I_{dif}$. It is a characteristic of the species being reduced and independent of the concentration of that species. Equation (29-1) applies only to reversible electrode reactions and how closely a system follows the equation can be taken as a measure of the degree of reversibility. Furthermore, this equation can only be applied to simple systems, for example, where all the reactants and products are soluble and where none of the concentration terms in the Nernst equation is of second power or higher.

The Ilkovič Equation 29.5

The relationship between the diffusion current and parameters pertaining to the system under consideration has been derived by Ilkovič and the resulting equation is named after him:

$$I_{dif} = 607nCD^{1/2}m^{2/3}t^{1/6} \qquad \text{(at 25°C)} \qquad (29\text{-}2)$$

Here I_{dif} is the average diffusion current in microamperes during the life of a drop, n the number of electrons transferred in the electrode reaction, C the concentration of the reacting species in millimoles per liter, and D the diffusion coefficient of the reacting species in square centimeters per second. The remaining two factors are characteristics of the capillary, m being the milligrams of mercury flowing from the capillary per second and t the drop time in seconds. The numerical factor 607 derives from some constants and conversion factors. From the equation it can be deduced that the diffusion current is proportional to the concentration of the reacting species if the parameters pertaining to the electrode are kept constant.

Multiple Polarographic Waves 29.6

When more than one reducible species is present in the solution, the resulting polarogram consists of successive waves due to the reduction of each species in turn. However, well-separated waves are only obtained if the half-wave potentials of the various species differ by at least 0.15 V. Such a situation is illustrated in Figure 29-4. Points A and B are the half-wave potentials of the two reducible substances. $I_{dif,A}$ and $I_{dif,B}$ are the corresponding diffusion currents. To apply the equation of the wave I must be measured from the base line of each individual wave, as is done for $I_{dif,B}$. The dashed line indicates

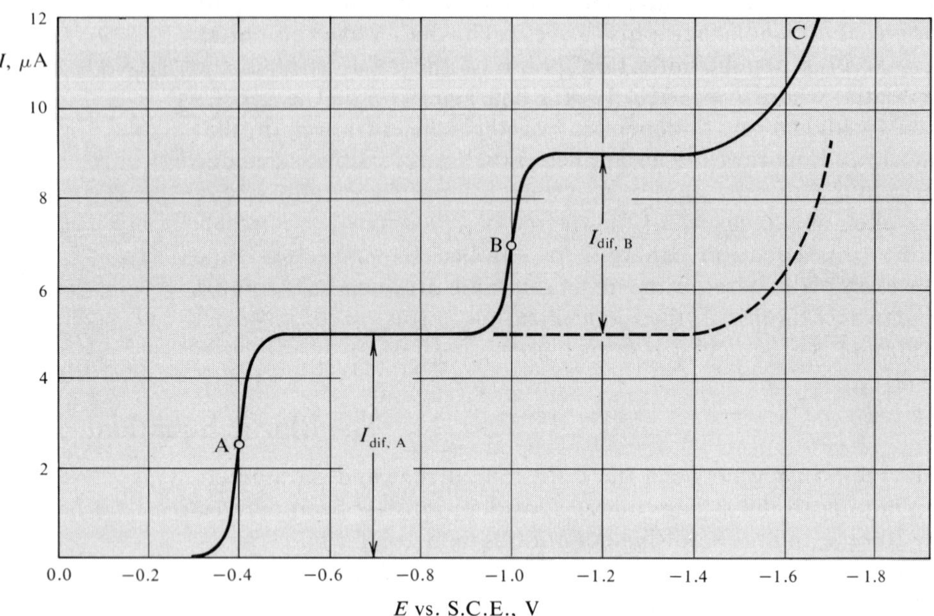

Figure 29-4. Typical plot of current (in microamperes) versus applied potential (in volts versus S.C.E.) for multiple polarographic waves.

the shape of the polarogram that would result if the substance *B* were absent. The increased current at point *C* corresponds to the reduction of hydrogen ion and represents the beginning of the hydrogen wave.

29.7 Dropping Mercury Electrode

The dropping mercury electrode, often abbreviated d.m.e., is a short length of fine-bore (~0.04 mm inside diameter) glass capillary tubing, connected by flexible tubing to a reservoir of mercury, which can be raised or lowered to adjust the mercury pressure. Each drop, when it reaches a certain size, falls from the capillary tip. The rate at which drops form and fall is determined by the capillary bore and the mercury pressure and is usually adjusted to one drop every 3 to 7 seconds.

The effective electrode, which is the mercury drop, is continually renewed; therefore, its surface does not become fouled by reaction products as in the case of a fixed electrode. Since all drops form identically and thus have the same history, reproducibility is achieved. The high kinetic overpotential of hydrogen on mercury allows the reduction of many metal ions that otherwise could not be reduced under the prevailing conditions. Many metal ions are reduced to the metal, which dissolves in the mercury; this removal leads to a less negative half-wave potential than would otherwise be expected. Since the solution is stirred by the fall of the drops, each new drop comes into contact with fresh solution, and changes in the solution density are avoided. Consequently, convection is minimized as a transport mechanism.

One disadvantage of the dropping mercury electrode is the fluctuation in the current with the growth and fall of the drops. The extent of concentration overpotential, and hence of the current, is related to the electrode area. As the drop grows, this area increases and then suddenly becomes quite small as the drop falls; an oscillating current results. The usual technique is to record the average current during the life of the drop.

The potential region over which the dropping mercury electrode can be used in an aqueous solution extends from about +0.4 to −2 V versus S.C.E. The positive limit is determined by the potential at which the mercury of the electrode is oxidized anodically. This value may be less positive than +0.4 V if the solution contains an agent that forms either a soluble complex or a precipitate with the ions of mercury. The negative limit, as previously considered, is determined by the potential at which hydrogen ion or a component of the supporting electrolyte is reduced.

If it is desired to study the potential–current relation of a solution at a potential more positive than +0.4 V versus S.C.E., it becomes necessary to use a noble metal, such as platinum, as the electrode, but then the difficulties mentioned in Section 29.2 are encountered. The situation can be improved somewhat by employing a small platinum electrode rotating at high speed (600 rpm).

Polarographic Maxima 29.8

In many cases, an extremely large, erratic current, known as a polarographic maximum, is observed at the potential where a wave starts to level off (see Figure 29-1). The maximum interferes in the measurement of the half-wave potential and the wave height but may be avoided by the addition of a so-called maximum suppressor to the electrolyte solution. Gelatin (about 0.005% in the electrolyte) and surface-active agents (e.g., dyes and detergents) are used for this purpose.

29.9 Applied Polarography

As was pointed out, the half-wave potential is a characteristic of a reacting species and, under given conditions, is independent of the concentration of a substance. Consequently, its value may be determined for an unknown substance and the substance then be identified by comparing the experimentally established value with tabulated values.

When applying polarography to quantitative analysis, one might proceed as follows. The parameters of the electrode are experimentally determined and then diffusion current of the species is measured. These data and the value for the diffusion coefficient taken from tables are employed together with the Ilkovič equation to calculate the concentration of the sought-for species. Generally, a result, thus obtained, has an intolerably large error due to the accumulation of small errors in the individual factors. In practice, the factors are combined into a single constant, k, so that $I_{dif} = kC$. The value of k is determined under the actual, experimental conditions prevailing from the diffusion current of a solution of exactly known concentration. This value is then used to calculate the concentration of an unknown from its diffusion current measured under identical conditions. Alternatively, a calibration curve of diffusion current versus concentration may be established employing a number of solutions of known concentrations. Another technique is the so-called method of standard addition, which is best understood from an example.

Example 29-1. Exactly 100 ml of a lead(II) solution (brought to the correct pH and with an appropriate supporting electrolyte added) is introduced into a polarographic cell. Oxygen is removed by bubbling nitrogen gas. The limiting current measured at −0.60 V versus S.C.E. is found to be 12.5 S.D. (scale divisions). (Calibration of the meter in amperes is unnecessary.) Then a volume of exactly 10 ml of 0.0100 F $Pb(NO_3)_2$ is added. After passage of nitrogen gas and establishment of the original temperature, a current of 16.8 S.D. is observed at the same potential. Under identical conditions the supporting electrolyte alone yields a current of 0.5 S.D. Calculate the concentration of lead in the original solution.

Correction for the residual current gives 12.5 − 0.5 = 12.0 S.D. and 16.8 − 0.5 = 16.3 S.D. for the two diffusion currents. Let x = molar concentration of Pb^{2+} in the original solution. Then

$$12.0 = kx$$

After the standard addition

$$16.3 = k\,\frac{100x + 10 \times 0.0100}{110}$$

On the elimination of k by division of one equation by the other, the result is

$$\frac{12.0}{16.3} = \frac{110x}{100x + 0.100}$$

$$x = 2.02 \times 10^{-3}\ M\ Pb^{2+}$$

Since diffusion currents change by 1 to 2% per degree centigrade, it is important to keep the temperature constant during a series of related polarographic measurements.

A few practical examples taken randomly may illustrate the diverse possibilities offered by polarography. In a supporting electrolyte 1 F in potassium chloride, cadmium, lead and zinc have their half-wave potentials at −0.64, −0.44, and −1.00 V versus S.C.E., respectively. In the same medium copper exhibits two waves, one at +0.04 and another at −0.22 V versus S.C.E. The first wave corresponds to the reduction Cu(II) → Cu(I) and the second to the reaction Cu(I) → Cu. The waves of the four elements are sufficiently separated to allow their consecutive determination provided the concentration ratios are not excessively large. Since aluminum is polarographically inactive, the four elements mentioned above can be determined polarographically when present as impurities or minor constituents in pure aluminum metal or aluminum alloys. The large separation of the half-wave potential of cadmium and zinc permits the simple determination of cadmium impurities in pure zinc metal or zinc salts. In these materials it is also possible to determine lead and copper. In an ammonia–ammonium chloride medium cobalt can be oxidized to cobalt(III), which then gives waves at −0.28 and −1.30 V versus S.C.E. [Co(III) → Co(II) → Co]. Since in this medium nickel gives a wave at about −1.1 V, the first cobalt wave may be employed to determine cobalt in trace amounts in nickel metal or nickel salts, thus solving an otherwise very difficult analytical problem in a rather simple fashion. Unless special conditions are established, the alkali metals, the alkaline earths, aluminum, and a few other elements do not yield polarographic waves. Consequently, it is possible to determine small amounts of polarographically active metals in the presence of extremely high concentrations of the inactive components. In addition to its application to inorganic analysis, polarography is becoming progressively important for the determination of reducible or oxidizable organic substances.

29.10 Amperometric Titrations

Closely related to polarography is a titration technique known as the amperometric titration, where polarographic arrangements in connection with a graphical procedure are used to detect the end point.

The method may be exemplified by the titration of cadmium ion with EDTA. Consider a solution of a cadmium salt in an ammonium nitrate–ammonia supporting electrolyte of pH 10. A polarogram of this solution is represented by the uppermost line of Figure 29-5a. On addition of some standard

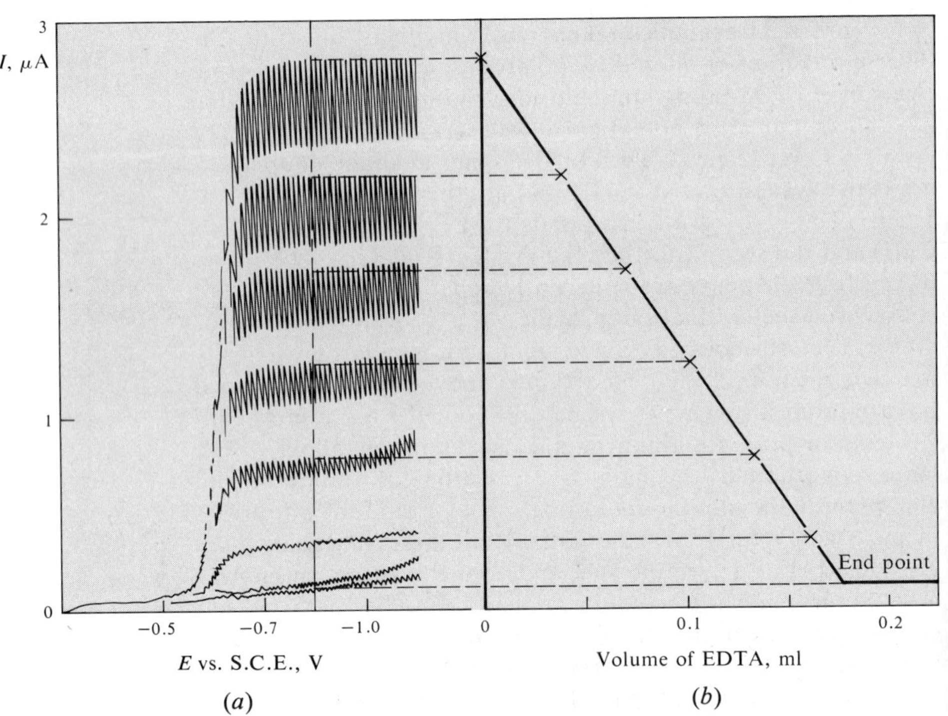

Figure 29-5. Relation of polarograms (a) of cadmium in ammoniacal buffer of pH 10 to the amperometric titration curve (b) for cadmium with EDTA at −0.9 V versus S.C.E.

EDTA solution, the soluble cadmium–EDTA complex is formed. The cadmium ion complexed by the EDTA is reduced at a much more negative potential than that complexed by water or ammonia. Consequently, the diffusion current, which is proportional to the concentration of the "free" cadmium ion, becomes smaller. The polarogram then obtained is also shown in Figure 29-5a. As more EDTA is added, the diffusion current decreases further. At the equivalence point, virtually all the cadmium is complexed by EDTA, the diffusion current is essentially zero, and the polarogram shows only the residual current. On addition of EDTA beyond the equivalence point, no appreciable change is observed in the polarogram. It is not necessary to record a polarogram after each addition of titrant. The current can be read at some point on the plateau of the cadmium wave, for example, at −1.0 V versus S.C.E. The data obtained are plotted versus the volume of titrant to yield the titration curve shown in Figure 29-5b.

Generally, experimental curves will depart from strict linearity since the solution volume increases during the titration. This dilution effect can be corrected for by multiplying each reading by the factor $(V + v)/V$, where V is the initial volume of the solution and v is the volume of titrant added to the point of the observation. A plot of such corrected current versus the volume of titrant will consist of two straight lines. Alternatively, the effect of the dilution may be made negligible by employing a titrant 10 or more times greater in concentration than the species being titrated. Such concentrated titrant solutions are then delivered from a micro buret.

To locate the amperometric end point, only the changes in the current are needed. This makes simplification possible. The drop time and the temperature are unimportant as long as they are constant during a titration. The galvanometer need not be calibrated, and the readings in scale divisions may be plotted. The difficulties encountered in polarography with solid electrodes are without moment, and a rotating platinum electrode may be substituted for the dropping mercury electrode, often to advantage, since a larger diffusion current is obtained and thereby greater sensitivity. The current associated with the species being titrated may be superimposed on waves of substances not responding to the titrant.

The precipitation titration of lead(II) ion with chromate ion to form lead chromate may serve as an example of the effect of half-wave potentials and the selected potential value on the shape of the titration curve obtained. Chromate and lead(II) ions have half-wave potentials of about +0.1 and about −0.4 V versus S.C.E., respectively. Two possibilities for the titration exist: A potential can be selected at which only chromate is reduced (i.e., a value on the plateau the chromate wave, say, −0.2 V versus S.C.E.) or at which both lead and chromate are reduced (i.e., on the plateau of the lead wave, say, −0.6 V versus S.C.E.). Corresponding titration curves at these two potentials are shown in Figure 29-6 as curves A and B, respectively. The shape of curve A derives from the fact that up to

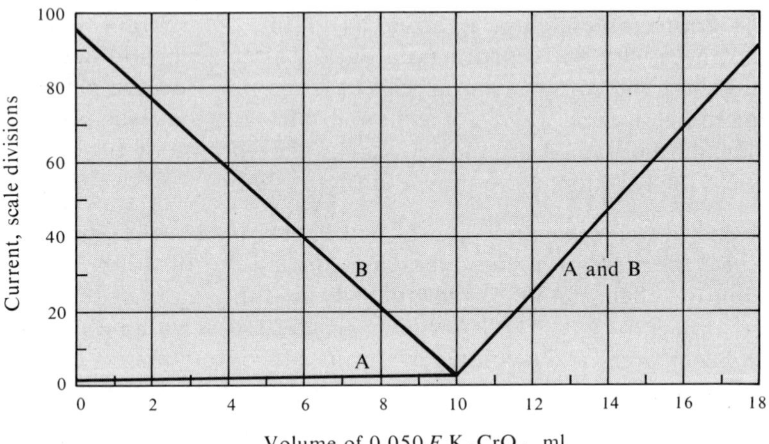

Figure 29-6. Titration curve for the amperometric titration of 50 ml of 0.01 *F* $Pb(NO_3)_2$ with 0.05 *F* K_2CrO_4: curve A at −0.20 V versus S.C.E.; curve B at −0.60 V versus S.C.E.

the equivalence point no species is present that is reducible at the applied potential of −0.2 V versus S.C.E. Chromate, although reducible at this potential, when added reacts with lead to form lead chromate, which precipitates, and is no longer available for an electrode reaction. Beyond the equivalence point, free, reducible chromate is present and causes the flow of a current that increases as the concentration of chromate increases. In the case of curve B, lead is present from the start and is reduced at the applied potential of −0.6 V versus S.C.E. As its concentration decreases via its removal as insoluble lead chromate, the current decreases progressively. Beyond the equivalence point the situation is identical to that leading to curve A. For both curves, the minimum value of the current corresponds to the residual current and a small current caused by the minute amounts of dissolved lead chromate.

29.11 Biamperometric Titrations

Such titrations are performed with the arrangement shown in Figure 29-7. Two identical platinum electrodes dip into the solution to be titrated and are connected with a milliammeter and a potentiometer circuit that allows application of a

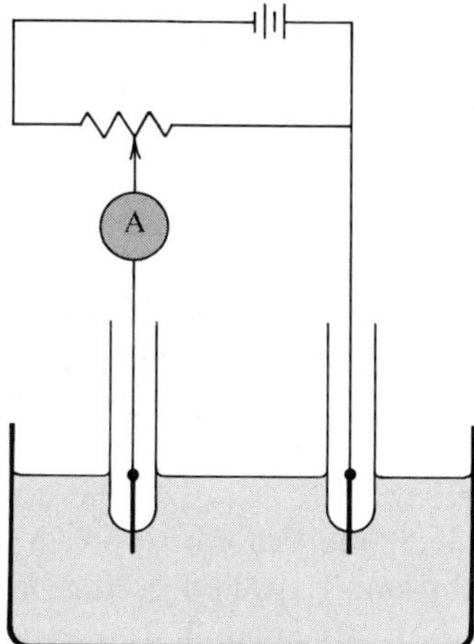

Figure 29-7. Experimental arrangement for titration with two polarized electrodes.

constant d.c. voltage, usually less than 0.5 V. The solution should be stirred well and at a constant rate. The principles involved in such a titration are best illustrated by the discussion of an example.

Consider the titration of a solution containing iodine and an excess of iodide ion with a thiosulfate solution (Figure 29-8). At the start of the titration, iodine (present as the triiodide ion) is reduced at the cathode and iodide is oxidized at the

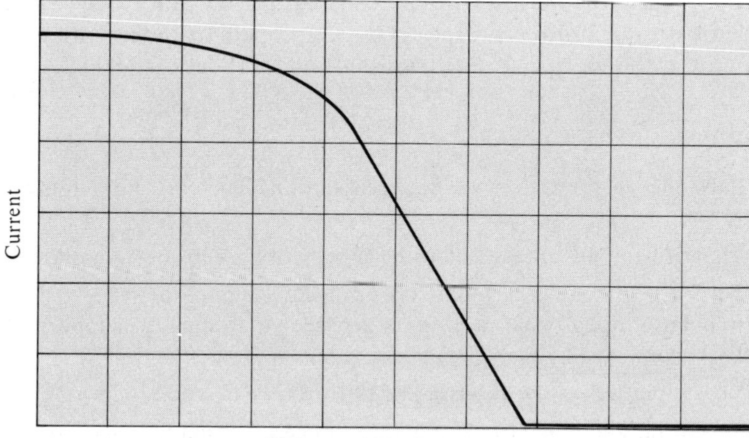

Figure 29-8. General shape of the curve for the titration of iodine in the presence of an excess of iodide with thiosulfate employing two polarizable electrodes.

anode. Since the reactions at the two polarizable electrodes are the reverse of each other, the value of the decomposition voltage is zero, and therefore a current flows even at the small voltage applied. The electrode reactions occur to an equal extent and the composition of the solution remains unchanged. As thiosulfate is added, the iodine concentration decreases. At first, the change in the current is small. However, as the equivalence point is approached, the current decreases linearly with the volume of thiosulfate solution added. At the equivalence point, the concentration of iodine is essentially zero and practically no current flows, since in the absence of iodine no combination of electrode reactions having a decomposition voltage less than the small applied voltage is possible. Beyond the equivalence point, no current flows, since the irreversible thiosulfate–tetrathionate reaction has a sizable kinetic overpotential and therefore the decomposition voltage is larger than the applied voltage.

Because of the peculiar shape of the titration curve which reaches (or in other cases starts from) a "dead stop" at the end point, the method is also known as a dead-stop titration. The technique can be used in a number of titrations in which the following redox couples constitute at least one of the reactant–product pairs: Br_2/Br^-, I_2/I^-, Ti^{4+}/Ti^{3+}, Ce^{4+}/Ce^{3+}, Fe^{3+}/Fe^{2+}, and MnO_4^-/Mn^{2+}. Such titrations offer the advantages of simple circuitry and of a simple electrode system not even requiring a reference electrode. Since only current *changes* are of interest, the meter need not be calibrated. The dead-stop technique is frequently employed to detect the end point in the Karl Fischer titration of water (Section 25.11).

29.12 Questions

29-1. Elaborate on the difference between limiting current and diffusion current.

29-2. What are the advantages of using a mercury electrode in polarography?

29-3. In polarography what are the purposes of adding a supporting electrolyte?

29-4. What is plotted on the abscissa and the ordinate of a polarogram?

29-5. What values obtained from a polarogram can be used for the purpose of (a) quantitative analysis and (b) qualitative analysis?

29-6. What are the possible mechanisms for the transport of electricity in an electrolytic cell? Which of these are undesirable in polarography?

-7. What electrode reaction in aqueous solution limits a polarogram on (a) the negative side and (b) the positive side?

-8. What happens to the current when the dropping mercury electrode changes from being the cathode and becomes the anode?

-9. How much difference in the values of the half-wave potentials is generally necessary so that two polarographic waves are clearly distinguishable?

0. Discuss the advantages and disadvantages of polarographic determinations compared with amperometric titrations.

1. How can the addition of a complexing agent affect the value of a half-wave potential?

2. How does the half-wave potential depend on concentration?

3. Why should oxygen be removed in a polarographic determination and how is it done?

4. Sketch the titration curves when CrO_4^{2-} is titrated amperometrically with Pb^{2+} at (a) -0.1 and (b) -0.6 V versus S.C.E. Compare the curves with those for the reverse titration shown in Figure 29-6.

Problems 29.13

-1. Into a polarographic cell are placed successively exactly 50 ml of an ammoniacal buffer (pH 10), exactly 5 ml of a 0.0100 *M* cadmium(II) solution, and exactly 5 ml of a solution having an unknown cadmium concentration. Following each addition, nitrogen gas is passed to remove oxygen, and the current measured at -1.0 V versus S.C.E. is found to be 1, 18, and 25 arbitrary scale divisions, respectively. Calculate the molar concentration of cadmium in the unknown solution.

Answer: 0.0054 *M*

-2. A 10.0-ml portion of a solution 1.0×10^{-3} *M* in a dipositive metal ion and containing supporting electrolyte is placed in a polarographic cell and deaerated. The polarographic study, in which this cation is reduced to the metal, requires 10 minutes, during which time the average current passed is 8.0 μA. Calculate the percentage decrease in the concentration of this ion in the cell solution ($1\ \text{F} = 9.65 \times 10^4$ C).

-3. A volume of 25.0 ml of a solution containing an unknown amount of lead(II) was polarographed and a diffusion current of 4.5 μA found. A volume of 5.0 ml of 5.0×10^{-3} *F* $Pb(NO_3)_2$ was added to the original solution. The diffusion current was then found to be 9.7 μA. Calculate the concentration of lead in the original solution.

-4. A volume of 50.0 ml of a 2.0×10^{-3} *F* $Cd(NO_3)_2$ solution was polarographed. The polarogram, which took 12 minutes to perform, showed an average current of 9 μA. Calculate the weight of cadmium reduced to the metal and the percentage decrease in concentration.

30 ELECTROGRAVIMETRY AND ELECTROSEPARATION

If a product of an electrolytic reaction is plated (that is, is deposited) quantitatively on an electrode, the amount of that product can be established by weighing the (dry) electrode before and after the electrolysis. This type of quantitative analysis is called electrogravimetric analysis or electrogravimetry.

During the electrolysis the product (or products) is separated from the solvent and other components of the solution. Consequently, electrolysis may be performed not only to determine a constituent of a multicomponent mixture but also with the sole intent of achieving a separation. The technique is then called electroseparation. The principles underlying electrolysis have been discussed in preceding chapters. The present chapter deals with their application to the two techniques mentioned.

Requirements for Electrogravimetric Analysis 30.1

For an electrogravimetric procedure to be feasible, the deposition of the substance of interest must be complete, and the deposit must be of known composition and must adhere firmly so that the electrode can be rinsed, dried, and reweighed without losses. In addition, the electrode must be inert; that is, it must not undergo any change in its weight in the electrolysis process. All these factors have their counterpart in conventional gravimetric determinations (Chapter 6) and need no elaboration. In practice, the obtaining of a pure deposit is frequently complicated by the presence of other substances that may be codeposited with the substance of interest. In these cases the conditions of the electrolysis must be adjusted and controlled so as to secure selectivity.

Completeness of an Electrodeposition 30.2

The Nernst equation can be used to calculate the concentration of undeposited sought-for material in the solution and hence to establish the completeness of the electrodeposition. However, this calculation requires knowledge of the electrode potential of the "working electrode," that is, of the electrode at which the process of interest occurs. The potential of this electrode can be measured by comparing it with that of a third electrode acting as a reference electrode. For this purpose a saturated calomel electrode may be connected with the electrolysis cell by a salt bridge and the voltage of the cell consisting of the cathode and reference half-cells be measured by means of a potentiometer circuit. From this voltage the potential of the cathode can be computed.

Even so, however, a concentration thus calculated is seldom more than a rough approximation because of the concentration overpotential established by the depletion of ions in the vicinity of the working electrode. This overpotential is not due to the potential measurement, which is performed with the potentiometer under zero-current condition, but develops because the cathode is also part of the operating electrolysis cell.

Although the concentration overpotential can be decreased by vigorous stirring of the electrolyte, elevating the temperature, and using large electrodes, the concentration calculated corresponds to that around the electrode, and is always smaller than that in the bulk of the solution.

Example 30-1. In the electrogravimetric determination of copper the reference electrode technique described above is used. The voltage of the cell consisting of a calomel electrode ($E = +0.24$ V) and the cathode of the electrolysis cell is found to be 0.05 V with the calomel electrode as the positive pole. What is the concentration of copper(II) remaining undeposited in the acidic cell solution? If the initial concentration of copper(II) was 1.0×10^{-2} M, how complete is the deposition?

Substitution in equation (25-4) yields

$$E_c = +0.24 - 0.05 = +0.19 \text{ V}$$

From the Nernst equation for the copper(II)–copper couple,

$$+0.19 = +0.34 + \frac{0.059}{2} \log[Cu^{2+}]$$

$$\log[Cu^{2+}] = \frac{(0.19 - 0.34) \times 2}{0.059}$$

and

$$[Cu^{2+}] = 10^{-5}\, M$$

$$\%\ \text{undeposited copper} = \frac{10^{-5} \times 10^{2}}{10^{-2}} = 0.1\%$$

Hence, the deposition is 99.9% complete.

Selectivity of Electrodepositions 30.3

For the purposes of electrogravimetry, only a single substance should be deposited at a time. Factors that determine whether a selective electrodeposition can be effected include the standard electrode potentials of the species present, their concentrations, overpotentials, the efficiency of the stirring, temperature, pH, and the presence and concentration of complexing agents.

Consider the electrogravimetric determination of copper in a solution that is 1.0×10^{-1} *M* in copper(II) ion, 1.0×10^{-1} *M* in antimony(III), and 1.0 *M* in hydrogen ion. Antimony(III) is present as the antimony ion in an acidic solution and has the following electrode reaction:

$$SbO^+ + 2H^+ + 3e \rightleftharpoons Sb + H_2O \qquad E^0 = +0.21\ V$$

The Nernst equation for this half-cell takes the form

$$E_c = +0.21 + \frac{0.059}{3} \log[SbO^+][H^+]^2$$

$$= +0.21 + 0.020 \log 10^{-1} = +0.19\ V$$

The potential of the copper(II)–copper half-cell is given by

$$E_c = +0.34 + \frac{0.059}{2} \log[Cu^{2+}]$$

According to the last equation (substituting $[Cu^{2+}] = 1.0 \times 10^{-1}$), copper starts to deposit on the cathode when the cathode is at +0.31 V. As the electrolysis proceeds, the concentration of copper(II) ion in the cell solution decreases and the cathode must be made more negative (that is, less positive) in order for the electrolysis to continue. When the concentration of copper ion falls to 1.0×10^{-2} *M*, the cathode has to be at least at +0.28 V; similarly, at a copper concentration of 1.0×10^{-4} *M*, the potential has to be at least +0.22 V. These calculations, however, neglect concentration overpotential. At current densities encountered in practice the copper(II) ion concentration in the vicinity of the cathode may be several orders of magnitude lower than that in the bulk of the solution. As a consequence, the cathode potential will be substantially lower than that calculated from the bulk concentration and may even be as low as +0.18 V. This is already 0.01 V lower than the potential calculated above for the antimonyl couple; hence, antimony will start to deposit with the copper. Of course, efficient stirring will decrease the concentration overpotential greatly, but still some overpotential will persist, and thus it will be impossible to deposit all the copper in a pure form.

In contrast, the difference in the values of the standard electrode potentials of copper(II) and nickel(II) is so large that copper can be deposited quantitatively before nickel begins to plate. As a consequence copper can be determined even in the presence of considerable amounts of nickel, as can be shown by the following example.

Example 30-2. Assume an electrolyte solution is $0.10\,F$ in all three solutes: copper(II) sulfate, nickel(II) sulfate, and sulfuric acid. What will be the copper concentration in the solution when nickel is just starting to plate? $E^0_{Cu^{2+}/Cu} = +0.34$ V, $E^0_{Ni^{2+}/Ni} = -0.25$ V.

The potential at which nickel starts to plate is given by

$$E_c = -0.25 + \frac{0.059}{2}\log(1.0 \times 10^{-1})$$

This potential, by the statement of the problem, must be the same as that of the copper(II) half-cell, and the two Nernst equations can be equated:

$$-0.25 + \frac{0.059}{2}\log(1.0 \times 10^{-1}) = +0.34 + \frac{0.059}{2}\log[Cu^{2+}]$$

$$[Cu^{2+}] = \sim 10^{-21}\,M \text{ when Ni begins to plate}$$

It can be appreciated that the greater the difference in the electrode potentials of two metals, the more complete will be the separation that can be achieved by electrolysis. Actually the formal potentials of the redox systems under the prevailing solution conditions are the ones that govern the possibility of an electroseparation. Thus the addition of complexing agents can greatly influence the selectivity (Section 24.10). For example, antimony(III) forms a far more stable tartrate complex in weakly acidic solution than copper(II); the addition of tartrate causes a greater shift to negative potentials for the antimony(III) and thus creates more favorable conditions for the selective deposition of copper in the presence of antimony.

30.4 Techniques of Electrogravimetry

Two principal techniques for electrogravimetry are employed: electrolysis at constant (or controlled) current and electrolysis at constant (or controlled) potential. The former method, which has long been routinely employed, involves adjustment of the *applied voltage* to give a desired current, which is maintained during the electrolysis by occasional manual adjustments. In the latter method, which has received extensive attention only in recent decades, the *potential* of the working electrode is controlled, that is, kept constant at an appropriate value, preferably by an electronic device.

Constant-Current Electrolysis 30.5

The name of this technique stems from the fact that the current is adjusted at the start of the electrolysis and is then maintained roughly at that level during the course of the electrolysis. The usual electrolysis arrangement is shown schematically in Figure 30-1. The mode of its function is already known from

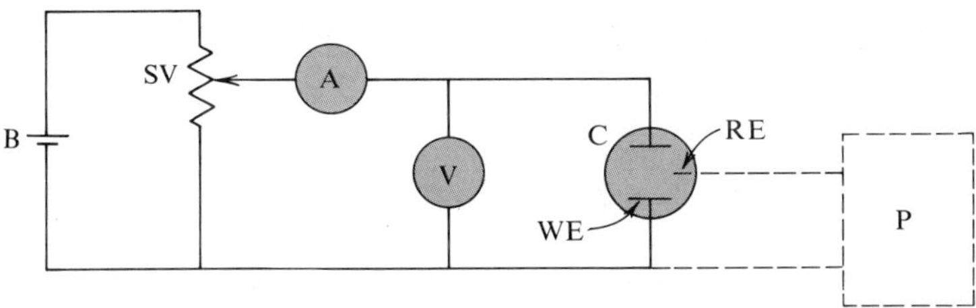

Figure 30-1. Schematic diagram of assembly for controlled-current and controlled-potential electrolysis. A, ammeter; B, battery or other voltage source; C, electrolysis cell; WE, working electrode; SV, variable resistance to adjust applied voltage. Controlled-potential electrodeposition additionally employs the following (dashed lines): RE, reference electrode; P, potentiometer circuit (e.g., that shown in Figure 22-2).

previous discussions. The constant-current technique is widely applied to the routine determination of components in many important alloys and ores. Most metals are deposited cathodically; lead in some cases is deposited anodically as lead(IV) oxide. According to the composition of the mixture and depending on which metal is to be determined, acidic, neutral, or alkaline solutions may be used and complexing agents such as tartrate, cyanide, or EDTA may be added. Constant-current electrolysis has low selectivity; the reason for this shortcoming will become obvious from the discussion of an example, the determination of copper. At the start of the electrolysis the current is adjusted to a certain level. As the electrolysis proceeds, the concentration of copper decreases progressively and with less copper available the current drops. For a while the original level can be restored by increasing the applied voltage. Sooner or later, however, such an increase will no longer be effective in bringing more copper ions to the cathode; then the current level can only be maintained by forcing another electrode reaction to take place. If another reducible metal is present it will plate and the selective determination of the copper will become impossible. If no such metal is present, hydrogen ion will be reduced to hydrogen gas. This gas evolution may seem to present no problem because the hydrogen does not plate and thus cannot affect the weight of the copper deposit. However, the concurrent reduction of copper and hydrogen ions causes a flaky, poorly adherent copper deposit that may lead to losses of copper. Similar considerations hold in other cases and are valid analogously for a deposition on the anode. Of course, the situation can be improved by adequate stirring of the solution, increasing the temperature and lowering the current density via the use of electrodes with large surface areas. But these factors are subject to limitations.

It thus becomes obvious that it is not the current that should be controlled but the potential of the working electrode. This potential, in case of a cathodic deposition, should be as negative as possible, to assure as large an electrolysis current as feasible, but at the same time should be limited to such a value as to exclude the occurrence of an undesired electrode reaction. This limiting of the potential can be achieved by adding a depolarizer (Section 30.6) or by controlling the electrode potential (Section 30.7) rather than the cell current.

30.6 Depolarizers

A depolarizer is a substance that undergoes a reaction at a certain electrode potential and thereby limits the potential of the working electrode. The depolarizer acts as a "potential buffer" and prevents undesired electrode reactions from occurring. For example, the reduction of hydrogen ion to hydrogen gas during the electrodeposition of copper can be avoided by adding nitric acid as the depolarizer. The nitrate ion is reduced to nitrite ion ($NO_3^- + 2H^+ + 2e \rightarrow NO_2^- + H_2O$) or ammonium ion ($NO_3^- + 10H^+ + 8e \rightarrow NH_4^+ + 3H_2O$) at a potential more positive than that at which the discharge of hydrogen occurs. As long as the supply of nitrate ion lasts, the cathode potential is limited by the reduction of nitrate and little or no hydrogen gas forms. Nitrate ion here functions as a cathodic depolarizer.

In a similar manner, hydrazine, which is oxidized to nitrogen in acidic or neutral media, may be used as an anodic depolarizer, for example, in the deposition of lead(IV) oxide, to prevent the evolution of oxygen or the deposition of undesired metal oxides.

Controlled-Potential Electrolysis 30.7

Increased selectivity for an electrodeposition can be obtained by controlling the potential of the working electrode. An experimental arrangement is shown schematically in Figure 30-1. The slide wire is adjusted so that the potential of the working electrode has the appropriate value. The manual adjustment is tedious and in routine practice an electronic controlling device, known as a potentiostat, is employed. This improvement increases markedly both the complexity of the circuitry and the cost of the apparatus.

By careful selection and control of the electrode potential, electrogravimetric determinations can be achieved that are impossible with controlled-current electrolysis. A metal, for example, may be determined in the presence of a second one when the two formal potentials differ by only about 0.2 to 0.3 V. By weighing the (dried) electrode between each change in potential, two or three components of an alloy can be determined consecutively. Changing the pH, adding complexing agents, or both, between steps may further improve the selectivity.

Electroseparation and the Mercury Cathode 30.8

In the foregoing considerations, the final goal of the electrolysis was to single out one element at a time for the purpose of its immediate determination. Of course, by the nature of the process it was thereby separated from the solution and from any other components present in it. Electrolysis, as already mentioned, may also be applied with the sole purpose of effecting a separation. It has the advantage, for example, over separation by precipitation or extraction that no reagents are introduced that may interfere in a later analytical procedure. After the electrolytic separation has been accomplished, the substances remaining in solution may be determined by any appropriate method. The deposited metals may be redissolved and then determined in a suitable way. Although platinum or other electrodes used in electrogravimetry may be employed, the mercury electrode has received special attention.

A mercury cathode is of little value in electrogravimetry because of the inconvenience of weighing a liquid, and because of the volatility and high density of mercury. These factors are of no moment in electroseparations. Mercury is of special usefulness because of the high kinetic overpotential of hydrogen on this metal (Section 28.7). It offers the additional advantage that many metals are soluble in mercury through amalgam formation. In this way, many more metals can be separated at a mercury cathode. A major use of the mercury cathode is to separate large amounts of iron before the determination of aluminum or alkaline earth metals. Where the metals amalgamated with the mercury must be recovered, the electrolysis vessel may be fitted with a drain tube and stopcock. The controlled-current electrolysis technique may be used, with the solution being well stirred. Greater selectivity can be secured by employing the controlled-potential technique. In this case, polarographic measurements can be used to special advantage to ascertain the potentials most suitable for selective separations.

30.9 Questions

30-1. Explain why the mercury cathode is of little value in electrogravimetric analysis but of great value in electroseparations.

30-2. Why is it impossible to accomplish the electrogravimetric determination of metals such as sodium and barium?

30-3. Why is it necessary, in most cases, to use a third electrode as the reference when determining the potential of the working electrode?

30-4. Will the actual potential of the working electrode, as measured against a third electrode, be identical with that calculated by the Nernst equation using the concentration in the bulk of the solution? Explain.

30-5. Why is the reduction of hydrogen ion to hydrogen gas to be avoided in the electrogravimetric determination of copper?

30.10 Problems

30-1. Two metal ions, one dipositive and one tripositive, are to be separated by electrolysis. It is desired to deposit the dipositive cation until its solution concentration is $1.0 \times 10^{-6}\,M$. If the standard potential of the dipositive ion is 0.15 V more positive than that of the tripositive one, what is the maximum concentration of the latter ion allowed in the solution before it starts to deposit with the dipositive metal?

Answer: $3 \times 10^{-2}\,M$

30-2. Calculate the minimum difference in standard potentials necessary so that the concentration of a unipositive metal cation can be decreased by electrolysis to $1.0 \times 10^{-6}\,M$ without deposition of any of a dipositive metal cation that is present at a concentration of $1.0 \times 10^{-2}\,M$.

Assume that there is no overpotential for the deposition of either metal.

Answer: 0.41 V

Calculate the minimum concentration of Zn^{2+} required for the deposition of zinc on a zinc electrode from a solution of pH 7.50. Assume the kinetic overpotential of H^+ on zinc is -0.50 V. $E^0_{Zn^{2+}/Zn} = -0.763$ V.

A solution, 1.0 M in Pb^{2+} and 1.0×10^{-2} M in H^+, is electrolyzed with bright platinum electrodes. Calculate the molar concentration of Pb^{2+} in the solution when hydrogen starts to evolve. The kinetic overpotential for hydrogen is -0.09 V on platinum and -0.67 V on lead. Assume that no concentration polarization is present. $E^0_{Pb^{2+}/Pb} = -0.126$ V.

What kinetic overpotential for hydrogen evolution would be necessary to allow lead to be deposited from a solution 1.0×10^{-3} M in Pb^{2+} and 1.0×10^{-1} M in H^+? $E^0_{Pb^{2+}/Pb} = -0.126$ V.

A solution is 0.0010 M in silver ion and contains gold(III) ion. At approximately how large a concentration of gold(III) ion would the two metals deposit simultaneously upon electrolysis? $E^0_{Ag^+/Ag} = +0.799$ V, $E^0_{Au^{3+}/Au} = +1.50$ V.

Two metal ions, one dipositive and one tripositive, are to be separated by plating the former metal and leaving the latter in solution. What difference is necessary between the standard potentials of the two ions, assuming that the concentration of the tripositive ion is 0.010 M?

31 COULOMETRY AND COULOMETRIC TITRATIONS

Current Efficiency 31.1

Coulometry is a method of analysis in which the amount of sought-for substance is calculated from the quantity of electricity (that is, coulombs = amperes × seconds) necessary to react electrochemically and completely with that substance. The calculation is based on Faraday's laws (Section 22.5) and requires the knowledge of the current efficiency. The latter is defined as $100N_r/N_t$, where N_r is the total number of coulombs used to promote the desired reaction and N_t is the total number of coulombs passed. Consider the electrolysis of a solution containing a copper(II) salt. If all the electrons flowing to the cathode combine with copper(II) ion to plate copper metal, the current efficiency is 100%. If, however, hydrogen ion is reduced to hydrogen gas or if nitrate ion, acting as a depolarizer, is reduced, the current efficiency in regard to the copper(II) reduction is less than 100%. When such is the case, the exact percentage is generally variable and irreproducible; as a consequence, 100% efficiency is usually required in a coulometric method.

Measurement of Coulombs 31.2

The number of coulombs passed through an electrolysis cell can be measured in various ways. Where the electrolysis is performed at constant current, the current in amperes need only be multiplied by the elapsed time in seconds. Where an electrode potential is maintained constant, the current decreases with time; empirical evaluation of the area under the current versus time curve is then possible. However, the most satisfactory technique is the use of a coulometer.

The silver coulometer, commonly used, is simply an electrolysis cell filled with a silver nitrate solution placed in series with the cell in which the coulometric analysis is to be accomplished. The (dry) cathode of this coulometer is weighed before and after the electrolysis. From the weight of silver deposited, either the number of coulombs passed may be calculated or, more directly, the equivalents involved in the electrolysis. Of course, the electrolysis in the coulometer must proceed with

100% efficiency, which in this case is not difficult to attain. The hydrogen–oxygen (or gas) coulometer is also widely used. Water containing, for example, potassium sulfate to reduce the electrical resistance is electrolyzed between two platinum electrodes. The total volume of hydrogen and oxygen produced is measured and related to the quantity of electricity passed.

Electromechanical devices known as current integrators are progressively replacing the older-type coulometers. These devices operate on a principle similar to that of the watt-hour meter used to measure home consumption of electricity. The number of coulombs can be read directly from the meter.

31.3 Coulometric Determinations

Both of the techniques of electrolysis described in Section 30.4, constant-current and controlled-potential electrolysis, are applicable in coulometric methods. Constant-current electrolysis is seldom employed, because with this technique it is very difficult to avoid the occurrence of undesired reactions, and thus to attain 100% current efficiency. However, the technique is the basis for coulometric titrations (Section 31.4). The difficulty to maintain full current efficiency is greatly reduced in the controlled-potential technique, which is therefore the method of choice in coulometry. The circuit described in Figure 30-1 is used with the addition of a coulometer in series with the electrolysis cell. Instead of noble metal electrodes, a mercury cathode is often used advantageously, since the high kinetic overpotential of hydrogen on mercury reduces the possibility of hydrogen discharge and thereby facilitates achieving 100% current efficiency for the reduction of interest. The electrolysis in controlled-potential coulometric determinations is stopped when the cell current has fallen to a quite small, negligible value. Such coulometric determinations succeed, unlike electrogravimetric methods, even when no physical separation occurs at the working electrode, but simply a change in oxidation state. However, in such cases it is necessary to separate the anode and cathode compartments so that the reaction at the working electrode does not occur in reverse direction at the counter electrode.

31.4 Coulometric Titrations

In a coulometric titration, the titrant substance is generated with 100% current efficiency by a suitable electrode reaction secured by the presence of an excess of a suitable titrant precursor in the cell solution. Hence, the "addition rate" of the titrant is controlled by the current. If the current is held constant, the time may be measured and in the plot of a titration curve serves as the analog of the volume of the titrant solution in a conventional titration.

As an example of a coulometric titration, consider the determination of arsenic. The sample solution containing arsenic(III) is placed in a beaker, made about 0.1 *M* in iodide ion, and then some starch indicator is added. The platinum anode is inserted into this solution. The anodic half-cell thus formed is connected to the cathodic half-cell (consisting, for example, of a potassium chloride solution and a platinum wire) via a salt bridge. Separation of the anode and cathode compartments is necessary in order to avoid reversal at the cathode of the titration reaction taking place at the anode. The solution in the beaker containing the anode is stirred, and a constant current is passed. The iodine (or triodide ion) formed at the anode by the oxidation of iodide ion reacts rapidly and stoichiometrically with the arsenic(III). When all the arsenic is converted to arsenic(V), the first trace of iodine now generated will undergo the starch-iodine reaction, thus signaling the end point visually. From the elapsed time and the value of the (constant!) current, the amount of arsenic present in the sample solution can be readily calculated.

Most techniques for the detection of an end point can be utilized with coulometric titrations. Instrumental methods, notably the amperometric, dead-stop, and potentiometric techniques, are usually preferred. Since a small current and a small time interval and hence a small number of coulombs can be measured more readily and precisely than a small volume of a solution, coulometric titrations are especially valuable for the determination of extremely small amounts of material. The method also allows the generation of titrants that are difficult to prepare or to use in ordinary titrimetry, including substances readily oxidized by air (e.g., Ti^{3+}, Cu^{+}).

Questions 31.5

1. Why is a current efficiency of 100% required in coulometry?
2. What is the special value of the mercury cathode in coulometry?
3. What are the advantages of coulometric titrations over conventional titrimetry?

31.6 Problems

31-1. An As_2O_3 (*197.02*) sample is dissolved in the usual way and a volume of exactly 1 ml of this sample solution is diluted to exactly 250 ml. A 5.00-ml volume of this dilute solution is analyzed by a coulometric titration using iodide ion as the titrant precursor at a constant current of 4.00 mA. The end point corresponded to 300 seconds. Calculate the mg of As_2O_3 represented by each milliliter of the original sample solution. (1 F = 9.65×10^4C.)

Answer: 30.6 mg of As_2O_3 per milliliter

31-2. Electrically generated Br_2 is used for the coulometric titration of $HAsO_2$. The end point is detected when polarized platinum electrodes become depolarized. A current of 1.00 mA was passed for 35.6 seconds to react with all the $HAsO_2$ in a one-tenth aliquot of the sample. What amount, in milligrams, of $HAsO_2$ (*107.9*) was present in the total sample?

31-3. Electrically generated Br_2 is used for the coulometric titration of Sb(III). The end point is detected when polarized platinum electrodes become depolarized. A current of 1.350 mA was passed for 467.0 seconds before the end point was reached. What amount of antimony(III), expressed as milligrams of Sb_2O_3 (*291.5*), was present in the sample?

31-4. A gas coulometer (H_2 and O_2) is placed in series with an electrolysis cell which is filled with copper(II) sulfate solution. After electrolysis, the copper (*63.54*) deposit weighed 235.1 mg and the coulometer contained 180.0 ml of gas at 722 mm of mercury pressure and 27.0°C. Calculate the current efficiency in the electrolysis of copper. Assume that one mole of gas occupies 22.4 liters at 760 mm of mercury and 0.0°C.

Answer: 80.0% efficiency

32 ELECTROLYTIC CONDUCTANCE AND CONDUCTOMETRIC TITRATIONS

Introduction 32.1

Electricity is conducted through a solution, that is, through an electrolytic conductor, by the motion of charged ions under the influence of an applied electrical field. The conductance is defined as the reciprocal of the resistance of the solution and is the summation of the contributions to the conductances of all ions in the solution. Each of these contributions is usually called an ionic conductance. The ionic conductance of an ion depends on its charge and the rate at which it migrates under the influence of the electrical field. Thus, a dipositive cation contributes twice as much to the conductance as a monopositive cation if both ions migrate at the same speed. The migration rate is influenced by (1) the magnitude of the applied electrical field, (2) the charge and size of the solvated ion, (3) the temperature, (4) the viscosity and dielectric properties of the solvent, and (5) the attractive forces between the ion of interest and other ions present in the solution.

In a concentrated solution, an ion is surrounded by other ions to which it is attracted and which retard its movement toward a charged electrode. If such a solution is progressively diluted, the attractive forces diminish and vanish at infinite dilution; thus, the ionic conductance increases with dilution and reaches a limiting value at infinite dilution. Frequently the approximation is made that the value of the ionic conductance at low finite concentrations (usually below 0.1 M) is equal to that at infinite dilution.

Units of Conductance 32.2

Conductance, as the reciprocal of resistance, is expressed as reciprocal ohms (abbreviated either mho or Ω^{-1}) and is directly proportional to the cross-sectional area (in square centimeters), a, and inversely proportional to the length (in centimeters), l, of a conductor.

$$\text{Conductance} = \frac{1}{R} = \frac{\kappa a}{l} \qquad (32\text{-}1)$$

The proportionality constant, denoted by the Greek letter kappa, κ, is called the specific conductance or, in the case of electrolytic conduction, the electrolytic conductivity, and has units of mhos per centimeter.

The equivalent conductance of an ion is designated by the Greek letter lambda, λ, and is defined as the electrolytic conductivity of a hypothetical solution that contains one equivalent of the ion per milliliter of solution. Since equivalents per milliliter equal equivalents per liter divided by 1000 it follows that

$$\lambda = \frac{\kappa}{C^*/1000} \qquad \text{mhos-cm}^2/\text{equivalent} \qquad (32\text{-}2)$$

C^* is the concentration in equivalents per liter and is here defined as the molarity multiplied by the absolute numerical value of the charge on the ion.

Solution of equation (32-2) for κ and substitution in (32-1) yields

$$\text{conductance} = \frac{1}{R} = \frac{a\lambda C^*}{l \times 1000} = K\lambda C^* \qquad (32\text{-}3)$$

where the proportionality constant K equals $a/(l \times 1000)$. Each ion contributes independently to the conductance of the solution; hence,

$$\text{conductance} = \frac{1}{R} = K(\lambda_1 C_1^* + \lambda_2 C_2^* + \cdots)$$

$$= K\sum_i \lambda_i C_i^* \qquad (32\text{-}4)$$

The equivalent conductance at infinite dilution is designed λ^0. Values for some common ions are listed in Table J in the Appendix.

32.3 Application of Ohm's Law to Solutions

When a d.c. voltage is applied to two electrodes dipping into an electrolyte solution, Ohm's law is not obeyed. The causes for this fact have been considered in a previous section, and include the establishment of back emf, kinetic overvoltage, and concentration overvoltage.

If the polarity of the applied voltage is changed rapidly (that is, if an a.c. source is used) the situation is different. For a rather simplified picture, assume the polarity during the first half of a cycle is such that one of the electrodes is the cathode. Then cations are attracted to this electrode and anions are repelled; by the motion of the ions the electric current is carried through the solution. However, when an electrode reaction starts to take place, the second half of the cycle is reached and the polarity is reversed; that is, the electrode becomes the anode. Now cations are repelled and anions are attracted. Again before any appreciable reaction can occur the polarity changes, and so forth. Consequently, the ions carry the electrical current by a reciprocating motion, that is, merely by an oscillation, with no deposition taking place. Thus no back emf or overvoltage develops and Ohm's law is obeyed.

Measurement of Resistance 32.4

The measurement of the resistance and therefore the conductance of an electrolyte solution is usually performed with a modified Wheatstone bridge circuit, shown in Figure 32-1. A

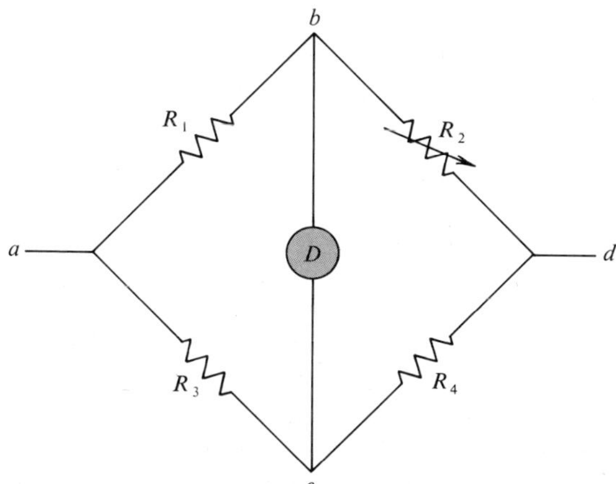

Figure 32-1. Wheatstone bridge circuit. See the text for a detailed explanation.

voltage is applied across points a and d. One (or more) of the resistors, R_1, R_2, R_3, and R_4, is variable and its resistance is accurately known at the various settings. By varying this resistance a setting can be found at which no current flows through the detector, D, and the bridge is said to be balanced.

Since no current passes through the detector when the bridge is balanced, points b and c must be at the same potential and consequently $I_1R_1 = I_3R_3$. Similarly, it follows that $I_2R_2 = I_4R_4$. Division of one equation by the other yields $I_1R_1/I_2R_2 =$

I_3R_3/I_4R_4. Since no current passes through the detector, the current I_1 passing through R_1 must equal the current I_2 passing through R_2; that is, $I_1 = I_2$. Similarly, $I_3 = I_4$. Hence, for the balanced bridge, $R_1/R_2 = R_3/R_4$. If three of the resistances are known, the fourth is established and is thereby measured.

In the measurement of electrolytic conductance, it is customary to apply a pure sine-wave signal (10^2 to 10^3 cycles/sec) across points a and d.

32.5 Conductance Cells

A conductance cell often consists of a pair of platinized platinum electrodes rigidly mounted in a glass vessel. For any given cell, the values of a and l are fixed and the ratio l/a is called the cell constant, K_{cell}. Although this constant might be calculated from the cell geometry, it is commonly evaluated by filling the cell with a solution of known electrolytic conductivity and measuring the resistance. The value of K_{cell} is then known from the relation $K_{cell} = \kappa R$. Since conductance varies by about 2% per degree centigrade, it is necessary to control the cell temperature by use of a constant-temperature bath.

32.6 Conductivity Determinations

Conductivity determinations receive limited application in inorganic analysis. One of the most common uses is to check the quality of water intended for process or boiler-feed use. This measurement of conductivity is not specific for any particular impurity, but serves as a rough measure of the total ion content of the water. Carbon in steel and other metals and alloys is frequently determined by heating the sample in a tube furnace in a stream of oxygen. The exit gases are passed through a solution of an alkali metal or alkaline earth hydroxide. The decrease in the conductance of this solution can be correlated with the amount of carbon dioxide absorbed and this with the carbon content of the sample.

32.7 Conductometric Titrations

Any reaction in which the conductance of the reactants differs markedly from that of the products can, in principle, serve as the basis of a conductometric titration. In such a titration the conductance of the solution is measured after the addition of each increment of titrant solution and the values obtained are plotted versus the volume of titrant. The end point corresponds to a break in this curve and is located by the extrapolation of two linear segments. The conductance of the solution corresponds to the sum of the contributions of all ions present. Ions that do not enter into the titration reaction give a background conductance on which the changes in conductance due to the titration reaction are superimposed. This background does not invalidate a conductometric titration unless it is so large that the small changes due to the titration are indiscernible. High concentrations of inert electrolytes are therefore undesirable in conductometric titrations.

Since the end point is obtained from *changes* in conductance, knowledge of the absolute value of the cell constant is not required and, in addition, the conductance may be expressed in arbitrary units. However, dilution effects must either be minimized or corrected for (Section 29.10); the temperature should be kept constant throughout the titration.

Conductometric Acid–Base Titrations 32.8

Consider the titration of hydrochloric acid with sodium hydroxide. The reaction equation for the present purposes may be expressed

$$(\underset{350}{H^+} + \underset{76}{Cl^-}) + (\underset{50}{Na^+} + \underset{198}{OH^-}) \rightarrow H_2O + (Na^+ + Cl^-)$$

where the number below an ion is its λ^0 value.

The conductance associated with the hydrogen ion starts at a value proportional to its equivalent conductance (350) and decreases linearly to zero at the equivalence point. The conductance of the chloride ion remains constant throughout the titration at a value proportional to its equivalent conductance (76).

Hydroxide ion is not present in significant amounts until the equivalence point is passed, and then the conductance associated with this ion increases linearly to a value proportional to its equivalent conductance (198) when a 100% excess of titrant is added. The contribution to the conductance of the sodium ion starts at zero, is proportional to its equivalent conductance (50) at the equivalence point, and is twice that value (100) at the 100% excess point.

These relations are shown graphically as dashed lines in Figure 32-2. The point by point summation of these lines yields

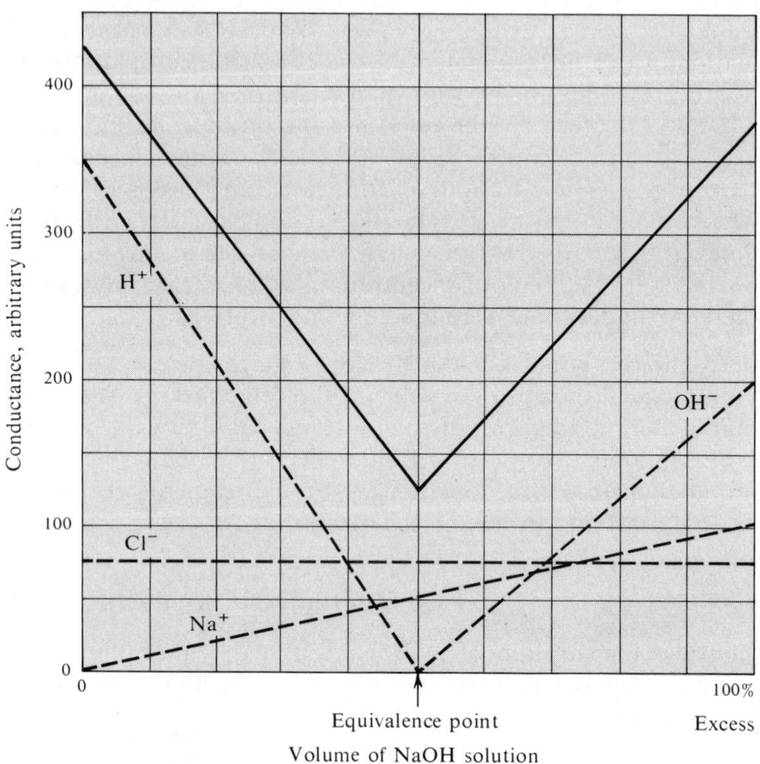

Figure 32-2. Conductometric titration of hydrochloric acid with sodium hydroxide. The contribution of each ion to the conductance shown as a dashed line.

the theoretical titration curve (solid line). Only three points need be calculated to establish this curve: the starting point, equivalence point, and 100% excess point. This applies whenever the substances reacting are strong electrolytes. With weak electrolytes more points are needed and the calculations are involved.

In the conductometric titration of a weak acid, such as acetic acid, with sodium hydroxide (Figure 32-3, curve A), the conductance at the start is small, because only a small portion of the acid is dissociated. As sodium hydroxide is added, the conductance decreases at first because strongly conducting hydrogen ions are removed and acetate ions are formed that repress the dissociation of the acid and thereby prevent the formation of more hydrogen ion. As the titration proceeds the conductance increases nearly linearly up to the equivalence

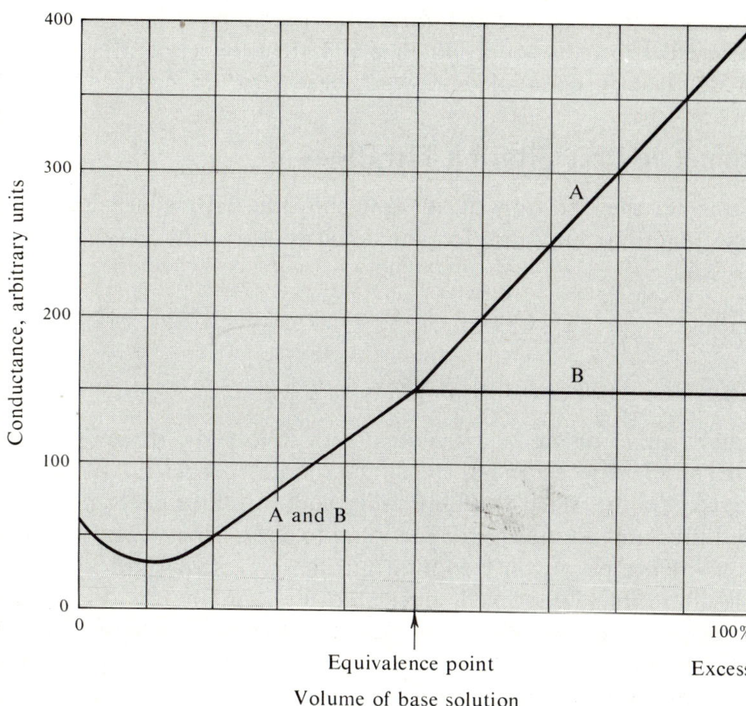

Figure 32-3. Curve A: conductometric titration of acetic acid with sodium hydroxide. Curve B: conductometric titration of acetic acid with ammonia.

point. This increase is caused by introduction of conducting sodium ions by the titrant solution and the formation of acetate ions. Beyond that point, the conductance increases more rapidly with the addition of an excess of titrant and thereby introduction of highly conducting hydroxide ions. Since the slope of the titration curve changes only slightly at the equivalence point, it is difficult to locate the end point precisely. Analogous considerations apply to the titration of a weak base with a strong acid.

When a weak acid, such as acetic acid, is titrated conductometrically with a weak base, such as ammonia, the titration curve (Figure 32-3, curve B) up to the equivalence point is similar to that for the titration of a weak acid with a strong base. Beyond that point, however, the conductance increases only slightly, since the ammonia is only partially dissociated. Further, its protolysis ($NH_3 + H_2O \rightleftharpoons NH_4^+ + OH^-$) is repressed by the ammonium ions present. Since the break in this curve is sharper than that with a strong base as titrant, the end point can be located far more precisely. This is quite different from the situation in a pH titration (potentiometric or visual) where a weak acid–weak base titration is infeasible.

32.9 Conductometric Precipitation Titrations

Consider the titration of sodium chloride with silver nitrate. The reaction equation for the present purposes may be expressed

$$\underset{50}{(Na^+} + \underset{76}{Cl^-)} + \underset{62}{(Ag^+} + \underset{71}{NO_3^-)} \rightarrow \underline{AgCl} + (Na^+ + NO_3^-)$$

where the number below an ion is its λ^0 value.

Since only strong electrolytes are involved, three points are sufficient to establish the shape of the theoretical titration curve. At the starting point, only sodium chloride is present and the conductance is proportional to 50 + 76 = 126. At the equivalence point, only sodium nitrate is present (and negligible amounts of dissolved silver chloride) and the conductance is proportional to 50 + 71 = 121. At the 100% excess point, sodium nitrate and silver nitrate are present and the conductance is proportional to 50 + 71 + 62 + 71 = 254. The titration curve is shown in Figure 32-4.

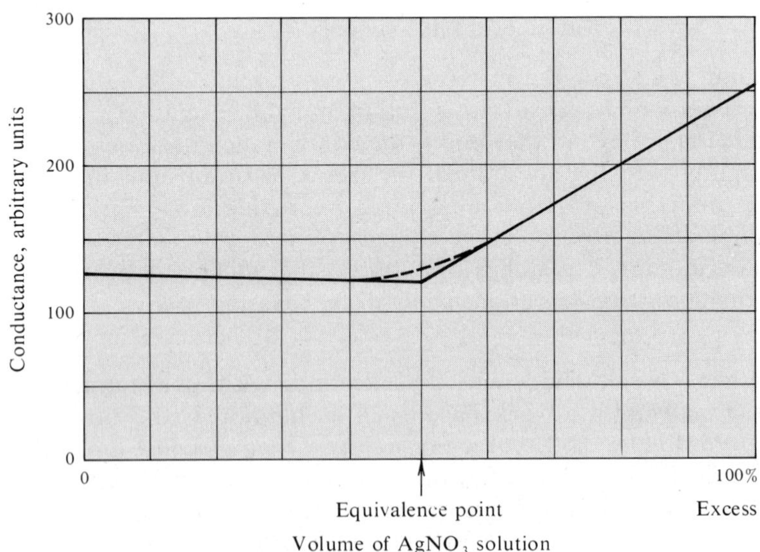

Figure 32-4. Conductometric titration of sodium chloride with silver nitrate.

If silver chloride were appreciably more soluble, then the conductance in the vicinity of the equivalence point would be increased measurably since the concentration of silver and chloride ions could not be neglected. In such a case, the two straight line-portions would be connected by a curved section (dashed line in Figure 32-4), but the end point could still be located precisely by extrapolation of the straight-line segments.

Remarks 32.10

Conductometric titrations can be applied to many chemical systems where an ionic reaction fails to go to completion. The dissociation of the principal product causes curvature in the titration plot in the region around the equivalence point. However, the dissociation is repressed before and beyond this point by a common-ion effect, thus yielding straight-line segments, which are extrapolated to establish the end point. In addition, conductometric titrations can be performed with dilute solutions, often 0.001 *F* or lower. These advantages are shared with other "linear" titrations, including amperometric titrations (Section 29.14) and some photometric titrations (Chapter 38).

Questions 32.11

How does the charge of an ion influence its equivalent ionic conductance?

Explain why the equivalent conductance of a solution does not increase proportionally when the concentration of the solution is increased.

Does Ohm's law apply to a solution of an electrolyte? Explain.

What is the merit of determining a cell constant indirectly rather than calculating it?

Explain the necessity of maintaining a constant temperature in conductometric titrations.

What type of acid–base titration can be performed more satisfactorily conductometrically than as a pH titration? Elaborate.

Problems 32.12

A mixture of exactly 50 ml of 0.0200 *F* HCl and 50 ml of 0.0200 *F* acetic acid is titrated conductometrically with 1.000 *F* NaOH. The following data were secured during the titration, with conductance expressed in arbitrary units:

ml NaOH	0.0	0.40	0.80	1.20	1.60	1.80	2.20	2.60	2.80
Conductance	37.0	28.0	16.0	12.0	14.5	15.5	20.0	27.5	31.5

Plot the experimental points of the titration curve and extrapolate the line segments to locate the two end points. (a) State the values of the end points and (b) the equivalence points. (c) In your own words, attempt an explanation of the shape of this titration curve.

Answers: (a) 1.0 ml, 2.05 ml; (b) 1.0 ml, 2.0 ml

32-2. A 1.0255-g sample of a water-soluble organic acid is weighed out, dissolved, and diluted with water to a volume of exactly 250 ml. A 25.00-ml aliquot is transferred to a conductivity cell and without further dilution is titrated with 0.1000 F NaOH. The following data are obtained, where R is the measured resistance in ohms and V is the volume of NaOH solution in milliliters.

R	V	R	V	R	V	R	V
29.4	0.00	32.5	1.80	23.8	3.60	16.2	5.40
32.6	0.20	31.7	2.00	23.0	3.80	15.1	5.60
34.1	0.40	31.1	2.20	22.3	4.00	14.5	5.80
38.0	0.60	30.1	2.40	21.7	4.20	13.8	6.00
36.9	0.80	29.1	2.60	21.0	4.40	13.1	6.20
35.8	1.00	27.7	2.80	20.4	4.60	12.6	6.40
34.9	1.20	26.7	3.00	19.6	4.80	12.0	6.60
34.1	1.40	25.7	3.20	18.5	5.00	11.5	6.80
33.3	1.60	24.7	3.40	17.3	5.20	11.1	7.00

Convert the resistance values to conductance, correct for dilution, and plot the titration curve. From the curve, conclude whether the acid is monoprotic or diprotic. Calculate the formula weight of the acid.

Answers: diprotic acid; FW = 427

32-3. The specific conductance of 0.100 F HCl is 0.0394 mho. What is the equivalent conductance of the solution?

Answer: 394 mhos

32-4. What is the equivalent conductance, in mhos, at infinite dilution and 25°C of a solution of:

(a) HCl
(b) NaCl
(c) Na_2SO_4
(d) Ag_2SO_4
(e) $NaKSO_4$
(f) H_2SO_4

32-5. Draw the schematic figures for the conductimetric titration curves for the following titrations.

(a) HNO_3 with NaOH
(b) H_2SO_4 with $BaCl_2$
(c) Ag_2SO_4 with $BaCl_2$
(d) $AgNO_3$ with KCl
(e) $FeCl_3$ with NaOH

33 REVIEW QUESTIONS ON ELECTROANALYTICAL METHODS AND REDOX TITRATIONS

Compare the relative advantages and disadvantages of electrometric titration methods involving linear plots of data (conductometric and amperometric titrations) with those involving logarithmic plots (potentiometric redox and acid–base titrations).

Which electroanalytical methods are based on concentration overpotential? In which should such overpotential, as a nuisance, be minimized as much as possible? In which is such overpotential absent?

Which methods do not require complete reaction or complete consumption of the sample solution taken for the determination?

Which methods are capable of the greatest sensitivity, that is, respond to very small concentrations of the substance to be determined?

Which electroanalytical methods require the least amount of equipment? Which a considerable amount?

How do you measure a voltage without drawing a current? In which methods is this technique used?

Which methods operate at constant current? Which at constant potential?

In which electroanalytical methods is it necessary to know the absolute value of the (a) current, (b) voltage, and (c) resistance? In which is it only necessary to know (d) differences in parameters a, b, c, or (e) a function proportional to one of these parameters?

Which methods require that the temperature be known precisely? Which require only that the temperature be constant during the period of measurements? Which do not require temperature control as long as changes are not extreme?

Which electroanalytical methods require close control of pH and under what circumstances?

How does the presence of a complexing agent affect (a) a redox titration, (b) polarographic measurements, and (c) a determination by controlled-potential electrogravimetry?

Compare the formal potential of a redox indicator with the pK_a value of an acid–base indicator, and elaborate on the parallels in the use of these indicators, respectively, in redox and acid–base titrations.

What is an inert electrode? Name some methods in which it finds application.

33-14. Which electroanalytical methods require the presence of a high concentration of inert salts and for what purpose? In which method should a high concentration of inert electrolytes be avoided?

33-15. Which electrochemical methods can be used for analytical purposes by the direct measurement of voltage, current, etc., and can be modified to titration methods based on the measurement of differences in one of the parameters?

33-16. In which analytical methods can theoretical predictions as to feasibility, performance, conditions, etc., be based on the application of the Nernst equation?

33-17. Which electroanalytical techniques and measurements may be employed to obtain data or perform operations for an *immediate* purpose other than the detection or determination of a substance?

33-18. In which methods does mercury play an important role and why does it offer special advantage?

33-19. In which methods and techniques of measurement is connection of half-cells via a salt bridge a common practice?

33-20. In which electroanalytical methods is the measurement of time involved?

33-21. What is a junction potential? How is it avoided or at least minimized? In which methods does its occurrence impose a limitation on the accuracy and precision of the measurement?

33-22. What ways and means can you suggest to test whether an electrode reaction is reversible or irreversible?

33-23. In a certain supporting electrolyte two metals have their polarographic waves essentially coinciding. Elaborate on the possibility and mechanism of resolving the coincidence by the addition of a complexing agent.

33-24. Name some redox reactions that are autocatalytic.

33-25. In which redox titrations and electrometric determinations must oxygen be excluded? Explain why and discuss how the exclusion can be achieved in practice.

33-26. When is a small electrode area desirable in an electroanalytical process? Explain why. When is a large electrode area desirable? Explain.

33-27. How would you proceed to make a silver coulometer work on a titrimetric basis?

33-28. By what means can you decrease concentration overpotentials?

33-29. Some substances can be used as primary standards for both a redox and an acid–base titration. Would such a substance have the same equivalent weight in both titrations? Name a few substances that would fall in this category.

. Why is sulfuric acid or sodium hydroxide added to the electrolyte in a gas coulometer, although only water is decomposed?

. What solutions should not be electrolyzed with a gold or platinum foil as the anode? Explain why. What electrode material would be suitable in such cases?

34 LAWS OF THE ABSORPTION OF LIGHT

General Remarks 34.1

The student from his study of qualitative analysis is aware that phenomena involving light can be used for analytical purposes. The color developed in a solution, the color of a flame, and the scattering of light associated with the appearance of a turbidity can serve for the detection and identification of substances. These phenomena and others have their counterparts in quantitative analysis. The analytical methods based on light and its interaction with matter can be called optical methods. Here the terms "light" and "optical" are not restricted to the involvement of visible radiation but are used in the broadest sense of the word to include all radiant, electromagnetic energy. Most optical methods can be grouped into those involving the absorption of light (absorptiometry, including colorimetry and photometry, Chapters 34 through 38, and absorption flame photometry, Chapter 40) and the emission of light (emissiometry, including spectroscopy and emission flame photometry, Chapter 40). Some diverse optical methods can be treated together (Chapters 39 and 41).

Matter and radiant energy interact by several mechanisms. Some will be briefly discussed in the presentation of the relevant methods. In many cases, however, a full understanding of the processes requires more knowledge of physics and physical chemistry than the student has presently at his command; he will be exposed to these facts in other courses. However, such knowledge is not required for an appreciation of the operation of an instrument, the working of a procedure, and its application to analysis. Historically, several of the analytical methods were developed before their theoretical basis was elucidated and indeed the methods subsequently were of importance in the elaboration of underlying theory.

Many of the terms employed in connection with optical methods are often used with different nuances, in either broad or restricted senses, and for one and the same phenomenon or property more than one term may be assigned. Different authors prefer different terms and often apply them with slightly different connotations. It is therefore appropriate to define initially some of the important terms. The student is urged to review these definitions as his study of the optical methods of analysis proceeds and as he thereby gains a fuller appreciation of the implication of the terms. It is also recommended that the student review the relevant chapters on the properties of light in an elementary textbook on physics.

34.2 Definitions

Light. Light, for the present purposes, is defined as a form of radiant, electromagnetic energy, and is assumed to propagate as a transverse wave, that is, a wave in which the vibration is in a plane perpendicular to the direction of propagation. The term light is here not restricted to visible radiation.

Wavelength. Wavelength, denoted by the Greek letter lambda, λ, is the distance along the direction of propagation between two points that are in phase on adjacent waves. The wavelength may be expressed in angstroms (Å), millimicrons ($m\mu$), or microns (μ), depending on its magnitude. These units of length may be related to more familiar metric units:

$$1\,\text{Å} = 1 \times 10^{-1}\,m\mu = 1 \times 10^{-4}\,\mu = 1 \times 10^{-7}\,\text{mm} = 1 \times 10^{-8}\,\text{cm} = 1 \times 10^{-10}\,\text{m}$$

The use of the nanometer (nm) for the millimicron ($m\mu$) is being encouraged internationally ($1\text{ nm} = 1\ m\mu = 1 \times 10^{-9}$ m); the micron and millimicron are here used since they are more frequently encountered at this time.

Frequency. Frequency, denoted by the Greek letter nu, ν, is the number of wave cycles per second. Frequency in one sense is more fundamental than wavelength. The frequency of (monochromatic) light remains constant no matter the medium in which it is propagated or transmitted. In contrast, the wavelength varies inversely with the velocity of light in the medium; that is, $\nu = c_{\text{light}}/\lambda$, where c_{light} is the speed of light in the medium. The higher the refractive index of the medium, the smaller is the velocity of light in that medium.

Monochromatic Light. Monochromatic light is a term used to denote light of a single frequency or, less precisely, of a single wavelength. *Compound light* or heterochromatic light denotes light of many frequencies (wavelengths).

Spectrum. The overall spectrum of radiant, electrometric energy extends from cosmic, gamma, and X-rays (<1 to ~ 100 Å), through the ultraviolet (~ 10 to $\sim 380\ m\mu$), the visible (380 to 780 mμ), the infrared ($\sim$0.78 to $\sim$300 μ), and microwaves ($\sim$ 0.1 to $\sim$ 1000 cm), to radio waves ($>$ 1 m). In analytical chemistry the term spectrum is commonly incorporated into the following connotations. An *emission* spectrum is the array of all wavelengths obtained from compound light emitted by a radiation source by separation into the individual wavelengths. An *absorption spectrum* is the array of lines or, more frequently, bands of those wavelengths that are removed from compound light when it passes through an absorbing medium. The array is commonly displayed as a curve in a plot of absorbance versus wavelength.

Color. Color is the psychophysical sensation experienced when light of one or more wavelengths in the visible region is viewed by the eye. It should be emphasized that the term light, for some purposes, is used in a most restrictive sense to denote radiant energy in the visible region only. Since the eye is not capable of distinguishing whether visible light is monochromatic or compound, the term monochromatic light, as here used, is not necessarily identical with "one" color or "pure" color.

Radiant Power. Radiant power (also known as radiant flux), denoted by P, is the rate at which radiant energy is transported by a beam of light. In simple terms, the radiant power is the intensity of the light.

Transmittance. Transmittance, denoted by T, is the ratio of the radiant power, P, transmitted by a sample to the radiant power, P_0, incident on the sample. Hence, $T = P/P_0$. (Terms formerly used in place of transmittance include transmittancy and transmission.) The *percentage transmittance* is given by $T \times 100$ and may be denoted by $\%\,T$.

Absorbance. Absorbance, denoted by A, is defined as the negative logarithm (to the base 10) of the transmittance; that is, $A = -\log T = \log(1/T) = \log(100/\%\,T)$. (Terms formerly used in place of absorbance include optical density, extinction, and absorbancy.)

Absorptivity. Absorptivity, denoted by a, is a measure of the ability of a material to absorb light. Since analytical chemistry is here concerned mainly with solutions, absorptivity is more closely defined as the ratio of the absorbance to the product of the actual concentration of the absorbing species and the length of the light path in the absorbing medium. Consequently, a is the specific absorbance, that is, the absorbance per unit concentration and unit thickness. By convention, the concentration, c, is expressed in grams per liter and the length of the light path, b, in centimeters. Hence $a = A/bc$ and a has units of liters per gram-centimeter. (Older terms for absorptivity include specific extinction, extinction coefficient, specific absorption, and absorbancy index.)

Molar Absorptivity. Molar absorptivity, denoted by the Greek letter epsilon, ϵ, is the absorptivity with the actual concentration of the absorbing species expressed in moles per liter, C, and the length of light path in centimeters. Hence, the molar absorptivity has units of liters per mole-centimeter. (Older terms for molar absorptivity include molecular extinction coefficient and molar absorbancy index.)

As will be shown later, absorptivity and molar absorptivity depend on the wavelength of the light and the nature of the absorbing species.

34.3 Laws of the Absorption of Light

Suppose the sample under study is a solution of a light-absorbing solute in a nonabsorbing solvent placed in a rectangular glass cell. Suppose further that a beam of parallel, monochromatic light strikes the cell at a right angle to one of its sides. It will be observed in general that only a portion of the radiant power of the incident light beam will be transmitted, that is, will leave the cell on the opposite side. A portion of the radiant power is lost by reflections at the air–glass and glass–solution interfaces. A second portion is lost by scattering caused by dust particles and inhomogeneities in the solution and the glass, and a third portion is consumed in an absorption process by the solute.

For analytical purposes, as will become more evident in the subsequent development of the topic, interest is centered on the decrease in the radiant power caused by absorption. Consequently, it is necessary to allow for the light lost by scattering and reflection. How this is done will be considered later. For the presentation of the laws of the absorption of light, which follows, scattering and reflection are assumed not to occur.

Lambert's Law. Lambert's law, also known as the Bouguer–Lambert law, describes the relation between the radiant power of the incident light, P_0, and that of the transmitted light, P, as a function of the length of the light path at a constant concentration. Suppose light of 80 arbitrary energy units is incident on a solution of given concentration and thickness. Assume that for the particular solution (reflection and scattering being negligible) the amount of light absorbed is such that light of only 40 units emerges. Then the transmittance of the sample is 40/80 = 0.50 or 50%. The situation is shown in Figure 34-1. If the transmitted beam of 40 units falls upon an identical "layer" of solution, again 50% of the radiant energy will be absorbed and the transmitted beam will have 40 × 0.5 = 20 units. After passing through a third identical layer the energy will be 10 units, and so forth. Thus the law may be stated: The transmitted radiant power decreases in a geometric progression as the length of light path increases in an arithmetic progression. The mathematical expression of this law takes the form†

$$\log \frac{P}{P_0} = -kb \qquad (34\text{-}1)$$

where k is a proportionality constant and b is the length of the light path. As described above, the ratio of the radiant power

†The student familiar with calculus will understand the following derivation of Lambert's law. The law, which is a particular application of the general law of decay (radioactive decay, decay of population, etc.), states: The decreases in radiant power of the transmitted light is proportional to the radiant power of the incident light and to the increase in the length of the light path in the absorbing medium.

Let dP be the change in radiant power when light of radiant power P passes through an infinitesimal thickness of the sample, db. The mathematical formulation of the stated law is

$$dP = -k'P\,db$$

where k' is a proportionality constant and the minus sign indicates that the radiant power decreases as the light path is increased.

Rearrangement and integration between the limits P_0 and P corresponding to b equal to zero and b (that is, integration across the entire thickness of the sample, b) proceeds as follows:

$$\frac{dP}{P} = -k'\,db$$

$$\int_{P_0}^{P} \frac{dP}{P} = -k' \int_0^b db$$

$$\ln \frac{P}{P_0} = -k'b$$

Conversion to Briggsian logarithms (i.e., base 10) involves a mere change in the proportionality constant:

$$\log \frac{P}{P_0} = -kb$$

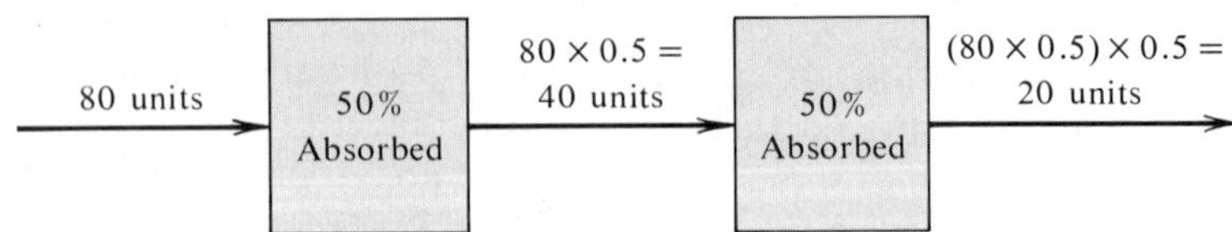

Figure 34-1. Arrangement of absorbing layers to demonstrate derivation of Lambert's law.

of the transmitted light to that of the incident light, P/P_0, is the transmittance, T, and is related to the absorbance by the equation $-\log T = A$. Thus, the law may be written in a number of forms that are fully equivalent:

$$\log T = -kb, \qquad \log \frac{1}{T} = kb,$$

$$\log \frac{100}{\%T} = kb, \qquad A = kb \tag{34-2}$$

Beer's Law. Beer's law describes the relation of the radiant power of the incident and transmitted light as a function of the concentration of the absorbing species for a light path of constant length.

Reference is again made to Figure 34-1. Each "layer" contains solution of the same concentration and consequently contains the same number of absorbing species (which are assumed not to react with each other). If the absorbing particles of the second layer were transferred to the first layer, the concentration there would be doubled. Light now transmitted by the first layer would have an intensity equal to that which has passed through two layers of the solution of original concentration. Consequently, the law may be stated as: The radiant power of a beam decreases in a geometric progression as the concentration increases in an arithmetic progression. The mathematical formulation of the law takes the form†

$$\log \frac{P}{P_0} = -k'c \tag{34-3}$$

where k' is a proportionality constant and c the concentration.

As in the case of Lambert's law, the ratio of the radiant powers may be expressed as transmittance or absorbance, and hence the law may be written in a number of forms that are fully

†The derivation of the mathematical expression parallels that given for Lambert's law with the starting expression being $dP = -k'' P dc$.

equivalent:

$$\log T = -k'c, \qquad \log \frac{1}{T} = k'c,$$

$$\log \frac{100}{\% T} = k'c, \qquad A = k'c \tag{34-4}$$

Lambert–Beer's Law. Lambert–Beer's law, as the name implies, is the combination of the two laws and considers the relation between the radiant power of the incident and transmitted light as a function both of the length of the light path and of the concentration of the absorbing species. Its mathematical formulation takes the form

$$\log \frac{P}{P_0} = -k''bc \tag{34-5}$$

where k'' is a proportionality constant, b the length of the light path, and c the concentration.

Note that the unit of the radiant power, whatever it may be, cancels in the fraction; hence, the left side of the equation is a dimensionless number. Thus it follows that the value (and units) of the proportionality constant depends on the units in which b and c are expressed. The light path, b, is usually expressed in centimeters. The concentration is usually expressed either in grams per liter or in moles per liter and is then denoted by c and C, respectively. The proportionality constant, k'', then becomes the absorptivity, a, and molar absorptivity, ϵ, respectively (see the definitions of these terms given in Section 34.2). Thus Lambert–Beer's law may be written in a number of forms which are fully equivalent:

$$\log \frac{P}{P_0} = -abc, \qquad \log T = -abc, \qquad \log \frac{1}{T} = abc,$$

$$\log \frac{100}{\% T} = abc, \qquad A = abc \tag{34-6a}$$

$$\log \frac{P}{P_0} = -\epsilon bC, \qquad \log T = -\epsilon bC, \qquad \log \frac{1}{T} = \epsilon bC,$$

$$\log \frac{100}{\% T} = \epsilon bC, \qquad A = \epsilon bC \tag{34-6b}$$

It should be emphasized that the physical content of the law is the same irrespective of whether the concentration is expressed in grams per liter or moles per liter. One form of expression is convertible to the other via the formula weight of the absorbing species.

The proportionality constant of Lambert–Beer's law, whether given as a or ϵ, is a characteristic of the absorbing species, and the numerical value will, in general, depend on the wavelength at which it is measured. Lambert–Beer's law is strictly obeyed only with monochromatic light. The implications of this restriction will be considered later.

34.4 Illustrative Examples

A few illustrative examples will advance the appreciation of Lambert–Beer's law and will acquaint the student with calculations based on it.

Example 34-1. The transmittance of a sample is 30.0%. What is its absorbance?

$$A = \log \frac{100}{\% T} = \log \frac{100}{30.0} = 2 - \log 30 = 2 - 1.477 = 0.523$$

Example 34-2. A sample shows an absorbance of 0.70. What is the transmittance and per cent transmittance?

$$-\log T = A = 0.70$$
$$T = 10^{-0.70} = 10^{0.30-1} = 10^{0.30} \times 10^{-1} = 2.0 \times 10^{-1} = 0.20$$
$$\% T = 0.20 \times 100 = 20\%$$

Example 34-3. A solution containing 2.5 mg per 100 ml of a light-absorbing solute of formula weight 200 shows a transmittance of 20% when measured in a cell with a light path of 1.0 cm.

(a) Calculate the absorptivity of the solute.

$$a = \frac{-\log T}{bc} = \frac{-\log \frac{20}{100}}{1.0 \times \underbrace{2.5 \times 10^{-3}}_{\text{for conversion of mg to g}} \times \underbrace{10}_{\text{for conversion of 100 ml to 1 liter}}}$$

$$= \frac{-\log 0.20}{2.5 \times 10^{-2}} = \frac{0.70}{2.5 \times 10^{-2}} = \frac{70}{2.5} = 28 \text{ liters/g-cm}$$

(b) Calculate the molar absorptivity of the solute.

$$C = \frac{2.5}{200 \times 100} M$$

$$\epsilon = \frac{-\log T}{bC} = \frac{-\log \frac{20}{100}}{1.0 \times \frac{2.5}{200 \times 100}} = \frac{0.70 \times 2.00 \times 10^4}{2.5}$$

$$= 5.6 \times 10^3 \text{ liters/mole-cm}$$

The result can also be obtained immediately by multiplying the value found for the absorptivity by the formula weight:

$$28 \times 200 = 5.6 \times 10^3 \text{ liters/mole-cm}$$

(c) What per cent transmittance will be observed when the solution is measured in a 2.0-cm cell?

For the 1.0-cm cell, by Lambert–Beer's law

$$-\log 0.20 = a \times 1.0 \times c$$

and for the 2.0-cm cell,

$$-\log T = a \times 2.0 \times c$$

Division of one equation by the other yields

$$\frac{-\log 0.20}{-\log T} = \frac{1.0}{2.0}$$

$\log T = 2.0 \times \log 0.20 = 2.0 \times (0.30 - 1) = 2.0 \times (-0.70) = -1.40$. Hence,

$$T = 10^{-1.40} = 10^{0.60-2} = 4.0 \times 10^{-2} = 0.040$$

and

$$\% T = 4.0\%$$

It should be noted that doubling the length of the light path does *not* halve the transmittance, because of the logarithmic relationship existing between C and T. It is the absorbance that is directly proportional to the light path.

Since the path length is here simply an integer, as in common practicc, the calculation may be simplified.

As previously established,

$$\log T = 2.0 \times \log 0.20$$

Hence,

$$\log T = \log(0.20)^{2.0}$$

and

$$T = 0.20^{2.0} = 0.040$$

or

$$\% T = 4.0\%$$

(d) What absorbance will be obtained when 0.50 mg of the solute is present in 50.0 ml of solution and the measurements are performed in a 3.0-cm cell?

The absorptivity has been calculated in part (a). The concentration here is

$$\frac{0.50 \times 10^{-3}}{50.0} \times 10^3 = 0.010 \text{ g/liter}$$

Hence,

$$A = abc = 28 \times 3.0 \times 0.010 = 0.84$$

34.5 Questions

34-1. What relationship exists between the absorbance of a solution and the concentration of the absorbing species?

34-2. Would doubling the concentration of a solution cause the transmittance to decrease by a factor of 2, other conditions being unchanged? Explain your answer.

34-3. By what processes is light generally lost when a beam passes through a cell containing an absorbing solution? Which of these processes is of primary concern for analytical applications?

34-4. What is monochromatic light? Is monochromatic light identical with light viewed by the human eye as a single or pure color?

34-5. On what does the numerical value of absorptivity depend?

34-6 Does the concentration of the absorbing species influence the absorptivity?

34-7. What is the relation between absorbance and transmittance?

34-8. Within what wavelength region does visible light fall?

34-9. Would the introduction of some scattering dust particles into a solution change the absorbance of the solution? Would it change the molar absorptivity of the absorbing solute? Explain your answers.

34-10. With reference to Figure 34-1 assume that the two layers contain solutions of different substances and exhibit different absorbances. Show that the total absorbance will be the sum of the absorbances of the two layers. Will the overall transmittance be the sum of the two transmittances?

34-11. How would the absorbance value for a solution change if the concentration were doubled and the path length were halved?

34.6 Problems

34-1. Calculate the absorbance value of each of the following samples from the measured % T value.

	% *T*	
(a)	20.0	(*Answer:* $A = 0.699$)
(b)	50.0	
(c)	35.0	
(d)	65.0	
(e)	43.0	
(f)	61.0	

Calculate the % T of each of the following samples from the measured absorbance value.

	A	
(a)	0.235	(*Answer:* 58.2 % T)
(b)	1.72	
(c)	0.689	
(d)	0.055	
(e)	0.830	
(f)	0.425	

The absorbance of a solution, containing 4.0 mg of a light-absorbing solute (*150.0*) in a volume of 250.0 ml is measured in a 5.00-cm cell and found to be 0.400. Calculate the molar absorptivity of the solute.

Answer: 7.5×10^2 liters/mole-cm

Calculate the molar absorptivity for a particular solute from the following data: formula weight, 200.0; % T, 50.0; c, 1.3 mg/liter; cell path length, 2.00 cm.

A solution containing 10.0 g of a light-absorbing solute (*125.0*) in 250.0 ml shows 25.0% T in a cell having a 5.00-cm light path. Calculate the absorbance of the solution and the absorptivity and molar absorptivity of the solute.

A solution contains 4.0 mg of a light-absorbing compound (*50.0*) in 300.0 ml. The solution shows 35.0% T when measured in a 5.00-cm cell. Calculate the absorbance of the solution and the absorptivity and molar absorptivity of the solute.

A solution containing 5.0 mg of a light-absorbing substance in 250.0 ml shows 50.0% T when measured in a cell with a 3.00-cm light path. What will be the % T of a solution containing 1.0 mg of the substance in exactly 1 liter when measured in a 10.0-cm cell?

The molar absorptivity of a substance is 2.3×10^4 liters/mole-cm. A solution of the substance is placed in a 1.00-cm cell and the transmittance is found to be 0.200. Calculate the molar concentration of the solution.

Answer: 3.0×10^{-5} M

A solution shows 64.0% T when measured in a 2.00-cm cell. Calculate the % T when the same solution is measured in a 1.00-cm cell.

A solution containing 10.0 ppm of $KMnO_4$ (*158.0*) shows an absorbance of 0.80 in a 2.00-cm cell at 520 mμ. Calculate the molar absorptivity of $KMnO_4$.

35 COLORIMETRIC METHODS OF ANALYSIS

Colorimetric Determinations 35.1

One of the simplest forms of colorimetry† operates as follows. The sample solution and a series of standard solutions containing the sought-for species in different, known concentrations are individually placed in identical clear glass tubes. The standards are arranged in the order of their concentration and the sample solution is compared with the standards for a color match. The concentration of the sample solution is evaluated from the concentration of the most closely matching standard. This procedure is simple and rapid, but its accuracy is poor.

Somewhat better results are obtained with a technique of colorimetry that involves a pair of Hehner cylinders. These are identical, graduated cylinders having an optically flat bottom and on their side, near the bottom, a drain tube fitted with a stopcock. The sample solution is placed in one cylinder and a standard solution in the other. The two cylinders, side by side, are then placed on a stand on a homogeneously illuminated plate of frosted glass (or simply on white paper) and are viewed from above. A judgment is made as to which cylinder presents the darker color and then liquid is allowed to drain via the stopcock from that cylinder. The light path is thereby decreased and the color viewed becomes progressively lighter. Eventually a color match is attained. When this condition prevails, the radiant powers of the light beams transmitted by the two solutions are identical. Since the radiant power of the light beams entering the two solutions is also identical, the absorbances of the two solutions must be identical.

Hence, by application of Lambert–Beer's law,

$$\mathrm{A} = ab_x c_x = ab_s c_s$$

†A remark on terminology seems appropriate. In physics, colorimetry refers to the measurement or specification of colors. In the common usage of chemical analysis a colorimeter, unless otherwise specified, is a device for comparing the color of a sample solution with that of a standard for the purpose of establishing a concentration. Consequently, devices such as the Duboscq colorimeter may more appropriately be termed color comparators.

where the subscripts x and s refer to the sample solution and the standard solution, respectively. The absorptivity, a, of course, has no subscript because the same absorbing species is present in both solutions. Canceling a and rearranging yields

$$c_x = \frac{b_s}{b_x} \times c_s$$

The concentration of the standard solution, c_s, is known. Further, b_s and b_x are proportional to the respective volume graduations because the diameters of the cylinders are identical. These graduations are read when color match has been achieved and the values substituted in the fraction b_s/b_x. The concentration of the absorbing species in the sample solution can then be calculated. Hehner cylinders are rarely employed in present-day chemical analysis, but the description of their use serves as an instructive introduction to the application of Lambert–Beer's law in visual colorimetry.

35.2 The Duboscq Colorimeter

The Duboscq colorimeter, illustrated in Figure 35-1, is a more elaborate arrangement for individually varying the length of the light path in two solutions so that a color match may be secured visually. Two identical light beams emerging from a homogeneously illuminated frosted glass plate are passed up through two flat-bottomed cells containing the sample and the standard solution, respectively. The light beams enter vertically mounted glass plungers having optically flat ends. These rods extend to an optical system that presents the two light beams to an eyepiece, where they are viewed in juxtaposition as semicircles. Each cell is mounted on a support that can be raised or lowered individually by a rack and pinion movement. The length of the light path through each of the solutions can then be varied separately until a color match is secured. When the colors are matched the dividing line between the two half-fields viewed in the eyepiece virtually disappears. The length of the light path for each solution is then read on the calibrated scale attached to each rack. The calculation of the result is identical to that for the Hehner cylinders. In practice the color match is attained several times and the results are averaged.

35.3 Limitations of Visual Colorimetry

The human eye is quite capable of detecting minute differences in the intensities of the colors of two solutions if these solutions

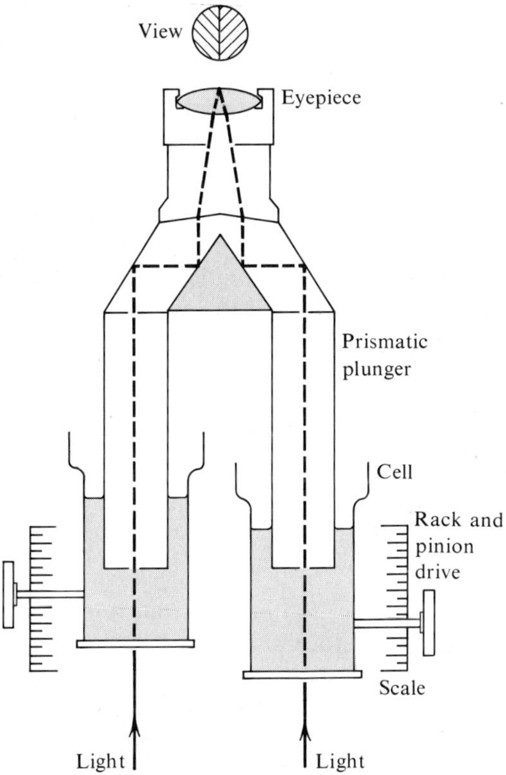

Figure 35-1. Schematic diagram of a Duboscq colorimeter.

exhibit colors of the same shade (that is, hue) and if they are viewed concurrently. If, however, a second color is present in one of the solutions the situation becomes different. This is due to the fact that the eye is unable to "analyze" colors, that is, for example, to differentiate whether a certain green is a single color or whether it is a mixture of yellow and blue. (Note the difference between eye and ear; the latter is able, for example, to isolate the sound of a violin from the total acoustic impression provided by a full orchestra.) The practical consequence of this fact is best understood by the discussion of an example. Manganese is readily determined colorimetrically by oxidation to permanganate and comparing the color developed with that of a standard. This method is quite sensitive because of the high absorptivity of permanganate and is applicable to the analysis of steel. Suppose a steel also contains chromium, which under the conditions of the determination is oxidized to dichromate. The standard solution shows only the purple color of permanganate. In contrast, the sample solution presents a mixture of the colors of permanganate and of dichromate (red-orange). This color mixture to the eye is a new color. Consequently, in a Duboscq comparator, no color match (neither of color shade nor of color intensity) can be attained, whatever adjustment of the two path lengths is made. One remedy to overcome the difficulty is to add to the standard solution an amount of dichromate equal to that present in the sample solution. However, such an addition is impractical because it requires the concentration of dichromate in the sample solution be known. Another and far better approach is to place, in both light beams of the comparator, a filter that "cuts out" the color of dichromate. This filter has a color of its own, and what is viewed now is a new color which is a mixture of the colors of the permanganate and the filter. Since the intensity of the filter color is the same in both beams, variation of the path lengths and thereby of the intensity of the permanganate color can now lead to a color match detectable by the eye.

The use of such a filter confers an additional advantage. As has been mentioned before and as will be elaborated in more detail, Lambert–Beer's law is strictly valid only for monochromatic light. The lower the degree of monochromacity, the lower the concentration up to which the law is followed with reasonable closeness. The color filter cuts out from the compound light a narrower wavelength band and thus provides light of greater monochromacity.

36 PHOTOMETRIC METHODS OF ANALYSIS

Photometric Determinations 36.1

Photometric determinations are based on the measurement of the ratio of the radiant power of the light entering a sample to that emerging from it.† Although visual photometry is possible, a photoelectric detector replaces the eye in current analytical practice. Photometric methods employing *white* light are feasible, but light of a restricted wavelength region is used almost exclusively. Where restricted spectral regions are secured by filters, the instrument may be termed a filter photometer and measurements employing it as filter photometry. Where light of a narrow wavelength range is obtained by using a portion of a prism or grating spectrum, the instrument is known as a spectrophotometer and measurements employing it as spectrophotometry. Where a distinction between filter photometry and spectrophotometry is unnecessary, the general terms photometer, photometry, and photometric determination are appropriate.

The constructional and operational details of a photometer are the concern of the practice of chemical analysis and consequently fall outside the scope of this volume. However, the essential components of a photometer and their function must be described briefly because this knowledge is a requisite for an understanding of the basic aspects of photometric methods.

The principal components of a typical spectrophotometer are shown schematically in Figure 36-1. The light emerging from the source is collimated to a parallel beam, which passes through a monochromater (here, a prism and slits) and enters the cell, where absorption may take place. The beam emerging from the cell strikes a photoelectric detector, which generates an electrical signal proportional to the radiant power of the

†A photometer in the strict sense of physics is a device that measures the radiant power of light, regardless of any previous interactions of that light. In chemical analysis, unless otherwise qualified, the term photometer refers to an apparatus that allows the measurement of the absorbance or transmittance of a system. Consequently, for the usual term "photometric determination" the more exact expression "absorption photometric determination" should be understood.

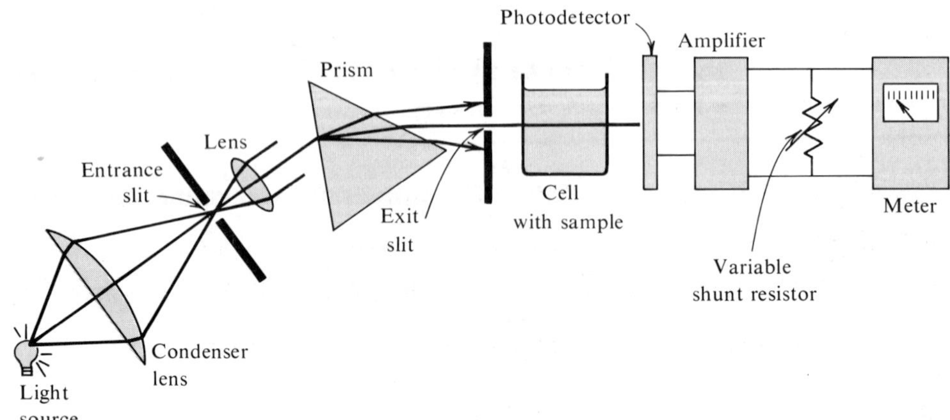

Figure 36-1. Schematic arrangement of the principal components of a spectrophotometer. (For simplification, the narrow beam of parallel light present after the second lens is shown as a single ray. The prism can be rotated so that light of any desired wavelength may pass through the exit slit, which is usually adjustable to allow variation.)

incident light beam. The detector output, amplified if necessary, is fed to a meter, the sensitivity of which can be varied by a shunt resistor.

The performance of a photometric determination employing such an instrument can be described in terms of a sample solution consisting of a single absorbing species in a nonabsorbing solvent. The cell is first filled with the solvent, placed in the light beam, and an appropriate wavelength is set. The needle of the meter is brought to the 100% T position by adjusting the resistor. Then the cell is emptied, filled with the sample solution, and positioned in the light beam exactly as before. Or, more conveniently, the sample is placed in the beam in a second, identical cell. The meter now indicates a value of less than 100% T, since a portion of the radiant power is absorbed. The value read corresponds to the per cent transmittance of the sample solution. If the cell length and the absorptivity are known, the concentration of the absorbing species can be calculated by application of Lambert–Beer's law.

Several consequences of this mode of operation should be appreciated. As previously mentioned (Section 34.3), processes other than absorption—scattering and reflection—may reduce the radiant power of the light beam. Scattering, due to inhomogeneities or dust particles in the solution, may be made negligible by appropriate treatment (thorough mixing and filtration). Reflection is taken care of in the following manner. As noted above, the meter reading is proportional to the radiant power of the beam striking the detector. With the solvent in the cell, the radiant power of the beam is reduced by reflection and the initial adjustment of the meter to 100% T in effect compensates for reflection losses. This compensation is fully successful if the indices of refraction of the solution and the solvent are identical or sufficiently so. Fortunately, this condition is usually met in photometric practice.

From an understanding of the manner in which reflection losses are compensated, it will be appreciated that any absorbance due to the solvent itself is also compensated concurrently. Consequently the restriction to a nonabsorbing solvent, previously assumed, is unnecessary. This additional compensation is fully achieved only in quite dilute solutions, because the concentration of the solvent in the sample solution must not depart markedly from that of the pure solvent.

The possibility of compensating for impurities in reagents and solvents used for the treatment of the sample will be discussed in Section 36.7.

In principle the comparison of the radiant powers of the beam transmitted by the solvent and the sample solution under identical conditions corresponds to the measurement of the absorbance of the solute as if it alone were distributed homogeneously throughout the space actually occupied by the sample solution.

A further important conclusion is that, if two or more absorbing solutes are present in the solution, the absorbance value (but not the transmittance value!) obtained will be the sum of the individual absorbances due to the various absorbing species.

Transmittance and Absorbance Curves 36.2

In order to select a wavelength appropriate for a photometric determination, it is necessary to establish the wavelength region in which significant absorption occurs. This selection is made from an inspection of a curve obtained by plotting either transmittance or absorbance versus wavelength.

A transmittance curve is obtained in the following manner. A cell, filled with the solvent, is placed in the light beam and the instrument at a wavelength of, say, 400 mμ is adjusted to read

100% T. Then the solvent is replaced by the solution and the transmittance read. This process is repeated at, say, 410, 420, 430, ... mμ. The interval between the wavelength settings need not be constant; indeed, the values selected will depend on the shape of the curve and the degree to which finer details of the curve must be secured. The values of % T found are plotted versus the wavelength values and a smooth curve is drawn through the points. Either by substitution in the formula $A = -\log T$ or by reading the absorbance scale provided additionally on most instruments, the corresponding absorbance values are obtained and plotted versus the wavelength values to obtain the absorbance curve. A typical transmittance curve and its equivalent absorbance curve are shown in Figure 36-2.

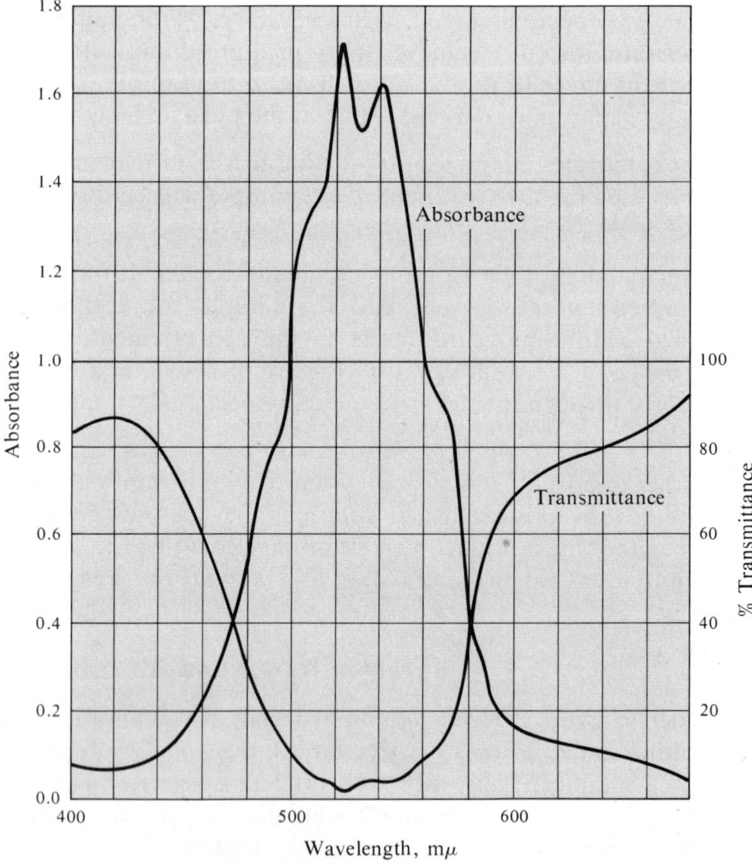

Figure 36-2. Absorbance and transmittance curves for $KMnO_4$ in water (40 mg/liter, 1-cm cell) for the region 400 to 680 mμ.

It will be seen that an absorbance maximum corresponds to a transmittance minimum and vice versa.

The attention of the student is directed to the following relation that may be a source of confusion. The transmittance scale is commonly limited by the values zero and unity and the per cent transmittance scale by zero and one hundred. In contrast, the absorbance, as is obvious from its definition, extends from zero to infinity, since $-\log 1 = 0$ and $-\log 0 = \infty$.

Calibration Curves 36.3

It may seem feasible to obtain the result of a photometric determination as follows. The absorbance of the absorbing solute is secured by experiment; the length of the light path in the cell used is measured and the absorptivity of the solute is taken from a table of listed values. From these data, the concentration of the solute is then calculated by application of Lambert–Beer's law. Unfortunately, such a procedure seldom affords an acceptable result. Although the absorptivity at a given wavelength is a characteristic of a substance, different absorptivity values will be obtained with different photometers, principally because the degree of monochromacity will vary from instrument to instrument. Even with a single instrument and at an unchanged wavelength setting different absorptivity values may be obtained for the same substance at different slit widths, since the degree of monochromacity decreases as the slit width increases. It is possible to determine the absorptivity of a substance for a given instrument at a given wavelength setting and slit width and then to employ the value obtained for the calculation of results with the absorbance measurement secured under identical operating conditions. But there still remains the problem of determining the length of the light path. For rectangular cells the distance between the inner surfaces of two opposite cell faces may be taken as the path length provided the beam consists of parallel light and traverses perpendicularly to the two faces. These conditions, however, are not always met. The assignment of a definite path length becomes even more difficult when cylindrical cells are employed.

Consequently, it is more convenient to use a calibration curve, which is obtained as follows. A series of standard solutions containing the species of interest in different but known concentrations is prepared and the transmittance or a function related to it is plotted versus the concentration. Then the transmittance of the sample solution is determined under identical instrument settings and the concentration of the sought-for substance is obtained from the calibration curve. Various methods of plotting a calibration curve are feasible; several are considered below.

36.4 Plot of Transmittance versus Concentration

Since transmittance and concentration are related by a logarithmic function, a straight line is not obtained on plotting these two variables on ordinary (linear) graph paper (see Figure 36-3). This fact does not hinder the practical use of the

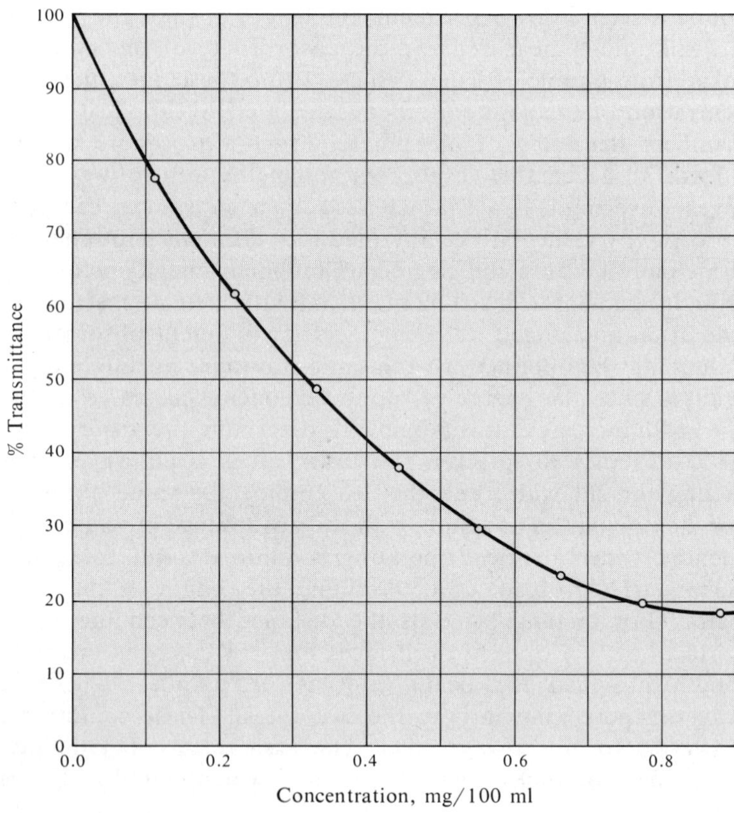

Figure 36-3. Typical calibration curve, per cent transmittance versus concentration. (See also Figure 36-4.)

plot as a calibration curve, but it is difficult to decide from such a curve whether Lambert–Beer's law is obeyed and to judge whether the conditions for a photometric measurement might be improved. Consequently, such a plot is seldom used in analytical practice.

Plot of Absorbance versus Concentration 36.5

If Lambert–Beer's law is obeyed, a straight line is obtained on plotting either transmittance versus concentration on semilogarithmic graph paper or, what amounts to the same thing, absorbance versus concentration on linear graph paper. The latter method is usually preferred, because it is easier to interpolate on an equal-increment scale. For these two plots the slopes of the lines correspond to $-ab$ or $+ab$, respectively, in the expression of the law. Indeed the slope of the straight-line portion can be read and be used to calculate the concentration directly.

The absorbance versus concentration curve corresponding to the curve in Figure 36-3 is shown in Figure 36-4. Beer's law is

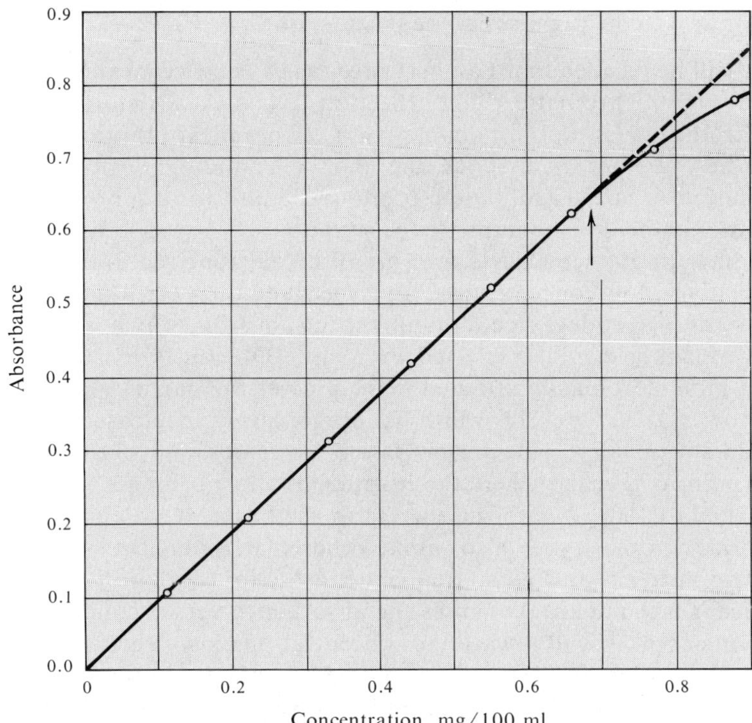

Figure 36-4. Typical calibration curve, absorbance versus concentration. Arrow indicates point at which deviation from Lambert–Beer's law becomes pronounced. (Same data as Figure 36-3.)

obeyed up to a concentration of about 0.67 mg/100 ml. Beyond that point deviation from a straight line becomes clearly evident. The nonlinear portion can still be used as a calibration curve, but the curvature serves as a warning that the working limits for the system are close and that caution need be exercised in the interpretation of the results (see below).

36.6 Deviations from Lambert–Beer's Law

An absorbance versus concentration calibration curve is usually a straight line at a sufficiently low concentration range. At some higher concentration, however, the curve starts to deviate from a straight line and may bend either toward the ordinate or the abscissa. Various factors may operate singly or together to produce a deviation. Frequently, however, the deviation is not caused by a true failure of the law but is, rather, due to the fact that the conditions prevailing do not confirm to the premises of the law.

It will be recalled that Lambert–Beer's law is strictly valid only for monochromatic light (Section 34.3). Strict monochromacity is not attainable with a practical photometer. Where a filter is used as the monochromating device, the limitations are usually quite pronounced and for a given filter little can be done to improve the situation. Consequently, with a filter photometer deviations are often encountered even at a relatively low concentration. In a spectrophotometer the prism or grating produces a continuous spectrum from which a small wavelength region is selected by a slit. Obviously the degree of monochromacity attained is the greater the narrow the slit. However, as the slit width is progressively decreased, the radiant power of the beam passed decreases and eventually a point is reached where the instrument can no longer be adjusted to 100% T with the solvent in the beam. Consequently, monochromacity is also limited here. Insufficient monochromacity is the most frequent reason for deviations from Beer's law and always causes the absorbance versus concentration curve to bend toward the concentration axis. Fortunately, as will be seen in Section 36.9, the requirement of strict monochromacity can often be relaxed in practical photometry.

In the normal absorption process radiant energy is consumed by the absorbing species, which is brought to an excited state. As this species returns to the ground state, the absorbed energy is dissipated as heat. In some cases, however, the absorbed energy, or a part of it, is emitted as light. This phenomenon is known as fluorescence (see Chapter 39) and is undesirable in photometric determinations because large deviations from straight-line calibration curves result. In theory fluorescence could be compensated for by having the same degree of fluorescence in reference and sample solutions. This condition, however, is difficult to achieve because fluorescence is extremely sensitive to impurities and other solution conditions.

Sometimes it is held that deviations from the law are the result of chemical effects, especially where dissociation–association equilibria are involved, since the position of the equilibrium will depend on the concentrations of the participating species. In many cases, however, the deviations are not true ones but rather result from inappropriate plotting of the calibration curve. An example will illustrate the point in question.

Suppose the formal concentration, C_{In}, of an acid–base indicator of the weak-acid type is to be determined photometrically. The dissociation equilibrium of the indicator and the relevant material balance may be written

$$HIn \rightleftharpoons H^+ + In^-$$
$$C_{In} = [HIn] + [In^-]$$

Suppose further that the anion In^- is the only species that absorbs at the wavelength selected. If the photometric measurements are made on strongly alkaline solutions a straight-line calibration curve is obtained when absorbance is plotted versus C_{In}, because in this medium dissociation is essentially complete, practically no HIn is present, and consequently $C_{In} \simeq [In^-]$. Next suppose that the photometric measurements are made under less alkaline conditions where, say, the indicator is only about 50% dissociated. Under these conditions the exact degree of dissociation (at a certain pH maintained by a buffer) will depend on the formal concentration of the indicator. The higher this concentration, the lower is the degree of dissociation. Nevertheless, if the absorbance were plotted against the molar concentration of the absorbing species, $[In^-]$, a straight line would be obtained. However, the conventional plot is one of absorbance versus C_{In}, since it is the latter quantity which is of analytical interest. Because an equality no longer exists between C_{In} and $[In^-]$, a straight line cannot be expected. The two quantities are related via the expression for the acidity constant, and, in fact, this constant can be computed from such an apparent deviation from Lambert–Beer's law. In a photometric determination, therefore, it is necessary to clearly differentiate between the absorbing species and sought-for substance and, where the two are not identical, the relationship between their concentrations under the prevailing conditions becomes of utmost importance.

Finally, it may be seen that deviations (usually erratic ones) from a straight-line plot may result from temperature changes during the measurements. Variations in temperature cause expansion or contraction of the solution and thereby changes in concentration. The novice sometimes fails to appreciate that the solution under measurement is exposed to a light beam of considerable intensity and the radiant power absorbed by the solution is converted to heat. Thus, significant warming can occur when the solution is allowed to remain in the cell compartment of the instrument for a protracted period of time. Such warming can result in evaporation losses, especially if volatile nonaqueous solvents are employed. Additionally the warming frequently causes the formation of gas bubbles due to the release of absorbed gases (commonly air). The bubbles cling to the cell walls, and if they are in the light path, highly erratic analytical results are obtained. Temperature effects become especially serious when the absorbing species participates in a temperature-sensitive equilibrium.

36.7 Blank and Reference Solutions

Although photometric determinations can be performed with highly concentrated solutions of substances of low absorptivities, the principal analytical application is to rather dilute solutions. Often the substance to be determined is transformed by a suitable chemical reaction to a species that absorbs highly. The reagents and solvents used in this process are seldom of adequate purity. The impurities present will have the greater influence on the result the smaller the concentration of the substance to be determined. For example, in the photometric determination of small amounts of silica in water, silica impurities may be present in the reagents and are introduced from the glass vessels used in the operation. It is necessary to compensate for these impurities. Such compensation is secured by using a "blank" instead of the solvent in the initial setting of the meter to 100% T. A blank is commonly prepared in a manner identical with the treatment of the sample, but without its addition. If the sample is a liquid, an equal volume of the solvent is usually added in the preparation of the blank.

A blank of the type described above does not compensate for any species present in the original sample that absorbs at the wavelength used in the photometric measurement of the substance of interest. In this situation and preferentially where higher accuracy is desired, comparison is made with a solution of a standard. Such a solution is commonly prepared from material closely resembling the sample material in composition, but having an exactly known content of the component to be determined. For example, in the photometric determination of manganese in steel, a standard steel of known manganese content is used. Solutions of the sample and standard are prepared in an identical manner. A cell filled with water is placed in the light beam when the meter is set to 100% T. Then the transmittances (or absorbances) of the sample and standard solutions are measured. From the two values obtained the concentration of the manganese in the sample solution is calculated and thus the manganese content of the steel sample. The closer the transmittances of the sample and standard solutions the better the result of the determination.

Another way of obviating the interference by absorbing impurities in the sample is application of the technique of standard addition. A known amount of the substance to be determined is added to one aliquot of the sample solution. This aliquot and one of the original solution are then treated in an identical fashion. The transmittances of the two final solutions are measured with the solvent or a blank being employed to set 100% T, and the data obtained are used to calculate the unknown concentration. The calculations involved when operating with a standard or with the standard addition technique are illustrated in Section 37.3.

Photometric Error 36.8

As with any other experimental measurement, the transmittance or absorbance value finally obtained in a photometric determination is limited in accuracy and precision. The reliability of the result depends on the quality of the instrument, the conditions under which the determination is made, the reproducibility of the instrumental setting, etc., and, of course, on the care of the operator. It is of special interest to consider how the precision of the photometric measurement, that is, the so-called photometric error, affects the result of the determination. The relationship between this photometric error and the precision of the concentration value obtained is complex. The mathematical relationship can be obtained by differentiation of the expression of Lambert–Beer's law, but the derivation is beyond the scope of this textbook. Such a formula, however, was used to calculate the data leading to the curve shown in Figure 36-5. This plot allows the relative

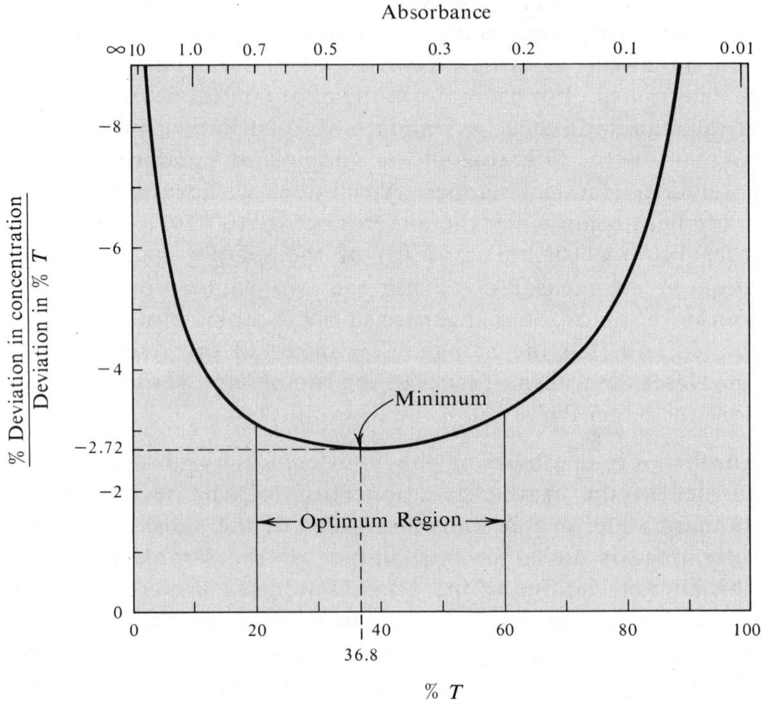

Figure 36-5. Plot of the relative deviation in concentration caused by a 1% photometric error versus per cent transmittance (lower abscissa scale) and absorbance (upper abscissa scale). Calculated from $100\ \Delta c/c\ \Delta T = 0.434/T \log T$, for $\Delta T = 1\%$.

deviation in the concentration caused by an absolute photometric error of 1% T to be read. The curve has an interesting shape. It can be seen that the relative deviation in concentration is a minimum at a transmittance of about 37%. Here a photometric error of 1% T, that is, of 1 division on a 100-division transmittance scale, causes a relative deviation of 2.7% in the concentration. The same photometric error at 80% T produces a relative deviation of 5.6% in concentration. For a good instrument the actual photometric error encountered is often only 0.2% T, which at the optimum condition of 37% T yields a concentration value precise within 0.5%.

It can be appreciated that, where possible, the conditions should be adjusted so that in the measurement of the sample solution, the transmittance falls between about 10 and 80% or better between 20 and 60%; that is, the absorbance lies between 0.7 and 0.2. How these conditions can be secured will be considered in Section 36.9.

The above considerations apply fully only to simple instruments. Each type of instrument has its particular point or range of optimum performance, which either is given in the manufacturer's instructions or may be established experimentally. An attempt should be made to operate as close as possible to the optimum condition or within the range prescribed.

Optimum Conditions for Photometric Methods 36.9

Before the transmittance or absorbance is read, photometric methods may involve a number of preliminary steps. Consequently, the selection of the optimum conditions must be based on the evaluation of many facts. Two groups of steps may be differentiated: (1) those involving chemical operations and (2) those related to instrumental considerations.

Only in rare cases is the substance to be determined a highly absorbing species per se. Chemical reactions of various types are employed to produce such a species. Some examples of the approaches encountered are considered in Chapter 37. Often the substance to be determined is a minor, or even a trace constituent, so that separation and enrichment steps are necessary, involving such techniques as precipitation, extraction, distillation, and chromatography. The principles of such methods are discussed in other chapters of this textbook. Here the discussion is directed to the second group of steps, instrumental considerations.

One of the basic factors to be considered is the selection of an appropriate wavelength. To delineate some of the relevant aspects reference is made to the absorbance curve shown in Figure 36-6. This curve is a hypothetical one and is drawn to emphasize various features relevant to the discussion. The photometric determination will have the highest sensitivity if the wavelength used corresponds to the largest absorbance peak, that is, to 450 mμ for the assumed curve. However, the choice must often be tempered by practical considerations. The situation involving the presence of some other substance that absorbs at that wavelength needs no elaboration. The

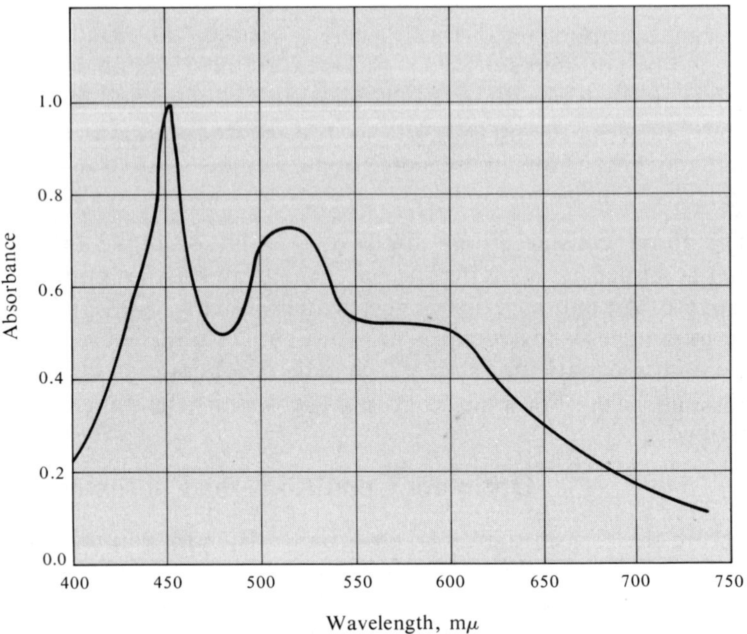

Figure 36-6. Hypothetical absorbance curve.

major difficulty encountered is the inability to reproduce in actual determinations the exact wavelength setting used in the earlier establishment of the calibration curve. A slight change in the wavelength setting results in a considerable change in the absorptivity and consequently in the slope of the calibration curve. This difficulty is avoided if a reference solution is used and the wavelength setting is left untouched during all the relevant photometric measurements, or if the calibration and determination are performed consecutively without changing the setting.

An additional point is of importance in relation to peaks. For each wavelength in the definite band passed by the slit the absorptivity of the substance is different and the light has a low degree of monochromacity. Consider in Figure 36-6 the flat peak around 525 mμ or especially the long shoulder stretching between 550 and 600 mμ. Along the shoulder the absorptivity at each wavelength is essentially the same. Thus the substance, so to speak, does not differentiate between light of 560 and, say, 590 mμ. For the substance either of the wavelengths are absorbed to the same extent, and consequently light at any wavelength range along the shoulder is monochromatic for the substance, but the light is not monochromatic in terms of wavelength. Thus selecting a wavelength band covering a well-rounded peak or a flat shoulder will yield quasi-monochromatic light; then the strict requirements set forth in the derivation of the Lambert–Beer law can be relaxed. For the hypothetical substance considered in Figure 36-6, selecting a wavelength along the shoulder would resolve the problem of monochromacity but, of course, the sensitivity would be reduced by a factor of about 2.

As concluded in Section 36.8, the photometric error will have a relatively small effect on the precision of the concentration value if the transmittance of the sample solution falls between about 20 and 60% *T*. It is important to consider this fact in establishing a photometric procedure, so that the most favorable portion of the calibration curve will correspond to the optimal range of the error curve.

When the transmittance of the sample solution falls outside the optimum range the situation can be remedied as follows. If the transmittance is too small, either the sample solution may be diluted to provide a lower concentration of the absorbing species or the photometric measurement may be made in a cell affording a shorter light path or at another wavelength at which the absorptivity is smaller. The last remedy is less convenient because it requires that a new calibration curve be established at the new wavelength. If the transmittance is too great, the photometric measurement may be made with a cell that affords a longer light path. If such a cell is not available or its use does not bring about the desired result, change may be possible to another wavelength where the absorptivity is higher. The only other and less convenient approaches are to prepare the final solution anew either using a large amount of sample or diluting to a smaller volume. Evaporation of solvent from the final solution is of value in some cases.

Ultraviolet and Infrared Spectrophotometry 36.10

The treatment so far has used examples that involve the visible region of the spectrum. The principles and laws discussed, however, are applicable to the absorption of electromagnetic waves of any wavelength. Of special importance are photometric measurements in the ultraviolet (below about 400 mμ) and infrared (above about 780 mμ). The procedures involved if the species to be determined absorbs in the ultraviolet region are identical with those described for the visible range. But some modifications of the instrument are necessary. Most glass is not transparent to shorter wavelengths, and prisms, gratings, lenses, and cells must be fashioned from special glass or quartz.

Similar conditions exist for the infrared region. Glass prisms have insufficient dispersing power for such light, and glass becomes progressively less transparent with increasing wavelength. Consequently, prisms and cells fashioned from sodium chloride are commonly employed. In addition, solvents, including water, have strong absorption bands in the infrared region. Where no suitable solvent is available, either due to strong absorption or to inadequate solvent properties, the sample can be mixed with a solid, often potassium bromide, which is transparent at the relevant wavelength region range. The finely powdered, homogeneous mixture is pressed to a flat disk, which is mounted in the light beam. A disk made from the diluent only can be used to set the instrument to 100% *T*.

Infrared absorbance measurements are most often employed for organic substances to obtain spectra that allow deductions as to the structures. There is also a difference in the mechanism of absorption. In the visible and ultraviolet region the absorption of energy involves excitation of electrons to higher energy levels. In contrast, in the infrared region the absorption involves an increase in the vibrational and rotational energy of the absorbing species. Vibrations are excited in molecules by the stretching and bending of chemical bonds and account for the frequent use of this technique for the elucidation of structural details.

36.11 Questions

36-1. Define absorbance and transmittance.

36-2. In your own words, state what is meant by the terms spectrum, monochromacity, and compound light.

36-3. What is stated by Lambert's law? Beer's law?

36-4. Explain why the degree of monochromacity of the light is of importance in photometric determinations.

36-5. Describe in your own words the difference between colorimetry and photometry as the terms are used in chemical analysis.

36-6. What is the approximate range of the visible portion of the electromagnetic spectrum in angstroms? In millimicrons? In microns?

36-7. What is a "blank" in a photometric determination? What is its function?

Explain what difficulties may be encountered if the wavelength selected for a spectrophotometric determination corresponds to a sharp peak in the absorbance curve?

What are the advantages of a plot of absorbance versus concentration rather than per cent transmittance versus concentration as a calibration curve for a photometric determination?

When the concentration of an absorbing species is doubled and other conditions are kept constant, what is the effect on the transmittance? On the absorbance?

List the causes of some of the so-called deviations from Lambert–Beer's law.

What are the advantages of using a standard sample as a reference solution in a photometric determination?

Why are unsatisfactory results obtained when a practical photometric determination is based on the use of a tabulated absorptivity value?

Is light of "single color" or "pure color" identical with monochromatic light? Explain.

Absorbance data may be presented as a plot of the logarithm of the absorbance versus wavelength. In such a plot all curves for a single substance will show the same shape, regardless of its concentration, and will only be displaced from each other with regard to the ordinate. By a mere parallel shift, two such curves may be superimposed. Derive this finding as a mathematical consequence of Lambert–Beer's law. (*Hint:* Express $\log A$ in terms of the expression for Lambert–Beer's law.)

In a *double-beam* spectrophotometer the beam energy from the monochromator is split into two beams by an optical system. One beam passes through a cell filled with the solvent and the second through an identical (matched) cell filled with the sample solution. The emerging light beams strike identical photodetectors and the *ratio* of the two electrical signals is presented on a meter dial. Based on Figure 36-1, sketch your conception of such a double-beam instrument. Consider its operation in obtaining data for a transmittance or absorbance curve and in a photometric determination. What might be its advantages and disadvantages over a single-beam instrument?

37 APPLIED COLORIMETRIC AND PHOTOMETRIC DETERMINATIONS

Only relatively few substances are so highly colored or, more generally, have such a high absorptivity that their direct colorimetric or photometric determination is possible. Of the inorganic cations, copper(II) and nickel(II), for example, form colored aquo ions, but the color is so feeble that relatively high concentrations are required for a reasonably successful determination. However, the principal domain of photometry is the determination of small amounts and low concentrations. Therefore, it is usually necessary to transfer the cation or any other species to be determined into a strongly absorbing entity. Sometimes oxidation (or reduction) will yield a strongly absorbing species. For example, manganese is oxidized to permanganate ion and chromium to chromate ion. However, the most satisfactory approach for obtaining strongly absorbing species is to employ a chromogenic agent, that is, a color-forming agent.

Chromogenic Agents 37.1

For inorganic analysis, chromogenic agents are usually complexing agents that form a strongly absorbing complex with the species of interest. Only a limited number of inorganic substances are suitable as chromogenic agents for quantitative purposes. Common examples include ammonia, which forms an intensely blue complex with copper(II), and thiocyanate ion, which forms an intensely red complex with iron(III) and a red-orange one with molybdenum(V). Most chromogenic agents are organic substances. Several requirements must be fulfilled if a chromogenic agent is to be suitable for a given photometric purpose. The agent itself should not absorb light in the wavelength region in which the complex absorbs. If it does absorb slightly, compensation is possible by a proper reagent blank; where significant absorption by this agent occurs, special measures are required (see the consideration of dithizone in Section 37.2).

The absorptivity of the colored complex formed should be high and remain essentially constant when changes in conditions take place within a reasonable range. Such conditions include temperature, pH, and the presence of inert salts.

The color should develop rapidly and once obtained should be stable at least for a period of time sufficient to permit the photometric (or colorimetric) measurement. The complex formed should have a high value for its stability constant. As was explained in Section 36.7, a straight-line relationship, according to Lambert–Beer's law, is obtained only if the absorbance is plotted versus the concentration of the absorbing species, but in practice the concentration of the sought-for substance is usually plotted on the abscissa. As long as the conversion of the species to be determined to the colored complex is essentially complete, a straight-line plot will still be obtained. If this complex dissociates, however, deviation from a straight line will result. Fortunately, the equilibrium can often be shifted favorably by the addition of an adequate excess of the chromogenic agent. But such an excessive addition can obviously only be employed if the reagent itself does not absorb appreciably at the wavelength used in the determination.

High selectivity is a further desirable feature in order that interferences be few in number and hence that the necessity of prior separations or resort to masking be minimized. From a practical viewpoint, it is also preferable that the solution of the chromogenic agent be stable in storage so that its frequent preparation is avoided.

Many of the complexes used for photometric determinations are readily extractable from an aqueous solution into an organic solvent nonmiscible with water. This behavior offers an advantage in permitting the concentration of small amounts of the colored complex and often confers added selectivity on the determination.

37.2 Examples of Photometric Determinations

For inorganic substances, an extremely large number of photometric (and colorimetric) procedures are known and additional ones are developed continually. Only a few representative examples can be given here to illustrate the principles involved.

Manganese is often determined by oxidation to permanganate ion with potassium periodate or ammonium persulfate as the oxidant. The absorbance of the permanganate ion can be measured at about 530 mμ (the molar absorptivity is 2.3×10^3 liters/mole-cm). The method is used routinely for the determination of manganese in steel and other alloys. Iron, if present, is decolorized by the addition of phosphoric acid, which yields a colorless iron(III) complex (Section 25.5).

Nickel can be determined as the red complex formed with dimethylglyoxime, $CH_3C(:NOH)C(:NOH)CH_3$. The complex can be extracted into chloroform and determined at 375 or 327 mμ (molar absorptivities of 3.5×10^3 and 5.0×10^3 liters/mole-cm, respectively). This separation improves the selectivity since, for example, copper may be kept in the aqueous phase by masking it as the copper(I)–thiosulfate complex. The sensitivity is excellent since less than 2 μg of nickel present per milliliter of the aqueous solution can be determined accurately. Alternatively, the nickel–dimethylglyoxime complex can be oxidized (e.g., with bromine) in an alkaline aqueous medium to an intensely red-colored entity, which allows direct photometric measurements at 465 mμ (molar absorptivity, $\sim 1.5 \times 10^4$ liters/mole-cm).

Titanium, in small amounts in alloys, rocks, minerals, etc., is frequently determined in a strongly acidic medium by the addition of hydrogen peroxide and photometric measurement of the orange-colored peroxotitanate ion at 410 mμ (molar absorptivity, 5×10^2 liters/mole-cm).

Copper may be determined photometrically in many materials, including blood serum, as the yellow- to brown-colored copper(II) complex formed with the diethyldithiocarbamate anion, $(C_2H_5)_2NCS_2^-$. The complex is extractable into 1-butanol, chloroform, carbon tetrachloride, or other organic solvents. The selectivity of the extraction is often augmented by the addition of citrate and EDTA, which mask many cations. Photometric measurements are made at 436 mμ (molar absorptivity in carbon tetrachloride, 1.3×10^4 liters/mole-cm).

So-called isopolyacids are obtained by the condensation of two or more molecules of a simple oxygen acid to form an aggregate; their structure involves a central ion linked to surrounding acid anhydride molecules through oxygen. If two or more different acids are involved, the resulting condensed acid is known as a heteropoly acid; the molybdophosphoric acids, $P_2O_5 \cdot xMoO_3 \cdot yH_2O$ (x = 12 or 24 commonly) are well-known examples. Heteropolyacids and their anions are of importance in chemical analysis, especially in photometric and colorimetric methods and in solvent-extraction procedures.

37.2 EXAMPLES OF PHOTOMETRIC DETERMINATIONS

The yellow-colored polymolybdophosphate complex (also known as phosphomolybdate) is frequently employed for the determination of phosphate in rocks and fertilizers. Photometric measurements are made at 380, 400, or 420 mμ. Some heteropolyacids are reduced readily to blue-colored species known as heteropoly blues. With molybdophosphoric acid, the product is known as phosphomolybdenum blue. The development of such a more deeply colored species affords greater sensitivity and allows the determination of extremely small amounts of phosphorus. The reduction is effected with, for example, tin(II) or ascorbic acid.

Silicon and arsenic form, respectively, silicophosphoric and arsenophosphoric acids; these are also reducible to heteropoly blues. These reactions are frequently employed for the determination of small amounts of these elements. For example, silica in water is often determined photometrically in this manner.

An important method for the photometric or colorimetric determination of small amounts of ammonia, known as Nessler's method, may be mentioned. When conducted as a visual comparison method, the ammonia-containing sample solution and a series of ammonia standards are placed in special flat-bottomed test tubes known as Nessler tubes. A solution of potassium tetraiodomercurate(II) is added. The overall reaction occurring may be expressed as

$$\underset{\text{Nessler's reagent}}{2HgI_4^{2-}} + 2NH_3 \rightarrow \underset{\text{intense orange}}{\underline{NH_2Hg_2I_3}} + NH_4^+ + 5I^- \qquad (37\text{-}1)$$

(The colored product may be formulated in other ways.) A suspending agent is added, usually gum arabic, to retard flocculation and settling of the colloidal precipitate. The solutions are all diluted to the same volume and the visual comparison effected promptly. Alternatively, standards and a sample solution, prepared similarly, may be measured photometrically. For the determination of small amounts of nitrogen, a Kjeldahl digestion can be performed (see Section 14.9) and after the addition of sodium hydroxide the ammonia is distilled (or diffused) and "nesslerized."

Dithizone (i.e., diphenylthiocarbazone), $C_6H_5NHNHCSN:NC_6H_5$, forms colored complexes with many metal ions. The reagent itself as well as the complexes are only slightly soluble in water but are readily soluble in carbon tetrachloride. In this solvent, the "free" dithizone is green, but many of the complexes are red or orange. Dithizone is one of the most sensitive reagents for metals and is frequently employed in trace analysis, notably in the separation and determination of lead, mercury, bismuth, and copper. Unfortunately, the reagent absorbs at the wavelength maximum of the metal complexes and vice versa, although to a different degree. Consequently, it is necessary either to remove the excess of the reagent or to add it in an exact, reproducibly measured amount. In the latter case, either the increase in the absorbance may be measured at a wavelength at which the metal complex absorbs strongly or the decrease in absorbance at a wavelength at which dithizone itself absorbs strongly. If dithizone is represented by HDz, the extraction of a metal–dithizone complex from an aqueous solution can be represented by

$$\underset{\text{in water}}{M^{n+}} + \underset{\text{in solvent}}{n\text{HDz}} \rightleftharpoons \underset{\text{in solvent}}{\text{MDz}_n} + \underset{\text{in water}}{n\text{H}^+} \qquad (37\text{-}2)$$

From this equation it can be appreciated that the extraction efficiency of a metal will depend on the pH and the stability of the metal–dithizone complex. The stronger a complex, the lower the pH at which it can be effectively extracted. This behavior confers some degree of selectivity on the use of dithizone in practical analysis. (Solvent extraction is considered further in Section 45.4.)

Various oxidants are determined by allowing them to react with essentially colorless substances to form colored products. The absorbance measured is proportional to the amount of the oxidant. Redox indicators (Section 24.9) often find use for this purpose. Probably the best known of these procedures is the colorimetric or photometric determination of "active" chlorine in water in the parts per million range using *o*-tolidine, $(4\text{-}NH_2\text{-}3\text{-}CH_3C_6H_3)_2$. The oxidation product of this compound in acidic solution is yellow. Since solutions of chlorine are unstable, artificial standards are often used when a visual comparison is applied. These standards are made of yellow-colored plastic or glass or are obtained by mixing solutions of copper(II) sulfate and potassium dichromate. As a photometric method with measurements at 438 mμ, the sensitivity limit is about 0.01 ppm.

Absorption photometry is also possible in the solid and gaseous states. Of special analytical interest is absorption flame photometry. Discussion of this topic is deferred to Chapter 40.

37.3 Calculations in Photometric Determinations

Some illustrative examples of calculations involving Lambert–Beer's law have already been given in Section 34.4. Some additional examples will further promote an understanding of the principles involved in applied photometric determinations.

Example 37-1. A substance S is determined in the following way. A 0.7520-g amount of the sample material containing S is dissolved, a chromogenic agent is added, and the solution is diluted to a volume of exactly 250 ml. As a reference, a standard known to contain 0.345% of S is used; a 0.3500-g amount of the standard is dissolved, the chromogenic agent is added, and the solution diluted to a volume of exactly 50 ml. A 1-cm cell is used for all photometric measurements. The instrument is adjusted to 100% transmittance with the pure solvent in the cell; then the transmittance is read for the sample solution and for the standard solution; the values found are 52.5 and 45.0%, respectively. Calculate the per cent of S in the sample material.

For the reference solution, the following equation applies:

$$-\log 0.450 = ab \times \frac{0.3500\text{ g} \times 0.345\%}{50.0\text{ ml}}$$

And for the sample solution where x is the per cent of S in the sample material:

$$-\log 0.535 = ab \times \frac{0.7520\text{ g} \times x}{250\text{ ml}}$$

By division,

$$\frac{-\log 0.450}{-\log 0.525} = \frac{0.3500 \times 0.345}{50.0} \times \frac{250}{0.7520 \times x}$$

and

$$\frac{0.653 - 1}{0.720 - 1} = \frac{0.802_5}{x}$$

Solving,

$$x = 0.802_5 \times \frac{0.280}{0.347} = 0.648$$

Hence, the per cent of S in the sample material is 0.648%.

Example 37-2. A 0.500-g sample of a substance, after appropriate treatment, yielded a solution of exactly 100 ml. The solution showed a transmittance of 5.0% in a 1-cm cell. What volume of this solution in milliliters must be further diluted to exactly 250 ml to obtain a solution having a transmittance of 40%, which is close to the optimum value (see Section 36.3), when measured in the 1-cm cell?

Let C and C' represent the original and final concentration, respectively. Then

$$-\log 0.050 = abC$$

$$-\log 0.40 = abC'$$

The relation between the two concentrations is obviously

$$C' \times 250 = C \times x$$

where x is the milliliters of the initial solution to be diluted to exactly 250 ml. Hence,

$$C' = \frac{C \times x}{250}$$

Insertion of this expression for C' in the expression of Lambert–Beer's law, division of one expression by the other, and substitution of the values for the logarithms yields

$$\frac{0.70 - 2}{0.60 - 1} = \frac{a \times 1 \times 250}{a \times 1 \times x}$$

$$x = \frac{250 \times 0.40}{1.3} = 77 \text{ ml}$$

Example 37-3. What transmittance will be read in a 5.00-cm cell for a 0.00200% solution of a substance of formula weight 250 and possessing at the wavelength used a molar absorptivity of 5.00×10^3 liters/mole-cm?

$$\log T = -bc = -5.00 \times 10^3 \times 5.00 \times \frac{0.00200 \times 1000}{100 \times 250} = -2.00$$

Hence

$$T = 0.010$$

Example 37-4. The concentration of a substance S in a solution is determined photometrically by the technique of a standard addition. The transmittance of the solution is found to be 45.1%. Then to the solution a known amount of S is added corresponding to an increase of 0.0040 g of S per liter of the solution. The transmittance is now found to be 29.5% under identical conditions. What is the concentration of S in the initial solution assuming Lambert–Beer's law is obeyed?

Let x be the concentration of S in grams per liter in the initial solution. Then

$$-\log 0.451 = ab \times x$$

The concentration of S after the "standard addition" is $x + 0.0040$. Hence,

$$-\log 0.295 = ab \times (x + 0.0040)$$

Division and substitution of the value of the logarithms yields

$$\frac{0.654 - 1}{0.470 - 1} = \frac{x}{x + 0.0040}$$

$$\frac{0.346}{0.530} = \frac{x}{x + 0.0040}$$

On rearrangement,

$$0.184x = 0.0040 \times 0.346$$

And solving,

$$x = 0.0075_2$$

Hence, the concentration of S in the sample solution is 0.0075 g/liter.

37.4 Questions

37-1. What is a chromogenic agent and what properties should it have for advantageous use in photometry?

37-2. Explain why inorganic complexing agents are seldom used as chromogenic agents.

37-3. When a photometric determination is performed at a wavelength corresponding to a sharp peak in the absorbance curve, what special precautions must be taken?

37-4. What are the advantages of combining a solvent extraction with a photometric determination?

37-5. Outline in your own words some general principles and possibilities for transferring a substance to be determined into a strongly absorbing species.

37-6. Elaborate on the possibilities of using a system that does not fully comply with Lambert–Beer's law for a spectrophotometric determination.

37-7. How can oxidation and reduction reactions be used in photometric analysis? Elaborate on the principles and give examples.

37-8. What might be the consequences of having a turbidity present in a solution subjected to a photometric measurement?

37-9. A pair of cells is badly matched and cell A has a light path 1% longer than cell B. Elaborate on the effect on the results if cell A is used for the blank in the standardization but B is used for the blank in the actual determination. Will the result be high or low?

37-10. How would you cope with a permanent absorbing background present in a solution to be subjected to a photometric determination?

37-11. Silver ion is found to interfere in the determination of ammonia by Nessler's method. Explain.

37-12. Care should be exercised in the cleaning, storage, and use of photometric cells. Explain.

37-13. Elaborate on the possibility of determining the formal solubility of a sparingly soluble electrolyte MA when an accurate photometric method is known for the determination of the cation M.

37.5 Problems

37-1. What is the absorbance when the transmittance is 69.9%?

Answer: 0.156

2. What is the per cent transmittance corresponding to an absorbance of 0.025?

Answer: 94.4%

3. Calculate the molar absorptivity of a substance of formula weight 273 showing in a 0.0100% solution a transmittance of 24.6% in a 2-cm cell.

Answer: 831 liters/mole-cm

4. To 50 ml of an aqueous $CuSO_4$ solution, an excess of aqueous ammonia is added to form the copper(II)–ammine complex, and the solution is diluted to exactly 100 ml with water. A volume of 50 ml of a standard $CuSO_4$ solution known to contain 2.0 mg of copper per milliliter is analogously treated and diluted. Portions of the two solutions are placed in the cells of a Duboscq comparator. When a color match is secured, the lengths of the light paths in the sample and standard solution are found to be 43 and 50 mm, respectively. Calculate the copper concentration in the original sample solution.

Answer: 2.3 mg/ml

5. A standard contains 0.620% of a substance S. When a 0.5000-g amount of the standard is dissolved, the color developed, and the solution after dilution to exactly 100 ml is measured photometrically in a 1-cm cell, a transmittance of 19.3% is observed. A similar material having an unknown content of per cent S is to be analyzed. A 0.6402-g amount of it is analogously treated and dissolved to a volume of 50 ml, and under the same conditions a transmittance of 55.3% is observed. What is the percentage of S in the unknown?

Answer: 0.087% S

6. What is the molarity of a solution which shows in a 2-cm cell the same transmittance as a 3.0×10^{-4} *M* solution of the same substance in a 5-cm cell?

Answer: 7.5×10^{-4} *M*

7. Solution *A* shows a transmittance of 20.0% in a 1-cm cell. Solution *B* of the same substance shows a transmittance of 60.0% in a 1-cm cell. What volume in milliliters of solution *B* must be added to 100 ml of solution *A* to obtain a solution which shows a transmittance of 40.0% in a 1-cm cell?

Answer: 171 ml

8. The aqueous solution of a substance shows a transmittance of 28.7% in a 1-cm cell. A volume of 40.0 ml of this solution is diluted with 100 ml of water. What transmittance will the dilute solution show when measured in a 5-cm cell?

Answer: 16.8% *T*

9. In the determination of inorganic ions it is difficult to attain a molar absorptivity greater than 2×10^4 liters/mole-cm when operating in the visible region. The largest cells that most instruments will accommodate without modification have a 5-cm light path. Such instruments, with meticulous attention to possible sources of error, have a minimum detectable absorbance value of about 0.005. For an in-

organic ion calculate the molar concentration of a solution that will give a significant response under these favorable conditions.

Answer: 5×10^{-8} *M*

37-10. A solution containing 4.65 mg of a light-absorbing substance (*486.0*) in a volume of 250.0 ml shows 34.8% transmittance when measured in a 1.00-cm cell at 432 mμ. When a solution, containing 1.02 mg of the same substance in 100.0 ml, is measured in a 2.00-cm cell at 521 mμ, 44.0% transmittance is observed. How much smaller is the molar absorptivity at 432 mμ than at 531 mμ? Express the result as a percentage of the molar absorptivity at the latter wavelength.

37-11. In an acid–base titration, 30.0 ml of a 1.0×10^{-2} *F* NaOH solution is found to be equivalent to 40.0 ml of a solution of potassium tetroxalate, $KHC_2O_4 \cdot H_2C_2O_4 \cdot 2H_2O$. In the titration of 20.0 ml of this tetroxalate solution under acidic conditions, a volume of 5.0 ml of a $KMnO_4$ solution is required. Calculate the per cent transmittance of a solution containing 0.50 ml of this $KMnO_4$ solution per liter when measured in a 3.00-cm cell. The molar absorptivity of $KMnO_4$ is 6.7×10^3 liters/mole-cm at the wavelength used in the measurements.

37-12. Of a standard steel sample containing 1.67% w/w chromium, an amount of 0.5000 g is dissolved in acid, the chromium is suitably oxidized to dichromate, and the solution is diluted to a volume of 250.0 ml. A volume of 10.0 ml of this solution is diluted with water and acid to 100.0 ml; the resulting solution shows 40.7% transmittance in a 1.00-cm cell. When an amount of 0.7500 g of an unknown steel is dissolved, oxidized, and diluted to 200.0 ml, the resulting solution shows 61.3% transmittance under identical instrument settings. What is the chromium percentage in the unknown steel?

37-13. A soluble complex, MY, dissociates according to the reaction $MY^+ \rightleftharpoons M^+ + Y$, where M^+ is a metal ion and Y is a complexing agent. At 500 mμ, the species Y and MY^+ do not absorb light, but a 1.0×10^{-3} *M* solution of M^+ in a 3.00-cm cell shows 50.0% transmittance. When 2.0×10^{-4} mole of MY is dissolved in water and diluted to exactly 100 ml, the solution shows 76.0% transmittance at 500 mμ. Calculate the stability constant of the complex.

37-14. The molar absorptivities of compound A are 341 liters/mole-cm at 510 mμ and 692 liters/mole-cm at 625 mμ. The molar absorptivities of compound B are 722 liters/mole-cm at 510 mμ and 488 liters/mole-cm at 625 mμ. A solution containing only A and B as light-absorbing species shows 32.7% transmittance at 510 mμ and 16.2% transmittance at 625 mμ when measured in a 5.00-cm cell. Calculate the molar concentrations of A and B in the solution.

37-15. Iron(III) gives a red complex with 5-sulfosalicylic acid; this is a sensitive reaction used for the photometric determination of iron. A 0.100-g sample of material to be analyzed is dissolved in 20.00 ml of water and a volume of 5.00 ml of 1% 5-sulfosalicylic acid solution is added. When this mixture is measured in a 1.00-cm cell against water, the % T reading is 54.3. When 10.00 ml of water and 5.00 ml of the reagent solution are mixed and measured in a 5.00-cm cell, the reading is 90.1% T. An amount of 0.200 g of a standard with a content of 0.15% Fe is dissolved in 30.00 ml of water and a volume of 5.00 ml of reagent is added. When measured in a 2.00-cm cell the transmittance reads 60.2% T. Calculate the per cent Fe in the sample.

38 PHOTOMETRIC TITRATIONS

A titration to a photometric end point, or simply a photometric titration, is performed by measuring the absorbance of the solution after each addition of titrant and then plotting the absorbance values versus the milliliters of titrant. The volume of titrant solution required to reach the end point is evaluated from the titration curve thus obtained. In comparison with the conventional titration technique a photometric titration obviously requires more effort and additional instrumentation. However, these disadvantages are often more than counterbalanced because of the higher accuracy and precision obtained and because in many cases a titration can be performed photometrically under conditions and with systems where a visual titration is not possible. Some of these conditions will be made evident by discussion and examples; thereby, also, the field of application and the advantages of photometric titrations will become clear.

Two types of photometric titrations may be differentiated: (1) titrations employing an added indicator and (2) self-indicating titrations. With these two types not only are titration curves of different shapes obtained, but there are also some other points that warrant their separate treatment.

Titrations with Indicator Added 38.1

Consider the titration of hydrochloric acid with sodium hydroxide with preneutralized methyl orange as the indicator and assume that the wavelength is selected so that only the yellow (base) form of the indicator absorbs. The titration curve resulting is shown as curve A of Figure 38-1. At the starting point and up to the vicinity of the end point the absorbance is zero, because neither of the species initially present nor those formed during this portion of the titration absorb. Then the indicator begins to change its color and the nonabsorbing red (acid) form of the indicator disappears and the yellow absorbing (base) form appears. Consequently, the absorbance increases. After all the indicator has been transformed to the base form, further addition of nonabsorbing titrant does not

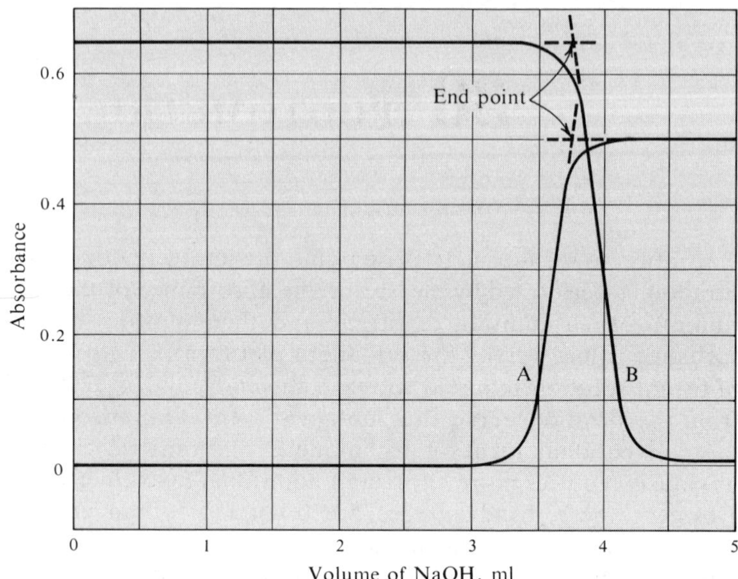

Figure 38-1. Titration curves for the photometric titration of HCl with NaOH using methyl orange as indicator at wavelengths at which the yellow (base) form (curve A) and red (acid) form absorb (curve B), respectively.

cause any changes in absorbance. The question arises as to what point is to be taken as the end point. Methyl orange has a color transition range of pH 3.2 to 4.4. Since the pH at the equivalence point of the titration is 7.0, the indicator must be completely transformed to the base form. Consequently, the point at the upper end of the step in curve A is to be taken as the end point. It should be noted that a slight curvature exists in this area but that the end point is obtained by extrapolation of the two essentially straight-line portions as shown by dashed lines in the figure. Alternatively, if the wavelength is selected so that only the red (acid) form of the indicator absorbs, curve B would result. Here the *first* change in direction is extrapolated and taken as the end point because the pH of the equivalence point (pH 7.0) is already passed when the indicator begins to change color.

Such a photometric acid–base titration has advantages over a visual titration where the sample solution is extremely dilute or where it is colored by nonreacting material to an extent that the eye is unable to discern a color change. A photometric titration is extremely valuable for the titration of weak acids and bases where the color change of the indicator is gradual, and consequently precise visual location of the end point becomes difficult.

Photometric end-point detection employing an added indicator is, of course, not restricted to acid–base titrations. The technique may equally well be applied to other titrations (with the obvious exception of precipitation titrations; see, however, Section 39.2) whenever the adverse conditions mentioned above prevail and replacement of the eye by a photodetector is advantageous.

Self-Indicating Titrations 38.2

Many titrations exist where the system during the titration changes color without an indicator being added. Only few of these systems, however, show a sufficiently abrupt and intense change in color to be suitable for visual indication. Notable examples are titrations with permanganate.

Consider the titration of iron(II) in acidic, phosphoric acid–containing medium with permanganate. In a visual titration, permanganate is added until the first drop in excess imparts a pink color to the solution. This small excess causes no appreciable error if a large amount of iron is being determined. In contrast, in the titration of a small amount of iron, the excess of permanganate necessary to indicate the end of the titration may amount to an error of several per cent. Of course, an indicator blank may be applied, but such a correction becomes difficult or impossible when the sample solution is colored, as is frequently the case with natural waters or solutions obtained from dissolution of alloys. Here a photometric end point is especially expedient. Assume that a wavelength is selected at which permanganate absorbs strongly (535 mμ) and at which no other species absorbs significantly. The titration curve resulting is shown as curve A in Figure 38-2. Up to the equivalence point the absorbance remains zero. Beyond that point the absorbance increases linearly with the amount of permanganate added. The best line through the few points secured after the end point is drawn and extrapolated to intersect with the zero line; the intersection is taken as the end point. Curve B illustrates the situation when a colored but, of course, nonreacting, impurity is present. This "background" absorbance is to be added to the titration curve. Neither the shape of the curve is thereby

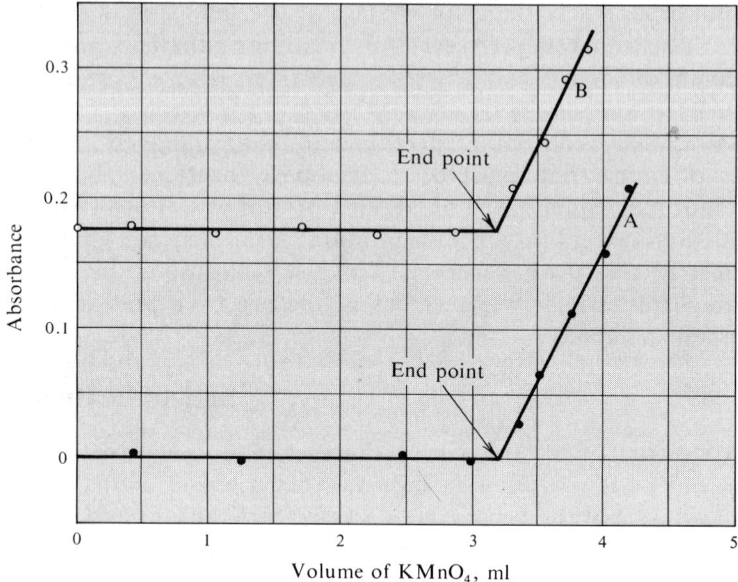

Figure 38-2. Titration curves for the photometric titration of an acidic, dilute Fe(II) solution with $KMnO_4$, at a wavelength at which only MnO_4^- absorbs (ca. 530 mμ). Curve A, pure solution; curve B in the presence of foreign material absorbing at the selected wavelength but not reacting with $KMnO_4$.

changed nor its position relative to the abscissa. All that happens is a shift of the curve parallel to the ordinate.

As mentioned before, only a few titration systems show color changes as abrupt as those involving permanganate. In most cases the color change is too gradual for visual self-indication. But many such systems lend themselves readily to photometric titrations. As an example, consider the titration of copper(II) in ammoniacal solution with EDTA using a yellow filter. The titration curve for this case is shown in Figure 38-3. Initially a high absorbance reading is encountered, because the copper(II)–ammine complexes absorb strongly at the wavelength region transmitted by the filter. On progressive addition of EDTA, the copper is transferred to the more weakly absorbing copper(II)–EDTA complex and the titration curve descends in a linear fashion. Beyond the equivalence point, the addition of nonabsorbing EDTA causes no change in absorbance and a horizontal line results. Extension of the two straight lines yields an intersect that is taken as the end point.

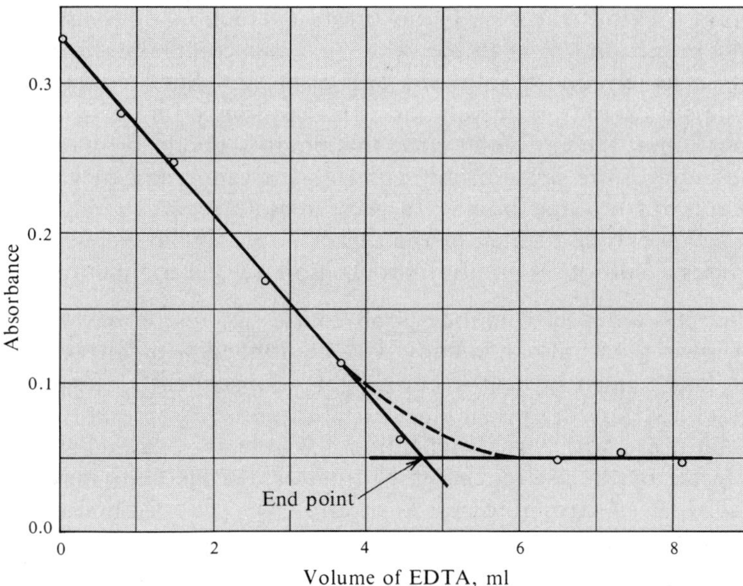

Figure 38-3. Titration curves for the photometric titration of an ammoniacal copper(II) solution with EDTA using a yellow filter.

Some General Considerations 38.3

The two segments in the curve for a self-indicating titration will only be straight if no distortion results from the dilution of the solution during the titration. Such distortion is obviated if the concentration of the titrant solution is far greater (50 to 100 times) than that of the species being titrated. The volume of titrant then required is so small that the resulting dilution has no significant effect. Of course, a buret that is capable of delivering a small volume precisely (that is, a microburet) is required. Another possibility is to correct the absorbance for dilution by multiplying each reading by the factor $(V + v)/V$, where V is the initial volume of the solution to be titrated and v is the volume of titrant solution added, both expressed in the same unit, usually milliliters.

Dilution effects are usually of no moment if an indicator is used because the region of interest on the titration curve is close to the end point and restricted to a very small range of volume.

Titration curves, especially when self-indication is employed, may show a pronounced curvature in the vicinity of the end point. This is the case if the titration reaction does not go to completion, for example, where a weak complex is formed or extremely dilute solutions are involved. Such curvature, unless it extends extremely far, is no hindrance for the precise location of the end point since this point is obtained by extrapolation of the straight-line segments corresponding to values removed from that point. In these remote regions either the substance to be titrated or the titrant is present in excess and causes a shift of the equilibrium position toward completion.

The photometric technique is obviously not restricted to the visible region but can equally well be applied with ultraviolet or, less frequently, near-infrared light. Consequently, the technique extends the possibilities for titrations considerably beyond those with visual indication. Where the wavelength is selected by the use of a monochromator, the methods may be called a spectrophotometric titration, if the distinction is desired.

For the occasional performance of a photometric titration any filter photometer or spectrophotometer may be employed. The titration is performed outside the instrument in a beaker as usual. But after each addition of titrant a portion of the solution is withdrawn, placed in a cell, and the absorbance measured. The portion is reunited with the main portion and the next increment of titrant added, and so forth. More convenient, of course, is a photometer modified to accommodate the titration vessel in the cell compartment. This lighttight compartment has a hole for insertion of the buret tip and provision for stirring the solution. Magnetic stirring is quite convenient. Special devices are available for performing photometric titrations and are called phototitrators. Indeed, automatic titrations are possible by having the change in absorbance at the end point generate an electrical signal to operate a relay that stops the flow of titrant solution.

38.4 Questions

38-1. In your own words define the following terms: self-indication, titration curve, and dilution effect.

38-2. Would you expect extraneous light (remaining at a constant level) to have an effect on accuracy and precision of a photometric titration? Explain your answer.

38-3. Substances A and B are nonabsorbing and react completely to form a stable absorbing species AB_2. Both a photometric determination and photometric titration are possible. Elaborate on the theoretical and practical advantages and disadvantages of the two methods for the determination of either A or B.

Sketch the photometric titration curves for the following self-indicating systems where the numbers given are the approximate effective absorptivities. Assume a 1:1 reaction in all cases. Neglect dilution effects, plot three points (starting point, equivalence point, and 100% excess titrant added), and connect them by straight-line segments.

	Species titrated	Titrant	Reaction product
(a)	5000	5000	0
(b)	0	5000	5000
(c)	0	0	5000
(d)	5000	0	5000
(e)	1000	5000	0
(f)	5000	5000	5000
(g)	1000	3000	8000

Bismuth ion, EDTA, and the bismuth–EDTA complex are colorless; the copper–EDTA complex is colored. Bismuth ion can be titrated in acidic medium with EDTA if some copper(II) is added as the "indicator." Elaborate on the shape of the titration curve. Could a mixture of bismuth and copper(II) be titrated photometrically so that both metals are determined? Explain.

A self-indicating photometric titration and an amperometric titration (Chapter 29) have titration curves similar in shape. Compare the two techniques in their application, limitations, and advantages; name cases where one would succeed, the other fail; what is plotted on the ordinate and abscissa to obtain the titration curves; name interferences for one which, however, would be of no consequence in the other.

A photometric titration curve has the shape of an inverted "V" that rests on the abscissa. What can you deduct from this fact as to the absorptivities of titrant, species titrated, and reaction product?

Some organic acids show considerable differences in their ultraviolet absorbance spectra when dissociated and not dissociated and thus lend themselves to photometric titration. Would you expect that such acids are very strong or very weak? Explain your answer.

39 TURBIDIMETRY, NEPHELOMETRY, AND FLUORIMETRY

Three diverse methods, closely related to photometric and colorimetric determinations either in principle or instrumentation, or both, may be considered briefly. Two of this methods, turbidimetry and nephelometry, are based on the scattering of light. The third method, fluorimetry, is based on the measurements of fluorescent light emitted by some substances on appropriate excitation.

Turbidimetry 39.1

When a turbid solution, that is, a suspension of solid particles in a liquid, is brought into the light path of a photometer, less radiant power reaches the photodetector than if the clear (nonabsorbing) liquid were in the light path. This reduction results from the scattering of light due to reflection on and refraction by the suspended particles. The scattered light is dispersed in all directions, and consequently the radiant power of the beam directed toward the detector is diminished. Some true absorption of light by the particles may also occur. A turbidimetric measurement can be performed in the same manner as a photometric measurement and the result be expressed in absorbance units. A relation between the absorbance and "concentration" of the suspended material can be derived that is analogous to the Lambert–Beer law. However, this formula holds, if at all, only for a small degree of turbidity and a very short light path, because the relationship among number, size, and shape of the particles and the amount of light scattered is complicated. Consequently, the calibration curves obtained for turbidimetric determinations usually are far from linear.

It is not difficult to understand that for a given amount of material suspended in a given volume of liquid, the scattering due to reflection increases with a decrease of particle size. But this reasoning holds only to a certain particle size below which no reflection occurs. A body can only reflect light when its size is at least one half of that of the wavelength of the light striking it. Consequently, blue light is scattered more than, say, yellow or red light and thus affords higher sensitivity in turbidimetry. These facts, coupled with the knowledge that a uniform particle size can seldom be obtained, suggests the rather involved situation that exists.

It is usually extremely difficult to adjust the conditions in a turbidimetric determination so that the particle size is reproducible. Some of the factors exerting a decisive influence include the manner, order, and rate of mixing the reactants, agitation, temperature, the presence or absence of inert electrolytes, and the concentration of the solutions mixed. Large particles are

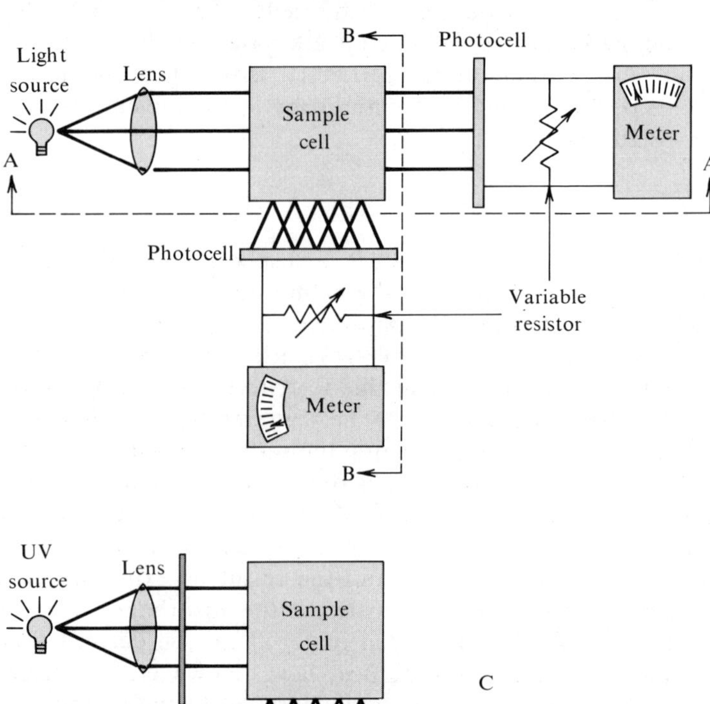

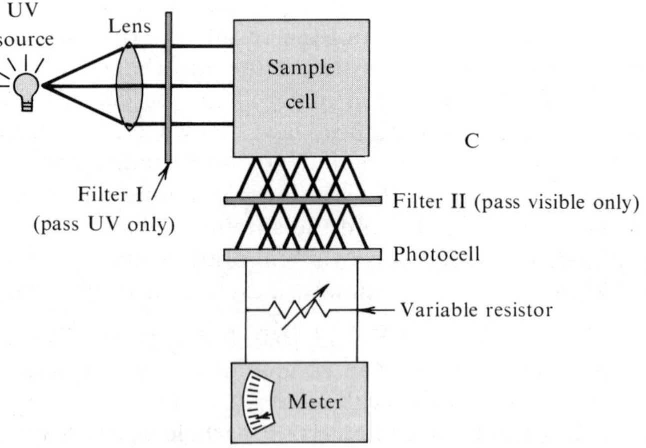

Figure 39-1. Schematic arrangement of the components of a turbidimeter (A), a nephelometer (B), and a fluorimeter (C), viewed from above.

unwanted, in contrast to the requirements for a gravimetric determination, because they cause inhomogeneities by settling. Protective colloids such as gum arabic and gelatin are frequently added to reduce flocculation and settling. In a turbidimetric determination, low accuracy and precision are usually accepted in exchange for speed and simplicity.

As can be seen in A at the left of Figure 39-1 the arrangement of the components in a turbidimeter is the same as in a simple photometer. Indeed, any photometer can be used as a turbidimeter. Although monochromacity is desirable it is not mandatory, and operation with a filter of moderate quality or even with white light is possible. Turbidimetric measurements can also be made visually with a colorimeter.

Possibly the most frequent application of turbidimetric methods in inorganic analysis is the determination of chloride and sulfate in water via the formation of silver chloride and barium sulfate, respectively. Here sufficient accuracy often is secured by comparison of the turbidity developed in the sample with that of standards. These procedures are satisfactory at concentrations as low as a few parts per million. The determination of ammonia by Nessler's method, considered in Section 37.3, is a hybrid between a colorimetric (or photometric) and a turbidimetric method.

Turbidimetric Titrations 39.2

Turbidimetric titrations are performed in a manner analogous to photometric titrations (see Chapter 38). The "absorbance" is plotted versus the volume of titrant solution added. The "absorbance" increases as the precipitate forms and the curve levels off (or declines slightly) when all the sought-for substance is precipitated. Thus the end point corresponds to an abrupt change in slope. The change, for a given amount of substance titrated, occurs at the same volume of titrant added, regardless of the size of the particles formed. The particle size only influences the slope of the curve before the leveled portion. In addition, the end point is obtained from the two lines drawn through a series of points, thus affording an averaging process that reduces the influence of random errors. Consequently, the accuracy and precision attained in a turbidimetric titration are often significantly better than those obtained in the corresponding turbidimetric determination.

39.3 Nephelometry

In a turbidimetric measurement the ratio of the intensities of the transmitted and incident light is measured. It is, however, possible to directly measure the intensity of the scattered light, or more exactly, a portion of that light, and thereby to gain sensitivity. The technique employing this principle is known as nephelometry. The instrument used is termed a nephelometer, and the schematic layout of its principal components is shown in B of Figure 39-1. The photodetector is usually, but not necessarily, mounted at 90° to the incident beam.

Most of the requirements discussed in connection with turbidimetry also apply here and the difficulties encountered are similar. The calibration curves here are also far from linear, unless very low concentrations of the sought-for substances prevail. A nephelometric determination for small quantities of a sought-for substance commonly yields better results than its turbidimetric counterpart because the direct measurement of a small light intensity is more accurate and precise than that of a small change in the intensity of a strong light.

However, this advantage is rarely of such moment that the purchase of a special instrument is justified. Consequently, nephelometry does not play a significant role in common routine analysis, but the technique is an important tool for the evaluation of size and shape of macromolecules and the determination of their molecular weight.

39.4 Fluorimetry

When radiant energy in the visible or ultraviolet is absorbed by a substance, electrons are raised to higher energy levels. Usually the electrons return to the ground state immediately and the energy is released as heat and dissipated in the absorbing medium. With some substances, however, a portion of the absorbed energy is reemitted as light. If the light emission is prompt (usually about 10^{-8} second), the phenomenon is called fluorescence. If the emission is delayed, the process is a form of phosphorescence. The emitted light contains less energy and consequently is of longer wavelengths than the exciting light. This statement represents Stoke's law, for which, however, rare exceptions exist. The return of the electrons to the ground state may proceed in several steps, each representing a discrete, quantized energy difference.

However, molecules give rise to band spectra because of the many vibrational and rotational transitions associated with each electronic transition. Consequently, fluorescent spectra, as absorbance spectra, consist of connected bands. In fact, a plot of fluorescence intensity versus wavelength is often a mirror image of the absorbance curve of the same species obtained under comparable conditions.

The intensity of the fluorescent light depends on various factors. For analytical purposes it suffices to state that at low concentrations (10^{-4} to 10^{-7} M) and within narrow limits, the intensity of the fluorescence is proportional both to the intensity of the exciting light and the concentration of the fluorescing species. In many cases the intensity of the fluorescence is influenced by the temperature, the pH, and the presence of substances that either increase the intensity (activators) or decrease it (quenching agents or quenchers).

Fluorimeters of various design exist, but all operate on the same principle. The light from the excitor lamp (often an ultraviolet-radiating mercury vapor lamp) passes into the sample. The fluorescent light is emitted in all directions; its intensity is measured at an angle to the incident beam, usually at 90°, using a photodetector and meter. In simple instruments, such as that shown schematically in C of Figure 39-1, two filters are used. The filter between the source and the cell isolates the exciting light of the desired wavelength(s). The filter between the cell and the photodetector excludes any exciting light and only allows passage of the fluorescent light. More elaborate instruments, known as spectrofluorometers, are capable of greater selectivity and sensitivity. A monochromator is used to single out a particular wavelength of the fluorescent light for measurement; sometimes a second monochromator is employed to secure monochromatic exciting light.

In a fluorimetric determination the concentration of the species to be determined is obtained from a calibration curve, which is a plot of the intensity of the fluorescent light versus the concentration of the sought-for substance. Substances that quench fluorescence may be determined by the use of a calibration curve that is obtained by addition of known amounts of the quencher to a definite volume of a solution of a suitable fluorescent substance at a definite concentration.

Fluorimetric methods are of considerable value for the determination of trace amounts of substances where photometric and other methods lack adequate sensitivity or selectivity.

The higher sensitivity of fluorimetric determinations when compared with photometric ones is due to the fact that in fluorimetry the strength of a small signal is measured rather than a small difference in two large signals. Where a substance to be determined is itself not fluorescent, it may often be converted to a fluorescent species by reaction with a suitable fluorogenic agent. For inorganic analysis, such agents are usually complexing agents. For example, beryllium in the submicrogram range is determined fluorimetrically after reaction with the organic reagent morin. 8-Quinolinol forms fluorescent chelates with various metals including aluminum, gallium, zinc, and magnesium. The gallium chelate, for example, can be extracted into chloroform. In this solvent, "free" 8-quinolinol shows negligible fluorescence. In this way gallium can be determined fluorimetrically down to 3×10^{-8} g/ml of solvent.

Fluorimetry is not restricted to solutions. The determination of trace amounts of uranium is frequently accomplished by melting in a mold a known amount of the sample with a fixed amount of sodium fluoride. A flat button is obtained on cooling. Similar buttons are prepared containing known amounts of uranium. Under ultraviolet light, a yellow-green fluorescence is visible even to the eye. With a suitable fluorimeter even 10^{-9} g of uranium can be determined with an error of about $\pm 10\%$.

39.5 Questions

39-1. Many fluorimeters can readily be used as nephelometers. Discuss the modifications required when switching from fluorimetry to nephelometry.

39-2. Elaborate on the fact that a turbidimetric titration commonly yields better results than a turbidimetric determination.

39-3. Fluorescence can be detected by the eye if the sample is located and irradiated in a black box having a tube attached through which the observation is made. Elaborate (a) on the possibility of a fluorescence analog to visual colorimetry and (b) on the use of an acid–base indicator fluorescent only in its base form in the titration of the acidity of a highly colored sample (e.g., red wine).

39-4. In the turbidimetric or nephelometric determination of a suspended material that is colored, would you recommend that the incident beam should be of a wavelength that is strongly absorbed? Explain your answer.

39-5. The continuous turbidimetric monitoring of smoke or dust in an air stream is possible. Speculate on the instrumental arrangements.

. The intensity of the fluorescence excited in a fluorescent substance is proportional to the intensity of the exciting light, but only up to a certain intensity. Explain why. When operating within the proportionality range an increase in the sensitivity of a determination can be achieved. How? Compare this situation with that in a photometric determination. How would an increase of the intensity of the light beam affect the sensitivity of a determination by this technique?

40 SPECTROSCOPY AND FLAME PHOTOMETRY

Atomic Spectra 40.1

Matter in any form, when exposed to sufficient energy, can be excited and may then emit light. Of special interest here is the excitation of atoms and atomic ions. When light emitted from excited atoms is passed through a narrow slit and dispersed by a prism or grating, the spectrum obtained under certain conditions consists of discrete lines. Such a spectrum, called either a line spectrum or, from its origin, an atomic spectrum, is obtained, for example, by exposing a material to the heat of a flame, by passing a high-voltage electric spark through a gas, or by striking an electric arc between electrodes, one of which either consists of the substance to be excited or is impregnated with it.

The energy provided by such sources disrupts the chemical bonding to leave atoms or atomic ions and then brings orbital electrons to excited states, that is, to higher energy levels. On return of the electrons to a lower level or to the ground state, the energy taken up is emitted as light. Only definite, quantized transitions are possible, and each transition corresponds to the emission of light of a specific wavelength. Since certain transitions are more probable than others, the intensities of the various lines differ. The greater the energy provided by the exiting source, the higher the possible level to which the electrons are excited, and the shorter the wavelengths of the light emitted. Consequently, with an increase in the energy supplied by the source, more spectral lines are observed, especially at shorter wavelengths.

The wavelengths of the observed spectral lines are a characteristic of the emitting element and thus spectra may be employed for qualitative purposes. The intensity of the lines can be correlated with the amount of substance present and thus may serve quantitative purposes. Two techniques, emission flame photometry and spectroscopy, may be differentiated. Although both are based on atomic spectra, their separate treatment is warranted because they differ markedly with respect to the energy levels excited, instrument design, and practical procedures. Absorption flame photometry involves the absorption of light by a vapor of metal atoms. Although this technique is an absorption method, it is best considered in this chapter because of its relation to emission flame photometry.

40.2 Emission Flame Photometry

It has been known since antiquity that salts of certain metals when introduced into a flame will impart a color to it. Flame tests have long been applied in qualitative inorganic analysis. However, early attempts to employ flame colors for quantitative purposes failed, mainly because of difficulties in introducing the sample appropriately and evenly into the flame. These difficulties were resolved about 1929 by Lundgårdh, who developed the burner-atomizer, which since then has been modified and perfected. The technique of emission flame photometry thus developed is quite attractive because liquid samples can be employed directly. In addition, flame spectra are simple, that is, only comparatively few lines are produced; consequently, the use of a relatively inexpensive instrument is permitted; furthermore, the technique is rapid and shows high sensitivity for some metals.

The operation of a flame photometer becomes obvious from Figure 40-1, which shows the principal parts schematically. The sample solution is sucked up by an atomizer operated by one of the flame-producing gases and as a fine spray is fed into the flame. The light emitted is collected by a parabolic mirror and a lens and passed through a monochromator (prism or

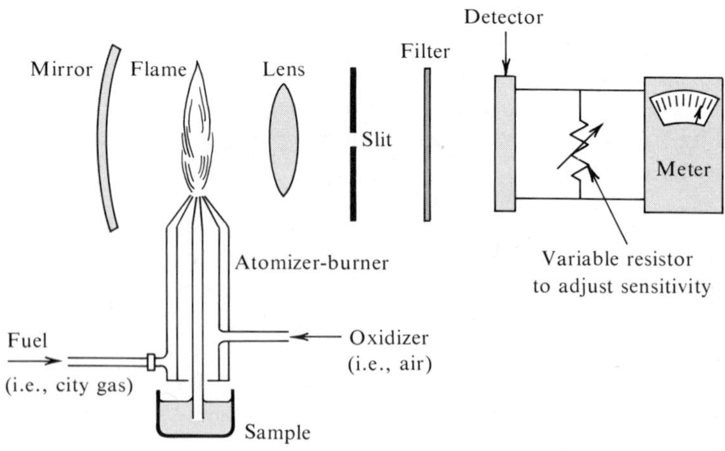

Figure 40-1. Schematic arrangements of the components of an emission flame photometer.

grating) or simply a filter. The light of the appropriately selected wavelength strikes a photodetector and the magnitude of the electrical signal developed is read on a meter. Not shown in this figure are pressure regulators and flow meters, which are installed in the gas lines to permit reproducible adjustment of flow rates and the ratio of fuel to oxidizer gas. In addition to simple city gas–air flames (1900°C), flames of higher temperature are used, including gas–oxygen (2800°C), hydrogen–oxygen (2800°C), and acetylene–oxygen (3000°C).

The temperatures involved in emission flame photometry are low in comparison to those prevailing in an arc or spark (see Section 40.4). Consequently, only spectral lines corresponding to relatively low excitation energies are produced. Thus the flame spectrum of an element is simple and lines in the visible and near ultraviolet are the only ones excited in an ordinary gas flame. For many elements the energy available in a flame is insufficient to excite any line; this confers a favorable degree of selectivity on flame photometry.

Although a prism or grating monochromator may be used, a filter suffices for many analytical purposes, because of the simplicity of the flame spectra. Special filters are often employed that have a narrow "band pass" and therefore allow the selection of a single prominent wavelength.

A major advantage of flame photometry is that it permits the use of a small volume of sample solution. Often a single milliliter of the solution supplies the flame with material long enough to permit changing filters several times and to secure readings for several elements. The filters are frequently mounted on a disk so that they may be changed simply by its rotation.

As the student can conclude from his experience with the flame tests for sodium and potassium, the sensitivity of emission flame photometry is high for these and several other metals. Thus 0.01 ppm of sodium or 1 ppm of barium can be detected readily and somewhat higher concentrations can be determined with an accuracy of a few per cent or better. This high sensitivity is especially valuable for sodium and potassium and certain other metals for which other analytical methods are either tedious or inapplicable to minute amounts.

Calibration Curve. A calibration curve may be employed to obtain the concentration of the element to be determined. Such a curve is established with solutions containing the sought-for substance in known concentration and under identical instrument settings, including the rate of sample introduction, gas pressure, and gas ratio. A correction may be necessary for the flame background, which is the radiation emitted over a wide spectral range due to the combustion of the fuel and the hot or burning solvent or other substances present in the sample. The reading of the output meter is plotted versus concentration.

Internal Standard. An internal standard offers another approach for establishing the unknown concentration. Lithium is frequently used for this purpose because it shows a readily excitable spectral line and is rarely present in practical samples. First, the ratio of the intensity of the lithium line to that of the line of the element to be determined is established experimentally at various concentration levels for both. Then in the actual determination a known concentration of lithium is established in the sample solution and the ratio of the line intensities is measured again. From the two ratios and the known concentration of the lithium, the concentration of the sought-for substance is obtained by proportion, or graphically via a calibration curve.

Standard Addition. Standard addition is a further possibility. In this technique known amounts of the sought-for substance are added to aliquots of the sample solution. These aliquots and an untreated one are fed to the flame and for each solution the reading is taken. Either a graphical procedure may be employed or the unknown concentration may be calculated as in other cases of standard additions (for example, see Chapter 29 on polarography and Section 36.7 on photometric determinations).

Sources of errors exist with emission flame photometry as with other analytical methods. The following discussion will aid in understanding the causes of such errors and the possibilities of remedies. The excitation of electron transitions in a metal with subsequent emission of light takes place only if the metal is in the gaseous state. Since the metal to be determined is present as a dissolved salt, several processes must take place to have the metal eventually present as an atomic vapor in the flame. The sequence of events can be presented as follows:

liquid sample → atomization to droplets of liquid → evaporation of the solvent and formation of a residue →

decomposition of the residue, disruption of bonds, and evaporation to atoms	→	excitation of atoms in the gas phase	→	loss of excitation energy with emission of light

Unless each of these processes proceeds in an appropriate fashion, a faulty result may be obtained.

Chemical Interference. Chemical interference is one type of disturbance possible and comes about as follows. The energy available in a flame is sufficient to vaporize only metals of high volatility. To vaporize the metal the salt must be decomposed. The rate at which this process proceeds is low for some salts and high for others. Consequently, the anions present in the solution play an important role in flame photometry. The highest line intensities are usually obtained with chlorides and nitrates. Phosphates and sulfates are less suitable. When, for example, sulfate is added in increasing amounts to a solution containing calcium chloride, the line intensity for calcium decreases progressively, but eventually a point is reached where the addition of more sulfate has no further effect on the line intensity. Where the reduced intensity still affords sufficient sensitivity, sulfate ion (or the relevant interfering anion) can be added deliberately to the sample solution to an extent where there is no further effect on the line intensity. Obviously the same addition must also be made to the standard solutions used to establish the calibration curve.

The addition of organic solvents or complexing agents, or both, to the sample solution frequently reduces or even obviates chemical interferences and indeed may sometimes increase the sensitivity of a method. When huge amounts of interfering anions are present, a separation may become necessary. Such a separation is often achieved by simply passing the sample solution through a column of an anion exchange resin in the chloride or nitrate form (see Chapter 47).

Excitation Interference. Excitation interference is a peculiar phenomenon occurring with metals of the first group of the periodic table. An alkali metal may show higher line intensity in the presence of another alkali metal than if present alone in the solution. This effect can be explained in the following manner.

The desirable sequence of events is to elevate an electron in the metal atom to a higher energy level and have it return at once to the ground state with simultaneous emission of light. Once returned to that state the atom is ready to repeat the process. Consequently, it is desirable to have as high as possible a number of atoms in the ground state. With alkali metals, because of their low ionization energy, a substantial portion of the atoms dissociate according to $M \rightleftharpoons M^+ + e$. The *ions* are not able to participate in the emission mechanism. If another alkali metal is added, it too ionizes. Thereby the electron concentration in the flame is increased and the above equilibrium is shifted to the left. Consequently, more *atoms* of the metal to be determined are present and the line intensity is increased.

Radiation Interference. Radiation interference needs only brief mention. If another element is present that emits light of a wavelength identical or very close to that of the metal to be determined, the monochromating device may not be able to discriminate between the lines. The interfering element must then either be removed or be added in identical quantities to the solutions used in establishing the calibration curve.

Flame Temperature. Flame temperature has a great influence on a determination. If the temperature is too low to cause dissociation of the salt, to effect vaporization, and to excite the atoms of the metal, no lines, or only very weak ones, are obtained. But too high a flame temperature may also have detrimental effects. Under this condition electrons may not only be elevated to higher energy levels but may completely dissociate, leaving an atomic ion behind. Then immediate return to a lower energy level or to the ground state is not possible and a loss of intensity of the emitted light is the consequence.

Emission flame photometry represents an extremely valuable approach for the determination of readily excitable elements in small samples and at low concentrations, especially where many samples must be analyzed routinely. The determination of sodium, potassium, and calcium in such diverse samples as blood serum, urine, soil extracts, and industrial and natural waters is standard procedure in many laboratories. With flames of a higher temperature (notably acetylene–oxygen), magnesium, iron, and certain other elements may also be determined routinely.

40.3 Absorption Flame Photometry

This technique was introduced for analytical purposes by Walsh about 1955 under the designation atomic absorption spectrophotometry. It has achieved rather rapidly a definite place in analytical practice, because of its high sensitivity, selectivity, and relative simplicity. The underlying principles and the relation to flame emission photometry will become clear from the following discussion.

In the metal vapor in a flame, only a rather small portion of the metal atoms are in the excited state; most of the atoms are in the ground state. Consequently, the intensity of the light emitted at any given time is small compared with the number of potentially excitable atoms present. If a strong beam of compound light is passed through such a flame, the atoms in the ground state are capable of absorbing energy from this light and having electrons moved to higher orbitals. Of course, only light of a definite wavelength will be absorbed. This wavelength corresponds to the exact energy needed to bring about the quantized electron transition and is the same as that emitted by the heat-excited atoms.

Upon return of the light-excited atoms to the ground state, the absorbed energy is emitted as light. However, this light is not emitted solely in the direction of the exciting beam, but rather in all directions in space. Consequently, if the spectrum of the exciting beam, after passage through the flame is displayed, since there is less intensity at the wavelength absorbed, a line dark in comparison to the otherwise bright and continuous spectrum is observed. If the intensity of the light at the wavelength, for example, of sodium were measured with and without sodium present in the flame, an absorption photometric process could be established completely analogous to that in absorption photometry discussed in Chapter 36.

In contrast to the rather broad bands shown in the absorbance spectra of composite ions, atoms absorb at sharply defined wavelengths. Consequently, only an extremely minute portion of the compound exciting light is used and the radiant energy corresponding to these wavelengths present in the exciting beam is quite insufficient for practical application. It is therefore necessary to use an exciting beam that contains a high intensity of light of the required wavelengths. For sodium, for example, such a beam can be obtained from a "sodium lamp," that is, a nearly evacuated glass envelope, in which a small amount of sodium metal is vaporized and highly excited by an electric discharge. The emitted light has sufficient intensity for the purpose at hand and offers the additional advantage that it is, so to speak, tailored to the element to be determined, because the particular wavelength is specific for the element and unable to excite any other element.

40.3 ABSORPTION FLAME PHOTOMETRY

A flame absorptiometric determination of sodium would appear to proceed as follows. An intense narrow beam of light from the sodium lamp is passed through the empty flame (that is, no sample is fed into it) and falls on a photodetector. The electrical signal created by the detector is displayed on a meter, and the meter needle by means of a shunt resistor is brought to read 100% T. Now the sample solution containing sodium is fed into the flame. Then a portion of the light is absorbed and the meter shows a lower reading. This process, in principle, is fully analogous to absorptiometry as described for solutions in Chapter 36. However, the actual operation is not that simple. The light incident on the photodetector consists of several portions: (1) light from the exciting beam (less whatever intensity was absorbed), (2) light from the heat excitation of the sodium, and (3) light from the flame background. The latter two portions are undesirable and should be excluded. Fortunately, this exclusion is achieved comparatively easily. All that is needed is to chop the exciting beam before it enters into the flame. This chopping is achieved by placing into the beam a rapidly rotating disk with slots. The chopped exciting beam now creates a fluctuating signal in the detector output while the flame emission and the background will essentially cause a steady signal. It is quite simple to electronically separate the two signals and have only the component corresponding to the exciting beam (after rectification) displayed on the meter.

Thus the flame background is readily excluded, which represents a distinctive advantage over flame emission photometry. The fluctuating signal created has an additional benefit in that such a signal can be amplified more readily and by less expensive means.

Since the exciting light presents the wavelengths of the particular element to be determined, high selectivity is obtained and radiation interference is hardly of any consequence in flame absorption photometry. As was mentioned above, most of the atoms in the flame are present in the ground state and thus ready to interact with the exciting beam. Consequently, the method is also more sensitive than emission flame photometry, except possibly for the alkali metals, which are most readily excited by heat.

A disadvantage of the absorption flame photometric method, beside the more involved (and thus more expensive) apparatus required, is the necessity of essentially a separate light source for each element to be determined. However, progress is being made in the fashioning of multielement sources.

For alkali metals the type of lamp described above is in use. For other elements so-called hollow cathode lamps are employed. Their construction is shown schematically in Figure 40-2. The glass envelope, E, with a quartz window, Q, contains some argon at very low pressure. Across the ring-shaped

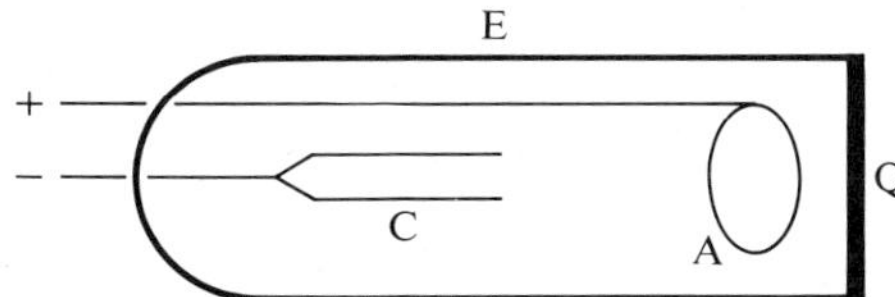

Figure 40-2. Schematic presentation of a hollow cathode lamp.

anode, A, and the cathode, C, is placed a high electrical voltage. Argon atoms are ionized and the positive ions formed are attracted by the cathode. These ions are accelerated and hit the inside of the cathode with great force. This cathode is either made of the metal of interest or lined with it. When the accelerated argon ions impinge on the surface, metal atoms are knocked out, vaporized, and excited. As a consequence, a strong beam containing wavelengths of the lines of the particular elements is emitted.

The operation of the atomizer, the flame, regulation of the gases, etc., are essentially identical with those in emission flame photometry. The evaluation of the results via a calibration curve or by means of standard additions proceeds similarly.

Emission Spectrography 40.4

The low energy provided by a flame is sufficient to excite only a limited number of elements. To obtain the spectral lines for elements that are not excited in a flame it is necessary to provide greater excitation energy. This is achieved by the use of a d.c. or a.c. arc or a high-voltage spark. In a d.c. arc, for example, the electrons emitted by the cathode bombard the anode and the temperature there is raised to possibly 6000° K. The anode may either be fashioned from the material to be analyzed or be impregnated with it. Spectra thus obtained contain a multitude of lines, the majority of which lie in the ultraviolet region. The number of lines obtained depends on the excitation conditions, the elements present, and the temperature. For transition metals, up to 1000 lines have been identified under certain conditions.

With these facts at hand it can readily be appreciated that the requirements for the apparatus are much greater than under the simple conditions of flame emission photometry. Lines of different elements are frequently close together and monochromators of considerable resolution are mandatory. But even then, for the identification or determination of an element, characteristic lines must be selected that are free of coincidences.

The principal parts of a typical spectrograph and their arrangement are shown schematically in Figure 40-3. The light emitted

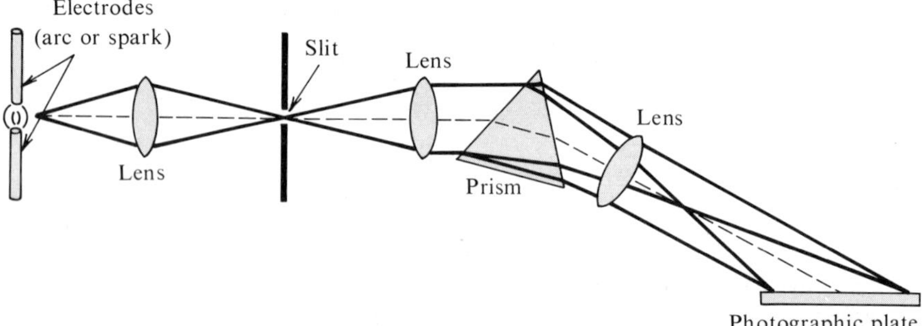

Figure 40-3. Schematic arrangement of the components of a spectrograph.

from the arc (or spark) is focused by a lens on the entrance slit of the monochromator, passes through a collimating lens (that is, one that renders the light parallel), then through the prism, and by a further lens is finally projected on a photographic plate or film. This plate or film can be moved up or down so that several spectra can be recorded consecutively but separately. In a grating spectrograph the prism is replaced by a grating, which is usually ruled on a concave mirror.

If the material to be analyzed is electrically conducting, for example, an alloy, it may be shaped to rods, which are used as electrodes. Alternatively, a graphite electrode may be used with a small crater at its tip. The powdered sample material, often with graphite added to make it electrically conducting, is introduced into this crater or a sample solution is placed there and evaporated. The arc or spark is then fired and the spectrum photographed. With a d.c. arc it is necessary to have the sample material as or at the anode.

From the location of lines at certain wavelengths it is readily possible to evaluate the spectrum qualitatively. Since at the extreme temperature conditions chemical bonds are completely disrupted, only elements can be detected, and no deductions as to their combination can be made. Several possibilities exist for the application of spectrography to quantitative analysis.

For the analysis of an alloy, for example, the following procedure may be employed. First a spectrogram is obtained from the sample. Then the plate or film is moved and spectrograms are obtained for one or more standard alloys of similar and exactly known composition, under identical instrument settings, including the gap between the electrodes, electrical current, and time of exposure of the plate to the spectrum. The plate is then developed and fixed under specified and reproducible conditions, and dried. For each element to be determined one or several lines are selected and their intensities, that is, the degree of blackening on the plate, are compared for the sample and standard. For this comparison an instrument known as a densitometer may be used that functions analogously to an absorption photometer and measures the absorbance of the spectral line on the plate in relation to a clear portion of the emulsion. From the knowledge of the ratio of the line absorbances for sample and standard and the known content of the element to be determined in the standard, the percentage of a sought-for element in the unknown can be calculated. The method of an internal standard can be used in a manner analogous to that mentioned for flame photometry. It has the advantage that the results are relatively independent of small variations in the exciting source and the photographic development and fixation. The internal standard method is most readily applicable to sample solutions, but it can be used with powdered samples as well.

The method of "last lines" or "persistent lines" is also often employed with sample solutions. This technique is based on the following. The spectral lines of an element differ in intensity since the various electron transitions vary in their probability. If a spectrum is obtained from each solution of a dilution series it will be noted that first the weakest lines disappear, and finally the most intense, most persistent lines. To establish the "last lines" for an element a measured volume of a stock solution of exactly known concentration is brought onto the electrode, and by sparking or arcing for a definite time the spectrum obtained. A portion of the stock solution is then diluted by a known ratio and a spectrum is again secured under identical conditions. This process is repeated until the last lines have disappeared. Then the concentration at which this disappearance took place is established. To analyze a sample the spectra of a dilution series are obtained in identical fashion. From the dilution ratios employed and the concentration at which the last lines disappear it is possible to calculate the concentration of the sought-for substance in the original sample solution. The technique may be employed for the concurrent determination of more than one component in a mixture. The method can also be applied to powdered samples with graphite serving as the diluent.

Other approaches are also used to obtain quantitative results, but the methods described above will suffice to provide an understanding of the principles involved. With close attention to operating details and the use of a good spectrograph, the result of a determination may be precise and accurate to within $\pm 4\%$ (relative).

Modern instruments exist in which the photographic recording is eliminated by the use of photodetectors. A single detector may be employed that is so moved that it slowly scans the spectrum. The detector output is amplified and fed to a strip chart recorder. There the spectrum is displayed as an array of sharp peaks, the locations and heights of which can be related to wavelengths and intensities of the corresponding lines. Alternatively, several photodetectors are used and each mounted exactly at the location of a particular spectral line. Each output is fed to a separate counting device, which after appropriate calibration allows the content of the relevant component to be read directly. With such emission spectrometers a number of elements can be determined concurrently in a relatively short period of time. Such instruments are expensive, but they pay for themselves where numerous samples must be analyzed routinely and rapidly for many constituents, notably in the process control of metallurgical melts.

40.5 Questions

40-1. Elaborate on the difference between line spectra and continuous spectra.

40-2. What parameters in flame emission photometry govern the number of spectral lines excited and the intensity of a single line?

40-3. Elaborate on the difference in principles of emission and absorption flame photometry.

40-4. What methods can you name that are used to determine the concentration of a sought-for element in emission spectrography?

5. In the determination of cations, what influence will anions exhibit in emission spectrography and in emission flame photometry?
6. Explain why only relatively few spectral lines are excited in a flame.
7. Contrast emission spectrography with wet analysis methods (gravimetry, visual titrimetry, and solution absorption photometry) with respect to advantages and disadvantages, the useful concentration range of the sought-for substance in the sample, speed, instrumental requirements, necessity of separations, ease of sample preparation, and accuracy and precision.

41 X-RAY METHODS

Introduction 41.1

X-rays are electromagnetic radiation in the wavelength range 0.1 to 100 Å. They are emitted during certain types of radioactive decay and can be produced in an X-ray tube (see Figure 41-1a). Such a tube consists of an evacuated glass envelope containing a wire filament (cathode) and a "target" (anode). A voltage difference up to several thousand volts is established between the filament and the target. The filament is heated and the electrons then emitted are attracted by the anode and greatly accelerated in the strong electric field. The high-velocity electrons on striking the target are capable of knocking out electrons from the inner orbitals (for example, the K or L shell) of the target atoms. Each electron thus removed is then replaced by one from a higher orbital, and the energy lost by this second electron is emitted as an X-ray.

The X-ray spectrum obtained, after suitable dispersion, consists of "white" noncharacteristic radiation constituting a continuous spectrum, with superimposed sharp lines characteristic of the element used as the target and related to the atomic number by Moseley's law:

$$\frac{c}{\lambda} = a(Z - \sigma)^2 \tag{41-1}$$

Here c is the velocity of light, λ the wavelength, a a proportionality constant, and Z the atomic number. σ is a constant with a value dependent on the series of lines, that is, dependent upon whether the line originates from an electron falling from the L to the K shell (K_α lines), the M to the K shell (K_β lines), the M to the L shell (L_α lines), etc.

X-ray line spectra are simple, that is, they consist of only a few lines and are similar for all elements, but the actual wavelengths of each of these lines varies from element to element. The X-ray lines are not significantly influenced by the state of the element (gas, liquid, or solid) or whether it is present as the free element or chemically bonded in a compound.

Lenses and prisms in the usual sense cannot be used with X-rays. In an X-ray spectrograph (Figure 41-1), in which a

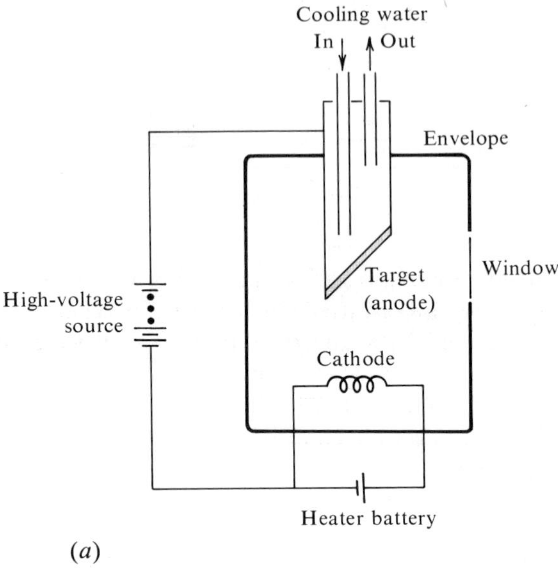

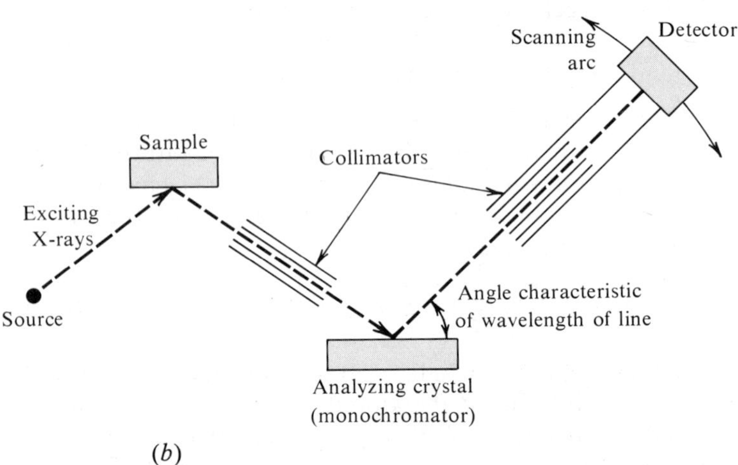

Figure 41-1. (a) X-ray tube; (b) X-ray spectrograph.

sample undergoes excitation by X-rays, the collimator necessary to obtain a parallel beam is a long narrow gap between two metal blocks or several gaps between closely spaced metal plates or concentrically arranged tubes. The dispersion of the emitted spectrum is effected by a crystal, which acts as a diffraction grating since the interatomic distances in the crystal are of the same order as the wavelengths of the X-rays. The spectrum can be recorded on a photographic plate. Alternatively, a detector can scan the arc, as shown in the figure, and the signal created is fed to a recorder. Radiation detectors including the Geiger counter and related devices, based on the ionizing effect of X-rays on a rarefied gas, may be used (see Section 42-2 and an elementary textbook on chemistry or physics). Photoelectric counters receive special attention. Such a counter is a photomultiplier tube, the window of which is coated with a fluorescent material. If the material is struck by an X-ray, fluorescence occurs in the visible region and its intensity is measured by the photomultiplier in connection with a relevant circuit and meter. Scintillation counters have a single crystal in place of the coating.

X-ray methods of interest to analytical chemists are based on emission (especially secondary "fluorescent" emission), diffraction, and absorption.

X-Ray Emission and Fluorescent Methods 41.2

If the substance to be analyzed is made the target in an X-ray tube, a spectrum is produced containing the characteristic lines of all the elements present. The wavelength of a line (or simply the angle between the beam of thc line and the incident beam) can be measured and used for qualitative purposes. Measurement of the intensity of the lines serves for a quantitative evaluation. This technique of X-ray emission analysis is restricted because tubes that can be dismantled are needed to introduce the sample material as the target. Since the tube is operated under vacuum, readily volatized substances cannot be analyzed. In addition, upon bombardment with the electrons the target reaches high temperatures even when cooled by the flow of water through airtight tubes; many substances cannot stand such conditions.

Another approach, X-ray fluorescence, is of greater general applicability. X-rays can be produced not only by bombarding a substance with electrons but also by irradiating the substance with short-wavelength X-rays. The secondary, or fluorescent, X-rays thus produced are less intense than those obtained by electron bombardment by a factor of about 1000. Consequently, a primary X-ray beam of high intensity and detectors of high sensitivity are mandatory. The primary components of an X-ray spectrograph and their arrangements are shown in Figure 41-lb. Solid or liquid samples may be used. The sample containers are fashioned from polymers, such as polyethylene or aluminum because these materials show low absorptivities for X-rays. Elements with an atomic number of 12 and higher can be analyzed in air; the elements with atomic number 5 through 11 emit X-rays of short wavelengths that are absorbed in air. These lower elements require special equipment that allows operation in vacuum or in a helium atmosphere.

A major domain of X-ray fluorescence methods is the rapid analysis of alloys, minerals, and other solid materials. The method is nondestructive. Depending on the quality of the instrument and care of the operator, even minor constituents can be determined with an accuracy and precision of 1 to 2% or even better.

41.3 X-Ray Diffraction Methods

When a monochromatic X-ray beam strikes a crystalline substance, diffraction takes place. Depending on the wavelength of the X-rays, the interatomic spacing of the layers in the crystal, and the position of the crystal relative to the beam, extinction will occur in some directions and reenforcement in others. Thus, the radiation leaving the crystal will, upon striking a photographic plate, create an arrangement of spots, called a "diffraction pattern." The pattern is characteristic of the crystal system and can be used to identify a crystal. For the purposes of analytical chemistry, often a powdered sample, with random distribution of the crystallites, is placed in the X-ray beam and the diffraction pattern recorded on a narrow strip of photographic film as a series of lines.

Study of the position of the lines (and their resolution) serves to estimate the ratio of crystalline to amorphous material and, with the aid of known spacing values, makes possible the identification of the crystalline phase or phases present. Thus, for example, X-ray powder diffraction can allow a decision as to whether a sample is a mixture of sodium chloride and potassium sulfate or of potassium chloride and sodium sulfate. Wet methods of analysis would reveal only the presence of sodium, potassium, chloride, and sulfate ions, but would not establish the pairing of the ions in the solid sample. Simple mixtures can also be analyzed semiquantitatively by measurement of the relative intensities of X-ray diffraction.

In general, however, the application of X-ray diffraction to analytical chemistry is limited. The method is applied for other purposes, including the determination of stresses and imperfections in materials such as alloys or plastics, and the measurement of particle size. The method is one of the principal tools in crystallographic studies.

X-Ray Absorption Methods 41.4

When a beam of X-rays, either polychromatic or monochromatic, is passed through a sample, the X-rays interact with the matter and energy is absorbed. Measurement of the intensity of the beam with and without the sample in it allows a determination of the absorbance in a photometer arrangement, as in the ultraviolet and visible regions. But here either a Geiger or a scintillation counter is used as the detector. In general, the higher the atomic number of an element, the greater is its absorptivity for X-rays. Thus lead, having a high atomic number, may be rapidly determined by X-ray absorption in gasoline to which it was added as tetraethyllead; the other elements present are principally carbon and hydrogen, which have low atomic numbers.

Electron Probe Microanalysis 41.5

Electron probe analysis is a relatively recent technique (first practical application by Castaing in 1951) in which the specimen is bombarded with a thin electron beam. The X-ray spectrum excited contains the wavelengths characteristic of the elements present in the minute area struck by the electron probe. The intensities of these wavelengths are proportional to the mass concentration of these elements. The operation of an electron probe microanalyzer can best be explained by reference to the admittedly much simplified schematic diagram shown in Figure 41-2. The electron beam from the electron "gun" by passage through two (or more) magnetic fields, that is, through "magnetic lenses," is focused and demagnified onto a region about 1 μ in diameter of the electron-opaque specimen. The X-rays generated are diffracted by a crystal and the wavelengths of interest are isolated by a slit and fall onto a detector (Geiger counter, scintillation counter, etc.). To direct the probe to the exact point of interest, observation via a mirror system (not shown in the figure) and a visual optical microscope is often used. The whole arrangement (except the optical microscope) is enclosed and under high vacuum. The specimen can be translated relative to the electron probe and point by point analysis is thus possible. In this way the distribution of a selected element can be established rapidly.

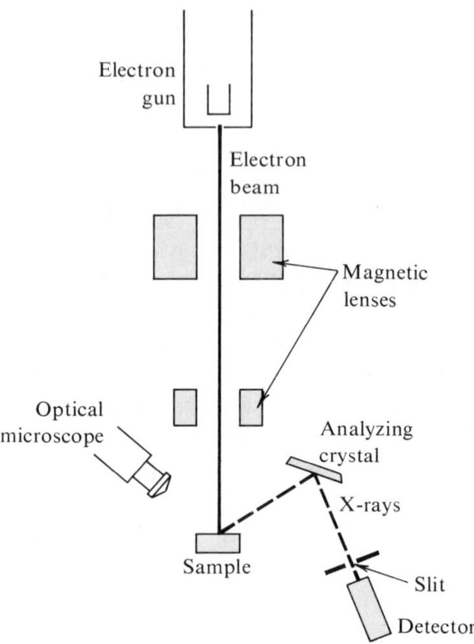

Figure 41-2. Schematic diagram of an electron probe analyzer.

The electron probe microanalyzer is quite elaborate and complex, but it allows the analysis of exceedingly minute segregations, inclusions, etc., at the surface of materials. Only surface layers can be investigated because the electron beam does not penetrate deeper than about 1 μ. In a region about 1 μ in diameter it is possible to detect elements present at concentrations of a few per cent, and the actual amount of an element excited is often only 10^{-14} to 10^{-15} g. Where appropriate standards are available, the determination of elements is possible with a relative error of only a few per cent, which is amazingly small for the effective sample size.

41.6 Questions

41-1. What relation exists between the wavelength of an X-ray and the atomic number of the element emitting the ray?

41-2. In an X-ray spectrograph, what is used as a collimator and as a monochromator?

41-3. Why is a simple X-ray spectrograph not suited for measuring the X-ray fluorescence of elements with atomic numbers less than 12?

41-4. What are "white" X-rays?

Compare the application of X-ray methods with other nondestructive techniques with respect to application, instrumentation, speed of analysis, and precision.

X-ray lines are designated by a capital English letter and a Greek letter subscript. To what do these two letters refer?

By what means can X-rays be produced?

42 ANALYTICAL METHODS BASED ON RADIOACTIVITY

Radioactivity and Radioactive Decay 42.1

In this chapter it is assumed that the student has an understanding of the fundamentals of the phenomena of radioactivity on the level as they are presented in elementary textbooks on physics and chemistry. Only a brief recapitulation of some basic facts can be given here.

Radioactive isotopes in their spontaneous decay eject one or more energetic particles or radiate energy, including alpha particles (high-speed, dipositive helium ions, He^{2+}), beta particles (negaton emission, high-speed electrons), positrons, and gamma rays (electromagnetic radiation of very short wavelength). Where the decay involves electron capture, X-rays are produced. Formerly, since only a few elements are naturally radioactive, analytical methods based on radioactivity had limited applicability. With the development of more intense sources of high-energy particles, notably the nuclear reactor, and with the resulting availability of artificially produced radioactive isotopes of a large number of elements, the possibilities for the analytical use of radioactivity have increased enormously.

The rate of disintegration of a radioactive isotope follows the so-called exponential decay law (as does the absorption of radiant energy; see Section 34.3). The law states: The rate of disintegration is proportional to the number of nuclei of active isotope present. In the notation of calculus, this takes the form

$$\frac{dN}{dt} = -\lambda N \qquad (42\text{-}1)$$

where N is the number of nuclei of the radioactive isotope present, t the time, and the Greek letter lambda, λ, a proportionality constant known as the decay constant. The value of this constant is a characteristic of the isotope. The minus sign enters the equation because the number of nuclei of the isotope decreases with time. The disintegration rate, dN/dt, is usually expressed in curies (1 Ci = 3.7×10^{10} disintegrations per second) or a submultiple, such as millicuries (1 mCi = 1×10^{-3} Ci). The specific activity of a sample is usually placed on a weight basis and expressed in curies or millicuries per gram of sample.

Rearrangement and integration of equation (42-1) between $t = 0$ and $t = t$ yields

$$N_t = N_0 e^{-\lambda t} \tag{42-2}$$

where N_t and N_0 are the number of nuclei of the radioactive isotope present at time t and time zero, respectively; e is the base of natural logarithms.

It is also convenient to consider radioactive decay in terms of the half-life of the isotope, that is, the time required for the number of nuclei of the isotope present in a sample to be reduced to one half of the original number. The relation between the half-life, denoted by the Greek letter tau, τ, and the decay constant can be readily derived. From the definition of the half-life, $N_t = N_0/2$; substitution in equation (42-2) yields

$$\tfrac{1}{2}N_0 = N_0 e^{-\lambda\tau}$$
$$\tfrac{1}{2} = e^{-\lambda\tau}$$

Conversion to logarithms to the base 10 yields

$$\log \tfrac{1}{2} = -\lambda\tau \log e$$
$$-0.3010 = -0.4343\,\lambda\tau$$
$$\lambda = \frac{0.693}{\tau} \tag{42-3}$$

Combination of the integrated and differential forms of the decay law, equations (42-1) and (42-2), yields

$$\frac{dN}{dt} = -\lambda N_0 e^{-\lambda t} \tag{42-4}$$

The rate of disintegration at time $t = 0$ is given by

$$\left(\frac{dN}{dt}\right)_{t=0} = -\lambda N_0 \tag{42-5}$$

Consequently,

$$\frac{dN}{dt} = \left(\frac{dN}{dt}\right)_{t=0} e^{-\lambda t} = \left(\frac{dN}{dt}\right)_{t=0} e^{-0.693t/\tau} \tag{42-6}$$

Conversion of this equation to the logarithmic form yields

$$\log\left(\frac{dN}{dt}\right) = \log\left(\frac{dN}{dt}\right)_{t=0} - \frac{0.301}{\tau}\,t \tag{42-7}$$

A semilogarithmic plot of disintegration rate (logarithmic axis) versus time (linear axis) should therefore yield a straight line, the slope of which is $-0.301/\tau$. Thus, the half-life of a radioactive isotope can be established by measurement of the disintegration rate at various times. A sample containing more than one radioactive species will give a plot showing a number of essentially straight-line portions if the half-lives of the isotopes are sufficiently different.

Detection and Counting of Radioactivity 42.2

A number of devices are available for the detection and counting of the energetic particles and radiant energy produced in radioactive decay. Which device is best applied will depend on the type of emission to be counted, its intensity, and whether or not differentiation is sought between emissions of different energies.

The Geiger–Müller counter (or simply Geiger counter) consists of a glass tube lined with a metal cylinder and a central wire which is insulated from the cylinder. The tube is filled with appropriate gases at low pressure. A voltage of at least 600 V is applied between the wire and cylinder. When a high-speed particle or ionizing radiant energy (gamma or X-rays) enters the tube, often through a permeable window, some of the gas is ionized, a discharge between tube and wire occurs, and the voltage drops momentarily. The electrical "pulse" thus produced is amplified and recorded by counting devices. In the Geiger–Müller counter, the impulse size is essentially independent of the energy of the incident particle or ray; thus the number of disintegrations is counted. Somewhat similar counters operating at higher voltages have an impulse response that is proportional to the energy of the incident ionizing particle or ray and are called proportional counters.

A scintillation counter consists of a fluorescent material mounted on a photomultiplier tube. The material shows fluorescence in the visible region with an intensity proportional to the energy of the incident particles or ray. Since the photomultiplier responds linearly to the intensity of the fluorescence, the response of the entire scintillation counter is proportional to the energy of the incident particles or rays. The output of the counter is amplified and fed to a counting device.

The detailed description of counting devices is beyond the scope of this chapter. When the number of counts exceeds about 50 per second, mechanical counters are replaced by so-called scaling circuits connected to electromechanical counters; the entire unit is called a scalar. If a so-called discrimination circuit (discriminator) is placed between the amplified output of a proportional or scintillation counter and a scalar, it is possible to count selectively pulses exceeding a preset size.

In the selection and use of counters, the geometry of the sample must be considered since radioactive decay occurs randomly in all directions. Often it suffices to place the sample and standards in the same orientation and position with respect to the counter. It is also necessary to subtract any background count caused by low-level radioactivity of the instruments and their surrounding and by cosmic rays. Adequate shielding of the counting chamber by lead reduces the background markedly.

42.3 Activation Analysis

In this technique the elements to be determined are converted to radioactive isotopes by bombardment of the sample with high-energy particles; the activity thus produced is measured. Activation by various charged particles is feasible, but for such particles the capture possibilities are few, because they are readily deflected by the electrical field of the orbital electrons or of the nucleus itself. Consequently, activation by neutrons, which are uncharged, is usually preferred. Although a neutron source can be secured by appropriate use of a cyclotron, van de Graaff generator, or radioactive isotope source, a nuclear reactor is preferred because of the high neutron flux it provides. On the capture of a single neutron by a nucleus, an isotope of the same element is formed with the atomic weight increased by one unit. The sample, after sufficient exposure (minutes to weeks), is removed from the reactor and the disintegration rate is recorded over a period of time. From the semilogarithmic plot (see above) then constructed it is possible to ascertain half-lives and thereby to detect what elements are present in the sample. Also quantitative data can be obtained from the plot.

In principle, the concentration of an element might be calculated from the knowledge of the decay constant and decay pattern of the relevant radioactive isotope, the particular geometry and efficiency of the counter, and the pertinent data relating to the possible nuclear reactions. However, this is not the procedure adopted in practice. Uncertainties in the counter geometry and other factors make it preferable to irradiate and then count standards under identical conditions and to use the data thus secured to make quantitative conclusions about the composition of the sample.

In the study of complex materials by activation analysis, chemical separations may be effected before the various counting operations. Often some of the nonradioactive form of an element is added as a "carrier" to allow the separations to proceed on a reasonable scale or to permit determination of the yield during the chemical processing.

Activation analysis is especially valuable in trace analysis. In favorable cases samples in the milligram range can often be analyzed containing 1 nanogram (i.e., 10^{-9} g) or less of a detectable element. The sensitivity of activation analysis for a given element obviously depends on the intensity of the particle source and the sample exposure time. It also depends on the ability of the element to capture a particle and on the half-life of the radioactive isotope produced.

Tracer Methods 42.4

By introduction of a radioactive tracer (that is, a radioactive isotope) of one of the elements present, a sample is said to be "labeled," "spiked," or "tagged." Only a very small amount of the isotope need be added to secure a disintegration rate that can be counted easily. Frequently the amount added is negligible with respect to the amount of the element present in the sample. Tracer methods receive diverse application in chemistry and other fields of science and technology. For example, they allow the course of a complex chemical separation scheme to be followed readily and the efficiency of the process to be evaluated. They also allow the determination of physical constants, such as solubilities. Probably the tracer methods of greatest value in quantitative chemical analysis are based on the isotope dilution technique.

Isotope Dilution Methods 42.5

The principle of isotope dilution may be readily appreciated by consideration of an example. Assume that a certain substance is to be determined in a mixture. Exactly m grams of a radioactively labeled form of the substance having a specific activity of A millicuries per gram is added to the sample solution taken. The substance of interest is then precipitated and a known weight of the isolated precipitate is subjected to counting. The specific activity of the precipitate is thereby established as A' millicuries per gram. Simple reflection reveals that the value of A' will decrease as the amount of the (unlabeled) substance in the sample solution is increased and as the amount of labeled substance added is decreased. The required relation, therefore, is

$$A' = \frac{m}{m + x} A \qquad (42\text{-}8)$$

where x is the grams of sought-for substance in the sample solution taken. Solution for x yields

$$x = m\left(\frac{A}{A'} - 1\right) \qquad (42\text{-}9)$$

It is important to realize that the precipitation need not be quantitative! In a similar way other separation processes may be employed, including solvent extraction and ion exchange. The only requirement is that the amount of the separated material taken for counting be established (by gravimetry, photometry, etc.).

42.6 Questions

42-1. Define the terms curie and specific activity.

42-2. Explain why a higher "yield" in radioisotopes is secured on irradiation with neutrons than with charged particles.

42-3. What two principal methods of analysis exist that employ radioactivity?

42-4. Explain the term half-life.

42-5. Suggest possible experimental procedures for the use of the radioactive tracer technique in the determination of (a) a solubility product, (b) the degree of coprecipitation of a substance, and (c) the completeness of a separation by precipitation and by solvent extraction.

42-6. The following isotope dilution technique can be applied. Two *equal* aliquots of the sample solution are taken and "tagged" by addition of *different*, but known amounts of the radioactively labeled substance to be determined. The sought-for substance (the amount originally present plus the labeled portion added) is precipitated in each aliquot and the specific activity of each of the two precipitates is determined. From the ratio of the two values, the amount of sought-for substance originally present in an aliquot can be calculated without knowing the specific activity of the labeled substance added. Demonstrate this fact mathematically.

43 MISCELLANEOUS INSTRUMENTAL METHODS

Some additional instrumental methods that receive attention in quantitative analysis will be considered briefly. In an introductory course, it is only possible to delineate the background of these methods and to hint at their usefulness in the practice of analytical chemistry. Several of these methods are predominantly applied to specific problems, often related to organic chemistry, and will be discussed more closely in other courses.

Mass Spectrometry 43.1

Mass spectrometry involves the acceleration of gaseous (positive) ions in an electrical field and then their separation in a magnetic field into ion beams according to the mass/charge ratio of the ions. Each ion beam may be detected and its intensity measured. The instrument involved is known as a mass spectrometer. The use of this technique in the separation of isotopes and the establishment of their atomic masses is well known. It is also possible to measure the relative abundance of an ionic species in the established gas phase. Organic molecules, on cleavage in a beam of electrons, yield ionic fragments. The mass spectra (that is, the fragmentation patterns) thus obtained are sufficiently different for the analysis of mixtures, even for closely related molecules. For example, mixtures of 10 or more closely related hydrocarbons can be analyzed by mass spectrometry, often with an accuracy and precision of 1%. The use of a stable isotope of unusual mass number (e.g., ^{2}H, ^{13}C, ^{15}N, ^{18}O) in conjunction with a mass spectrometer offers one technique of so-called tracer analysis. For example, in the determination of an element by the isotope dilution approach, a known amount of the trace isotope is incorporated in some way into the sample or its reaction products (Section 42.4). Mass spectrometric measurement of the relative amounts of the isotopes of the element present allows calculation of the amount or percentage of that element in the original sample.

Spark source mass spectrometry is applicable for the study of inorganic solids. It employs a vacuum spark in which a high-voltage radiofrequency discharge (20 to 100 kV) is produced between two closely spaced electrodes of the material to be analyzed. The constituents of the sample that are vaporized and ionized are accelerated and separated. In this way information on over 50 elements at contents of 100 ppm to a few ppb can be secured; the accuracy is ±10 to 25% when suitable standards are used. Organic materials are ashed before examination by this technique.

43.2 Refractometry

When a beam of light passes obliquely from one medium to another of a different "density" to light, the beam changes direction; that is, the light is refracted. It will be recalled from physics that the index of refraction, n, is given by $n = \sin i / \sin r$, where i and r are the angles of incidence and refraction, respectively. Values for this index for the passage of light from air (or vacuum) into a substance are recorded in the literature, usually for 20°C and the D line of the sodium atomic spectrum (5993 Å), denoted by n_D^{20}. Instruments for measuring refractive indices are known as refractometers. Measurement of this index allows the identification of a substance and often serves as a criterion of its purity. By the use of a calibration curve, the composition of a transparent, homogeneous binary mixture is readily determined. In favorable cases, the method can be extended to ternary mixtures.

43.3 Polarimetry

Some substances, predominantly organic ones, are optically active; that is, they rotate the plane of polarized light at a specified wavelength, and to a different degree depending on the wavelength. The direction and magnitude of the optical rotation are characteristic of the substance. The specific rotation, α, is defined as the optical rotation caused by a column 10 cm long containing 1 gram of the optically active substance per milliliter of solution. Values for the specific rotation are tabulated in the literature, usually for 20°C and the D line of sodium, denoted by α_D^{20}. Instruments for the determination of optical rotation are known as polarimeters. Measurement of this rotation serves for the identification of an optically active substance and as a criterion of its purity. The measurement also serves to determine the concentration of such substances in solution. This method is routinely applied to sugar solutions and is then known as saccharimetry.

43.4 Microscopy

The optical microscope is a valuable tool in the chemical laboratory, but predominantly for qualitative purposes. Even minute amounts of substances and materials can be identified by a study of their form, color, appearance, crystal habit, and optical properties under ordinary and polarized light, complimented in some cases by chemical tests performed on the microscope stage. The purity of some solid substances can be assessed on a microscope "hot stage" by following the rate of crystallization from the melt. Noble metals in their compounds may be reduced to the metal and melted into tiny spheres; measurement of the particle diameter with the known density of the metal serves for the estimation of amounts of the metal too small for conventional weighing. Minute amounts of evolved gases may be measured by the size of suitably formed bubbles. The number of particles of each discrete component of a mixture may be counted under the microscope; from this information and the total number of particles counted, a "number" percentage of each component can be calculated. In minerology, thin slices of rocks, for example, are cut, ground to transparency, and polished; areas of identical optical properties are measured, related to particular components, and an estimate of the composition established. However, all these methods serve for special purposes only and require special training and equipment. A fuller discussion of chemical microscopy is beyond the scope of this textbook.

44 REVIEW QUESTIONS ON OPTICAL AND MISCELLANEOUS INSTRUMENTAL METHODS

Name various units in which wavelength is often expressed.

Name and briefly describe some modes of interaction of radiant energy and matter that are of interest in analytical chemistry.

What is the difference in the mechanism of the absorption of radiant energy in the ultraviolet visible and infrared regions?

What difference might you expect in the materials used in an ultraviolet spectrophotometer and an infrared spectrophotometer for prisms, lens, cells, etc.?

What finally happens to the radiant energy absorbed by a solution containing a nonfluorescent light-absorbing solute?

Elaborate on the difference between fluorescence and phosphorescence.

What is reflection of light? Refraction? Diffraction?

Elaborate on the complexity of the processes leading to light scattering.

How may light scattering be used for analytical purposes?

What is the effect of light scattering upon a spectrophotometric determination?

What is polarized light and how may it be used in chemical analysis?

What is understood by the term "spectral line"?

In your own words define a "blank" as the term is used in optical methods and elaborate on the difference between a reference and a blank.

Elaborate on the method of "internal standards." Name optical methods where this technique is chiefly applied.

Elaborate on the advantages of photometric, nephelometric, and turbidimetric titrations over photometry, nephelometry, and turbidimetry, respectively.

Compare photometric methods with gravimetric methods and visual titrimetric methods with regard to accuracy, concentration range of sought-for substance, interferences, instrumental requirements, and speed.

Explain why it is feasible to use a filter in flame emission photometry, but why a monochromator is necessary in emission spectrography.

44-18. Elaborate on the differences in principles, but not on the differences in instruments, between absorption flame photometry and conventional solution photometry.

44-19. An atomic fluorescence determination employs an arrangement that is essentially that for atomic absorption (Section 40.3) except that the hollow cathode lamp or other appropriate source is mounted at rt angle to the flame-detector axis. The hollow cathode lamp for the element to be determined is positioned and the detector signal is read at the proper wavelength for a solvent blank, the sample preparation, and at least one standard preparation. Radiation is absorbed in the flame by the atoms of the relevant element and the fluorescent radiation (return to the atomic ground state) is emitted in all directions. From your knowledge of solution fluorimetry (Section 39.4) and of atomic flame processes (Sections 40.2 and 40.3) elaborate on the basis of the atomic fluorescence technique. For this technique and for atomic absorption, contrast the detector signal when the flame contains none of the relevant element and a small amount. From this contrast can you see any advantage for the atomic fluorescence method?

44-20. What optical methods can be considered essentially nondestructive when applied to a solid sample? What fields of science and technology might be interested in them?

44-21. Briefly define or explain the following terms and name optical methods or apparatus with which they are associated: atomizer, scintillation counter, visual comparison, spark, target, hot stage, densitometer, and flame background.

44-22. In a precipitation titration employing a radiometric end point, the activity of the supernatant liquid over the precipitate can be followed and the counts per minute plotted versus the volume of titrant solution added. For the simplest radiometric titrations the resulting curves consist of two intersecting straight lines. The position of the lines will depend on whether the sample solution is active or the titrant solution, or both. Sketch your concept of the three shapes for the curves. Based on other titrations involving linear plots of data that you have encountered, suggest possible advantages for radiometric titrations.

45 EXTRACTION METHODS

Distribution Law 45.1

When a solute is allowed to distribute itself between a pair of immiscible solvents in contact, the equilibrium established is governed by the distribution law: At constant temperature (and constant pressure in the case of a gas), the ratio of the equilibrium concentrations of a species distributed between two immiscible solvents is constant. In inorganic analysis, one of the solvents is usually water and the other an organic liquid.

The distribution equilibrium for a species A may be written

$$A_w \rightleftharpoons A_o \qquad (45\text{-}1)$$

where the subscripts w and o refer to the aqueous and organic phases, respectively. It is here assumed that the molecular weight of the species is identical in both phases; that is, no association of molecules occurs. The distribution law can then be expressed as

$$P = \frac{[A]_o}{[A]_w} \qquad (45\text{-}2)$$

where P is known as the partition coefficient or the distribution constant. Its value will depend on the temperature and the nature of the solute and the solvents involved. Sometimes this relationship is written as the reciprocal.

Where the two phases are equilibrated in the presence of an excess of a sparingly soluble solute, the expression of the partition coefficient reduces simply to the ratio of the solubilities of the compound in the two phases:

$$P = \frac{S_o}{S_w} \qquad (45\text{-}3)$$

Distribution Ratio 45.2

In analytical practice, the distribution ratio is often of greater importance and is defined by

$$D = \frac{C_{A,o}}{C_{A,w}} \qquad (45\text{-}4)$$

where C_A is the formal concentration of the species A. The relation between the partition coefficient and the distribution ratio in the extraction of a metal ion may be appreciated by consideration of a simple example.

Assume that a metal ion M^+ on reaction with a complexing agent HZ forms the uncharged complex MZ, which is the only species extractable into the organic phase. The two equilibria involved may then be displayed as

$$\begin{array}{c} M^+ + HZ \rightleftharpoons H^+ + (MZ)_w \\ \qquad\qquad\qquad\qquad \Updownarrow \\ \qquad\qquad\qquad\qquad (MZ)_o \end{array} \tag{45-5}$$

Inspection of this scheme reveals qualitatively that an increase in the concentration of HZ as well as a decrease in the concentration of the hydrogen ion shifts the complexation equilibrium to the right. As a consequence, the position of the extraction equilibrium will also be shifted, and more of the metal will be extracted as its complex.

The quantitative treatment proceeds as follows. The expression for the partition coefficient takes the form

$$P = \frac{[MZ]_o}{[MZ]_w} \tag{45-6}$$

When the extraction is performed, the interest is not in how much of the metal *complex* is extracted, but rather in how much of the metal ion is removed from the aqueous phase and thus the distribution ratio of the metal ion, D, must be considered:

$$D = \frac{C_{M,o}}{C_{M,w}} \tag{45-7}$$

The formal concentration of the metal in the aqueous phase is given by the following material balance:

$$C_{M,w} = [M^+]_w + [MZ]_w \tag{45-8}$$

Since only the complex is extracted into the organic phase,

$$C_{M,o} = [MZ]_o \tag{45-9}$$

The expression for the distribution ratio then takes the form

$$D = \frac{[MZ]_o}{[M^+]_w + [MZ]_w} \tag{45-10}$$

Division of the numerator and the denominator of this equation by $[MZ]_w$ yields

$$D = \frac{\dfrac{[MZ]_o}{[MZ]_w}}{\dfrac{[M^+]_w}{[MZ]_w} + \dfrac{\cancel{[MZ]_w}}{\cancel{[MZ]_w}}} = \frac{P}{\dfrac{[M^+]_w}{[MZ]_w} + 1} \tag{45-11}$$

In the aqueous phase, the expression for the stability constant of the complex MZ, as considered in Section 20.5, is

$$K_{st} = \frac{[MZ]_w}{[M^+]_w[Z^-]_w} \tag{45-12}$$

In addition, for the aqueous phase, the expression for the dissociation constant of the weak acid HZ can be written

$$K_a = \frac{[H^+]_w[Z^-]_w}{[HZ]_w} \tag{45-13}$$

By use of the expressions for these two constants, the term $[M^+]_w/[MZ]_w$ can be derived from

$$K_{st}K_a = \frac{[MZ]_w}{[M^+]_w\cancel{[Z^-]_w}} \times \frac{[H^+]_w\cancel{[Z^-]_w}}{[HZ]_w} \tag{45-14}$$

Hence,

$$\frac{[M^+]_w}{[MZ]_w} = \frac{[H^+]_w}{[HZ]_w} \times \frac{1}{K_{st}K_a} \tag{45-15}$$

Introduction of this equality into the expression for D, (45-11), yields

$$D = \frac{P}{\dfrac{[H^+]_w}{[HZ]_w} \times \dfrac{1}{K_{st}K_a} + 1} \tag{45-16}$$

At a given temperature, P, K_{st}, and K_a are constants for a given metal ion, solvent, and complexing agent. Inspection of the above equation reveals that the value of the distribution ratio can be increased either by decreasing the concentration of the hydrogen ion in the aqueous phase or by increasing the concentration of the complexing agent in that phase. An increase in the value of the distribution ratio means that the efficiency of the extraction of the metal is enhanced. Thus, the previous qualitative conclusions are verified and placed on a mathematical basis.

Although the example considered above represents a simple case, the conclusions drawn from it have general validity; in more involved cases, the mathematical effort becomes considerable.

45.3 Selectivity of Extraction

The complexation can be viewed as a competition between the hydrogen ion and the metal ion for the complexing agent, and, other factors being equal, the stronger the complex formed, the lower is the pH at which the extraction is still efficient. This fact is important when more than one metal ion is present because it offers the possibility of securing some degree of selectivity in the extraction process with a given complexing agent and solvent. If a second complexing agent is added, which competes with the agent HZ for the metal ion, the amount of MZ formed will be decreased. If the other complex formed is nonextractable under the conditions existing, the efficiency of the extraction will be decreased. Consequently the second complexing agent can function as a masking agent for the metal. Appropriate adjustment of the concentrations of the masking agent and the extractant can be of additional value in some cases.

45.4 Application of Extraction

Extraction can be applied in two ways: either to separate the desired species from other substances or to separate these substances and leave the desired species in the aqueous phase. Extracts, after dilution to known volume, can often be used for a photometric determination (Chapter 37). An extracted nonvolatile material is readily recovered by evaporation of the organic phase, and any organic complexing agent remaining can be destroyed by ashing.

Separations by means of solvent extraction have distinct advantages over the use of precipitation. In the extraction process, there is no analog to coprecipitation, used in the restricted sense of this term (Section 6.7). A substance that is not extracted alone under the prevailing conditions will not be "coextracted" with another substance. Hence, a separation by extraction is usually "cleaner" than that by a precipitation. Where an extraction is efficient, the technique can be used to concentrate trace amounts of a substance from a large volume of an aqueous solution. Separations can even be effected where the efficiency is low. All that need be done is to repeat the extraction with fresh portions of the solvent until only a negligible amount of the substance remains unextracted. Such a repeated extraction is illustrated by the following example.

Example 45-1. The partition coefficient of iodine between carbon tetrachloride and water has the value 85. Exactly 100 ml of an aqueous 0.0010 *M* iodine solution is extracted twice with 20-ml portions of the organic solvent. What percentage of the iodine initially present remains in the aqueous phase after each extraction step?

Let x = the molar concentration of iodine in the aqueous phase after an extraction. After the first extraction, the millimoles of iodine in the organic phase equals $100 \times 0.0010 - 100x$; hence, the molar concentration of iodine in that phase equals

$$\frac{100 \times 0.0010 - 100x}{20}$$

The expression for the partition coefficient yields

$$85 = \frac{100 \times 1.0 \times 10^{-3} - 100x}{20x}$$

$$x = 5.6 \times 10^{-5}\, M$$

Hencc,

$$\begin{matrix}\% \text{ iodine remaining} \\ \text{in the aqueous phase}\end{matrix} = \frac{5.6 \times 10^{-5} \times 100}{1.0 \times 10^{-3}} = 5.6\%$$

Similarly, for the second extraction step:

$$85 = \frac{100 \times 5.6 \times 10^{-5} - 100x}{20x}$$

$$x = 3.1 \times 10^{-6}\, M$$

$$\begin{matrix}\% \text{ iodine remaining} \\ \text{in the aqueous phase}\end{matrix} = \frac{3.1 \times 10^{-6} \times 100}{1.0 \times 10^{-3}} = 0.31\%$$

The application of the extraction technique to inorganic compounds per se is limited because only relatively few show sufficiently favorable partition coefficients. Notable examples are the extraction of elemental iodine into various solvents and of metal–halide complexes, including iron(III) as $HFeCl_4$, into ether. In most cases, organic extraction agents are employed.

Dithizone, in the extraction and photometric determination of metals, has already been considered in Section 37.3. This compound forms colored chelates with about 20 metal ions and hence has a low selectivity as far as complex formation is concerned. The chelates formed by dithizone with various metal ions differ in their stability. Since dithizone is a weak acid, some selectivity in the extraction can therefore be secured by appropriate adjustment of the pH. Additional complexing agents may be added for the purpose of masking. The concept of masking was briefly explained in Section 20.10. Thus, if potassium cyanide is added, lead(II), which does not form a cyano complex, can be extracted selectively into carbon tetrachloride, in the presence of copper(II), zinc(II), cadmium(II), mercury(II), silver(I), and certain other cations that form stable, nonextractable cyano complexes.

8-Quinolinol, which is also known by the trivial name oxine, offers another example of a reagent that forms extractable chelates with a large number of metal ions. Here, also, pH adjustment and the application of masking agents increase the selectivity considerably. Some further examples of extraction as a prelude to photometric determinations are considered in Chapter 37.

45.5 Leaching

Extraction in the broadest sense of this term also applies to the selective removal of one or more substances from a solid phase by treatment with a suitable liquid and is then termed leaching. For example, the components of a mixture of sodium perchlorate and potassium perchlorate may be separated by extraction with 1-butanol, which only dissolves the sodium perchlorate. From a mixture of the chlorides of calcium, barium, and strontium, only calcium chloride is extracted by a water-free mixture of ether and ethanol.

45.6 Questions

45-1. Elaborate on the significance of masking in connection with extraction equilibria.

45-2. Discuss the extraction end point in redox titrations involving iodine (Section 25.7) in relation to the extraction technique and especially to Example 45-1.

45-3. With the data of Example 45-1, calculate the molar concentration of iodine remaining in the aqueous phase if 5.0-ml and 50.0-ml portions of the organic solvent are employed in the two extraction steps. Discuss the influence of the volume of the organic phase on the amount of iodine unextracted.

45-4. Discuss the influence of pH on the extraction of chelates.

45-5. State why only relatively few inorganic compounds are extractable per se into organic solvents.

In practice, a buffer is sometimes added to the aqueous phase of an extraction system. Elaborate on the possible functions of the buffer.

A metal ion M reacts with HZ and HY to form the complexes MZ and MY, which are the only species extractable into a given solvent. If $K_{st, MZ} = K_{st, MY}$ and $P_{MZ} = P_{MY}$, and HZ is a far weaker acid than HY, which complex will be extracted preferentially at a given pH?

46 CHROMATOGRAPHY

Chromatography is a term for a group of techniques that effect the separation of substances by the selective alteration of the speed of their migration by contacting a mobile phase with a stationary phase. The mobile phase may consist of either a liquid solution or a gaseous mixture and the stationary phase may be either a solid or a liquid held immobile by a solid. Consequently, chromatographic processes may be differentiated according to the mobile phase–stationary phase relation as liquid–solid, gas–solid, liquid–liquid, and gas–liquid. As will be made clear in the development of the topic, the use of systems having a solid as the stationary phase is termed adsorption chromatography and of a liquid immobilized by a solid, partition chromatography.

Liquid–Solid and Liquid–Liquid Column Chromatography 46.1

The first successful application of the complete chromatographic technique may be credited to Tswett, although some aspects of the approach may have been realized earlier by other workers. In Tswett's initial experiments described by him in 1906, a long glass tube, constricted at its lower end, was vertically positioned and packed with finely divided calcium carbonate, which was then moistened with an organic solvent. A solution of various pigments from green leaves was applied to the top of the column and then solvent was passed through. The pigments moved down the column and were seen to separate progressively into distinct, colored bands. The assignment of the term chromatography by Tswett is thereby made clear (*chroma,* Greek = color; *graphein,* Greek = to write). In brief, such a chromatographic separation is explained by the different degrees to which the solutes are adsorbed on the solid, stationary phase, causing the bands to move down the column at different speeds. Such a process is therefore termed adsorption chromatography.

The passage of the solvent through the column is known as development of the chromatogram. By passage of sufficient solvent the bands may be washed progressively from the column. The solvent is then known as the eluting solvent or as the eluent. The operation is termed elution and the technique elution chromatography. The liquid emerging from the column, known as the eluate, can be collected as a number of fractions, which can be separately analyzed. To speed the elution, several different eluents can be applied in sequence. An alternative technique is to push from the tube the developed chromatogram, cut it into cylinderical sections, and analyze the substances retained in the various sections.

A solid support may be used to immobilize a liquid phase. The mechanism of the resulting chromatographic separation involves predominantly the partition of solutes between the two liquid phases, although adsorptive processes may play a secondary role. In the development of the chromatogram the solutes more soluble in the stationary phase move down the column at a smaller rate than the solutes more soluble in the mobile phase. The process may be viewed as a continuous solvent extraction involving a number of solutes differing in the value of their partition coefficient (Section 45.1) and is termed partition chromatography.

Both liquid–liquid and liquid–solid column chromatography are of great interest for the separation of organic substances—especially those with quite similar physical and chemical properties. Application of these techniques to inorganic substances is also feasible.

For adsorption chromatography, over 200 substances have received use as adsorbents; some common ones include alumina, calcium oxide and carbonate, magnesium oxide and carbonate, various silicates and active clays, silica, charcoal, starch, sugars, and powdered cellulose. For partition chromatography, the solid support is more frequently used to immobilize the more polar of the two solvents, usually an aqueous phase. Common supports receiving such use include silica gel, silicic acid, powdered cellulose, starch, and diatomaceous earth. Obviously the solid should not be dissolved by the solvents employed, and ideally neither the solid nor the solvent system should undergo unwanted reactions with the solutes to be separated.

46.2 Ion-Exchange Chromatography

Where the stationary phase interacts with a solute by actual ion exchange, a special form of chromatography, known as ion-exchange chromatography, is possible. This technique has attained an important position for the separation of inorganic ions with the commercial availability of ion-exchange resins. The subject is treated in Chapter 47.

Paper Chromatography 46.3

This technique first received extensive attention in the 1940s and rapidly become one of the most commonly used forms of chromatography. In this technique, usually one liquid phase is allowed to move along a strip or sheet of filter paper while a second liquid phase is immobilized by the paper. The process is best viewed as a form of liquid–liquid partition chromatography. Paper chromatography is of value for the separation and the subsequent identification or determination of micro amounts of both organic and inorganic substances.

Ascending paper chromatography may be performed with the simple arrangement shown in Figure 46-1. The solvent system, for example a mixture of 1-butanol and aqueous hydrochloric acid, is placed in the cylinder. The sample solution, often containing only microgram amounts of the solutes to be separated, is placed on the paper strip as a spot. (The exact

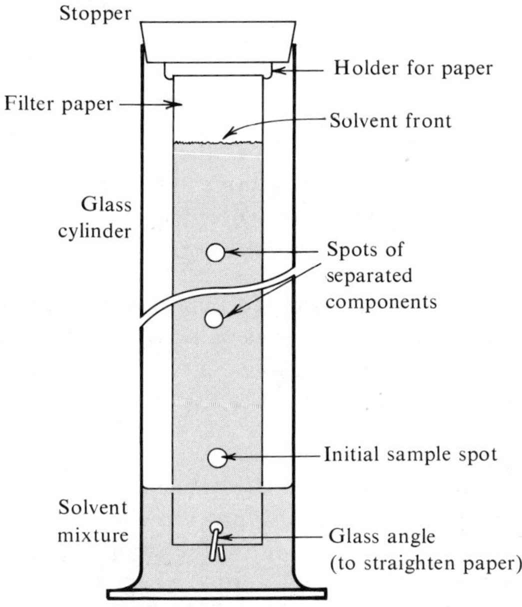

Figure 46-1. Simple arrangement for ascending paper chromatography.

volume spotted must be known if the chromatographic separation is to be followed by a determination.) The strip is fixed to the holder at the bottom of a stopper, weighted to prevent bending (e.g., by a glass angle or simply by a paper clip), and inserted into the cylinder so that the sample spot is well above the liquid level. The cylinder is closed tightly so that an atmosphere saturated with solvent vapors is maintained. By capillary action, the solvent mixture ascends the paper showing a sharp solvent front. Water is preferentially retained by the hydrophilic cellulose of the paper; consequently, the mobile phase will have a higher concentration of the organic solvent. Solutes, such as copper(II), bismuth(III), and lead(II) ions, will be partitioned between the two phases and will move up the paper at different speeds. After some hours, the development will be complete and the solutes will form discrete spots. The spots of colorless solutes and often even colored ones when very small amounts are involved are not directly visible. Location of the spots can be achieved by various means, including their fluorescence under ultraviolet illumination and reaction with chromogenic or fluorogenic reagents. For example, for the ions mentioned above, spraying of the dried developed chromatogram with a hydrogen sulfide solution yields the metal sulfides that appear as brown or black spots.

Of special value in paper chromatography is the R_F value of a solute, which is defined by

$$R_F = \frac{\text{distance between the center of the spot and the sample application line}}{\text{distance between the solvent front and the sample application line}} \qquad (46\text{-}1)$$

For a given solvent mixture and paper, the R_F value is a characteristic of the solute and is usually virtually independent of its concentration in the sample. From measured R_F values, inferences are often possible as to the substances present in the sample.

For quantitative purposes, a substance may be extracted from its spot and be determined appropriately. To secure larger amounts of the separated solutes, the sample solution may be applied as a streak or band across a wider strip or sheet of paper rather than as a spot. More sophisticated apparatus and arrangements are possible, including the use of specially impregnated paper.

Descending paper chromatography is another technique frequently applied. Here a trough containing the solvent mixture is located at the upper portion of the cylinder. The paper strip dips into the solvent, bends over the rim of the trough, and by means of a weight is kept stretched in a vertical position. The solvent descends under the influence of both capillary action and gravity. Obviously for identical papers the development is more rapid in the descending mode than in the ascending one and the distance that the solvent can travel is greater in the descending technique.

Thin-Layer Chromatography 46.4

This technique, often assigned the acronym TLC, did not attain prominence until the early 1960s. Commonly in this method a plate of glass is coated with a uniform layer of a water slurry of a finely powdered adsorbent, for example, silica gel or aluminum oxide, usually with an agent added that binds the adsorbent to the glass. The coated plate is dried in an oven and stored in a desiccator. The plates are spotted or streaked with the sample solution and the chromatogram is developed in a manner similar to that described for ascending paper chromatography. The advantages of thin-layer chromatography rest mainly in the possibility of using a great variety of chromatographic materials, including those that allow solvent systems to be used that would attack or destroy paper, and of securing smaller, better defined spots and zones and thereby superior separations. Flexible TLC plates are commercially available and are prepared by coating a metal foil or plastic film with the absorbent, using an appropriate binder and technique.

Gas–Solid and Gas–Liquid Chromatography 46.5

Gas chromatography involves the application of the principles described earlier to the separation and determination of substances volatile at the elevated temperatures employed. Consequently, the technique is applicable to many substances that are liquids or even solids at room temperature.

In gas–solid chromatography (also termed gas adsorption chromatography) a solid is packed into a column and the sample is passed through this column carried by an inert carrier gas. For example, activated carbon or silica gel may be the solid and nitrogen, helium, or argon the carrier gas. Different components are adsorbed on the column to different degrees and consequently pass through the column at different rates.

In gas-liquid chromatography (also termed gas partition chromatography) the column is packed with a granular solid (e.g., glass beads or crushed brick) coated with a nonvolatile liquid (e.g., silicone oils). The components of the sample carried by the gas stream partition between the liquid phase and the gas phase. Depending on the differences on their degree of "extraction" into the liquid, the solutes move through the column at different rates.

Gas chromatography usually involves the elution technique. The components as they are eluted sequentially by the carrier gas are detected or determined by a suitable method. If desired, they can be collected by appropriate means, for example, in a cold trap.

The principal components of a gas chromatograph are shown in Figure 46-2. To permit the use of a small oven, the configuration of the column is often other than linear; a spiral column is seen in the figure. In operation, the carrier gas is passed at a suitable, constant rate and the oven is maintained at the desired temperature by thermostatic control. The sample is injected by a syringe capable of delivering microliter volumes via a hypodermic needle through a port having a self-sealing silicone rubber disk. The port area is heated to assure prompt volatilization of the sample. Various devices find use for detecting the presence of something other than carrier gas in the column effluent. Ideally such a detector should have high sensitivity, be stable, give a signal that can be amplified and recorded, and respond linearly to the concentration of the component carried by the inert gas. A detector based on differential thermal conductivity is in common use and is shown in Figure 46-2. Identical platinum or tungsten wires are placed in the gas stream entering and leaving the column and within a thermostated block of metal. The wires are heated by an electric current. The electric conductivity of a wire depends on its temperature, and the latter varies according to the rate at which heat is lost to the surrounding gas. The two wires are made opposite arms of a conductivity bridge (Section 32.4). As long as the carrier gas alone is passing the two wires, the bridge will remain in balance.

When a component of the sample in the effluent gas reaches the detector, the heat loss of the wire exposed to it is altered and the bridge becomes unbalanced. The resulting bridge cur-

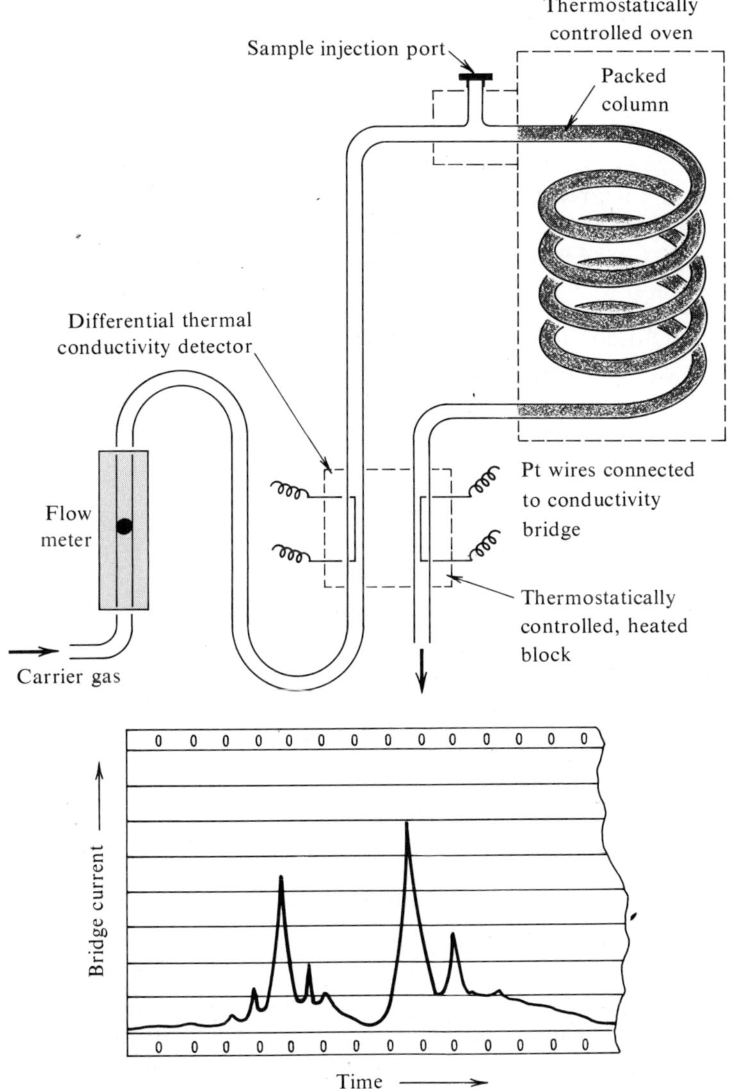

Figure 46-2. Left. schematic diagram of arrangement of the components of a gas chromatography. Right: gas chromatogram.

rent or bridge unbalance potential is detected, amplified, and fed to a chart recorder. The graph obtained, known as a gas chromatogram, is a plot of bridge current versus time. The time elapsed between injection of the sample and the appearance of a component in the effluent is known as the retention time for the substance. For given conditions (including the nature and velocity of the carrier gas and the packing, length, and temperature of the column), the value of the retention time is characteristic of a substance.

Inspection of the gas chromatogram at the foot of Figure 46-2 suggests that the sample consists of two major and at least five minor components. For quantitative purposes, the height of a peak or the area under it can be related to the amount of the component by the use of calibration curves for pure components or mixtures of known composition.

Gas chromatography finds its major application in the detection, identification, and determination of organic substances in complex mixtures. It is possible to collect components separately as they emerge from the column and to examine them by appropriate techniques, including elemental analysis, mass spectrometry, and infrared and fluorescence spectra.

46.6 Questions

46-1. In the various chromatographic techniques, what is the stationary phase? The mobile phase?

46-2. What is meant by the expressions "development of a chromatogram" and "elution"?

46-3. Define the terms "retention time" and "R_F value." In the relevant techniques, would a high affinity of a solute for the stationary phase correspond to a short or a long retention time? A small or a large value for the R_F value?

46-4. How is a liquid made the stationary phase in gas chromatography? In paper chromatography? In thin-layer chromatography?

46-5. What will be the maximum and minimum values attainable for an R_F value?

46-6. Why is it essential in gas chromatography that the sample be gaseous or be volatilized almost instantaneously? What would be the effect on a gas chromatogram if a liquid sample evaporated only slowly?

46-7. It is possible to obtain longer chromatograms in descending than ascending chromatography. Explain.

46-8. Paper chromatograms are usually developed with a mixture of two solvents, for example, water and an alcohol. However, development of some solutes can be effected with water alone. How do these two procedures differ with regard to the principles largely governing the separation of the various constituents of a sample?

46-9. Two-dimensional development in paper chromatography, also known as two-way paper chromatography, involves spotting a square sheet of paper near one corner and developing the chromatogram in solvent 1. The paper is dried and then developed with a different solvent 2 in a direction at right angles to the initial development. Draw your representation of the two-way chromatogram formed by a sample containing solutes *A* through *E*, each having the R_F values listed below in the two solvents, if each solvent front moves exactly 5 cm. Show each final spot as a small circle and mark the point of spotting the sample with an "X."

	Solute				
	A	*B*	*C*	*D*	*E*
R_F value, solvent 1:	0.15	0.35	0.40	0.75	0.85
R_F value, solvent 2:	0.70	0.15	0.40	0.25	0.50

What special merits of two-dimensional development can you see from this hypothetical example?

47 ION-EXCHANGE METHODS

The Nature of Ion Exchangers 47.1

Ion exchange can be defined as a reversible process in which ions are exchanged between a liquid and a solid. The solid is known as an ion exchanger and undergoes no substantial change. Although complex inorganic substances can function as ion exchangers, water-insoluble organic polymers are usually employed in analytical practice. An ion-exchange resin is a special type of a polyelectrolyte and consists of a three-dimensional polymeric hydrocarbon network to which are bonded a large number of electrically charged groups, such as the sulfonate group, $-SO_3^-$, or a quaternary ammonium group, $-N(CH_3)_3^+$. Ion-exchange resins are commercially available, usually as small, uniform beads, in diameters ranging from 0.04 to 1.2 mm.

Strongly Acidic Cation-Exchange Resins 47.2

Ion-exchange resins having sulfonic groups as the exchange sites are termed strongly acidic cation-exchange resins. When such a resin is treated with a strong acid (often 6 *F* hydrochloric acid), all the sulfonate groups are converted to the acid form (that is, the hydrogen form) and the resin may be represented by $(RSO_3^-)H^+$, where R is the resin network. The resin in this form behaves as an insoluble strong acid. The sulfonate group has less attraction for the hydrogen ion than for the sodium ion; consequently, if the resin in the hydrogen form is placed in a solution of sodium chloride, ion exchange occurs. The equilibrium established can be represented by

$$(RSO_3^-)H^+ + Na^+ \rightleftharpoons (RSO_3^-)Na^+ + H^+ \qquad (47\text{-}1)$$

For each equivalent of sodium ion, one equivalent of hydrogen ion is freed. The position of the equilibrium will depend on the acid strength of the resin and the concentration of the sodium ion in the solution in contact with the resin. For a strongly acidic resin, the exchange affinity for cations depends on the charge of the cation: Tripositive cations are held more firmly than dipositive cations, and these more firmly than unipositive ones. Thus, if the resin in the sodium form is placed in a dilute solution of an aluminum salt, ion exchange takes place and for each tripositive aluminum ion exchanged three unipositive sodium ions are freed. Within a given group of simple cations of equal charge, the affinity of the resin tends usually to be the greater, the greater the atomic number of the cation. Thus, for the second column of the periodic table, the order of the affinity is $Be^{2+} < Mg^{2+} < Ca^{2+} < Sr^{2+} < Ba^{2+}$. A fuller consideration of resin selectivity is beyond the scope of this textbook.

The affinity between an ion and an ion-exchange resin may be so great that the dissolution of a water-insoluble salt can be effected. For example, if a sufficient amount of a strongly acidic cation-exchange resin is placed into an aqueous suspension of barium sulfate, the latter is dissolved eventually. The process may be represented by

$$2(RSO_3^-)H^+ + \underline{BaSO_4} \rightleftharpoons (RSO_3^-)_2Ba^{2+} + 2H^+ + SO_4^{2-} \tag{47-2}$$

Note that in this example the exchange equivalency corresponds to two moles of hydrogen ion being displaced by one mole of barium ion.

47.3 Batch Operation with an Ion-Exchange Resin

When a strongly acidic cation-exchange resin is placed in a dilute solution of sodium chloride, exchange according to equation (47-1) takes place but the reaction does not go to completion. If more complete conversion of the sodium chloride to hydrochloric acid is desired, the equilibrated resin may be separated by filtration or decantation and a further portion of fresh resin be added to the solution. This so-called batch operation is tedious and receives little application in quantitative analysis.

47.4 Column Operation with an Ion-Exchange Resin

A column operation applied to ion-exchange processes is far more efficient than a batch operation. The resin is placed on top of a glass wool plug in a vertical tube, suitably a buret in many analytical applications. A strongly acidic cation-exchange resin can then be converted completely to the desired ionic form by passage of a sufficiently concentrated solution of the desired cation, usually by gravity flow through the resin column, which is often called the resin bed. Thus, to obtain the hydrogen form, 6 *F* hydrochloric acid is passed, followed by distilled water until the effluent is neutral. At no time should the liquid level be allowed to drop below the top of the resin bed; otherwise, air bubbles become entrapped and reduce the active surface available for ion exchange. Then the sample solution is placed on top of the resin bed and allowed to percolate through the column at an effluent rate of about 3 to 5 ml per minute per square centimeter of resin-bed cross section. Thus, in the conversion of sodium chloride to hydrochloric acid, the solution as it passes through the column will encounter fresh, unequilibrated layers of resin, and conversion to hydrochloric acid will be complete. Attention must be paid to certain details. The exchange capacity of the bed must be in excess of the amount of sodium ion to be exchanged. Further, the sodium chloride solution must be sufficiently dilute; otherwise, a concentrated solution of hydrochloric acid would result and the equilibrium would not favor full exchange of sodium for hydrogen. (Recall that the resin is converted to its hydrogen form by treatment with concentrated hydrochloric acid.) Eventually after prolonged use the column becomes exhausted and is then regenerated by passage of a concentrated solution of an appropriate ion. Commerical strongly acidic cation-exchange resins may be obtained with various exchange capacities, commonly within the range 0.6 to 2.3 meq of cation per milliliter of wet resin.

Strongly Basic Anion-Exchange Resins 47.5

Strongly basic anion exchange resins function analogously to the strongly acidic cation exchangers. Thus, the conversion of sodium chloride to sodium hydroxide by such a resin in its hydroxide form may be represented by

$$(RN(CH_3)_3^+)OH^- + Cl^- \rightleftharpoons (RN(CH_3)_3^+)Cl^- + OH^- \quad (47\text{-}3)$$

In practice, complete anion exchange is not as easily attained as cation exchange. Anion-exchange resins, especially in the hydroxide form, show lower chemical and thermal stability.

Ion-Exchange Methods in Analytical Chemistry 47.6

Some applications of ion-exchange methods in analytical chemistry may be briefly noted. A salt, preferably one with a high equivalent weight, can be employed for the standardization of a base or to prepare a standard acid solution. A weighed amount of the (dried) salt is dissolved in distilled water. The solution is then allowed to percolate through a cation-exchange column in the hydrogen form, and the acid formed is completely removed from the column by the passage of distilled water. The combined effluent, since it contains an amount of hydrogen ion that can be calculated from the amount of the salt taken, can be used directly to standardize a base. Alternatively, the effluent may be diluted to a known volume, to obtain a standard acid solution.

Cation exchange using a resin bed in the hydrogen form can be applied to determine the total cation content of a sample solution. On passage of an aliquot of the sample solution, the number of equivalents of hydrogen ion in the effluent is determined by titration with a standard base; thereby the number of equivalents of cations in the sample solution is established. If the sample solution contains free acid, and hydrogen ion is to be distinguished from other cations, an identical aliquot of the sample solution is titrated with the base without passage through the column. The volume of base required in this titration is subtracted from that required for the titration of the effluent. The difference corresponds to the equivalents of cations other than hydrogen ion. The method cannot be applied if the sample contains cations that precipitate as hydroxides during the titration.

Sodium hydroxide solutions containing carbonate may be passed through an anion-exchange column in the hydroxide form. The carbonate is exchanged for hydroxide, and the carbonate-free sodium hydroxide solution that emerges is suitable for use as a titrant.

Ion exchange offers a convenient and efficient method of removing from a sample solution cations that interfere in the determination of an anion, or conversely of removing anions that interfere in the determination of a cation.

A most significant analytical use of an ion-exchange resin is the concentration of a trace element to a point that it can be determined by an available technique [for example, the concentration of trace copper(II) in biological samples].

Ion-exchange resins are now commonly used to deionize water in the laboratory and in plant operations. Both cations and anions are removed by use of a mixed resin bed containing cation-exchange and anion-exchange resins in the hydrogen and hydroxide forms, respectively.

Ion-exchange chromatography utilizes the slight differences in the affinity of an exchange resin for different ions to effect their separation. Usually the ion mixture is placed on the resin bed. Then a solution is passed to move the ions down the column; those ions that have the highest affinity for the resin are the last to appear in the effluent. The eluting solution may contain one or more complexing agents that compete with the resin for the ions. The eluate is collected in a number of vessels, thus accomplishing complete or at least group separation of the ions. In this way it is possible to separate quantitatively ions that are quite similar in their chemical behavior, as, for example, the rare earth elements.

Some metals that form anionic complexes can be separated on *anion*-exchange columns. For example, cobalt(II), but not nickel(II), forms chloro complexes, such as $CoCl_4^{3-}$, in 12 F hydrochloric acid. A solution of cobalt(II) and nickel(II) 12 F in hydrochloric acid is placed on an anion-exchange column in the chloride form, and the column is washed with 12 F hydrochloric acid. The effluent contains only the nickel, since the cobalt is retained on the resin in the form of anionic chloro complexes. Then 0.1 F hydrochloric acid is passed; the chloro complexes of cobalt dissociate at this lower chloride concentration, and cobalt in cationic form emerges in the effluent.

Questions 47.7

Define the term ion exchanger.

What is the difference between a batch operation and a column operation? Which would be expected to be of the greatest value in analytical practice?

What will be the composition of the effluent upon percolation of a dilute solution of the following through a cation-exchange column in the hydrogen form: (a) KBr, (b) $Al(NH_4)(SO_4)_2$, (c) HNO_3, and (d) $NaCl + H_2SO_4$?

What will be the effluent obtained on the passage of a dilute solution of the following through an anion-exchange column in the hydroxide form: (a) NaI, (b) $BaCl_2$, (c) $Ba(NO_3)_2 + HNO_3$, and (d) HCl?

Write the equation representing the exchange reactions occurring when a solution containing $Ca(HCO_3)_2$ and NaCl is passed through a *mixed* bed with the cation and anion exchangers in the hydrogen and hydroxide forms, respectively.

What will be the effluent obtained when a dilute $BaCl_2$ solution is passed through a column of a cation-exchange resin in its hydrogen form? In its sodium form? In its iron(III) form?

Phosphate seriously interferes with the cation separations according to the hydrogen sulfide scheme. Elaborate on the possibility of employing ion exchange to overcome this interference.

Discuss the difference in the concept of equivalency in ion-exchange and acid–base titrations.

47.8 Problems

47-1. An ion-exchange column has an inside diameter of 2.0 cm and, in its use in a 3-minute period, a volume of 50 ml of effluent is collected. What is the flow rate of the column?

Answer: 5.3 ml/min for 1 cm^2 of bed cross section

47-2. In a column of 2.0-cm^2 cross section, a 10-cm bed of a cation-exchange resin in the hydrogen form is established. This resin in the hydrogen form is established. This resin has an exchange capacity of 2.4 meq/ml of wet resin. If the column is completely converted to the aluminum form, what is the weight of aluminum (*27*) retained?

Answer: 0.43 g

47-3. The sodium ion from 50 ml of a 1.5% solution of NaCl (*58*) is to be removed by a cation-exchange column in the hydrogen form having an exchange capacity of 5.0 meq/g of dry resin. Assuming that all the capacity of the resin is used, what weight of (dry) resin is required to exchange this amount of sodium ion?

Answer: 2.6 g

47-4. A volume of 25 ml of 0.10 F $Al_2(SO_4)_3$ is passed through a cation-exchange column in the sodium form. Assuming complete exchange of aluminum ion for sodium ion, what weight of sodium ion (*23*) is present in the effluent?

Answer: 0.34_5 g

47-5. A 0.3728-g sample of KCl (*74.56*) is dissolved in about 50 ml of distilled water and passed through a cation-exchange column in its hydrogen form. The column is washed with water and the combined effluent is titrated with a NaOH solution, requiring 50.00 ml. Calculate the formality of the base.

Answer: 0.1000 F

47-6. A mixture, containing only KCl (*74.56*) and $BaCl_2 \cdot 2H_2O$ (*244.3*), is dried and the content of water (*18.02*) found to be 10.00% w/w. An amount of 0.8000 g of the dried mixture is dissolved in water and passed through a cation-exchange resin in the H^+ form. The effluent and washings are collected in a 200.0-ml volumetric flask and diluted to mark. Assume 100% exchange efficiency and calculate what volume of 0.0250 F $Ba(OH)_2$ will be required to titrate a 25.00-ml portion of this acid solution.

47-7. A material is known to contain Na_2SO_4 (*142.04*), NaCl (*58.44*), $NaNO_3$ (*84.99*), and some inert material. A 2.000-g sample of this material is dissolved in water and diluted to mark with water in a 250-ml volumetric flask. A one-tenth aliquot is taken and passed through a cation-exchange resin in the H^+ form. The column is then rinsed with water. The combined effluent and washings require 25.77 ml of 0.1000 F NaOH. Another one-tenth aliquot requires 12.82 ml of 0.0800 F $AgNO_3$ in a precipitation titration. A one-fifth aliquot is reacted with an excess of $BaCl_2$ solution and yields 197.2 mg of $BaSO_4$ (*233.40*). Calculate the % w/w composition of the sample.

48 EVOLUTION METHODS

The separation by volatilization of one or more components of a sample can be undertaken for various reasons. Unwanted impurities or components may be removed, for example, ammonium salts and water by ignition of a precipitate. The weight of a sample before and after the volatilization of a single component may serve for the determination of the latter. Examples include the *indirect* gravimetric determination of water and carbon dioxide (Sections 6.1 and 6.4). Alternatively, the volatilized material may be recovered. Familiar examples include the preparation of distilled water and the distillation of ammonia (Sections 14.9 and 37.2). Some additional examples of evolution methods will be discussed in this chapter, with methods for water receiving detailed consideration because of their practical importance.

Water in Solids 48.1

Before the possibilities of the determination of water by evolution methods are considered, it is fruitful to discuss the various modes by which water may be associated with a sample. A broad classification into essential and nonessential water can be made.

Essential Water. Essential water is present in a stoichiometric ratio and is an essential part of the chemical composition of the substance. It may be present as water of crystallization or water of constitution, or as both.

Water of Crystallization. Water of crystallization is held by different substances with different strengths. Some substances lose such water by mere exposure to dry air. For example, sodium carbonate decahydrate effloresces (that is, it withers) readily to the monohydrate. Other substances require warming or even heating at elevated temperatures. For example, copper(II) sulfate requires about 150°C or higher to be converted to the anhydrous salt. Some water is lost at lower temperatures and relations exist between vapor pressure and temperature in hydrate equilibria; however, their discussion is beyond the scope of this textbook. It may be stated that it is frequently necessary to keep the ambient temperature as well as the humidity within a certain range if a specified hydrate is to be obtained and maintained.

Water of Constitution. Water of constitution does not show in the conventional formula of a substance as "H_2O" and is usually only volatilized at a high temperature. The loss of water by calcium hydroxide

$$Ca(OH)_2 \rightarrow CaO + H_2O \uparrow$$

offers a familiar example. The distinction between water of crystallization and water of constitution is made clear by the conversion of magnesium ammonium orthophosphate hexahydrate to the anhydrous salt and then to anhydrous magnesium pyrophosphate.

$$MgNH_4PO_4 \cdot 6H_2O \rightarrow MgNH_4PO_4 + 6H_2O \uparrow$$

$$2MgNH_4PO_4 \rightarrow Mg_2P_2O_7 + 2NH_3 \uparrow + H_2O \uparrow$$

Nonessential Water. Nonessential water is present in nonstoichiometric amounts and does not represent an essential part of the constitution of a substance and may be classified as occluded, sorbed and adsorbed water, and water present as a solid solution. Nonessential water loosely held by a material is often referred to as the "moisture" content of the sample.

Occluded Water. Occluded water is entrapped during the formation of crystals and cannot be removed by simple drying. Elevated temperatures are required and then the water is freed by rupture of the crystal. This rupture, known as decrepitation, is sometimes violent and in quantitative operations care should be exercised to avoid loss of material.

Sorbed Water. Sorbed water is held within the internal surfaces and crevices of substances possessing a porous, capillary structure. Familiar examples include starch, active carbon, natural and synthetic zeolites (certain magnesium aluminum silicates), and colloidal substances (such as low-temperature dried hydrous aluminum oxide). Sorbed water in some cases is strongly held, and its complete removal may require heating to a temperature of several hundred degrees and for a protracted period of time.

Adsorbed Water. Adsorbed water is usually held loosely as a monomolecular film on the surface of the material. Consequently, the amount, thus retained, increases with the degree of dispersion of the material. Such water can often be removed at room temperature by mere exposure to a dry atmosphere.

Solid Solutions. Solid solutions involving water are known. For example, glasses may hold water homogeneously dispersed in this way.

Hygroscopicity. Some substances (including liquids, e.g., concentrated sulfuric acid) take up water avidly from their surroundings. In some cases, the uptake of water by a solid can proceed so far that the substance dissolves (magnesium chloride is a familiar example). The water taken up by a substance may be incorporated as essential or nonessential water, or both. For example, water taken up by dehydrated barium chloride goes to reestablish the dihydrate; in contrast, that taken up by an ignited zeolite merely establishes some degree of water sorption. Hygroscopic substances are frequently a nuisance in analytical work and may require special precautions in their handling. However, some hygroscopic substances are useful as drying agents, that is, as desiccants.

Direct Gravimetric Determination of Water 48.2

If on drying (with or without the application of heat) of a sample the weight loss is solely caused by the volatilization of water, an *indirect* gravimetric determination of water succeeds. However, when other volatilizable substances are present or if

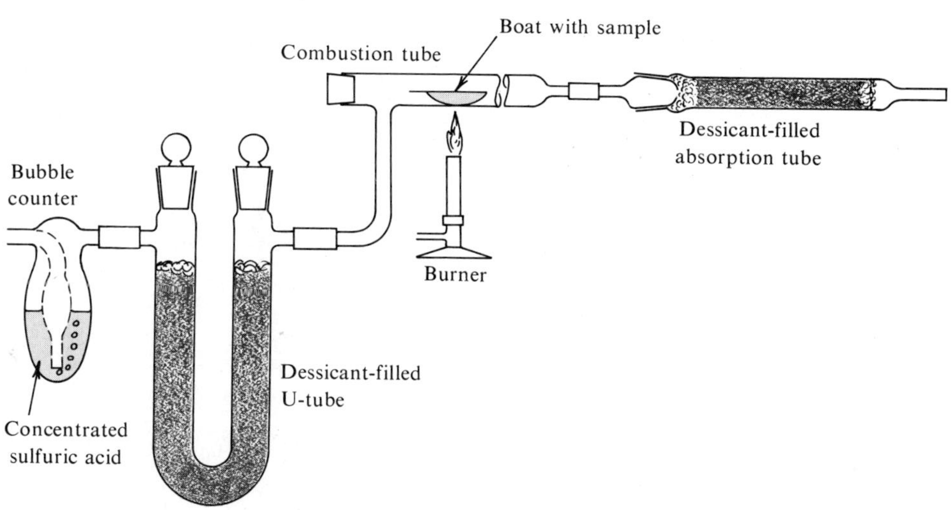

Figure 48-1. Apparatus for the direct determination of water.

the sample reacts with the oxygen of the air [e.g., iron(II) oxide in some minerals or ores is converted to iron(III) oxide], a *direct* determination of water must be undertaken. The water is volatilized in a suitable manner, collected selectively, and weighed. A simple assembly for this purpose is shown in Figure 48-1. The material is weighed into an elongated porcelain or metal vessel known as a boat, which is then placed into the tube and heated. A stream of air, dried by passage through a desiccant-filled tower, carries the evolved water to a U-tube packed with a water-absorbing material. The difference in the weight of this tube before and after the "run" is the weight of water evolved from the sample.

48.3 Water Determination by Azeotropic Distillation

Some liquids form mixtures that boil at a temperature lower than any of the components separately. The mixture having the lowest boiling point is called the azeotropic mixture or,

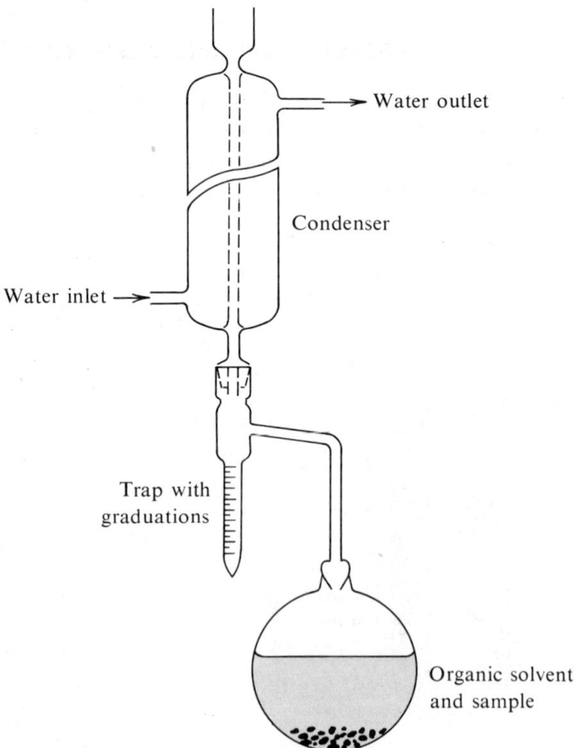

Figure 48-2. Apparatus for the determination of water by azeotropic distillation.

in short, the azeotrope. The composition of the azeotrope is fixed at a given atmospheric pressure. The alcohol–water azeotrope (ca. 96% v/v ethanol) and constant-boiling hydrochloric acid (Section 14.1) afford common examples. Azeotropic distillation is often applied to determine water firmly held by samples that would not tolerate extensive heating, such as coal and plant tissue. The assembly of Figure 48-2 is suitable. The sample is placed in a distillation flask, xylene is added, a condenser is fitted, and the flask is heated. The xylene–water azeotrope distills; the condensate flows down the condenser walls and is collected in a vertical tube. Upon cooling the components of the azeotrope separate, and water, having the higher density, accumulates at the bottom of the tube. Xylene overflows into the sample flask and is recycled. The tube is graduated and the volume of water collected can be read. Alternatively, and especially with small amounts of water, the determination of the water, thus collected, may be performed in other ways, notably by a Karl Fisher titration (Section 25.7).

Determination of Carbon Dioxide 48.4

The indirect determination of carbon dioxide suffers the same limitations as the analogous method for water and a direct evolution method is necessary in many cases. For this purpose the assembly shown in Figure 48-1 may be used, with the U-tube filled with a soda-lime mixture, which absorbs carbon dioxide. In this case the carried gas is not only predried but is also freed of carbon dioxide by an additional appropriate charge in the tower. If water is also evolved under the conditions of the run, it must be eliminated before the gas stream enters the soda-lime tube by allowing the stream to pass first through a tube containing a desiccant. Consequently, both water and carbon dioxide can be determined simultaneously.

C–H and N Analysis of Organic Samples 48.5

The possibility mentioned above for the simultaneous determination of carbon dioxide and water allows the determination of the carbon and hydrogen content of an organic material by its complete combustion. In essence, the assembly of Figure 48-1 is used with two absorption tubes, one for water and the other for carbon dioxide. The combustion tube contains a packing of oxidized copper wire, which is heated externally and assures the complete oxidation of evolved gases. Dry oxygen is frequently employed as the carrier gas to facilitate the combustion. To avoid interference by other gaseous substances formed, the exit end of the combustion tube contains various fillings. For example, silver wool may be inserted to trap halogens and sulfur. Nitrogen in organic samples, after Dumas, is determined as follows. The sample is mixed with copper(II) oxide and introduced into a tube that is already filled to about one half its length with this oxide. The tube is heated and combustion to water, carbon dioxide, and nitrogen occurs. The combustion products are swept by carbon dioxide as the carrier gas into a special gas buret known as a nitrometer. This device is a graduated tube filled with concentrated potassium hydroxide solution. This solution absorbs all the carbon dioxide and water but not the nitrogen gas. The volume of nitrogen trapped above the solution is read and related to the amount of nitrogen in the sample.

48.6 Distillation Methods

Although separation by distillation is largely the domain of organic chemistry, it finds some application in inorganic analysis. The volatility of most inorganic substances is low. Consequently, only a few can be distilled at moderate temperatures, even at reduced pressure. The separation of water via azeotropic distillation was considered in Section 48.3. The distillation of ammonia with water following a sample digestion by the Kjeldahl technique has been described in Section 14.9. Fluoride can be distilled as hexafluorosilicic acid from an acidic medium containing silicic acid, as was mentioned in Section 18.9. A further example is the distillation of the chlorides of arsenic, antimony, tin, and germanium. These chlorides distill from a concentrated hydrochloric acid solution at different temperatures and can thus be separated. Zinc metal is sometimes distilled at high temperature and thereby separated from less volatile components of alloys.

48.7 Determination of Arsenic by Volatilization

When a material containing arsenic is boiled in the presence of zinc metal in an acidic solution, complete reduction to gaseous arsine, AsH_3, occurs (e.g., $As^{3+} + 3Zn + 3H^+ \longrightarrow AsH_3\uparrow + 3Zn^{2+}$). The arsine can be carried by a gas stream into a glass tube having a small-bore section 1 to 2 cm in length. This portion is heated externally. Arsine is thermally decomposed and elemental arsenic deposits as a mirror on the glass walls. From the length of the mirrored section, the amount of arsenic in the sample is estimated by comparison of the results obtained with standards. Alternatively, the mirror is dissolved and the arsenic determined iodometrically (Section 25.8) or photometrically. This method, introduced by Marsh, is a standard procedure in forensic analysis when arsenic poisoning is encountered or suspected.

In the Gutzeit method, the reduction is performed in an assembly such as shown in Figure 48-3. The hydrogen gas evolved in the reaction between zinc and acid carries the arsine into the upper tube. Droplets and any hydrogen sulfide formed are held back by a cotton plug impregnated with lead acetate. The upper tube contains a moistened strip of paper impregnated with mercury(II) bromide. The latter reacts with the arsine and a coloration appears. The length of the colored

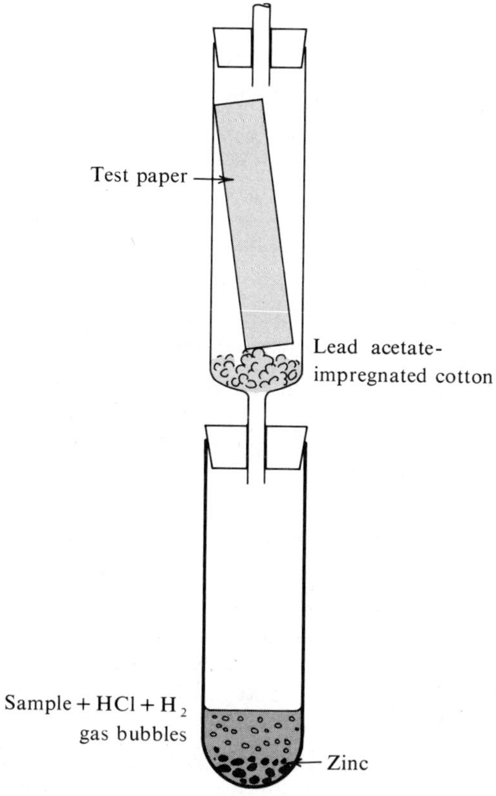

Figure 48-3. Apparatus for determination of arsenic by the Gutzeit technique.

portion is compared with that of strips obtained in the procedure with known amounts of arsenic under identical conditions. In this way trace amounts of arsenic can be determined rapidly with adequate accuracy.

48.8 Questions

48-1. What is the difference between essential and nonessential water? Describe in your own words and give examples other than those found in the text.

48-2. Is it possible for a substance to contain both essential and nonessential water? Given an example.

48-3. Define hygroscopicity and give examples of hygroscopic substances.

48-4. An air stream containing some water vapor is passed through an absorption tube filled with CaO. Write the equation for the reaction involved in the water absorption. Will the increase in weight give the correct amount for water?

48-5. A stream of dry air containing some carbon dioxide is passed through an absorption tube filled with CaO. Write the equation for the reaction involved in the absorption of the carbon dioxide. Will the increase in weight give the correct amount for carbon dioxide?

48-6. A stream of dry air containing some carbon dioxide is passed through an absorption tube filled with $Ca(OH)_2$. Write the equation for the reaction involved in the absorption of the carbon dioxide. Will the increase in weight necessarily give the correct amount for carbon dioxide? What else must be present in the absorption tube in order that the correct weight is obtained?

48-7. Name a solid substance, other than calcium oxide or hydroxide, suitable for the absorption of carbon dioxide.

48-8. What is the distinction to be made between dry and water-free copper sulfate?

48.9 Problems

48-1. What amount of H_2O (*18.0*), in milligrams, is volatilized when 0.543 g of $BaCl_2 \cdot 2H_2O$ (*244.3*) is heated to constant weight?

Answer: 80.0 mg

48-2. A material contains 50.0% $CaCO_3$ (*100.1*) and 50.0% $MgCO_3$ (*84.3*). On ignition, CO_2 (*44.0*) is completely volatilized. Calculate the per cent loss on ignition for a 0.2500-g sample of that material.

Answer: 48.1%

48-3. A material containing 3.0% $CaCO_3$ (*100.1*) and 97.0% FeO (*71.8*) is ignited. CO_2 (*44.0*) is evolved and oxygen (*16.0*) is absorbed, forming Fe_2O_3 (*159.7*). Calculate the per cent loss of ignition for this mixture.

Answer: −9.49%

When a 1.000-g sample of a material is dried, a loss of 16 mg of H_2O occurs. When 0.870 g of the same material (not dried) is *ignited*, a loss of 63 mg is observed. (a) Calculate the per cent H_2O content and (b) the per cent loss on ignition on a dry basis.

Answers: (a) 1.6%; (b) 5.7%

A material is known to contain 15.3% water. What will be the weight of an absorption tube when the water has been absorbed from a 1.520-g sample if the tube weighed 25.533 g before the start of the run?

Answer: 25.766 g

A substance known to contain *only* carbon, hydrogen, and nitrogen was analyzed by combustion. A microbalance precise to ± 1 or 2 μg was used. The following data were secured: Combustion boat + sample = 0.053256 g; boat empty = 0.050231 g; absorption tube + water = 7.634614 g and tube empty = 7.632564 g; absorption tube + carbon dioxide = 7.193111 g and tube empty = 7.184535 g. Calculate the composition of the substance in per cent C, H, and N. Give the simplest empirical formula possible for the substance.

Answers: 77.3_9% C, 7.5_1% H, 15.1_0% N; C_6H_7N (aniline)

49 ANALYSIS OF GASES

Introduction 49.1

The analysis of mixtures of gases is encountered in many areas of science and technology. Some common analytical problems include the evaluation of gaseous fuels, the analysis of combustion products, the study of respiration and metabolism, and the analysis of air pollutants and process gases from smelting and chemical operations.

Many approaches exist that allow the analytical resolution of gas mixtures. Some of the methods for individual gases considered in other sections have special application. Some examples may be cited.

Water vapor may be determined gravimetrically by absorption in a drying tube packed with a desiccant (Section 48.2).

Sulfur dioxide may be absorbed in an aqueous medium in a gas scrubber bottle. It can then be determined iodometrically (Section 25.7) or, after oxidation with hydrogen peroxide to sulfuric acid, gravimetrically as barium sulfate, by a precipitation titration with barium ion (Section 18.8), or by an acid–base titration.

Sulfur trioxide may be absorbed in an alkaline solution and be determined by any suitable method as sulfate.

Ammonia may be absorbed in water or dilute acid and be determined by an acid–base titration (Sections 14.9 and 14.10) or photometrically by Nessler's method (Section 37.2).

Carbon dioxide may be absorbed in a tube suitably packed (for example, with sodium hydroxide–coated asbestos) and be determined gravimetrically. Alternatively, it may be absorbed in a barium hydroxide solution and be determined by an acid–base titration or by measurements of the change in conductivity of the solution (Section 32.6).

Hydrogen chloride may be absorbed in water or in a base and be determined by an acid–base titration. Alternatively, it may be absorbed in a silver nitrate solution and the silver chloride precipitated be separated and weighed, or an excess of silver nitrate may be titrated with a standard chloride solution.

Physical measurements, including thermal conductivity, mass spectrometry, and photometry predominantly in the ultraviolet or infrared regions, are of value in the analysis of gas mixtures. Gas chromatography also offers salient possibilities for the analysis of gases (Section 46).

49.2 Gas-Volumetric Methods

The most explored methods for the analytical resolution of many common gas mixtures are based on the selective removal of one (or more) components of the mixture and measurement of the resulting change in volume or pressure. When not further qualified the term "gas analysis" usually implies these methods. In gas-volumetric methods, a known volume of the gas mixture is taken and a constituent is selectively removed, usually by an absorption process, and the new volume is measured at the same temperature and pressure. The process is repeated for other constituents using other absorbents. Since at constant pressure and temperature the amount of a gas is proportional to its volume, the composition of the gas in % v/v can be readily deduced. In manometric methods, the volume and temperature are kept constant and the change in pressure resulting from each absorption step is measured. The manometric method is capable of greater accuracy and is frequently employed when a small sample must be analyzed.

The principles underlying gas-volumetric methods can be appreciated by consideration of the resolution of a simple gas mixture: nitrogen, oxygen, and carbon dioxide. Simple forms of a gas buret and an absorption bulb are illustrated in Figure 49-1. Initially the gas buret, separated from the absorption bulb, is filled completely with a suitable confining liquid by opening both stopcocks and elevating the reservoir bulb; the lower stopcock is then closed. (For the assumed gas mixture, a sodium chloride solution acidified with sulfuric acid may serve as the confining liquid, since carbon dioxide and the other gases are only slightly soluble in it; mercury serves as the liquid of choice for the analysis of many gas mixtures.) The vessel containing the gaseous sample is next connected to the buret. The lower stopcock is then opened and a portion of the sample is drawn into the buret by lowering the reservoir bulb. The upper stopcock is then closed, and the reservoir is placed next to the gas buret and moved up and down until the level of the liquid in both is identical. The volume of the gas in the buret is read and recorded.

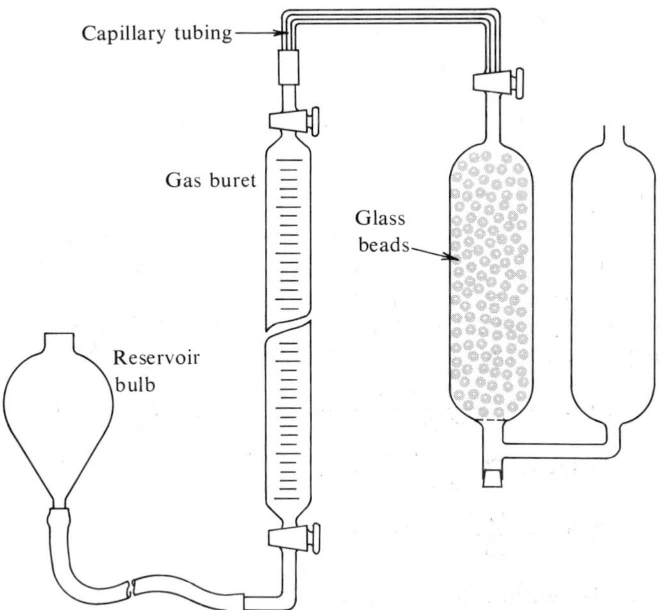

Figure 49-1. Schematic diagram of simple assembly for gas analysis.

An absorption bulb containing a concentrated potassium hydroxide solution (30 to 50%) is placed under suction until the solution is drawn through the capillary tubing to the inscribed mark and the stopcock is closed. This absorption bulb is then attached to the gas buret, as shown as Figure 49.1. The three stopcocks are opened and the gas mixture is forced into the absorption bulb by elevating the reservoir until the confining liquid reaches the area of the upper buret stopcock, which is then closed. The absorption bulb is then shaken gently for 1 minute or so. (The first of the two bulbs is often filled with glass beads in order that, when the solution is forced into the second bulb by the gas, the caustic-coated beads provide a greater area for the absorption of carbon dioxide.) The stopcocks are then opened and the sample is sucked back by lowering the reservoir bulb until the caustic solution is again aligned with the mark. The buret stopcock is closed. The reservoir bulb is again placed next to the buret and moved up and down until the liquid level in both is identical. The gas volume is read. The difference between the initial volume and the new volume (at constant temperature) corresponds to the carbon dioxide in the sample.

The potassium hydroxide–filled absorption bulb is replaced by one containing an absorbent for oxygen such as a potassium hydroxide solution of pyrogallol, that is, 1,2,3-$C_6H_3(OH)_3$, or an ammoniacal copper(I) chloride solution containing a coil of copper gauze. The absorption operation is then conducted as described for carbon dioxide, and the change in volume related to oxygen. The nitrogen of the sample remains in the buret after the two absorption steps.

Example 49-1. In the gas-volumetric analysis of a mixture of CO_2, O_2, and N_2 in the manner described above, the volumes (at constant temperature) in the gas buret initially and after the absorption of CO_2 and O_2 are 90.0, 44.0, and 18.0 ml, respectively. Calculate the composition of the gas mixture in % v/v.

$$\% \, CO_2 = \frac{90.0 - 44.0}{90.0} \times 100 = \frac{46.0}{90.0} \times 100 = 51.1\% \text{ v/v}$$

$$\% \, O_2 = \frac{44.0 - 18.0}{90.0} \times 100 = \frac{26.0}{90.0} \times 100 = 28.9\% \text{ v/v}$$

$$\% \, N_2 = \frac{18.0}{90.0} \times 100 = 20.0\% \text{ v/v}$$

In more elaborate equipment (e.g., Orsat apparatus), the various absorption units are attached to one "manifold" line and any one may be connected to the buret by opening the appropriate stopcock.

Hydrogen may be determined, usually after the absorption of the oxygen of the sample, by the deliberate addition of a known amount of oxygen. The mixture is then exposed to an electrically heated platinum spiral. A safe combustion occurs. The water formed condenses and the reduction in the volume of sample can be related to the hydrogen content of the sample.

49.3 Gasometric Methods

If a substance can be made to undergo a reaction that leads quantitatively to the liberation of a gas, its collection and gas-volumetric or manometric measurement can serve for the determination of the substance. For example, it is sometimes necessary to determine the amount of a metal as such present together with its oxide. Where the metal dissolves in an acid or base with the liberation of hydrogen, a weighed amount of the material is treated with acid (or sodium hydroxide in the case of aluminum or zinc), and the hydrogen that is evolved, usually with boiling of the solution, is collected over water. When a gas-volumetric finish is applied, the volume of hydrogen is measured at a known temperature and pressure and, after correction for the vapor pressure of water and reduction to standard conditions, the amount of hydrogen is calculated.

Example 49-2. A 1.500-g sample of an aluminum oxide sample known to contain some metallic aluminum (*26.98*) is dissolved in 10% NaOH solution. The solution is boiled and a volume of 148 ml of hydrogen is collected over water at 23.0°C at an atmospheric pressure of 756.1 torr. What is the percentage of Al in the sample? The vapor pressure of H_2O at 23.0°C is 21.1 torr. One (gram) mole of a gas occupies 2.24×10^4 ml at 0.00°C and 760.0 torr. The "effective" pressure of H_2 at 23.0°C = 756.1 − 21.1 = 735.0 torr. The measured volume, V_i, is reduced to standard conditions, V_{st}, by application of the perfect gas law (where the subscripts i and st refer to the measured and standard conditions, respectively):

$$V_{st} = V_i \times \frac{T_{st}}{T_i} \times \frac{P_i}{P_{st}} = 148 \times \frac{273}{273 + 23} \times \frac{735.0}{760.0} = 132 \text{ ml}$$

From the reaction equation

$$2\,Al + 2\,OH^- + 2\,H_2O \rightarrow 2\,AlO_2^- + 3\,H_2\uparrow$$

it can be seen that one mole of hydrogen gas corresponds to two-thirds mole of aluminum metal. Hence

$$\%\,Al = \frac{132}{2.24 \times 10^4} \times \frac{2}{3} \times \frac{26.98}{1.500} \times 100 = 7.1\%\text{ Al in sample}$$

Other common gasometric methods include the determination of ammonia by its reaction with sodium hypobromite ($2NH_3 + 3OBr^- \rightarrow N_2\uparrow + 3Br^- + 3H_2O$) and the determination of nitrite and nitrate by reaction with mercury and an excess of sulfuric acid (e.g., $2H^+ + 2NO_2^- + 2Hg + H_2SO_4 \rightarrow 2NO\uparrow + Hg_2SO_4 + 2H_2O$). In the Dumas determination of nitrogen in organic compounds, the sample is burned in a carbon dioxide atmosphere with a metal oxide as the source of oxygen. Carbon dioxide and other acidic gases are absorbed in potassium hydroxide solution; the nitrogen gas formed is collected above that solution and the gas volume is measured (Section 48.5).

Questions 49.4

. The gas laws discussed in previous courses are sometimes employed in calculations relevant to gas analysis. What is stated by the laws of Boyle, Charles, Gay-Lussac, and Dalton?

. Why should the water over which CO_2 is collected be previously saturated with that gas?

. A material contains sulfur as the only active ingredient. A sample of this material is combusted and the gases passed through a measured volume of standard base, the excess of which is titrated with acid to the phenolphthalein end point. Would it matter whether the sulfur is combusted to only SO_2 or to only SO_3 or to a mixture? Explain.

49-4. A gas stream contains N_2, O_2, SO_2, and SO_3. Suggest a gas analytical method to determine the two sulfur oxides separately.

49-5. What are the "standard conditions" to which reference is made when measuring gases?

49-6. Why are pellets or rings of glass or another inert material often present in absorption bulbs used in gas analysis?

49-7. Would you expect the correction for water vapor pressure to be larger when measurements of gas volumes are made over pure water than over concentrated KOH? Explain.

49-8. The "mole volume" of 22.4 liters is often used in gas analytical calculations. The values obtained are approximate. Why?

49-9. Gas analysis can be based on the following combustion technique. The known volume of a gas containing combustible components is mixed with a known volume of oxygen and the combustion initiated. The reaction mixture is then cooled and any CO_2 formed is determined. The data thus secured are in part used to calculate the composition of the sample. Write the reaction equations; indicate the volume of O_2 consumed, the contraction expected, and the volume of CO_2 produced for the combustion of one volume of each of the following substances: hydrogen; carbon monoxide; methane, CH_4; acetylene, C_2H_2; and ethane, C_2H_6.

Answer for acetylene: $2C_2H_2 + 5O_2 \longrightarrow 4CO_2 + 2H_2O$

O_2 consumption: 2.5 volumes

Contraction: 1.5 volumes [2acetylene + $5O_2$ volumes disappear and 4 volumes CO_2 appear per 2 volumes acetylene; thus $(2 + 5 - 4)/2 = 1.5$]

CO_2 produced: 2 volumes

49.5 Problems

49-1. A 0.320-g sample of limestone, containing $CaCO_3$ as the active ingredient and some inert material, is treated with acid and 78.5 ml of CO_2 developed is collected above water saturated with that gas. The temperature and pressure when measuring the gas volume are 23°C and 757 torr, respectively. Calculate the per cent $CaCO_3$ (*100.1*) in the limestone. Vapor pressure of H_2O at 23°C is 20.9 mm.

Answer: 97.9%

49-2. What minimum sample weight of a limestone containing 98% $CaCO_3$ (*100.1*) must be taken so that the CO_2 produced on ignition and collected dry over mercury will amount to at least 50 ml when measured at 20°C and 755 torr?

Answer: 0.21 g

49-3. What volume, in milliliters, of dry hydrogen gas will be evolved (measured at 23.5°C and 752 torr) when 86.2 mg of zinc metal (*65.4*) is treated with acid?

4. An organic material is combusted in air and the gas mixture containing CO_2, N_2, and O_2 after passing a drying tube is collected over mercury at 21°C and 752 torr. The gas volume is 60.2 ml. The gas is passed over concentrated KOH and the volume measured again and found to be 51.4 ml. How many milligrams of carbon (*12.0*) was present in the combusted material?

50 SAMPLING

Sampling 50.1

In practice, a chemical analysis is often made of a small sample taken from a large consignment of a material. If the sample is to be representative of the entire consignment, it must be selected in such a way that it possesses the characteristics of the bulk.† The manner in which this is accomplished is known as a sampling procedure. In the merchandising of an ore, it is often necessary to secure a representative sample from carloads, trainloads, or even shiploads. Obviously it is completely inadequate to take a single lump randomly. Since its composition may be that of accompanying rock (gangue) or that of an extremely rich portion of the ore, the value obtained for an element of principal interest might range from zero to a percentage far above the average. In practice, establishment of a representative sample often presents far more difficulties than the analysis itself. Law suits have resulted from inappropriate sampling. Indeed, in the modern period, sampling schemes are sometimes specified in purchase contracts, especially where the material involved is extremely inhomogeneous or variable in composition.

In general, sampling involves three steps: (1) the establishment of a gross sample, (2) its reduction to a subsample which can be brought to the laboratory, and (3) the preparation of the sample actually taken by the analyst for chemical analysis. There is no uniform, generally applicable method of sampling. The optimum technique will depend on the material of interest. The principal criterion is the degree of heterogeneity of the material.

---

†The discussion is here limited to the establishment of a representative "average" sample. However, this is not always the object of sampling. For example, to assure that the content of active chlorine of a municipal water is adequate to prevent the growth of microorganisms, the sample is not taken at the chlorinator, nor is an "average" sought; rather, the sample is best withdrawn from a far-distant, secondary distribution line. But even where a special sample is required, an appropriate sampling procedure is still necessary.

50.2 Establishment of a Sample

Liquids are frequently homogeneous and, if so, their sampling is simple. The student is warned, however, that attainment of homogeneity in thousands of gallons of a liquid product is far more difficult than the preparation of a simple solution in a beaker. Homogeneity cannot be blindly assumed in the sampling of a liquid. In practice, in the sampling of drums or tankcars, liquid is withdrawn at various levels and either a composite sample is established and analyzed or the samples taken at different levels are separately analyzed.

The establishment of a representative sample in a solid product is far more difficult, especially if there is considerable variation in composition and the particle size ranges from large lumps several pounds in weight to a fine dust! Statistical treatment is often of assistance to establish the optimum sampling conditions and to determine the minimum size of the sample and subsample so that they are representative of the consignment.

For bulk material, sample increments are often taken when the material is transferred. For example, with coal and ores, a shovelful or a wheelbarrow load may be taken from each truck or railway car or at regular intervals from the complete cross section of a conveyor belt. How often a sample increment should be taken and of what size will depend on the inhomogeneity of the material. The increments must then be placed together, reduced to a relatively uniform size, for example, by hammer or roll milling, and a subsample established.

Mechanical sampling devices are often used, but simple manual techniques are still widely applied. In coning the material is piled in a cone by placing each shovel of material in the center of a ring. Finally, the cone is flattened by pulling down material equally in all directions, quarters are established by the use of a board, and two diagonally opposite quarters are rejected. The remaining quarters are piled into a second cone, and so on. In quartering, the material is laid out evenly in a flat square, diagonals are drawn, and opposite quarters are rejected. The simple device known as a riffle consists of a trough divided into an even number of transverse slots. The alternate slots discharge into a left and right collecting bin. Thus, when the sample is distributed evenly over the trough of a simple riffle and allowed to fall into the bins, the sample is divided in half. The contents of either the right or left bin may then be riffled a second time, and so on. By application of one or more of these (or other) techniques with one or more reductions in particle size, a portion of the gross sample is secured as a subsample that is sent to the laboratory.

Metals and alloys may be sampled by sawing or drilling a large consignment. The "sawdust" or turnings are "quartered," etc., or, where appropriate, may be melted together to form a more homogenous subsample.

In the laboratory the subsample may be further quartered or coned to obtain a sample of a few grams that is taken for analysis. Since many materials cannot be brought into solution readily, the sample may be further reduced in particle size. For this purpose a ball mill or a mortar and pestle may be used.

A necessary requirement of the sampling procedure is that the composition of the sample either remains unchanged or changes only in a known way. This requirement is difficult to fulfill whenever the sample either contains loosely retained water (e.g., damp coal) or, as the particle size is reduced, absorbs water or is oxidized by the air. In such cases a separate sample of a few pounds is usually taken, brought to the laboratory under protective conditions, and the component undergoing change is separately determined. For example, the moisture content is often established with a separate sample. The other determinations are effected with the representative sample on a dry basis. The results may then be recalculated to an "as received" or "as is" basis in the manner considered in Section 6.4.

51 ATTACK OF THE SAMPLE

Most analytical methods require a solution of the sample in an aqueous medium. In practice, however, relatively few materials are completely or even substantially soluble in water. In the applications of "wet" methods mentioned in this textbook, it has been largely assumed that a sample solution has already been secured. In addition, the student in his parallel laboratory work mainly employs liquid samples and possibly a few solid samples that are brought into solution by a simple treatment. In the everyday practice of inorganic quantitative analysis, the situation is quite different, and many samples resistant to simple chemical attack are encountered.

Various measures for the treatment of solid samples may have been brought to the student's attention in a previous consideration of qualitative analysis. The measures commonly adopted in quantitative analysis are not basically different, but extreme care is required to avoid losses of material by spattering or volatilization. Further, it is necessary to avoid the introduction of any ions in the overall preparation of the sample solution that will interfere in or otherwise complicate a subsequent separation or determination. In some cases it is necessary to apply different dissolution techniques to separate samples to permit the determination of a number of constituents of a complex material. Where trace constituents of a sample are to be determined, it is imperative that significant amounts of these substances are not introduced from the reagents and vessels or as airborne contaminants.

Acid Attack of the Sample 51.1

Mineral acids, including hydrochloric, nitric, sulfuric, phosphoric, perchloric, and hydrofluoric acids, or mixtures of such acids, are frequently used to attack complex inorganic materials. Their use is advantageous, because any unwanted excess can usually be removed by volatilization. Any portion of the sample that is unattacked by suitable acid treatments is often separated by filtration and the washed residue is fused with a suitable flux (Section 51.2). The cooled fusion melt after solution may either be combined with the solution of the acid-soluble material or may be analyzed separately. In the attack of a sample by heating with a readily volatilizable acid, such as hydrochloric or nitric acid, it is necessary to avoid excessive loss of the acid or to make up for losses by occasional additions of acid.

Hydrochloric acid is a good solvent for many metals, metal oxides, and metal carbonates. With carbonates, care must be exercised to avoid loss of material by foaming or spattering during the evolution of carbon dioxide. The solution of a metal in a so-called nonoxidizing acid, such as hydrochloric acid, phosphoric acid, dilute perchloric acid, or dilute sulfuric acid, can be related to the more negative standard (or formal) electrode potential of the metal ion–metal redox couple as compared to that of the hydrogen ion–hydrogen gas couple (see Table G in the Appendix).

Nitric acid, as its anion manifests oxidizing properties, is known as an oxidizing acid, and attacks more metals than hydrochloric acid. Some metals, however, notably aluminum and chromium, are not attacked by concentrated nitric acid. This so-called passivity is associated with the formation of an insoluble, protective layer of the oxide (or of a more complex entity) on the metal surface. Nitric acid is favored for the dissolution of many alloys and offers the special advantage that certain metals (tin, antimony, and tungsten) may remain insoluble as oxides or acids, and can be separated immediately by filtration.

Many metals and alloys, notably the noble metals, require an acidic medium with stronger oxidizing power than nitric acid alone provides. Various mixed acids receive attention, especially aqua regia, which is a mixture of one volume of concentrated nitric acid with three volumes of concentrated hydrochloric acid. The formation of chloro complexes may aid in the dissolution of the noble metals by this mixture.

Sulfuric acid has certain advantages in the attack of samples. It shows strong dehydrating properties and possesses a high boiling point (~340°C); the latter fact offers the possibility of removing volatile acids by addition of sulfuric acid and subsequent heating.

Hot, concentrated perchloric acid is a powerful oxidant (Section 25.3) and attacks some alloys that are resistant to hydrochloric or nitric acid. Perchloric acid has the advantages of having a high boiling point (the azeotrope with water boiling at about 200°C) and of forming metal salts highly soluble in aqueous medium. Only potassium, rubidium, and cesium form sparingly soluble perchlorate salts (Section 7.2). Precautions must be taken to avoid contact of concentrated perchloric acid with organic matter, except under special, controlled conditions, because violent, even explosive reactions may result (see also Section 25.3).

Hydrofluoric acid receives special attention in the attack of silicate materials, including rocks, minerals, and glasses. On heating, fluosilicic acid and silicon tetrafluoride are volatilized (Section 7.3). The use of platinum or polytetrafluoroethylene (i.e., Teflon) ware is necessary. It is usually important to remove all the hydrofluoric acid preceding the analysis, because fluoride ion forms insoluble compounds or soluble complexes with many cations and also attacks glass. This removal is usually achieved by evaporation (in a hood!) of the sample solution with an excess of sulfuric acid or perchloric acid until white fumes appear.

Certain elements may be volatilized during the acid attack of a sample. This may be used to advantage as a means of separating the elements for their determination or to obviate their interference in subsequent determinations of interest. However, the loss in other cases may be unwanted and be a source of difficulty. What elements are volatilized will depend on the acid used and the conditions existing, including the nature of other components of the sample. Carbon dioxide and hydrogen sulfide are evolved from carbonates and sulfides. Silicon as fluosilicic acid and silicon tetrafluoride and boron as fluoboric acid are volatilized from a hydrofluoric acid medium (see above). From boiling hydrochloric acid, $AsCl_3$ and $GeCl_4$ are readily evolved and to some extent $SbCl_3$, $SnCl_4$, and $HgCl_2$, if adequate precautions are not taken (compare Sections 48.6 and 48.7). When a sample is heated with sulfuric acid, halides are evolved as the hydrogen halide and also other anions forming voltatile acids.

Attack of the Sample by Fusion with a Flux 51.2

Only a limited number of inorganic samples are brought completely into solution by acid attack. In such cases, the acid-insoluble residue or a further sample is subjected to fusion with a suitable flux. Fluxes usually melt in the region 180 to 850°C; however, with a few substances usually classified as fluxes, the attack involves sintering rather than melting. Depending on the flux and sample, either the finely powdered sample is intimately mixed with the flux and placed in a crucible, or a layer of flux is placed in the crucible followed by the powdered sample and covered by additional flux. The crucible is covered with a lid and heated until the flux melts. The elevated temperature is maintained until the sample material is largely or completely dissolved in the melt. The color and appearance of the melt often allows the progress and success of the fusion operation to be assessed. The cooled melt is dissolved in water or treated with acids. The success of the fusion depends on the temperature, the time allowed for the attack by the flux, and frequently on the degree of dispersion of the sample material. Fluxes are usually classified as alkaline or acidic and oxidizing or reducing in their action.

Acid–base behavior between solids or in the molten state at high temperature is explainable in terms of the Lewis theory (Section 16.4). Consider, for example, the fusion of silica with an alkali metal oxide, the latter being classified as an alkaline flux. The silicon dioxide is an electron-pair acceptor and acts as a Lewis acid. The metal oxide has an ionic crystal lattice (e.g., $Na^+O^{2-}Na^+$) and provides the oxide ion, which as an electron-pair donor acts as a Lewis base. The acid–base reaction in the fusion process leads first to the metasilicate ion, SiO_3^{2-}:

$$:\ddot{\underset{..}{O}}:Si:\ddot{\underset{..}{O}}: + :\ddot{\underset{..}{O}}:^{2-} \rightarrow \begin{matrix} :\ddot{\underset{..}{O}}:\underset{..}{Si}:O:^{2-} \\ :\ddot{\underset{..}{O}}: \end{matrix}$$

Further analogous reaction with the oxide ion yields the orthosilicate ion, SiO_4^{4-}. Some metal oxides on fusion act as Lewis acids and react with alkaline fluxes. Others are amphoteric in their behavior, notably aluminum, titanium, iron, and zinc, and react with both alkaline and acidic fluxes.

Alkaline fluxes are more commonly employed than acidic fluxes. For the decomposition of silicates, fusion with anhydrous sodium carbonate is useful. The carbonate ion at high temperatures is a good source of the oxide ion, carbon dioxide being liberated in the process. The over-all reaction of an acid-insoluble aluminum metasilicate on fusion with sodium carbonate may be written

$$Al_2(SiO_3)_3 + 4Na_2CO_3 \rightarrow 3Na_2SiO_3 + 2NaAlO_2 + 4CO_2\uparrow$$

The sodium metasilicate and sodium metaaluminate formed are soluble in water.

A mixture of potassium and sodium carbonate offers a flux melting lower than either carbonate alone. Other common alkaline fluxes include borax (sodium tetraborate) and sodium or potassium hydroxide.

Alkaline oxidizing fluxes convert metals to their higher oxidation states and thereby oxides that are more acidic and more readily undergo acid–base reactions with the alkaline component of the flux. Of such fluxes, sodium peroxide, Na_2O_2, is the most important. Milder oxidizing fluxes are obtained by mixing sodium carbonate with sodium or potassium nitrate or chlorate.

Of acidic fluxes, anhydrous potassium pyrosulfate is the most common. Either the available reagent-grade form of this salt is used or it is prepared by heating potassium hydrogen sulfate until volatilization of water ceases:

$$2KHSO_4 \rightarrow K_2S_2O_7 + H_2O\uparrow$$

The pyrosulfate ion is a source of sulfur trioxide:

$$S_2O_7^{2-} \rightarrow SO_4^{2-} + SO_3$$

To reduce the loss of this oxide by volatilization, the temperature should be just high enough to assure fusion of the salt and the crucible should be covered.

The sulfur trioxide acts as an electron-pair acceptor (Lewis acid) in its reaction with metal oxides, including those of titanium, iron, chromium, and aluminum. In this way, soluble metal sulfates are formed.

Reducing fluxes receive special use. A flux of sodium carbonate and sulfur converts arsenic, antimony, and tin to water-soluble thiosalts, such as Na_3AsS_4, Na_3SbS_4, and Na_2SnS_3.

The material of the crucible used in the fusion should ideally not be subject to attack by the flux. With platinum, loss of this expensive metal is one reason. From the viewpoint of chemical analysis, the introduction of contaminants is more serious. Since fluxes generally act more vigorously at higher temperatures, it is beneficial to conduct the fusion operation at the lowest possible temperature at which the sample is attacked, but the crucible is left unaffected. Frequently, however, attack of the crucible is tolerable if the contaminants thus introduced are of no consequence in the determinations of interest. For example, an iron crucible is often used for the sodium peroxide fusion of chrome ores. The crucible is attacked severely, but the iron thereby introduced does not interfere in the subsequent determination of chromate via a redox titration. When the same ores are analyzed for their iron content, nickel crucibles may be employed, since nickel does not interfere in the determination of iron.

For fusions with sodium carbonate, platinum crucibles are commonly used, although zirconium is receiving increasing attention as a suitable material. For fusions with sodium hydroxide, potassium hydroxide, or sodium peroxide, platinum should not be used because it is severely attacked, unless the fusion is performed with strict control of the temperature in an electric oven. Nickel, silver, gold, and zirconium crucibles are appropriate for fusions with sodium or potassium hydroxide. Sodium peroxide fusions are best performed in zirconium crucibles, but nickel or iron ones may be serviceable (see above). For fusions with potassium pyrosulfate, platinum crucibles are appropriate, but they are somewhat attacked if heated to too high a temperature. For this fusion vitreous silica is a suitable material. Porcelain crucibles are severely attacked in alkaline fusions but may be used if the contaminants, thus introduced, do not interfere in the intended analysis.

51.3 High-Temperature Attack by Halogens

Some materials, notably sulfides, are attacked by a current of chlorine or bromine at a high temperature. Sulfide is thereby converted to sulfur halides. Metals form either fused salts or, in the case of tin, antimony, arsenic, and certain other metals, are volatilized as their halides and may be absorbed in an appropriate solution and then be determined.

51.4 Attack of the Sample by Pyrohydrolysis

By passage of superheated steam over a finely divided sample held in a tube within a furnace at a suitable temperature in the range 500 to 1300°C, a few elements can be volatilized, notably fluorine, chlorine, and boron. The hydrofluoric, hydrochloric, or boric acid formed is condensed with the steam and determined by conventional procedures. Pyrohydrolysis affords a rapid separation of fluorine from many samples that present difficulties to attack by acids or fusion or that contain elements that complicate the determination of fluorine.

Questions 51.5

Explain or elaborate on the following terms: flux, oxidizing flux, nonoxidizing versus oxidizing acids, pyrohydrolysis, and aqua regia.

Methods for the attack of a sample are sometimes classified as wet attacks or dry attacks. Consider the approaches in this section in terms of these two categories.

State some factors that might deserve consideration in the selection of a mode of attack for an inorganic material.

What difficulties might result in the fusion of an inorganic sample with sodium carbonate in a porcelain crucible? Of a silicate sample?

Iron(III) oxide on fusion with potassium pyrosulfate yields iron(III) sulfate and potassium sulfate. Write a balanced overall reaction for this fusion. Elaborate on the possible acid–base reaction involved.

Magnesium mica (phlogopite), used in sheets for electrical insulation, has the approximate composition $KH_2Mg_3Al(SiO_4)_3$. On fusion of the ground mica sample with sodium carbonate, the products remaining in the crucible are magnesium carbonate, potassium metaaluminate, and sodium metasilicate. The cooled melt is treated with dilute hydrochloric acid. Carbon dioxide is thereby evolved and silica precipitated. Write balanced equations for the reactions taking place during this attack of the mica sample and the subsequent acid treatment.

52 CLASSIFICATION OF QUANTITATIVE ANALYSIS

In Chapter 1 no attempt was made to classify analytical chemistry, other than into qualitative and quantitative analysis, and only a brief comment was made on the distinction between a determination and an analysis. The student is now in a position to view the subject with some degree of perspective and to appreciate some general remarks on the various viewpoints from which quantitative analysis may be examined.

Organic versus Inorganic Analysis 52.1

In the present textbook, inorganic analysis has been the central topic. Organic analysis, to which the same principles apply, is concerned with the detection, identification, and determination of organic compounds. Elemental analysis of organic samples for carbon, hydrogen, and nitrogen by combustion methods (Section 48.6) or for nitrogen via Kjeldahl digestion (Section 14.9) is of great importance in organic chemical research.

Scale of Operations 52.2

Analytical procedures can be differentiated as to the absolute amount of the sample taken and the relative amount present of the constituent or constituents to be determined. With regard to the sample size, that is, to the scale of the analysis, the following classification is frequently employed: macro analysis, 0.1 g or greater; semimicro analysis, 0.01 to 0.1 g; micro analysis, 0.001 to 0.01 g; and ultramicro (or submicro) analysis, less than 0.001 g. This classification scheme is a loose one, and, since the limits of each category are really orders of magnitude rather than absolute numbers, the limits assigned by different workers will vary. However, this scheme has merit because the techniques and methods best employed may depend on the scale of analysis. For example, filtration by gravity is hardly possible in micro and ultramicro analysis, and the separation of a precipitate is accomplished either by filtration under suction or by centrifugation; in addition, the smaller the analytical sample, the more sensitive must be the balance used in its weighing.

The scale of analysis selected depends on various factors, including the amount of material available, the constituents present and their relative amounts, the expense of reagents and the efforts required in their preparation, the instruments available in the case of physicochemical methods, the frequency with which the determinations are encountered, and the time available. Some operations may be effected more rapidly and more expeditiously with small samples, for example, the washing and drying of a precipitate.

The above statements are directed to the scale of analysis, that is, to the sample size. The relative amount of a constituent of a sample is also of importance, and the following classification is appropriate: major constituents, 1 to 100% of the sample, and minor constituents, 0.01 to 1%, which together may be termed macro constituents, and constituents less than 0.01% of the sample, which are spoken of as trace (or micro) constituents. Again it is stressed that the classification is a loose one and is based on orders of magnitude. Often a trace constituent is determined using a sample of the order of 10 g. In such a case, one speaks of macro *analysis* with the particular trace determination performed by a micro or ultramicro *technique*.

52.3 Determination versus Analysis

The distinction between a determination and an analysis has been briefly drawn in Chapter 1. Applications of the principles of quantitative analysis introduced in this textbook and by experiments performed by the student in parallel laboratory work are largely simple determinations. Usually a single constituent is determined under uncomplicated circumstances: No extensive separation of interfering substances is required and the sample is readily brought into solution. In a practical analysis, the situation is often far more complicated: Establishment of a representative sample may present difficulties (Chapter 50), many constituents are present, interference must be considered and resolved, the material may be refractory to dissolution (Chapter 51), and more than one constituent must be determined.

52.4 Partial Analysis versus Complete Analysis

Sometimes only a few constituents of a complex material are of interest and need to be determined. This is known as partial analysis. In contrast, complete analysis ends only when all the constituents, as detected by suitable sensitive qualitative tests and methods, have been determined.

Proximate, Elemental, and Ultimate Analysis 52.5

Frequently two or more constituents show similar behavior in an analytical procedure and are determined together. This is known as proximate analysis. In the analysis of glass, for example, potassium and sodium may be weighed together as their sulfate salts and reported as total alkalies expressed as % Na_2O (Section 7.2). Similarly, in the analysis of rocks and ores, the sum of the oxides of iron and aluminum may be determined together and expressed as % R_2O_3 (Section 7.2). The percentage loss on ignition is usually proximate in character, because it includes moisture, carbonate, organic matter, and any other volatile substances present in the material. For materials that are largely water-soluble, the percentage of insoluble matter is frequently determined; this is another example of proximate analysis. When the percentages of various elements present in a material are determined, this is known as elemental analysis. The elemental analysis of a single substance may be termed ultimate analysis.

Nondestructive versus Destructive Analysis 52.6

Nondestructive analysis would involve establishment of the nature and content of the constituents of a material and at the finish having the sample present in its initial composition and amount. Such an ideal is closely approximated by some physical methods, notably methods based on the use of X-rays (Chapter 41) or radioactivity (Chapter 42). Where emission spectrography (Section 40.4) involves only a minute portion of a specimen, it is also appropriately termed nondestructive. However, nondestructive methods usually fail to provide all the answers sought by analysis; consequently, additional methods must be employed that are more or less destructive in nature.

Basic Steps in a Determination 52.7

In general terms, the determination of a constituent of a material involves four basic steps: (1) establishment of a sample suitable in character and amount; (2) its transfer by physical or chemical operations, or both, to a condition in which some property of the constituent can be measured; (3) performance of the measurement; and (4) expression of the value obtained in terms of the content of the constituent of interest by resort to stoichiometric relations, physical laws, standards, etc. Obviously the relative attention given to a particular step will vary from method to method and two or more steps may overlap in their conduct. The student will benefit by examining the performance of typical gravimetric, titrimetric, and photometric determinations from the standpoint of these four basic steps. In these three types of analytical methods, the measurements involved serve to establish the weight (that is, mass), volume, and absorption of light, respectively. It will be recognized that these measurements are actually physical in principle. Chemistry enters in the transfer of the sample to the form used in the measurement, in the interpretation of the result, and sometimes in understanding the phenomena underlying the property measured.

52.8 Classification by Methods and Techniques

The study and practice of quantitative analysis is often conveniently divided into two broad categories. Full appreciation of the labels variously assigned to these categories requires an appreciation of a large number of diverse methods. The first category, into which gravimetric and titrimetric methods are conventionally placed, is variously described as noninstrumental methods, classical analysis, chemical methods, or methods involving static properties. The second category, in which electrical and optical methods of analysis are major subjects, is variously labeled instrumental methods, physical or physicochemical methods, or methods involving dynamic properties. The implications of these labels will be clear from a brief discussion.

Gravimetric and titrimetric methods can be thought of as noninstrumental only by the arbitrary exclusion of the balance and the buret as instruments; hence, this label is best retired from use. It is a carryover from earlier decades when these two types of methods were the principal ones at the command of the analytical chemist and electrical and optical instruments were recent in their introduction and still rarely used in practical analysis. This also explains why these two types of methods, along with gas-analysis procedures involving the measurement of pressures or volumes (Chapter 49), are frequently termed classical analysis. Both gravimetric and titrimetric methods can obviously be labeled as chemical methods, since almost all gravimetric determinations involve either a chemical or electrochemical separation and titration involves a chemical reaction proceeding virtually to completion. Static properties are possessed by a system itself and basically do not involve the passage of energy into, through, or out of an analytical system for their external measurement. Common static properties include mass, volume, density, pressure, and chemical reactivity.

Electrical and optical methods can be thought of as instrumental ones, but, as observed above, the description is not recommended. If a crystal is placed in a light beam and the absorbance measured, this is clearly a physical procedure. In contrast, if the crystal is reacted with an acid and the absorbance of the resulting solution is measured, the overall procedure is appropriately described as a physicochemical one. Dynamic properties are exhibited when energy (or particles) pass into, through, or out of an analytical system. Methods dependent on optical, electrical, and radioactive phenomena all involve dynamic properties.

Although potentiometric, amperometric, and photometric titrations are usually treated in the context of physicochemical methods, as in this textbook, it is important to note that these are still chemical methods, but the end point is detected by a physicochemical technique. Electrogravimetric analysis involves an electrochemical separation, but a static property, mass, is still measured as in conventional gravimetric determinations.

Classification According to Application 52.9

It is also possible to discuss analytical principles, methods, and techniques from the standpoint of their application to a restricted field, for example, metallurgical analysis or clinical analysis. Another possibility is restriction to a group of related substances or to a single important product of commerce, for example, steel analysis, cement analysis, and the analysis of copper metal. In all such treatments, the principles are the same, but the emphasis given particular topics will differ.

Selection of an Analytical Method 52.10

The most critical question raised by the analytical chemist facing a new analytical problem is what method or methods should be selected. An adequate answer requires much knowledge and judgment, usually gained only by long experience, and may involve some trials, often with prepared samples somewhat similar in composition to the unknown. Such experience is not anticipated in the student or the nonspecialist in analytical chemistry, but some remarks may be of general interest.

The scale of analysis selected, as already noted in Section 52.2, will depend on the various factors, including the amount of material available and the value assigned to it. In practical analysis, the time factor is often of great importance. The best analysis is wasted if the data become available only when no longer useful. This consideration applies, for example, when analyses are used to control chemical processing and metallurgical operations or to aid in diagnosing clinical conditions. Consequently, it is frequently necessary to curtail accuracy for speed.

Interferences must be considered in the selection of a method; indeed, other factors being equal, a more *selective* method may be preferred over a more *sensitive* one. Often an assessment is required of how accurately and how precisely a result must be known to be useful. The attempt to achieve accuracy and precision in excess of that required for the immediate purpose of the determination or analysis may be costly in time and labor.

Sometimes the selection of a method is based on considerations other than the establishment of the optimum one. In many areas of analysis, so-called standard methods (actually standard procedures) are recognized either by their development or acceptance by professional societies and groups, by the force of purchase contracts, or by the requirements of industry or governmental codes. For example, in various countries so-called pharmacopoeias provide the methods required in the identification and analysis of officially recognized drugs. Often standard methods are so devised as to avoid the use of equipment and instruments that are not readily available in modest laboratories. By their nature, such methods may lag in the adoption of the latest developments. However, they are usually presented in great detail and with attention given to even minor operations; consequently, regardless of the accuracy, the results obtained by different analysts commonly agree quite favorably.

52.11 Questions

52-1. For liquid samples, analyses are sometimes placed on a volume basis, and, to fix the scale of analysis, 1 ml is taken as equal to 1 g. In volume units, what would be the approximate limits for macro, semimicro, micro, and ultramicro analysis? If a volume of 200 μl of blood is taken for a clinical analysis, what is the scale of analysis?

Is a component of 1 part per 1000 in a material a micro or a macro constituent?

Analytical methods are sometimes classified as "wet" or "dry," depending on whether or not a solution of the sample is necessary as a prelude to or as a necessary condition for the measurement. Name some methods and determinations that might fall into these two categories.

Physicochemical methods are sometimes stated to depend on the measurement of a mass-dependent property rather than mass itself. Comment on the implications of this statement.

A material containing loosely held water is heated at 110°C to constant weight for the purpose of establishing the percentage of water present. Is this a determination or an analysis? A proximate procedure? A chemical method? A nondestructive method? Elaborate on your answers.

APPENDIX

The data appearing in the following tables have been selected from various compilations and research publications. Principal sources include the following:

L. G. Sillén, and A. E. Martell, *Stability Constants of Metal-ion Complexes,* Special Publication No. 17, Chemical Society, London, 1964 [Tables A, B, C, and D].

G. Kortüm, W. Vogel, and K. Andrussow, Dissociation Constants of Organic Acids in Aqueous Solution, *Pure and Applied Chemistry*, **1**, 187–536 (1960) [Tables A and B].

W. Feitknecht and P. Schindler, Solubility Constants of Metal Oxides, Metal Hydroxides . . . in Aqueous Solution, *Pure and Applied Chemistry*, **6**, 130–199 (1963) [Table C].

K. Y. Yatsimirskiĭ and V. P. Vasil'ev, *Instability Constants of Complex Compounds* (D. A. Paterson, translator), Pergamon Press, Inc., Oxford, 1960 [Tables C and D].

R. Bates, *Journal of Research, National Bureau of Standards*, **66A**, 182 (1961) [Table E].

G. Charlot, D. Bezier, and J. Courtot, *Selected Constants, Oxydo-Reduction Potentials*, Pergamon Press, Inc., Oxford, 1958 [Table G].

E. Bishop, in *Comprehensive Analytical Chemistry*, C. L. Wilson and D. W. Wilson, editors, Elsevier Publishing Company, Amsterdam, 1960, Volume IB, pp. 151–184 [Tables A, B, and H].

A. Hickling and F. W. Salt, *Transactions of the Faraday Society*, **36**, 1226 (1940) [Table I].

Table A Dissociation Constants of Some Acids in Water at 25°C

Acid		K_a	Acid		K_a
Acetic	K_1	1.8×10^{-5}	Hypochlorous	K_1	2.8×10^{-8}
Arsenic	K_1	5.6×10^{-3}			
	K_2	1.2×10^{-7}	Iodic	K_1	1.8×10^{-1}
	K_3	3×10^{-12}	Nitrous	K_1	5×10^{-4}
Arsenious	K_1	1.4×10^{-9}	Oxalic	K_1	5.4×10^{-2}
Benzoic	K_1	6.3×10^{-5}		K_2	5.1×10^{-5}
Boric	K_1	5.9×10^{-10}	Phenol	K_1	1.1×10^{-10}
Carbonic	$K_1{}^a$	4.5×10^{-7}	Phosphoric	K_1	7.1×10^{-3}
	K_2	5.6×10^{-11}	(ortho)	K_2	6.3×10^{-8}
Chloroacetic	K_1	1.4×10^{-3}		K_3	4.4×10^{-13}
Chromic	K_2	3×10^{-7}	*o*-Phthalic	K_1	1.1×10^{-3}
Citric	K_1	7.4×10^{-4}		K_2	3.9×10^{-6}
	K_2	1.7×10^{-5}	Salicylic	K_1	1.0×10^{-3}
	K_3	3.9×10^{-7}		K_2	4×10^{-14}
Ethylenedinitrilotetracetic	K_1	1×10^{-2}	Sulfamic	K_1	1.0×10^{-1}
	K_2	2.1×10^{-3}	Sulfuric	K_1	1.1×10^{-2}
	K_3	6.9×10^{-7}	Sulfurous	K_1	1.7×10^{-2}
	K_4	7.4×10^{-11}		K_2	6.3×10^{-8}
Formic	K_1	1.8×10^{-4}	Tartaric	K_1	9.2×10^{-4}
Hydrocyanic	K_1	5×10^{-10}		K_2	4.3×10^{-5}
Hydrofluoric	K_1	6×10^{-4}	Thiocyanic	K_1	1.4×10^{-1}
Hydrogen sulfide	K_1	1.0×10^{-8}			
	K_2	1.2×10^{-14}			

[a] Apparent constant based on $C_{H_2CO_3} = [CO_2] + [H_2CO_3]$.

Table B Dissociation Constants of Some Bases in Water at 25°C

Base		K_b	Base		K_b
2-Amino-2-(hydroxymethyl)-1,3-propanediol	K_1	1.2×10^{-6}	Hydrazine	K_1	9.8×10^{-7}
			Hydroxylamine	K_1	9.6×10^{-9}
			Lead hydroxide	K_1	1.2×10^{-4}
Ammonia	K_1	1.8×10^{-5}	Piperidine	K_1	1.3×10^{-3}
Aniline	K_1	4.2×10^{-10}	Pyridine	K_1	1.5×10^{-9}
Diethylamine	K_1	1.3×10^{-3}	Silver hydroxide	K_1	6.0×10^{-5}
Hexamethylenetetramine	K_1	1×10^{-9}			

Table C Solubility Products of Some Common Electrolytes in Water at 25°C

Substance	Formula	K_{sp}
Aluminum hydroxide (amorphous)	$Al(OH)_3$	6×10^{-32}
Barium carbonate	$BaCO_3$	5.5×10^{-10}
Barium chromate	$BaCrO_4$	1.2×10^{-10}
Barium fluoride	BaF_2	1.0×10^{-6}
Barium iodate	$Ba(IO_3)_2$	1.5×10^{-9}
Barium maganate(VI)	$BaMnO_4$	2.5×10^{-10}
Barium oxalate	BaC_2O_4	1.7×10^{-7}
Barium sulfate	$BaSO_4$	1.3×10^{-10}
Bismuth sulfide	Bi_2S_3	$\sim 10^{-97}$
Cadmium sulfide	CdS	$\sim 10^{-27}$
Calcium carbonate	$CaCO_3$	4.8×10^{-9}
Calcium fluoride	CaF_2	4×10^{-11}
Calcium hydroxide	$Ca(OH)_2$	3.7×10^{-6}
Calcium oxalate	CaC_2O_4	2.3×10^{-9}
Calcium phosphate	$Ca_3(PO_4)_2$	$\sim 10^{-26}$
Calcium sulfate	$CaSO_4$	1.2×10^{-6}
Chromium(III) hydroxide	$Cr(OH)_3$	$\sim 10^{-30}$
Cobalt(II) hydroxide (pink, inactive)	$Co(OH)_2$	2×10^{-16}
Cobalt(III) hydroxide	$Co(OH)_3$	$\sim 10^{-43}$
Cobalt(II) sulfide	CoS	$\sim 10^{-23}$
Copper(I) chloride	$CuCl$	1.2×10^{-6}
Copper(II) hydroxide (inactive)	$Cu(OH)_2$	2×10^{-19}
Copper(I) iodide	CuI	5.0×10^{-12}
Copper(I) sulfide	Cu_2S	$\sim 10^{-48}$
Copper(II) sulfide	CuS	$\sim 10^{-36}$
Copper(I) thiocyanate	$CuSCN$	1.9×10^{-13} [a]
Iron(II) hydroxide (inactive)	$Fe(OH)_2$	8×10^{-6}
Iron(III) hydroxide (amorphous, inactive)	$Fe(OH)_3$	8×10^{-40}
Iron(II) sulfide	FeS	$\sim 10^{-19}$
Lead carbonate	$PbCO_3$	1×10^{-13} [b]
Lead chloride	$PbCl_2$	1.6×10^{-5}
Lead chromate	$PbCrO_4$	1.8×10^{-14} [b]
Lead fluoride	PbF_2	2.7×10^{-8}
Lead hydroxide	$Pb(OH)_2$	$\sim 10^{-20}$
Lead iodide	PbI_2	7.1×10^{-9}
Lead sulfate	$PbSO_4$	1.7×10^{-8}
Lead sulfide	PbS	2.5×10^{-27}

Table C (**continued**)

Substance	Formula	K_{sp}
Magnesium ammonium sulfate	$MgNH_4PO_4$	3×10^{-13}
Magnesium carbonate	$MgCO_3$	1×10^{-5}
Magnesium fluoride	MgF_2	6.6×10^{-9}
Magnesium hydroxide	$Mg(OH)_2$	1×10^{-11}
Manganese(II) carbonate	$MnCO_3$	1.8×10^{-11}
Manganese(II) hydroxide	$Mn(OH)_2$	1×10^{-13}
Manganese(II) sulfide	MnS	$\sim 10^{-13}$
Mercury(I) chloride	Hg_2Cl_2	1.3×10^{-18}
Mercury(I) chromate	Hg_2CrO_4	2.0×10^{-9}
Mercury(II) hydroxide	$Hg(OH)_2$	$\sim 10^{-26}$
Mercury(I) iodide	Hg_2I_2	4.9×10^{-29}
Mercury(II) sulfate	Hg_2SO_4	6.8×10^{-7}
Mercury(I) sulfide	Hg_2S	$\sim 10^{-47}$
Mercury(II) sulfide (black)	HgS	$\sim 10^{-52}$
Nickel hydroxide (inactive)	$Ni(OH)_2$	6×10^{-18}
Nickel sulfide	NiS	$\sim 10^{-21}$
Potassium tetraphenylboron	$KB(C_6H_5)_4$	3.2×10^{-8}
Silver arsenate	Ag_3AsO_4	1×10^{-22}[a]
Silver bromide	$AgBr$	5.2×10^{-13}
Silver carbonate	Ag_2CO_3	8.1×10^{-12}
Silver chloride	$AgCl$	1.8×10^{-10}
Silver chromate	Ag_2CrO_4	1.5×10^{-12}
Silver cyanide	$AgCN$	2.3×10^{-16}
Silver dicyanoargentate	$Ag[Ag(CN)_2]$	1.3×10^{-12}
Silver hydroxide	$AgOH$	2.0×10^{-8}
Silver iodide	AgI	8.3×10^{-17}
Silver sulfate	Ag_2SO_4	2.1×10^{-5}
Silver sulfide	AgS	6×10^{-50}
Silver thiocyanate	$AgSCN$	1×10^{-12}
Strontium carbonate	$SrCO_3$	1.1×10^{-10}
Strontium chromate	$SrCrO_4$	3.6×10^{-5}
Strontium fluoride	SrF_2	2.5×10^{-9}
Strontium sulfate	$SrSO_4$	3.2×10^{-7}
Zinc carbonate	$ZnCO_3$	1.4×10^{-11}
Zinc hexacyanoferrate(II)	$Zn_2Fe(CN)_6$	4.1×10^{-16}
Zinc hydroxide (amphorous)	$Zn(OH)_2$	2.5×10^{-16}
Zinc sulfide	ZnS	$\sim 10^{-24}$

[a]At 20°C.
[b]At 18°C.

Table D Stability Constants of Some Metal Ion Complexes in Water

Unless otherwise indicated, the values are for 25°C and ionic strength, μ, of 0.1. The value given to the right of the formula of a complex is the *overall* stability constant of that complex. Where known, the *stepwise* stability constants are given as log K values beneath the formula.

Complex	Overall stability constant
Ammonia	
$Ag(NH_2)_2^+$	1.7×10^7
3.31, 3.91	
$Cd(NH_3)_4^{2+}$	5.8×10^6
2.56, 2.01, 1.35, 0.84	
$Cu(NH_3)_4^{2+}$	2.0×10^{12}
4.06, 3.41, 2.80, 2.04	
$Hg(NH_3)_2^{2+}$	2×10^{19}
8.8, 8.7, 1.0, 0.8	
$Ni(NH_3)_2^{2+}$	4.2×10^7
2.71, 2.16, 1.64, 1.11	
$Zn(NH_3)_4^{2+}$	8.0×10^8
2.23, 2.30, 2.36, 2.01	
Cyanide	
$Ag(CN)_2^-$	1.1×10^{21}
$Cu(CN)_2^-$	1×10^{16}
$Fe(CN)_6^{4-}$	1×10^{24}
$Fe(CN)_6^{3-}$	1×10^{31}
$Hg(CN)_4^{2-}$	3.2×10^{41}
18.0, 16.70, 3.83, 2.98	
$Ni(CN)_4^{2-}$	1×10^{22}
Chloride	
$AgCl_4^{3-}$ ($\mu = 0.2$)	7.9×10^5
2.85, 1.87, 0.32, 0.86	
$HgCl_4^{2-}$ ($\mu = 0.5$)	1.2×10^{15}
6.74, 6.48, 0.85, 1.05	
Iodide	
CdI_4^{2-}	2.3×10^5
2.40, 1.26, 1.0, 0.7	
HgI_4^{2-} ($\mu = 0.5$)	7.2×10^{29}
12.87, 10.95, 3.67, 2.37	
Thiocyanate	
$Fe(SCN)_2^+$ ($\mu = 0.5$)	2.8×10^3
2.14, 1.31	
Thiosulfate	
$Ag(S_2O_3)_2^{3-}$ ($\mu = 4$)	5.2×10^{12}
7.36, 5.36	
Ethylenedinitrilotetraacetate (= Y^{4-}) (all values at 20°C)	
CaY^{2-}	5.0×10^{10}
CdY^{2-}	4×10^{16}
CuY^{2-}	6.3×10^{18}
FeY^{2-}	2×10^{14}
FeY^-	1×10^{25}
MgY^{2-}	4.9×10^8
ZnY^{2-}	3.2×10^{16}
Eriochrome Black T (= D^{3-}) (all values at 20°C)	
CaD^-	1.9×10^5
MgD^-	7.4×10^6
ZnD^-	2×10^{12}

Table E Recommended Standard Values of pH_s of Seven Standard Buffer Solutions

°C	Solution						
	A^a	B	C	D	E	F	G^a
	Tetroxalate	Tartrate	Phthalate	Phosphate	Phosphate	Borax	$Ca(OH)_2$
10	1.670	—	3.998	6.923	7.472	9.332	13.003
15	1.672	—	3.999	6.900	7.448	9.276	12.810
20	1.675	—	4.002	6.881	7.429	9.225	12.627
25	1.679	3.557	4.008	6.865	7.413	9.180	12.454
30	1.683	3.552	4.015	6.853	7.400	9.139	12.289
35	1.688	3.549	4.024	6.844	7.389	9.102	12.133
40	1.694	3.547	4.035	6.838	7.380	9.068	11.984

Composition of standard buffer solutions at 25°C:

A 0.04962 *F* solution of potassium tetroxalate, $KH_3(C_2O_4)_2 \cdot 2H_2O$

B Saturated solution of potassium hydrogen tartrate, $KHC_4H_4O_6$ (ca. 0.034 *F*)

C 0.04958 *F* solution of potassium hydrogen phthalate, $KHC_8H_4O_4$

D Phosphate buffer, 0.02490 *F* in potassium dihydrogen phosphate, KH_2PO_4, and 0.02490 *F* in disodium hydrogen phosphate, Na_2HPO_4

E Phosphate buffer, 0.008665 *F* in potassium dihydrogen phosphate, KH_2PO_4, and 0.03032 *F* in disodium hydrogen phosphate, Na_2HPO_4

F 0.009971 *F* borax (sodium tetraborate, $Na_2B_4O_7 \cdot 10H_2O$) solution

G Saturated solution of calcium hydroxide, $Ca(OH)_2$ (ca. 0.02025 *F*)

(For use of the pH_s values to two decimal places, solutions *A* and *C* may be made 0.050 *F*, solution *D* made 0.025 *F* in both salts, and solution *F*, 0.010 *F*.)

[a] Secondary standard.

Table F Common Acid–Base Indicators

Common trade name	pK_{In}	pH visual transition interval	Color[a]	
			Acidic	Basic
Cresol red		0.2–1.8	Red	Yellow
Thymol blue	1.6	1.2–2.8	Red	Yellow
Methyl yellow	3.1	2.4–4.0	Red	Yellow
Bromophenol blue	4.2	3.0–4.6	Yellow	Blue
Methyl orange	3.5	3.2–4.4	Red	Yellow-orange
Methyl orange + xylene cyanole FF, 40:56		(3.8–4.1)[b]	Violet	Green
Bromocresol green	4.9	3.9–5.4	Yellow	Blue
Methyl red	5.0	4.2–6.2	Pink	Yellow
Methyl red + methylene blue, 1:1		(~5.3)[b]	Red-violet	Green
Bromocresol purple	6.4	5.2–6.8	Yellow	Purple
Bromothymol blue	7.3	6.0–7.6	Yellow	Blue
Cresol red	8.4	7.2–8.8	Yellow	Red
Phenol red	8.0	6.8–8.2	Yellow	Red
Thymol blue	9.0	8.0–9.2	Yellow	Blue
Phenol-phthalein	(8.7)	(8.0–9.8)[c]	Colorless	Red-violet
Phenol-phthalein + methylene green, 1:2		(8.8)[d]	Green	Violet
Thymol-phthalein	(9.2)	(9.0–10.5)[c]	Colorless	Blue

[a]Colors in aqueous solution at lower and upper pH limits of the visual transition interval, respectively.

[b]Screened indicator, neutral gray at stated pH.

[c]Based on addition of 1 or 2 drops of a 0.1% indicator solution to 10 ml of aqueous solution.

[d]Screened indicator, pale blue at stated pH.

Table G Standard and Formal Electrode Potentials at 25°C (standard potentials shown in **boldface**)

Half-reaction equation	E^0 or E^f, V	Solution conditions for formal potentials
$S_2O_8^{2-} + 2e \rightleftharpoons 2SO_4^{2-}$	**+2.0**	
$H_2O_2 + 2H^+ + 2e \rightleftharpoons 2H_2O$	**+1.77**	
$MnO_4^- + 4H^+ + 3e \rightleftharpoons \underline{MnO_2} + 2H_2O$	**+1.69**	
	+1.70	1 F $HClO_4$
$Ce^{4+} + e \rightleftharpoons Ce^{3+}$	+1.60	1 F HNO_3
	+1.44	1 F H_2SO_4
	+1.28	1 F HCl
$\underline{NaBiO_3} + 6H^+ + 2e \rightleftharpoons Na^+ + Bi^{3+} + 3H_2O$	~**+1.6**	
$MnO_4^- + 8H^+ + 5e \rightleftharpoons Mn^{2+} + 4H_2O$	**+1.51**	
$2BrO_3^- + 12H^+ + 10e \rightleftharpoons Br_2 + 6H_2O$	~**+1.5**	
$\underline{PbO_2} + 4H^+ + 2e \rightleftharpoons Pb^{2+} + 2H_2O$	**+1.46**	
$Cl_2 + 2e \rightleftharpoons 2Cl^-$	**+1.359**	
$Cr_2O_7^{2-} + 14H^+ + 6e \rightleftharpoons 2Cr^{3+} + 7H_2O$	**+1.33**	
	+1.03	1 F $HClO_4$
	+1.00	1 F HCl
	+0.92	0.1 F H_2SO_4
$Tl^{3+} + 2e \rightleftharpoons Tl^+$	**+1.28**	
	+0.78	1 F HCl
$\underline{MnO_2} + 4H^+ + 2e \rightleftharpoons Mn^{2+} + 2H_2O$	**+1.23**	
$O_2(g) + 4H^+ + 4e \rightleftharpoons 2H_2O$	**+1.229**	
$ClO_4^- + 2H^+ + 2e \rightleftharpoons ClO_3^- + H_2O$	**+1.19**	
$2IO_3^- + 12H^+ + 10e \rightleftharpoons \underline{I_2} + 6H_2O$	**+1.19**	
$Br_2 + 2e \rightleftharpoons 2Br^-$	**+1.087**	
$VO_2^+ + 2H^+ + e \rightleftharpoons VO^{2+} + H_2O$	**+0.9994**	
$HNO_2 + H^+ + e \rightleftharpoons NO(g) + H_2O$	**+0.99**	
$NO_3^- + 3H^+ + 2e \rightleftharpoons HNO_2 + H_2O$	**+0.94**	
	+0.92	1 F HNO_3
$2Hg^{2+} + 2e \rightleftharpoons Hg_2^{2+}$	**+0.907**	
$Cu^{2+} + I^- + e \rightleftharpoons \underline{CuI}$	**+0.86**	

Table G (continued)

Half-reaction equation	E^0 or E^f, V	Solution conditions for formal potentials
$Ag^+ + e \rightleftharpoons Ag$	**+0.7994**	
$Hg_2^{2+} + 2e \rightleftharpoons 2Hg$	**+0.792**	
$Fe^{3+} + e \rightleftharpoons Fe^{2+}$	**+0.771**	
	+0.75	1 *F* $HClO_4$
	+0.73	1 *F* HNO_3
	+0.71	0.5 *F* HCl
	+0.70	1 *F* HCl
	+0.68	1 *F* H_2SO_4
	+0.64	5 *F* HCl
	+0.53	10 *F* HCl
	+0.46	2 *F* H_3PO_4
Benzoquinone + $2H^+ + e \rightleftharpoons$ hydroquinone	**+0.6994**	
	+0.696	1 *F* HCl
$O_2(g) + 2H^+ + 2e \rightleftharpoons H_2O_2$	**+0.69**	
$MnO_4^- + e \rightleftharpoons MnO_4^{2-}$	**+0.6**	
$MnO_4^- + 2H_2O + 3e \rightleftharpoons \underline{MnO_2} + 4OH^-$	**+0.57**	
$H_3AsO_4 + 2H^+ + 2e \rightleftharpoons H_3AsO_3 + H_2O$	**+0.559**	
	+0.577	1 *F* HCl
$I_3^- + 2e \rightleftharpoons 3I^-$	**+0.545**	
$\underline{I_2} + 2e \rightleftharpoons 2I^-$	**+0.536**	
$\underline{Ag_2CrO_4} + 2e \rightleftharpoons 2Ag + CrO_4^{2-}$	**+0.447**	
$UO_2^{2+} + 4H^+ + 2e \rightleftharpoons U^{4+} + 2H_2O$	+0.41	0.5 *F* H_2SO_4
	+0.31	1 *F* HCl
$Fe(CN)_6^{3-} + e \rightleftharpoons Fe(CN)_6^{4-}$	**+0.356**	
	+0.71	1 *F* HCl
$VO^{2+} + 2H^+ + e \rightleftharpoons V^{3+} + H_2O$	**+0.337**	
	+0.360	1 *F* H_2SO_4
$Cu^{2+} + 2e \rightleftharpoons Cu$	**+0.337**	
$\underline{Hg_2Cl_2} + 2e \rightleftharpoons 2Hg + 2Cl^-$	**+0.2680**	
	+0.3337	0.1 *F* KCl
	+0.2801	1 *F* KCl
	+0.2412	Satd. KCl
$\underline{AgCl} + e \rightleftharpoons Ag + Cl^-$	**+0.2224**	
$SbO^+ + 2H^+ + 3e \rightleftharpoons Sb + H_2O$	**+0.21**	
$SO_4^{2-} + 4H^+ + 2e \rightleftharpoons H_2SO_3 + H_2O$	**+0.17**	
$Cu^{2+} + e \rightleftharpoons Cu^+$	**+0.153**	
$Sn^{4+} + 2e \rightleftharpoons Sn^{2+}$	+0.14	1 *F* HCl

Table G (continued)

Half-reaction equation	E^0 or E^f, V	Solution conditions for formal potentials
$S + 2H^+ + 2e \rightleftharpoons H_2S$	**+0.14**	
$\underline{Hg_2Br_2} + 2e \rightleftharpoons 2Hg + 2Br^-$	**+0.1392**	
$TiO^{2+} + 2H^+ + e \rightleftharpoons$	+0.12	2 F H_2SO_4
$Ti^{3+} + H_2O$	−0.01	0.2 F H_2SO_4
$S_4O_6^{2-} + 2e \rightleftharpoons 2S_2O_3^{2-}$	**+0.09**	
$\underline{AgBr} + e \rightleftharpoons Ag + Br^-$	**+0.071**	
$2H^+ + 2e \rightleftharpoons H_2(g)$	**±0.0000**	
$Pb^{2+} + 2e \rightleftharpoons Pb$	**−0.126**	
$Sn^{2+} + 2e \rightleftharpoons Sn$	**−0.140**	
$\underline{AgI} + e \rightleftharpoons Ag + I^-$	**−0.152**	
$Ni^{2+} + 2e \rightleftharpoons Ni$	**−0.23**	
$V^{3+} + e \rightleftharpoons V^{2+}$	**−0.255**	
$Co^{2+} + 2e \rightleftharpoons Co$	**−0.28**	
$Tl^+ + e \rightleftharpoons Tl$	**−0.336**	
$Ti^{3+} + e \rightleftharpoons Ti^{2+}$	**−0.37**	
$Cd^{2+} + 2e \rightleftharpoons Cd$	**−0.402**	
$Cr^{3+} + e \rightleftharpoons Cr^{2+}$	**−0.41**	
	−0.38	1 F HCl
$Fe^{2+} + 2e \rightleftharpoons Fe$	**−0.440**	
$Cr^{3+} + 3e \rightleftharpoons Cr$	**−0.74**	
$Zn^{2+} + 2e \rightleftharpoons Zn$	**−0.7628**	
$Mn^{2+} + 2e \rightleftharpoons Mn$	**−1.190**	
$Al^{3+} + 3e \rightleftharpoons Al$	**−1.66**	
$Mg^{2+} + 2e \rightleftharpoons Mg$	**−2.37**	
$Na^+ + e \rightleftharpoons Na$	**−2.713**	
$Ca^{2+} + 2e \rightleftharpoons Ca$	**−2.87**	
$K^+ + e \rightleftharpoons K$	**−2.925**	
$Li^+ + e \rightleftharpoons Li$	**−3.03**	

Table H Redox Indicators

Common name	E^f, V	Color	
		Oxidized	Reduced
N-Phenylanthranilic acid	+1.08 at pH 0	Red-violet	Colorless
1,10-Phenanthroline [Fe(II) complex]	+1.06 at pH 0	Pale blue	Red
Xylene cyanole FF	+1.05 at pH 0	Orange	Green
Erioglaucine	+1.00 at pH 0	Orange-pink	Green-yellow
2,2′-Bipyridine [Fe(II) complex]	+0.97 at pH 0	Pale blue	Red
Diphenylaminesulfonic acid (and its salts)	+0.84 at pH 0	Blue-violet	Colorless
o-Dianisidine	+0.80 at pH 0	Red	Colorless
Diphenylamine and diphenylbenzidine	+0.76 at pH 0	Violet	Colorless
Variamine Blue B	+0.62 at pH 1.5	Blue	Colorless
Methylene blue	+0.52 at pH 3	Green-blue	Colorless
Neutral red	−0.325 at pH 7	Red	Colorless

Table I Cathodic Overpotential of Hydrogen on Various Materials[a]

Electrode material	Hydrogen overpotential (V) at current density (A/cm^2) of:			
	0.001	0.01	0.1	1
Aluminum	0.58	0.71	0.74	0.78
Bismuth	0.69	0.83	0.91	1.01
Cadmium	0.99	1.20	1.25	1.23
Chromium			0.67	0.77
Copper	0.60	0.75	0.82	0.84
Iron	0.40	0.53	0.64	0.77
Lead	0.67	0.97	1.12	1.08
Mercury	1.04	1.15	1.21	1.24
Nickel	0.33	0.42	0.51	0.59
Platinized Pt	0.01	0.03	0.05	0.07
Platinum	0.09	0.39	0.50	0.44
Silver	0.46	0.66	0.76	
Tin	0.85	0.98	0.99	0.98
Tungsten	0.27	0.35	0.47	0.54

[a] In 1 *F* HCl.

Table J Limiting Equivalent Ionic Conductances (aqueous solution, 25°C, rounded values)

Cation	λ^0, mhos-cm^2/equivalent	Anion	λ^0, mhos-cm^2/equivalent
H^+	350	F^-	55
Li^+	39	Cl^-	76
Na^+	50	Br^-	78
K^+	74	I^-	77
NH_4^+	73	ClO_4^-	67
Ag^+	62	OH^-	198
$\frac{1}{2}Mg^{2+}$	53	NO_3^-	71
$\frac{1}{2}Ca^{2+}$	60	HCO_3^-	44
$\frac{1}{2}Sr^{2+}$	59	$C_2H_3O_2^-$	41
$\frac{1}{2}Ba^{2+}$	64	$HC_2O_4^-$	40
$\frac{1}{2}Cu^{2+}$	54	$B(C_6H_5)_4^-$	18
$\frac{1}{2}Zn^{2+}$	53	$\frac{1}{2}SO_4^{2-}$	80
$\frac{1}{2}Pb^{2+}$	69	$\frac{1}{2}C_2O_4^{2-}$	74
$\frac{1}{3}La^{3+}$	70	$\frac{1}{2}CO_3^{2-}$	69

Table K 1968 Table of Relative Atomic Weights (^{12}C = 12)[a]

Element	Symbol	Atomic No.	Atomic weight	Element	Symbol	Atomic No.	Atomic weight
Actinium	Ac	89	[227][b]	Erbium	Er	68	167.26
Aluminum	Al	13	26.9815	Europium	Eu	63	151.96
Americium	Am	95	[243][b]	Fermium	Fm	100	[257][b]
Antimony	Sb	51	121.75	Fluorine	F	9	18.9984
Argon	Ar	18	39.948	Francium	Fr	87	[223][b]
Arsenic	As	33	74.9216	Gadolinium	Gd	64	157.25
Astatine	At	85	[210][b]	Gallium	Ga	31	69.72
Barium	Ba	56	137.34	Germanium	Ge	32	72.59
Berkelium	Bk	97	[247][b]	Gold	Au	79	196.967
Beryllium	Be	4	9.0122	Hafnium	Hf	72	178.49
Bismuth	Bi	83	208.980	Helium	He	2	4.0026
Boron	B	5	10.811[c]	Holmium	Ho	67	164.930
Bromine	Br	35	79.904[d]	Hydrogen	H	1	1.00797[c]
Cadmium	Cd	48	112.40	Indium	In	49	114.82
Calcium	Ca	20	40.08	Iodine	I	53	126.9044
Californium	Cf	98	[252][b]	Iridium	Ir	77	192.2
Carbon	C	6	12.01115[c]	Iron	Fe	26	55.847[d]
Cerium	Ce	58	140.12	Krypton	Kr	36	83.80
Cesium	Cs	55	132.905	Lanthanum	La	57	138.91
Chlorine	Cl	17	35.453[d]	Laurentium	Lr	103	[256][b]
Chromium	Cr	24	51.996[d]	Lead	Pb	82	207.19
Cobalt	Co	27	58.9332	Lithium	Li	3	6.939
Copper	Cu	29	63.546[d]	Lutetium	Lu	71	174.97
Curium	Cm	96	[247][b]	Magnesium	Mg	12	24.305
Dysprosium	Dy	66	162.50	Manganese	Mn	25	54.9380
Einsteinium	Es	99	[254][b]	Mendelevium	Md	101	[257][b]

[a] Permission of the International Union of Pure and Applied Chemistry.

[b] Value in brackets denotes the mass number of the isotope of longest known half-life (or a better known one for Cf, Po, Pm, and Tc).

[c] Atomic weight varies because of natural variation in isotopic composition: B, ±0.003; C, ±0.00005; H, ±0.00001; O, ±0.0001; Si, ±0.001; S, ±0.003.

[d] Atomic weight is believed to have the following experimental uncertainty: Br, ±0.001; Cl, ±0.001; Cu, ±0.001; Fe, ±0.003; Ag, ±0.001; Ne, ±0.003. (For other elements, the last digit given for the atomic weight is believed reliable to ±0.5.)

Table K (continued)

Element	Symbol	Atomic No.	Atomic weight	Element	Symbol	Atomic No.	Atomic weight
Mercury	Hg	80	200.59	Samarium	Sm	62	150.35
Molybdenum	Mo	42	95.94	Scandium	Sc	21	44.956
Neodymium	Nd	60	144.24	Selenium	Se	34	78.96
Neon	Ne	10	20.179[d]	Silicon	Si	14	28.086[c]
Neptunium	Np	93	[237][b]	Silver	Ag	47	107.868[d]
Nickel	Ni	28	58.71	Sodium	Na	11	22.9898
Niobium	Nb	41	92.906	Strontium	Sr	38	87.62
Nitrogen	N	7	14.0067	Sulfur	S	16	32.064[c]
Nobelium	No	102	[255][b]	Tantalum	Ta	73	180.948
Osmium	Os	76	190.2	Technetium	Tc	43	[99][b]
Oxygen	O	8	15.9994[c]	Tellurium	Te	52	127.60
Palladium	Pd	46	106.4	Terbium	Tb	65	158.924
Phosphorus	P	15	30.9738	Thallium	Tl	81	204.37
Platinum	Pt	78	195.09	Thorium	Th	90	232.038
Plutonium	Pu	94	[244][b]	Thulium	Tm	69	168.934
Polonium	Po	84	[210][b]	Tin	Sn	50	118.69
Potassium	K	19	39.102	Titanium	Ti	22	47.90
Praseodymium	Pr	59	140.907	Tungsten	W	74	183.85
Promethium	Pm	61	[147][b]	Uranium	U	92	238.03
Protactinium	Pa	91	[231][b]	Vanadium	V	23	50.942
Radium	Ra	88	[226][b]	Xenon	Xe	54	131.30
Radon	Rn	86	[222][b]	Ytterbium	Yb	70	173.04
Rhenium	Re	75	186.2	Yttrium	Y	39	88.905
Rhodium	Rh	45	102.905	Zinc	Zn	30	65.37
Rubidium	Rb	37	85.47	Zirconium	Zr	40	91.22
Ruthenium	Ru	44	101.07				

[a] Permission of the International Union of Pure and Applied Chemistry.

[b] Value in brackets denotes the mass number of the isotope of longest known half-life (or a better known one for Cf, Po, Pm, and Tc).

[c] Atomic weight varies because of natural variation in isotopic composition: B, ±0.003; C, ±0.00005; H, ±0.00001; O, ±0.0001; Si, ±0.001; S, ±0.003.

[d] Atomic weight is believed to have the following experimental uncertainty: Br, ±0.001; Cl, ±0.001; Cu, ±0.001; Fe, ±0.003; Ag, ±0.001; Ne, ±0.003. (For other elements, the last digit given for the atomic weight is believed reliable to ±0.5.)

APPENDIX

Table L Table of Base 10 Logarithms

Logarithms

Natural Numbers	0	1	2	3	4	5	6	7	8	9	Proportional Parts 1	2	3	4	5	6	7	8	9
10	0000	0043	0086	0128	0170	0212	0253	0294	0334	0374	4	8	12	17	21	25	29	33	37
11	0414	0453	0492	0531	0569	0607	0645	0682	0719	0755	4	8	11	15	19	23	26	30	34
12	0792	0828	0864	0899	0934	0969	1004	1038	1072	1106	3	7	10	14	17	21	24	28	31
13	1139	1173	1206	1239	1271	1303	1335	1367	1399	1430	3	6	10	13	16	19	23	26	29
14	1461	1492	1523	1553	1584	1614	1644	1673	1703	1732	3	6	9	12	15	18	21	24	27
15	1761	1790	1818	1847	1875	1903	1931	1959	1987	2014	3	6	8	11	14	17	20	22	25
16	2041	2068	2095	2122	2148	2175	2201	2227	2253	2279	3	5	8	11	13	16	18	21	24
17	2304	2330	2355	2380	2405	2430	2455	2480	2504	2529	2	5	7	10	12	15	17	20	22
18	2553	2577	2601	2625	2648	2672	2695	2718	2742	2765	2	5	7	9	12	14	16	19	21
19	2788	2810	2833	2856	2878	2900	2923	2945	2967	2989	2	4	7	9	11	13	16	18	20
20	3010	3032	3054	3075	3096	3118	3139	3160	3181	3201	2	4	6	8	11	13	15	17	19
21	3222	3243	3263	3284	3304	3324	3345	3365	3385	3404	2	4	6	8	10	12	14	16	18
22	3424	3444	3464	3483	3502	3522	3541	3560	3579	3598	2	4	6	8	10	12	14	15	17
23	3617	3636	3655	3674	3692	3711	3729	3747	3766	3784	2	4	6	7	9	11	13	15	17
24	3802	3820	3838	3856	3874	3892	3909	3927	3945	3962	2	4	5	7	9	11	12	14	16
25	3979	3997	4014	4031	4048	4065	4082	4099	4116	4133	2	3	5	7	9	10	12	14	15
26	4150	4166	4183	4200	4216	4232	4249	4265	4281	4298	2	3	5	7	8	10	11	13	15
27	4314	4330	4346	4362	4378	4393	4409	4425	4440	4456	2	3	5	6	8	9	11	13	14
28	4472	4487	4502	4518	4533	4548	4564	4579	4594	4609	2	3	5	6	8	9	11	12	14
29	4624	4639	4654	4669	4683	4698	4713	4728	4742	4757	1	3	4	6	7	9	10	12	13
30	4771	4786	4800	4814	4829	4843	4857	4871	4886	4900	1	3	4	6	7	9	10	11	13
31	4914	4928	4942	4955	4969	4983	4997	5011	5024	5038	1	3	4	6	7	8	10	11	12
32	5051	5065	5079	5092	5105	5119	5132	5145	5159	5172	1	3	4	5	7	8	9	11	12
33	5185	5198	5211	5224	5237	5250	5263	5276	5289	5302	1	3	4	5	6	8	9	10	12
34	5315	5328	5340	5353	5366	5378	5391	5403	5416	5428	1	3	4	5	6	8	9	10	11
35	5441	5453	5465	5478	5490	5502	5514	5527	5539	5551	1	2	4	5	6	7	9	10	11
36	5563	5575	5587	5599	5611	5623	5635	5647	5658	5670	1	2	4	5	6	7	8	10	11
37	5682	5694	5705	5717	5729	5740	5752	5763	5775	5786	1	2	3	5	6	7	8	9	10
38	5798	5809	5821	5832	5843	5855	5866	5877	5888	5899	1	2	3	5	6	7	8	9	10
39	5911	5922	5933	5944	5955	5966	5977	5988	5999	6010	1	2	3	4	5	7	8	9	10
40	6021	6031	6042	6053	6064	6075	6085	6096	6107	6117	1	2	3	4	5	6	8	9	10
41	6128	6138	6149	6160	6170	6180	6191	6201	6212	6222	1	2	3	4	5	6	7	8	9
42	6232	6243	6253	6263	6274	6284	6294	6304	6314	6325	1	2	3	4	5	6	7	8	9
43	6335	6345	6355	6365	6375	6385	6395	6405	6415	6425	1	2	3	4	5	6	7	8	9
44	6435	6444	6454	6464	6474	6484	6493	6503	6513	6522	1	2	3	4	5	6	7	8	9
45	6532	6542	6551	6561	6571	6580	6590	6599	6609	6618	1	2	3	4	5	6	7	8	9
46	6628	6637	6646	6656	6665	6675	6684	6693	6702	6712	1	2	3	4	5	6	7	7	8
47	6721	6730	6739	6749	6758	6767	6776	6785	6794	6803	1	2	3	4	5	5	6	7	8
48	6812	6821	6830	6839	6848	6857	6866	6875	6884	6893	1	2	3	4	4	5	6	7	8
49	6902	6911	6920	6928	6937	6946	6955	6964	6972	6981	1	2	3	4	4	5	6	7	8
50	6990	6998	7007	7016	7024	7033	7042	7050	7059	7067	1	2	3	3	4	5	6	7	8
51	7076	7084	7093	7101	7110	7118	7126	7135	7143	7152	1	2	3	3	4	5	6	7	8
52	7160	7168	7177	7185	7193	7202	7210	7218	7226	7235	1	2	2	3	4	5	6	7	7
53	7243	7251	7259	7267	7275	7284	7292	7300	7308	7316	1	2	2	3	4	5	6	6	7
54	7324	7332	7340	7348	7356	7364	7372	7380	7388	7396	1	2	2	3	4	5	6	6	7

Table L (continued)

Logarithms

Natural Numbers	0	1	2	3	4	5	6	7	8	9	Proportional Parts								
											1	2	3	4	5	6	7	8	9
55	7404	7412	7419	7427	7435	7443	7451	7459	7466	7474	1	2	2	3	4	5	5	6	7
56	7482	7490	7497	7505	7513	7520	7528	7536	7543	7551	1	2	2	3	4	5	5	6	7
57	7559	7566	7574	7582	7589	7597	7604	7612	7619	7627	1	2	2	3	4	5	5	6	7
58	7634	7642	7649	7657	7664	7672	7679	7686	7694	7701	1	1	2	3	4	4	5	6	7
59	7709	7716	7723	7731	7738	7745	7752	7760	7767	7774	1	1	2	3	4	4	5	6	7
60	7782	7789	7796	7803	7810	7818	7825	7832	7839	7846	1	1	2	3	4	4	5	6	6
61	7853	7860	7868	7875	7882	7889	7896	7903	7910	7917	1	1	2	3	4	4	5	6	6
62	7924	7931	7938	7945	7952	7959	7966	7973	7980	7987	1	1	2	3	3	4	5	6	6
63	7993	8000	8007	8014	8021	8028	8035	8041	8048	8055	1	1	2	3	3	4	5	5	6
64	8062	8069	8075	8082	8089	8096	8102	8109	8116	8122	1	1	2	3	3	4	5	5	6
65	8129	8136	8142	8149	8156	8162	8169	8176	8182	8189	1	1	2	3	3	4	5	5	6
66	8195	8202	8209	8215	8222	8228	8235	8241	8248	8254	1	1	2	3	3	4	5	5	6
67	8261	8267	8274	8280	8287	8293	8299	8306	8312	8319	1	1	2	3	3	4	5	5	6
68	8325	8331	8338	8344	8351	8357	8363	8370	8376	8382	1	1	2	3	3	4	4	5	6
69	8388	8395	8401	8407	8414	8420	8426	8432	8439	8445	1	1	2	2	3	4	4	5	6
70	8451	8457	8463	8470	8476	8482	8488	8494	8500	8506	1	1	2	2	3	4	4	5	6
71	8513	8519	8525	8531	8537	8543	8549	8555	8561	8567	1	1	2	2	3	4	4	5	5
72	8573	8579	8585	8591	8597	8603	8609	8615	8621	8627	1	1	2	2	3	4	4	5	5
73	8633	8639	8645	8651	8657	8663	8669	8675	8681	8686	1	1	2	2	3	4	4	5	5
74	8692	8698	8704	8710	8716	8722	8727	8733	8739	8745	1	1	2	2	3	4	4	5	5
75	8751	8756	8762	8768	8774	8779	8785	8791	8797	8802	1	1	2	2	3	3	4	5	5
76	8808	8814	8820	8825	8831	8837	8842	8848	8854	8859	1	1	2	2	3	3	4	5	5
77	8865	8871	8876	8882	8887	8893	8899	8904	8910	8915	1	1	2	2	3	3	4	4	5
78	8921	8927	8932	8938	8943	8949	8954	8960	8965	8971	1	1	2	2	3	3	4	4	5
79	8976	8982	8987	8993	8998	9004	9009	9015	9020	9026	1	1	2	2	3	3	4	4	5
80	9031	9036	9042	9047	9053	9058	9063	9069	9074	9079	1	1	2	2	3	3	4	4	5
81	9085	9090	9096	9101	9106	9112	9117	9122	9128	9133	1	1	2	2	3	3	4	4	5
82	9138	9143	9149	9154	9159	9165	9170	9175	9180	9186	1	1	2	2	3	3	4	4	5
83	9191	9196	9201	9206	9212	9217	9222	9227	9232	9238	1	1	2	2	3	3	4	4	5
84	9243	9248	9253	9258	9263	9269	9274	9279	9284	9289	1	1	2	2	3	3	4	4	5
85	9294	9299	9304	9309	9315	9320	9325	9330	9335	9340	1	1	2	2	3	3	4	4	5
86	9345	9350	9355	9360	9365	9370	9375	9380	9385	9390	1	1	2	2	3	3	4	4	5
87	9395	9400	9405	9410	9415	9420	9425	9430	9435	9440	0	1	1	2	2	3	3	4	4
88	9445	9450	9455	9460	9465	9469	9474	9479	9484	9489	0	1	1	2	2	3	3	4	4
89	9494	9499	9504	9509	9513	9518	9523	9528	9533	9538	0	1	1	2	2	3	3	4	4
90	9542	9547	9552	9557	9562	9566	9571	9576	9581	9586	0	1	1	2	2	3	3	4	4
91	9590	9595	9600	9605	9609	9614	9619	9624	9628	9633	0	1	1	2	2	3	3	4	4
92	9638	9643	9647	9652	9657	9661	9666	9671	9675	9680	0	1	1	2	2	3	3	4	4
93	9685	9689	9694	9699	9703	9708	9713	9717	9722	9727	0	1	1	2	2	3	3	4	4
94	9731	9736	9741	9745	9750	9754	9759	9763	9768	9773	0	1	1	2	2	3	3	4	4
95	9777	9782	9786	9791	9795	9800	9805	9809	9814	9818	0	1	1	2	2	3	3	4	4
96	9823	9827	9832	9836	9841	9845	9850	9854	9859	9863	0	1	1	2	2	3	3	4	4
97	9868	9872	9877	9881	9886	9890	9894	9899	9903	9908	0	1	1	2	2	3	3	4	4
98	9912	9917	9921	9926	9930	9934	9939	9943	9948	9952	0	1	1	2	2	3	3	4	4
99	9956	9961	9965	9969	9974	9978	9983	9987	9991	9996	0	1	1	2	2	3	3	3	4

INDEX

INDEX

INDEX

INDEX

INDEX

75 76 77 9 8 7 6 5